BIOLOGY AND ECOLOGY OF PIKE

BIOLOGY AND ECOLOGY OF PIKE

Editors

Christian Skov
DTU Aqua, Technical University of Denmark
Section for Inland Fisheries and Ecology
Silkeborg, Denmark

P. Anders Nilsson
Lund University
Department of Biology – Aquatic Ecology
Lund, Sweden
and
Karlstad University
Department of Environmental and Life Sciences – Biology
Karlstad, Sweden

CRC Press
Taylor & Francis Group
Boca Raton London New York

CRC Press is an imprint of the
Taylor & Francis Group, an **informa** business
A SCIENCE PUBLISHERS BOOK

Cover illustration: Reproduced by kind courtesy of Olof Engstedt (author of Chapter 10).

CRC Press
Taylor & Francis Group
6000 Broken Sound Parkway NW, Suite 300
Boca Raton, FL 33487-2742

Printed on acid-free paper
Version Date: 20171023

International Standard Book Number-13: 978-1-4822-6290-2 (Hardback)

Library of Congress Cataloging-in-Publication Data

Names: Skov, Christian, 1968- editor. | Nilsson, P. Anders, editor.
Title: Biology and ecology of pike / editors, Christian Skov, DTU Aqua, Technical University of Denmark, Section for Inland Fisheries and Ecology, Silkeborg, Denmark, P. Anders Nilsson, Lund University, Department of Biology & Aquatic Ecology, Lund, Sweden, and Karlstad University, Department of Environmental and Life Sciences & Biology, Karlstad, Sweden.
Description: Boca Raton, FL : CRC Press, 2017. | "A Science Publishers book." | Includes bibliographical references and index.
Identifiers: LCCN 2017034002 | ISBN 9781482262902 (hardback : alk. paper)
Subjects: LCSH: Pike. | Pike fisheries--Management.
Classification: LCC QL638.E7 B56 2017 | DDC 597.5/9--dc23
LC record available at https://lccn.loc.gov/2017034002

Visit the Taylor & Francis Web site at
http://www.taylorandfrancis.com

and the CRC Press Web site at
http://www.crcpress.com

To the ladies of our lives
Camilla, Felicia, Hanna, Dorte, Liv. And to Alexander.

Contents

Section III
Management and Fisheries

Chapter 1

Preface; Introduction to Pike and This Book

P. Anders Nilsson[*1] *and Christian Skov*[2]

The pike (*Esox lucius* Linnaeus, 1758), or northern pike as it is called in most of North America, is an iconic fish species that has fascinated mankind throughout history. Its large size and characteristic appearance have likely contributed to various tall stories, some of which are beautifully saved in Izaak Walton's "The Compleat Angler or the Contemplative man's Recreation" from 1653 (see e.g. Walton and Cotton 2005). The pike is there described as the tyrant of freshwaters, attacking everything from venomous frogs to birds to mules and people. There is also reference to anecdotes of pike being spontaneously generated from weed and glutinous matter with help of sunlight, and of specimens over 200 years old. Although such historical texts probably contain various degrees of misconception, they simultaneously nicely describe many aspects of pike as a fish species and fisheries target, and, perhaps most importantly, indicate how significant pike as a species has been and continues to be. Along similar lines, the pike has undoubtedly interested us for many years, from both personal and academic points of view, and it is with the abovementioned importance, significance and fascination in mind we have set the current book.

The last century shows an increasing and continuing interest in pike in the scientific literature (Fig. 1.1), warranting our summary of pike knowledge. Needless to say, it is impossible to cover all of those publications and knowledge

**Corresponding author*: [1]Lund University, Department of Biology – Aquatic ecology, Ecology Building, 22362 Lund, Sweden; Email: anders.nilsson@biol.lu.se
and
Karlstad University, Department of Environmental and Life Sciences – Biology, 65188 Karlstad, Sweden; Email: p.anders.nilsson@kau.se

[2]DTU Aqua, Technical University of Denmark, Section for Inland Fisheries and Ecology, Vejlsøvej 39 8600, Silkeborg, Denmark; Email: ck@aqua.dtu.dk

on pike in a single book. Also, the scientific interest in and knowledge of pike have previously been reviewed in synopses, books and journal special issues, such as Raat (1988), Craig (1996), Pierce (2012) and Farrell et al. (2008). Although these volumes remain substantially influential to the pike literature and knowledge, they leave temporal and geographical knowledge gaps to the understanding of pike. To complement and extend from these volumes, as well as to update the current state of pike research and knowledge, the current contribution focuses on pike from three main angles in three sections; pike individual ecology, pike populations and their communities, as well as management and fisheries of pike.

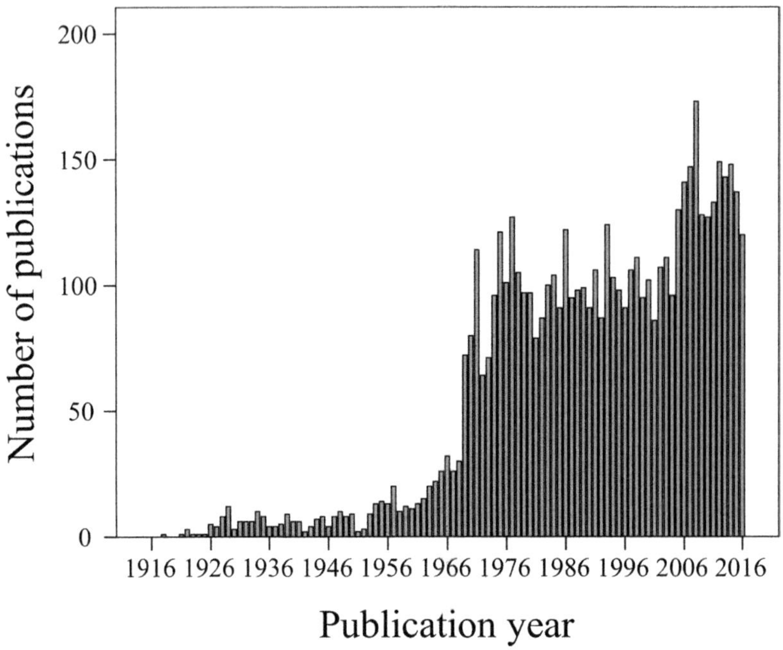

FIGURE 1.1 The annual (1916-2016) number of scientific publications including the search term "*Esox lucius*", extracted from The Web of Science.

Adaptive decisions are made by individuals, and as such decisions can affect, for example, pike individual success, growth and survival, they have bearing also for pike population densities and size structures, as well as for trophic processes and management strategies. This volume therefore starts with a section on pike *Individuals*, with chapters focusing on pike foraging behaviour and decisions, how pike cope with different environments, the bioenergetics and growth of pike, and pike spatial ecology. In chapter 2, pike foraging behaviour is described and scrutinized from a predation-cycle perspective, highlighting effects of pike and prey interactions, as well as pike risks of predation, kleptoparasitism and cannibalism, on pike foraging performance, prey preference, and functional responses. Pike individual success moreover depends on the environment they live in, and chapter 3 deals with crucial aspects of vegetation, water clarity, temperature, oxygen, salinity, and pH for different pike life stages, all with possible climate change in mind.

The consequences of foraging and environment should be translated into growth and reproduction to contribute to lifetime fitness, and pike bioenergetics, growth and physiological capacities are considered in chapter 4, in light of pike energy budgets, emphasising the special adaptations pike possess. The last chapter of the *Individuals* section of the book deals with the spatial ecology of pike, and reviews e.g. migration, dispersal and home-range aspects for pike in various habitats based on data from mainly telemetry studies, describing important behavioural aspects of pike adaptive spatial behaviour. The above aspects of pike individual ecology are directly linked to pike individual success, population growth and composition, as well as community dynamics, and can add mechanistic understanding to such higher-order processes.

The second section of the book deals with pike *Populations and the Communities* they are part of, as this organizational level is of utmost interest for e.g. conservation and biodiversity incentives. Here, attention is paid to pike population size and structure, pike population genetics, and the possible trophic effects from pike predation. Chapter 6, on pike population size and structure, provides insight into density-dependent and -independent factors affecting recruitment and survival in pike populations, and thereby crucial aspects on e.g. ecological bottlenecks for pike population density and composition. Pike population genetics is a topic of utmost importance for understanding e.g. evolutionary history and population distributions, and is also central for conservation, management and stocking incentives, and chapter 7 thoroughly informs about the possibilities and limitations to using population genetics for such purposes. The third and last chapter of the *Populations and Communities* section, chapter 8, focuses on consequences of pike on consumer-resource and trophic dynamics and interactions, and frames pike predation effects in both general and system-specific ecological terminology and concept, elaborating on the theoretical and empirical direct and indirect consequences of pike intra- and interspecific interactions on predation rates and trophic cascades and system responses.

The third book section is on important aspects of pike *Management and Fisheries*, with chapters on stocking for population enhancement, habitat restoration, stocking for lake restoration, recreational pike fisheries, commercial fisheries and aquaculture, as well as pike as an invasive species. Management of fish populations commonly involves various methods for population enhancement, and chapter 9 starts off the third book section by introducing and evaluating stocking of pike as a means of enhancing pike populations. One crucial aspect affecting pike population density and composition is reproduction and recruitment, and chapter 10 specifically addresses spawning habitat restoration as a key to pike management. As pike are apex predators in many systems, pike predation conveys the potential for top-down trophic cascades with consequences for lake system composition, and chapter 11 reviews and analyses attempts at using pike stocking for biomanipulation and lake restoration incentives. The pike can moreover be viewed as a natural resource, perhaps particularly for recreational fisheries, and chapter 12 approaches the sustainable management of pike for recreational fisheries from both biological and human dimensions. Pike are also a natural resource for human consumption, and commercial fisheries, stock assessment and aquaculture, across geography and history, are reviewed in chapter 13. In chapter 14, the final chapter of the book, the pike is viewed from a different angle; if pike are introduced to new systems, pike predation can have serious effects on prey species, trophic interactions and system composition and function, and the disastrous nature of pike as an invasive species is therefore evaluated and highlighted.

As the book is an edited volume, different researchers have authored the different chapters according to individual expertise. Apart from the chapter topics, that were decided and requested by the editors, the chapter authors were given free hands to interpret and summarize the chapter contents. Moreover, the book contents were communicated during a symposium and workshop on pike biology and ecology at the Department of Biology - Aquatic ecology at Lund University, Sweden, 11-13 October 2016. This workshop was convened by the editors, and funded by the Hans Kristiansson Memory Foundation and Aquatic Ecology at Lund University, to allow the chapter authors to network and get inspired by each others' work, as well as to present their chapter contents during a well-attended symposium that was open to the public.

We are happy to have taken the opportunity to edit this book. Pike biology and ecology has always been intriguing to us, clearly indicated by numerous and sometimes intense discussions on aspects of pike during the last decades as colleagues and friends. Editing this book has, as most such editing processes often turn out, been a long and winding road that during periods seemed endless. We are therefore thrilled to express our gratitude to all the chapter authors who, more or less willingly, early on accepted to take charge of the different chapters. We are also grateful to all the chapter reviewers: Colin Adams, Robert Arlinghaus, John Armstrong, Søren Berg, Christer Brönmark, Julien Cucherousset, Kristine Dunker, Erik Eichbach, Olof Engstedt, Jonna Engström-Öst, Nicolas Guillerault, Thrond Haugen, Lene Jacobsen, Martin Karlsson, Thomas Klefoth, Anna Kuparinen, Kai Lorentzen, Martyn Lucas, Petter Tibblin, Asbjørn Vøllestad and Klaus Wusyjack, who took their time to give constructive feedback to chapter authors. A special thank you goes to Kristine Dunker who kindly agreed to proofread and comment on the majority of the chapters. As we, Christian and Anders, have contributed equally to the production of this volume, the author order on the front of this book was decided by a roll of dice one fine Saturday evening in Grædstrup, Denmark.

One of our intentions with this volume is to highlight the importance of continuously summarizing and scrutinizing the knowledge on pike, to inspire ourselves, other researchers and further interest groups to update pike knowledge and to the greatest extent possible avoid future misconception on pike biology and ecology. Should we, for instance, view pike as ferocious, territorial, freshwater tyrants that always use a sit-and-wait, ambush tactic and consequently do not move? Should we, moreover, always view the pike as a cherished and desired fish species for recreational angling and commercial fisheries, that thereby deserves careful management and conservation, or may the pike be of great nuisance where it does not belong? The answers to some of these questions are given in this book while other answers are still not straightforward, why we strongly advocate continued formulation and testing of hypotheses on pike biology and ecology to further our understanding of this fascinating fish species on scientifically sound analyses. The scientific interest in pike (Fig. 1.1) indicates a continuing and strong interest in pike, an interest likely partly driven by challenged and revised views of pike biology and ecology. This reasoning strongly implies a need for continued and recurring review of the knowledge on pike, and we hence indeed look forward to updated volumes by other editors some 15-25 years from now. Meanwhile, we hope this book can be an inspiration to anyone with an interest in the biology and ecology of pike (Fig. 1.2).

FIGURE 1.2 Pike *Esox lucius* Linnaeus, 1758 (Photos: Henrik Baktoft, Olof Engstedt and Søren Berg).

REFERENCES CITED

Craig, J.F. 1996. *Pike: Biology and Exploitation.* London: Chapman & Hall.

Farrell, J.M., C. Skov and P.A. Nilsson. 2008. Preface to the International Pike Symposium: merging knowledge of ecology, biology, and management for a circumpolar species. *Hydrobiologia* 601:1-3.

Pierce, R.B. 2012. *Northern Pike: Ecology, Conservation, and Management History.* Minneapolis: University of Minnesota Press.

Raat, A.J.P. 1988. Synopsis of the biological data on the northern pike, *Esox lucius* Linnaeus, 1758. *FAO Fisheries Synopsis*: No. 30, Rev. 2, 178 p.

Walton, I. and C. Cotton. 2005. *The Compleat Angler, or the Contemplative Man's Recreation.* Darke County: Coachwhip Publications.

Section I

Individuals

Chapter 2

Finding Food and Staying Alive

P. Anders Nilsson[*1] and *Peter Eklöv*[2]

2.1 FORAGING BEHAVIOURS AND TRADE-OFFS

Finding food is central for individual animal growth and survival, and foraging performance hereby constitutes an important fitness component. When, where and how animals look for foraging opportunity can however also involve exposure to risks. For instance, foragers may have to leave shelter from predation to attain food, and the trade-offs between safety and food can vary across habitats and contexts (Werner and Gilliam 1984, Pettersson and Brönmark 1993, Brönmark et al. 2008). Moreover, individuals should according to optimal foraging theory choose food types to maximize energy intake per unit time (Stephens and Krebs 1986), although optimal diet theory can also be affected by prey behaviours and characters (Sih and Christensen 2001). Foraging decisions can furthermore be influenced by the social context in which they are made (Giraldeau and Caraco 2000). Altogether, there is general complexity in foraging activities of animals, and the pike (*Esox lucius* Linnaeus, 1758) is no exception. Even if pike are known as apex predators, consuming e.g. zooplankton, macroinvertebrates, amphibians, fish and even birds, they are particularly through their early, small-bodied life stages targeted by a variety of potential predators, e.g. insects, birds, and fish, including cannibals (Giles et al. 1986, Raat 1988, Billard 1996, Craig 1996, Engström 2001, Brown and McIntyre 2005, Pedreschi et al. 2015). Falling victim to predators or suffering injury from escaped attacks should be considered serious fitness costs, why pike are faced with a trade-off between finding food and staying alive.

**Corresponding author*: [1]Lund University, Department of Biology – Aquatic ecology, Ecology Building, 22362 Lund, Sweden; Email: anders.nilsson@biol.lu.se
and
Karlstad University, Department of Environmental and Life Sciences – Biology, 65188 Karlstad, Sweden; Email: p.anders.nilsson@kau.se

[2]Uppsala University, Department of Ecology and Genetics – Limnology, Norbyvägen 18D, 75236 Uppsala, Sweden; Email: peter.eklov@ebc.uu.se

Trading off safety against finding food should affect behavioural decisions. For instance, risks can displace optimal foraging away from maximising energy intake per unit time. The tradeoff may thereby affect overall food intake rates, affecting individual predator fitness, prey-specific risks of predation, and consequently consumer-resource interactions. This chapter focuses on the behavioural foraging decisions made by pike in different circumstances. Pike behaviour is evaluated in light of the predation cycle, from search to consumption and digestion of prey. Decisions made during the predation cycle affect food intake rates, and are therefore directly linked to the functional response, i.e. the per capita prey-dependent food intake rates of predators (Holling 1959). The functional response, in turn, scales individual foraging behaviours to individual fitness components, consumer-resource interactions and higher-order processes, and as behavioural decisions during the predation cycle can shape the functional response, decisions can have far-reaching consequences (Fryxell and Lundberg 1998, Abrams and Ginzburg 2000). Patterns and diversity of behavioural foraging decisions are hereby central for the understanding of not only pike success, but also the systems pike inhabit, why this chapter considers pike and the functional response, pike and the predation cycle, pike prey preference, as well as pike specialization and phenotypic plasticity.

2.2 PIKE AND THE FUNCTIONAL RESPONSE

Individual pike behaviours and efficiency during foraging should convey effects on per capita foraging rates, with potentials effects on both individual fitness components and consequences for populations and communities (Fryxell and Lundberg 1998, Stephens et al. 2007, chapters 6 and 8). A higher food intake rate should enhance growth, survival and reproduction fitness proxies (Wright and Giles 1987, Raat 1988, Wright and Shoesmith 1988, Engström-Öst et al. 2005), and efficient foraging should hence be adaptive and selected for. Food intake rate is commonly described and handled in terms of the functional response, and the functional response can elegantly map the effects of resource abundance on consumer populations and community responses by considering consumer per capita prey-dependent intake rates (Holling 1959, Fryxell and Lundberg 1998, Abrams and Ginzburg 2000), hereby bridging the order scales between individual foraging rates and higher-order processes. The most commonly used functional response model is the Holling type II

$$I = \frac{aN}{1 + ahN} \tag{2.1}$$

where per capita intake rate (I) is a function of 'attack rate' (a), prey density (N) and handling time (h) (Fig. 2.1).

The parameter a, in the literature often referred to as 'search efficiency', 'attack rate' or 'rate of encounter', works in concert with handling time (h) to shape the functional response. The parameter a is originally defined as the area or volume searched by a forager per unit time, and h is defined as the time the handling

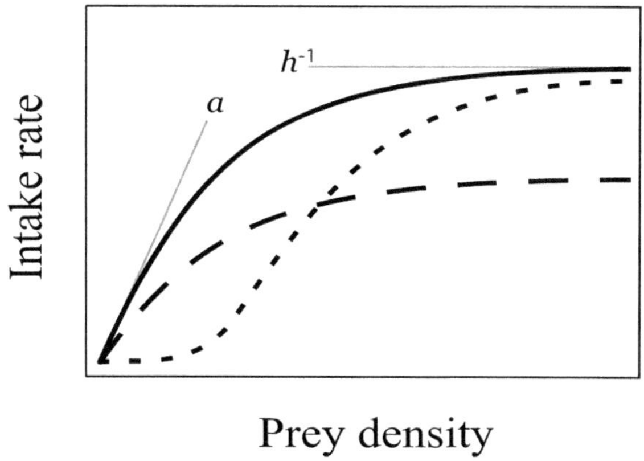

FIGURE 2.1 Schematic illustrations of a Holling type II functional response (solid curve), with indication of effects of parameters a and h (thin grey lines, equation 2.1), as well as the effects of predator-dependent only (long-dashed curve) and also prey-dependent (short-dashed curve) interference among foragers on the functional response shape (inspired by Nilsson et al. 2004).

of prey allocates from searching for prey. At low prey densities, intake rate I is to a lesser degree influenced by h, why intake rates follow a (Fig. 2.1). With increasing prey density, and thereby increasing intake rates, h has an increasing effect on the functional response as higher consumption rates imply more time allocated to manipulating and handling prey, making I asymptote at h^{-1} at high N (Fig. 2.1). Generally, high values of parameter a can lead to destabilizing consumer-resource dynamics, while high values of h reduce the impact of predators on prey, and thereby stabilize dynamics. The functional response, shaped by parameters a and h, can hereby affect individual predator and prey success but also affect higher-order ecological patterns (Fryxell and Lundberg 1998, Ranta et al. 2006, chapters 6 and 8). The assumptions of a are not entirely reflected in pike foraging behaviour, as pike can potentially be surrounded by prey without consuming them, or sometimes forage at high rates even if prey density is low. Moreover, pike may after consuming prey wait for days before resuming foraging (Ince and Thorpe 1976, Raat 1988), making it difficult to obtain reliable estimates of total handling time. This, in combination with prey evasive and avoidance behaviours as well as spatial distribution, may be a contributing factor behind the apparent lack of empirically derived functional response parameters for pike in the literature. The functional response is nevertheless an interesting angle of approach to describe pike-prey interactions and dynamics, as it is possible to empirically quantify separate predation-cycle components and combine them into parameters corresponding to a and h (Nilsson et al. 2006, Nilsson et al. 2007), and incorporate this in functional response models that evaluate effects of individual behaviours on predator-prey interactions and dynamics (Fryxell and Lundberg 1998, Nilsson 2001). We therefore below describe how pike may behave during the predation cycle, to return to the functional response and its consequences later in this chapter.

2.3 PIKE AND THE PREDATION CYCLE

Pike, or any predators, have to successfully progress through a number of critical steps to acquire food. These predation steps occur in a sequence where the predator has to *search* for prey, detect and *encounter* prey, identify prey as profitable and decide to *attack*, successfully subjugate and *capture* prey, *consume* prey, and finally *digest* prey, to eventually complete the predation cycle by anew searching for prey (Holling 1965, Webb 1986, Endler 1991, Fig. 2.2).

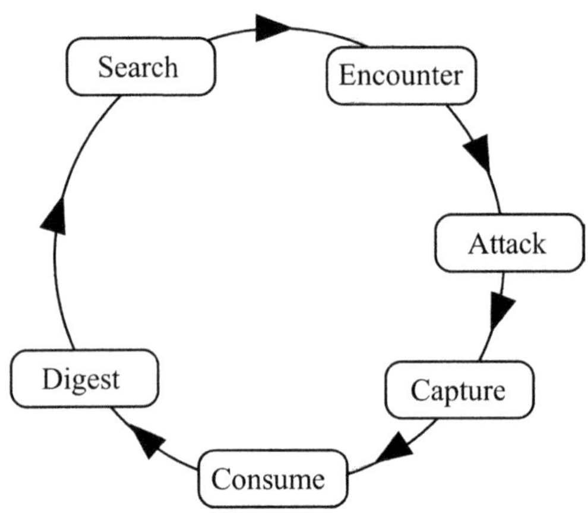

FIGURE 2.2 The predation cycle, illustrated by the sequential steps between predator search for prey and digestion of prey. The cycle is completed when the predator after digestion of prey anew initiates search (inspired by Holling 1965, Webb 1986, Endler 1991).

We here emphasise the importance to not confuse the names of the predation-cycle steps with the words used to describe the parameter *a* in the functional response above. Instead, we suggest to view the predation-cycle behaviours as a way of understanding parameters *a* and *h*, and thereby the functional response. We below describe the predation cycle from a pike point of view, with inclusion of prey adaptation to avoid pike predation.

2.3.1 Search

Being considered mainly sit-and-wait, ambush predators, pike may not show particularly conspicuous search behaviours when foraging (Webb and Skadsen 1980, but see below and chapter 5 for exceptions). Ambush predators, in comparison to predators with search strategies such as cruising or saltatory foraging, move very little in their search for prey encounter (O'Brien et al. 1990). Instead of moving around in search for prey, pike commonly rely on prey movement for encounter, by sitting and waiting, hid in structures such as vegetation, to ambush prey individuals as they swim by (Diana 1979). When adopting this search strategy, pike spend very little energy on the search part of their predation cycle, and as long as prey densities and behaviours allow for prey encounter, the sit-and-wait strategy should be considered adaptive. Prey fish may, in theory, behave adaptively to avoid encounter with ambushing pike by, for instance, avoiding habitats preferred by pike or avoiding actual positions held by foraging pike. However, structurally complex vegetation, where pike generally sit and wait for prey, also provide prey fish with foraging opportunity and shelter from actively searching predators, why a total avoidance of such habitat is not realistic. In fact, prey choose structurally complex refuges over open water regardless of the presence of predators since the benefits from the refuge shelter likely override potential negative effects from predation risk (Eklöv and Persson 1996). Moreover, as the strategy of pike is to not reveal their presence or position when in ambush search for prey, avoiding pike actual positions is unlikely a straightforward and successful prey adaptation to avoid pike encounters.

Although pike are generally considered stationary when foraging, individuals seem to shift actual positions between and over days, seasons and environmental conditions (e.g. Eklöv 1997, Nilsson 2006, Andersen et al. 2008, Knight et al. 2008, Kobler et al. 2008, Baktoft et al. 2012, chapter 5). This shifting of positions may be due to poor searching success in one spot triggering movement in anticipation of enhanced success elsewhere, a phenomenon that should be paid attention in future work to evaluate the nature and diversity of pike foraging modes. Regardless, moving could invoke potential risks for pike; leaving a hiding spot in search for enhanced foraging opportunity can expose pike individuals to encounter with potential predators, e.g. larger piscivorous fish. Moreover, as pike are cannibalistic, moving to change position in a landscape of cannibals in ambush should be quite risky, especially for smaller individuals. Pike can cannibalize on conspecifics up to 80% of their own body length (at least for smaller pike, Bry et al. 1992), while risk of cannibalism is markedly reduced for individuals above around 50 cm body length, above which substantial movement has been observed (Raat 1988, Grimm and Klinge 1996, Jepsen et al. 2001). Pike of most sizes nevertheless move (Jepsen et al. 2001, Baktoft et al. 2012, chapter 5), at least partly to change their foraging positions, but seem to end up closer to similar-sized pike than to larger after moving, supposedly avoiding the proximity of neighbouring cannibals (Nilsson 2006). Put in a simplistic way, pike search for prey can be summarized in relatively straightforward terms: simultaneously enhance prey encounters and avoid size-determined predator/cannibal encounters by hiding in ambush waiting for suitable prey to come close by.

2.3.2 Encounter

The predation-cycle stage of searching for prey may end when the predator gives up or when an encounter between predator and prey occurs. The onset of an encounter requires that the predator or the prey individual detects the other. Esocids and prey may use chemical, mechanical and/or visual cues and signals to detect each other, while use of the lateral line seems useful only at very short distances (e.g. Bleckmann 1986, Raat 1988, Stauffer and Semlitsch 1993, Nilsson and Brönmark 1999, Pettersson et al. 2000, Chivers et al. 2001, New et al. 2001, Brown 2003). The relative precision and efficiency by which pike and prey fish use cues and signals may dictate who detects whom first. From the pike point of view, it is of course important to detect prey before being detected, in order to, to the greatest extent possible, avert prey expression of anti-predator behaviours. As long as visibility conditions allow (see chapter 3), detection of prey at a distance should be most precise and efficient using visual cues, and as pike are camouflaged when hid in structured habitats in search for food, prey individuals have comparatively lower chances of detecting pike from longer distances. Light conditions, turbidity and water colouration (from e.g. humic substances) can affect the ease with which pike and prey use vision for detection and encounter, as well as avoid predation and choose habitat (Lehtiniemi et al. 2005, Andersen et al. 2008, Jönsson et al. 2013, chapter 3). Although chemical cues seem less important than visual stimuli to foraging pike, pike behave adaptively to the possibility that chemical cues from consumed prey can be avoided by new prey. As minnows (*Pimephales promelas* Rafinesque, 1820) can detect conspecific alarm signals in fecal deposits of pike, pike defecate away from their foraging spots, presumably to avoid prey detecting the area as dangerous, and thereby maintaining encounter probability (Brown et al. 1995).

The encounter rate between predator and prey is central for estimation of consumer-resource dynamics and predator-prey interactions, especially in functional response environments, and predator search efficiency and prey density are commonly used as proxies for the number of prey encountered per unit time, i.e. encounter rate (e.g. Holling 1959, Fryxell and Lundberg 1998). An interesting question in this context is how to define an encounter between pike and their prey. Although pike may occur in the vicinity of each other, making up smaller groups normally consisting of similar-sized individuals (Nilsson 2006), they could be viewed as solitary when searching for prey (Casselman and Lewis 1996, Nyqvist et al. 2012). From this point of view, each and all pike in close proximity to each other would experience a single individual prey swimming by as one encounter each. Obviously, the single individual prey could not be consumed by all pike. The prey individual, on the other hand, would upon detection of all pike in the group experience one predator encounter per pike, and even if it can only be predated once it would experience a comparably high risk of mortality. The encounter rates can hereby differ between individual pike and prey fish. Similarly, a single, solitary pike encountering a shoal of prey fish would numerically encounter each individual in the prey shoal, i.e. multiple simultaneous encounters. However, considering the predation-cycle progression towards consumption of prey, and that pike only on extremely rare occasions can catch more than single individual prey fish at the time, encountering a shoal of prey could mean only a single encounter with the possibility for predation-cycle continuation. Scrutinizing this argument may appear inconsequential, as pike only need to successfully capture one prey per foraging attempt, while prey need to in fact avoid all pike to survive. The argument however has bearing for the relative success of pike and prey. Prey can occur in dense shoals or schools for selfish, dilution reasons. The dilution effect suggests that the probabilistic risk for one of the individuals in a school of prey to fall victim to a predator decreases with the number of individual prey in the school;

as long as someone else in the school is caught you will probably not be caught yourself (Foster and Treherne 1981). A school of prey can hereby consist of a hypothetically large number of encounters for an individual pike, but in reality only one potentially effective encounter, while the prey per capita probability of mortality upon the same encounter is one divided by the number of individuals in the school. Simultaneously, a large number of prey individuals in the encountered school can cause confusion effects, where the schooling behaviour can reduce the possibility for pike to identify and discern single individual prey, decreasing the probability of the encounter leading to initiation of an attack by the pike (Landeau and Terborgh 1986).

2.3.3 Attack

Pike attacking prey is a spectacular event. Pike may show a variety of strike tactics and behaviours, but usually perform fast starts with extreme acceleration up to 130 ms^{-2} (Harper and Blake 1991, Frith and Blake 1995). This acceleration is accomplished by a combination of pike morphology, physiology and strike behaviour (Frith and Blake 1995, chapter 4). The exceptional acceleration ability commonly places foraging pike in an advantage over prey, as such acceleration capacity is quite rare and generally exceeds that of prey (Domenici and Blake 1997). Needless to say, the acceleration of pike takes place in a dense fluid and requires energy (Vogel 1994). This also means that the acceleration advantage relative to prey is valid only at close range. At the point of prey encounter at a distance exceeding individual strike distance (approximately half of the individual pike body length, Webb and Skadsen 1980, Raat 1988, Harper and Blake 1991), evasive prey behaviours have a high probability of success, and the pike is relatively unlikely to initiate a strike at all (Ranåker et al. 2012). If so, pike are faced with the decision of cancelling the predation-cycle progression, or, alternatively, move towards prey aiming to decrease attack distance. Pike may deliberately, very slowly and carefully approach and follow prey to obtain improved strike distance, position and/or angle (Hart and Connellan 1984, Nilsson et al. 1995, Nilsson and Brönmark 1999). Unless favourable strike circumstances are obtained, pike are less likely to successfully capture prey.

2.3.4 Capture

After encountering and choosing to attack prey, the next predation-cycle step for pike is to capture prey. Although being regarded as a very efficient attacker, pike capture success, i.e. attack followed by prey capture and subsequent successful consumption of prey (e.g. Turesson and Brönmark 2004), can be affected by a number of circumstances. For instance, prey individuals may differ in evasive and avoidance behavioural capacities, morphological defences, or body size, all of which could affect the probability of capture success (e.g. Nilsson et al. 1995, Christensen 1996, Nilsson and Brönmark 2000a, Clemente and Wilson 2016). In order to take the predation-cycle step from attack to capture of prey, pike must first efficiently direct their strike towards prey (Webb and Skadsen 1980). Failing to do so inevitably leads to a missed attack, consequently with zero probability of capture success. Pike can miss prey for several reasons. First, pike can make sloppy or unfocused attack attempts, for instance when recently fed or if prey behave in ways not triggering full strike attention. Moreover, as pike are very influenced by visual cues when making foraging decision, poor visibility conditions (see chapter 3) can reduce strike precision if conditions do not allow for e.g. accurate detection of distance and angle to prey. Interestingly, although deteriorated visibility conditions can reduce pike strike and capture success, extremely poor visibility can provide pike with a relative advantage over prey if pike and prey visual encounter occurs within the strike distance of the pike, leading to a higher relative probability of capture success (Jönsson et al. 2012). This relative advantage is presumably due to the reduced opportunity for prey to perform evasive behaviours during the brief moment between encounter and initiated strike.

If visual encounters occur at longer distance, the element of surprise as well as strike from short distance are no longer at work, and prey may be allowed time to behave adaptively according to the threat posed by pike. Prey fish can show a variety of evasive and cryptic behaviours, including shoaling/schooling, high swimming speed, acceleration/manoeuvrability, reduced movement, and habitat shift (Keenleyside 1979, Savino and Stein 1989, Magurran 1990, Christensen 1996, Webb and Fairchild 2001, Blake 2004). Some of these can be overridden by pike short-distance ambush attack also in high water visibility, while others are potentially at work in spite of pike attack and capture strategies. As mentioned above, confusion effects (Landeau and Terborgh 1986) via e.g. schooling could reduce pike ability to discern individual prey, and thereby diminish attack precision and consequently both capture rate and success. Moreover, as speed and manoeuvrability can jointly determine the outcome of escape performance in predator-prey interactions, both theoretically (Clemente and Wilson 2015) and empirically (Domenici et al. 2004), such prey behaviours should be involved also in the outcome of pike-prey interactions (Öhlund et al. 2015).

Pike are gape-size limited predators, and there are hence limits to the maximum prey to pike size ratio allowing prey ingestion. Prey size, and particularly prey body depth, can affect capture efficiency also below the size ratios limiting pike predation. Pike generally aim their strike at prey mid body, and prey with a deep mid-body morphology could make pike strike either anterior or posterior of mid body, resulting in a strike at body parts that move more during prey swimming (Webb and Skadsen 1980, Eklöv and Hamrin 1989). Striking relatively more moving body parts could reduce capture success by inducing missed attacks, and also reduce capture success as a first grip over the tail increases the handling and turning of prey before the pike is able to swallow the prey head first, which is the preferred mode of ingestion for *Esox* (Reimchen 1991, Nilsson and Brönmark 2000b). A prey adaptation on a similar note is the occurrence of false eyespots in prey fish, that can divert predator strikes towards eyespots (Kjernsmo and Merilaita 2013, Lönnstedt et al. 2013). Such false eyespots are sometimes formed and displayed just in front of the tail fin, and diverting predator strikes to this body end rather than the mid body or head means diverting strikes to less vulnerable body parts. In the case of pike as predator, it also enhances prey chances of first being missed and not struck at all, as well as if struck also enhances the escape potential during the prolonged handling time taken for pike to manipulate and turn prey to consume it.

2.3.5 Consume

After capturing a prey individual, a pike has to successfully manipulate and swallow it, i.e. consume it, to progress through the predation cycle. Although pike can grow very large and have a comparably large mouth, prey size may limit the consumption success for foraging pike. Pike generally swallow prey head first, so if a prey individual has a body depth greater than the gape width of a specific size of pike, the prey enjoys a size refuge from predation (Hambright 1991, Brönmark and Miner 1992, Nilsson and Brönmark 2000a). This means that capture of prey this deep cannot become successful, and the pike will eventually let go of the prey. Prey with body depths smaller than the gape size of a pike are vulnerable to predation, but deeper prey body depths increase handling time (e.g. Werner 1974, Hambright et al. 1991, Nilsson et al. 1995, Nilsson and Brönmark 2000a). Interestingly, with increasing handling time, prey not only increase probability of the pike losing its grip on prey, but also increases the probability of pike being detected and attacked by conspecifics. A long handling time increases the amount of visual, tactile and chemical cues produced and possibly attracting conspecifics, which increases the probability of conspecific encounter. Conspecific encounter can lead to conspecific attack in pike. If an attacking conspecific is of similar size to the pike handling prey, an attack may be aimed at the prey held in the mouth of the prey-handling pike, with possible kleptoparasitism as consequence. If the attacking conspecific is larger, and thereby a potential cannibal, an attack may instead be aimed for the prey-handling pike rather than the prey held in its mouth, leading to risk of cannibalism. Such cannibalistic attacks commonly also make the attacked pike let go of its prey, actually increasing the chance of survival for prey (Nilsson and Brönmark 1999). Consuming large, deep-bodied prey may thus be either impossible or risky for pike, with consequences for pike behaviour and efficiency during the consumption predation-cycle stage. To further emphasise the importance of the relation between pike gape size and prey size, Magnhagen and Heibo (2001) elegantly show that between-lake variation in prey size and morphology composition is reflected in lake-specific pike gape sizes; pike have relatively larger gapes in lakes with larger and deeper prey, i.e. pike adapt mouth morphology according to prey population composition. The study concludes possible genetic differentiation or phenotypic plasticity behind the finding, but also indicates a selective advantage of large gape sizes in pike populations.

Deep-bodied prey may enjoy a defence against gape-size limited pike predation, but this is not the only morphological prey-fish defence effective during the consumption predation-cycle stage. Prey fish such as perch (*Perca* spp.) and sticklebacks (*Gasterosteidae* spp.) have spines or spiny fin rays that complicate prey consumption for pike. Spines may substantially obstruct the swallowing if spines are caught in mouth parts of the pike. Spiny body parts of such fish are often directed or folded towards the posterior of the fish, and it should be essential for pike to swallow such prey head first as a tail-first attempt may be pointless even for small prey. Moreover, spiny prey could get stuck in the throat of the pike, potentially impairing gill ventilation and jeopardize not only successful foraging but also pike survival. Prey morphological defences can thus reduce the probability for pike to consume and ingest prey for digestion.

2.3.6 Digest

The final step of the pike predation cycle is digestion of food, and this step has to be dealt with to complete the predation cycle, as pike cannot resume foraging until there is enough stomach space to fit another prey item. Of course, if prey are very small compared to pike body size, several prey items may be ingested, while for larger prey digestion may limit how soon foraging can be resumed. Theoretically, taking several small prey fish instead of one large of the same total body mass could reduce digestion time, as several small prey should comprise a greater outer surface area on which digestive processes act (Jobling 1987, Bromley 1994, Andersen 1999). This, however, seems not to be the case at least for small compared to large crucian carp (*Carassius carassius* Linnaeus, 1758) prey, likely as a result of that small prey, closely packed together in the stomach, make up one collective outer surface for enzymes to work on that is comparable to the area of one large prey individual (Nilsson and Brönmark 2000b).

During early digestion of prey, pike mobility can be impaired for several potential reasons. The fast-start attack has probably filled the white, glycolytic muscles with lactic acid, and although pike have comparably low maintenance energy requirements the digestion process can in itself require energy (Johnson 1966, chapter 4), altogether reducing movement capacity. Furthermore, a very full stomach can distort the body shape of the pike and physically impair pike movement, and perhaps especially so if pike have swallowed e.g. spiny prey. Moreover, swallowing a large fish prey would not only fill the pike stomach, but also face the pike with the situation of an extra spine, from the prey, possibly reducing movement and swimming capacity. Reduced swimming efficiency implies lower opportunity for e.g. habitat change and predator avoidance, why the early stages of digestion can come with a cost to pike. Presumably, this cost may be the reason why pike individuals commonly regurgitate food upon stress from e.g. being caught or threatened (e.g. Treasurer 1988, personal observations). Although it would lead to losing a valuable and consumed food item, regurgitation should in theory reduce most of the movement-deteriorating aspects of having a full stomach, why it is reasonable to assume such a tradeoff behind the regurgitation behaviour.

The digestion of food in the stomach leaves gradually increasing room for further prey ingestion, and digestion is hereby the last factor affecting foraging resumption and thereby food intake rates for pike. The temporal delay that digestion of food can create on foraging is a part of the handling-time parameter h in the functional response, why digestion of food is important in terms of the predation cycle, pike foraging rates, pike and prey population composition, as well as higher-order community consequences. Before returning to the functional response and how it is affected by pike behaviours, we below devote some chapter space to pike optimal foraging decisions in light of pike foraging behaviours and intraspecific interactions.

2.4 PIKE PREY PREFERENCE

A sit-and-wait foraging strategy relies on prey movement for prey encounter. Movement strategies of prey may differ between sizes and/or species of prey, potentially leading to size-, density- and species-dependent encounter rates with prey. Such prey-dependent encounter rates could lead to patterns of prey selectivity in pike diets which may be the sole result of prey-dependent encounter probability if encounters are low and pike cannot afford to choose not to forage on encountered prey. Selectivity is in light of such processes not an active prey choice, in contrast to prey preference. Although selectivity can have far-reaching consequences in e.g. size-structured populations and communities (Ebenman and Persson 1988, Persson et al. 2004, chapter 8), the focus here lies on prey preference. Prey preference, here defined as a result of an active choice to attack or not depending on circumstances and prey characteristics, is governed by decision-making preceding attack, including the foraging value of prey individuals.

Pike should choose and attack prey so as to maximize energy intake over time, according to optimal foraging theory (Pyke et al. 1977, Stephens and Krebs 1986, Juanes 1994). This implies adaptive prey preference in pike. Pike commonly coexist with a number of potential prey fish species forming size-structured populations (Ebenman and Persson 1988, Raat 1988, Persson et al. 1996), introducing pike to a choice of size- and species-specific prey values. Pike should choose prey to maximise energy (e) over handling time (h) to optimise prey value (e/h) (e.g. Hart and Connellan 1984, Stephens and Krebs 1986, Nilsson et al. 2000). Large prey should include more energy than small, at first suggesting a possible e-maximising (and hence prey-size maximising) rule for pike. Handling time however increases non-linearly with prey size (S) for pike, generally following a power function

$$h = \alpha + \beta S^{\gamma} \tag{2.2}$$

where α is a positive y-axis intercept and β and γ are positive constants shaping the prey-size to handling-time relationship (Fig. 2.3). Assuming that prey size is an appropriate proxy for energy content of prey, prey value (V) could be estimated as

$$V = \frac{S}{h} \tag{2.3}$$

illustrating relatively lower prey values for very small and very large prey (Fig. 2.3).

As optimal foraging theory suggests to maximize energy intake per handling time, an optimal prey size (S^*)

$$S^* = \sqrt[\gamma]{\frac{\alpha}{\beta(\gamma - 1)}} \tag{2.4}$$

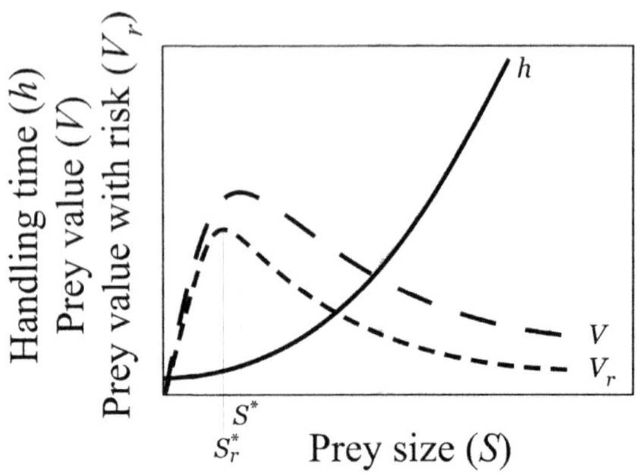

FIGURE 2.3 Schematic relationships between prey size (*S*) and handling time (solid curve, *h*), prey value (long-dashed curve, *V*), as well as prey value when handling time includes a risk (short-dashed curve, V_r). Vertical hairlines indicate optimal prey size in the absence (S^*) and presence (S_r^*) of risk. Inspired by Nilsson et al. (2000).

appears at a relatively small, but not smallest, prey to pike size ratio (Fig. 2.3), and pike should according to this prediction choose to attack prey of such sizes if given the opportunity. Individual pike choosing such small prey in natural settings could result from size-determined encounter rates, i.e. non-active selective processes, as size-structured prey fish populations generally consist of numerous young and small individuals (Ebenman and Persson 1988). As pike in controlled, experimental settings, with equal abundances of different prey sizes, repeatedly show preference for smaller prey (e.g. Beyerle and Williams 1968, Hart and Connellan 1984, Hart and Hamrin 1988, Wahl and Stein 1988, Hart and Hamrin 1990, Juanes 1994, Nilsson and Brönmark 2000a, Nilsson et al. 2000, Jönsson et al. 2013), it is however reasonable to assume that pike have adapted to the relatively higher value of small (but not smallest) prey, and forage accordingly.

The decreasing value of increasingly large prey above S^* originates from pike being gape-size limited predators. Prey body depth is the crucial measure of prey size, as pike after striking their prey, aiming at mid body, first lay prey down on the side and then handle and turn them to be able to swallow them head first, as this reduces handling time (Webb and Skadsen 1980, Raat 1988, Nilsson and Brönmark 2000b). It is therefore prey body depth, rather than length, that determines handling time for pike (Nilsson et al. 1995, Nilsson and Brönmark 2000a). As prey body depth approaches the gape-size limit of the pike individual, handling and manipulating prey becomes increasingly troublesome for the pike, why the rate of handling-time change increases with increasing prey to pike size ratio (Fig. 2.3), consequently reducing prey value, e.g. prey mass per unit time, above S^*. Prey body depth is hereby an important prey attribute affecting pike optimal foraging, and pike accordingly choose to forage on shallow-bodied prey phenotypes and species before deep-bodied prey of comparable body length (Mauck and Coble 1971, Nilsson et al. 1995, Nilsson and Brönmark 2000a). Pike thus actively choose to forage on small and preferably shallow-bodied prey according to time and energy budgets, and depending on the abundance and morphology composition of prey fish communities, such preferential predation can influence the effects of pike predation in aquatic systems (chapter 8).

The above reasoning on prey morphology- and size-determined handling times for pike does not consider e.g. specialized, inducible or constitutive morphological defence adaptations in prey (Harvell 1990). Prey fish can have spiny fins and rays that can affect pike foraging decisions and handling time beyond size-specifying body depth. For instance, pike choose soft-rayed rudd (*Scardinus erythrophthalmus* Linnaeus, 1758) before perch (*Perca fluviatilis* Linnaeus, 1758) when presented with prey within a narrow size range (Eklöv and Hamrin 1989). Moreover, the inducible morphological defence in crucian carp where induced individuals grow a deeper body (Brönmark and Miner 1992), also results in prey preference in pike, that choose shallow-bodied individuals if possible (Nilsson et al. 1995). Prey preference in pike is hereby influenced also by prey defence adaptations.

Prey preference in pike is thus far in this text described to be governed by prey size and the corresponding time it takes to handle prey, which is in line with the bases of optimal foraging theory (Stephens and Krebs 1986). Pike, as most foragers, however also behave in social contexts, and social foraging theory introduces further complexity to optimal foraging (Giraldeau and Caraco 2000). Pike social behaviours during foraging are relatively straightforward: avoid conspecifics as they can eat you or steal your food. Foraging under risk of such intraspecific interactions has consequences for optimal prey choice in pike.

2.4.1 Prey Preference under Risk

The reasoning behind equation 2.3 considers that pike should maximise prey mass per unit time to adhere to the assumptions of optimal foraging theory. This assumes that time is a cost in relation to energy contents of prey. For pike, handling time is however not a cost in time alone; time is also a cost in the form of risk (see also e.g. Gilliam 1990, Godin 1990). When handling prey after a successful attack, pike manipulate the prey in the mouth to be able to swallow it head first, and the movements involved with this manipulation creates cues that can attract conspecifics (Mathis and Smith 1993, Chivers and Smith 1998, Nilsson and Brönmark 1999). Attracting conspecifics implies attracting cannibals if approaching conspecifics are larger than the pike handling its prey (Bry et al. 1992). There are obvious fitness benefits of avoiding falling victim to cannibals, and one way of achieving avoidance is to reduce the time handling prey to reduce signals sent and thereby also probability of cannibalistic attack. For pike, also similar-sized conspecifics can pose a threat in the form of kleptoparasitism, depriving the victim of both the captured prey and the time and energy spent foraging until attack (Goss-Custard 1996, Nilsson and Brönmark 1999, Giraldeau and Caraco 2000, Krause and Ruxton 2002). In pike, these costs of falling victim to kleptoparasites can be accompanied by severe injury from attack, which should also be the case for unsuccessful cannibalistic attacks. Handling time in equation 2.2 is in pike reality hereby not only a cost of time *per se*, but also a potentially substantial risk of time-related intraspecific interaction. To minimize risk of interactions, pike should consider prey value as both time in itself and the time-related risk (r), making prey value under risk (V_r)

$$V_r = \frac{S}{h_r} = \frac{S}{\alpha + \beta S^{r\gamma}} \tag{2.5}$$

and the subsequent optimal prey size under risk of interaction (S_r^*) follow

$$S_r^* = \sqrt[r\gamma]{\frac{\alpha}{\beta(r\gamma - 1)}} \tag{2.6}$$

As long as risk is real, then $r > 1$ and $S_r^* < S^*$, why pike under risk of interaction should choose prey smaller than available and optimal in the absence of risk (Fig. 2.3), which is corroborated by experimental data (Nilsson et al. 2000).

Pike prey-size preference is under the influence of prey size, time constraints, and risk of intraspecific interactions. Choosing prey smaller than optimal from *e/h* perspectives can affect energy intake rates and budgets for pike, and effects on foraging rates can affect individual success and pike effects in communities. Moreover, in light of pike progression through the predation cycle as described above, risk of intraspecific interactions may reduce not only the prey size chosen by pike, but even reduce overall foraging propensity or efficiency. Factors affecting food intake in pike has direct links to the functional response, and we here return to the functional response to evaluate its behaviour under risk of intraspecific interactions.

2.5 PIKE INTRASPECIFIC INTERACTIONS AND FUNCTIONAL RESPONSES

Pike can show agonistic, kleptoparasitic and cannibalistic behaviours (Grimm 1981, Giles et al. 1986, Raat 1988, Bry et al. 1992, Craig 1996, Nilsson and Brönmark 1999). Exposure to such interaction could involve fitness costs, and pike individuals should hence strive to avoid intraspecific interactions. Avoidance of interaction could be accomplished by spatially and/or temporally avoiding conspecifics (Hawkins et al. 2005, Nilsson 2006); the risk of intraspecific interaction should disappear in the absence of conspecifics. If this is not possible, pike in the vicinity of conspecifics may avoid attracting the attention of conspecifics by, for instance, minimizing movement, by optimizing the time handling prey, or by altering foraging behaviours, as mentioned above in this chapter. Such avoidance behaviours are linked to decisions made during the predation cycle, and as predation-cycle efficiency has implications for the functional response, pike avoidance behaviours can affect the functional response. For instance, spatially avoiding conspecifics may simultaneously incur abandoning favourable foraging spots, reducing foraging potential for pike individuals. Moreover, minimizing movement by reducing foraging behaviours such as attacks on prey to avoid attracting conspecific attention should lead to reduced food intake rates (Nilsson et al. 2006). Furthermore, optimizing handling time by preferring smaller prey in the presence of conspecifics can imply reductions in prey encounter and energy intake, as a narrower prey-type window would reduce foraging opportunity (Mittelbach and Persson 1998). Behaviours such as these, adopted to avoid risk from conspecifics with effects on per capita intake rates, should act as interference on pike foraging. Interference among foragers reduces per capita prey-dependent food intake rates (see e.g. Triplet et al. 1999, Giraldeau and Caraco 2000, Skalski and Gilliam 2001), and interference can be incorporated in the functional response [equation (2.1)], for instance according to an extension of Hassel and Varley (1969)

$$I_m = \frac{aNP^{-m}}{1 + ahN} \tag{2.7}$$

where P is predator density and m is an interference parameter that decreases per capita foraging rates when $m > 0$. Depending on the type of behavioural interactions among foragers, interference can have different effects on the shape of the functional response (Fryxell and Lundberg 1998, Skalski and Gilliam 2001, Nilsson et al. 2004, Nilsson et al. 2007). For pike, it is possible that larger and/or aggressive individuals could suppress foraging propensity in neighbouring pike by constantly intimidating them from foraging. Exposure to intimidation could hereby lead to general hesitation to forage and consequently make completion of the predation cycle take longer time. If this effect on foraging propensity is not affected by prey availability, the parameter m would be a positive constant resulting in a reduced functional response according to the long-dashed curve in Fig. 2.1. The interference effect of conspecific presence and interaction could also be prey dependent (Abrams and Ginzburg 2000, Skalski and Gilliam 2001, Nilsson et al. 2003). This means that the parameter m is a function of prey density, leading to high interference effects at low prey density, but reduced effects of interference at high prey densities (short-dashed curve in Fig. 2.1). This scenario is possible if pike agonistic behaviours decrease with increasing prey availability, or if, for instance, prey encounters increase in safer microhabitats at higher prey densities, and hereby release foraging individuals from the imminent threat from conspecific presence.

Interference among foragers, regardless of mechanistic contribution to the shape of the functional response, should produce stabilizing effects in community dynamics. Such stabilization originates from the reduced functional and numerical response of predators to prey density, and subsequently dampened variation in predator density over time. In the case of intimidation interference (m is a positive constant), the stabilizing effects should be stronger than for prey-dependent interference as high prey densities would in the latter situation allow for higher per capita predation rates (Fig. 2.1) and subsequently lead to relatively higher numerical predator responses and comparably less stabilized consumer-resource dynamics (e.g. Fryxell and Lundberg 1998, Nilsson et al. 2007). Moreover, the behaviours behind the interference in pike can via shifts in prey-size preference and predation intensities, along with altered pike densities from cannibalism, have far-reaching trophic effects through top-down cascades from pike predation (e.g. Carpenter and Kitchell 1993, Nilsson 2001). These higher-order processes are mechanistically governed by individual foraging and interaction behaviours why they deserve mentioning in this chapter, while they are paid further attention in chapters 6, 8 and 11 of this volume.

The exact nature of pike intraspecific interference and its effects on the functional response, communities and trophic systems remain elusive, as pike per capita prey- and predator-dependent intake rates to our knowledge are missing from the literature. Until such experimental work has successfully quantified pike functional responses, their precise shape and higher-order consequences remain unknown, but it seems likely that both predator- and prey-dependent effects act in combination to determine per capita intake rates in pike. Moreover, as also environmental variables such as visibility conditions can affect pike predation efficiency and interference propensity (Nilsson et al. 2009, Jönsson et al. 2012, chapter 3), the links between individual behaviours and higher-order processes may be difficult to predict without further quantification of effects. Furthermore, pike populations should consist of individuals with diverse behavioural attributes also within sizes and ages, why many of the assumptions around pike success and effects in aquatic systems may underestimate important and interesting variation.

2.6 PIKE SPECIALIZATION AND PHENOTYPIC PLASTICITY

The predation-cycle approach to describe pike foraging behaviour, functional responses and their higher-order consequences is based on an optimality perspective that hitherto in this text does not consider diversity in adaptive traits. Such optimality perspective suggests that all individuals behave similarly according to prerequisites. This is not necessarily the case for pike. There is growing evidence that individual variation within populations can substantially affect acquisition of resources, competitive success and predator avoidance in various species (Bolnick et al. 2003, Eklöv and Svanbäck 2006, Sih et al. 2012, Svanbäck et al. 2015). Specifically, the trade-off between foraging and predation risk can drive individual specialization leading to divergence within subpopulations (Eklöv and Svanbäck 2006, Svanbäck et al. 2015). Moreover, changes in body size over ontogeny may cause individuals to experience differences in foraging opportunity and risks, leading to differences in individual decisions of finding and exploiting food resources (Svanbäck et al. 2015). Although often considered a specialist piscivore, strong individual diet specialization and plasticity, in both short-term diet and long-term trophic position, occur in pike (Beaudoin et al. 1999, Pedreschi et al. 2015). Specialization and plasticity may indicate within-population differences in foraging strategies and also a potential mechanism behind the development of different behavioural types of pike (Kobler et al. 2009). Different behavioural types of pike were found to relate to resource density and habitat structural complexity (Kobler et al. 2009). Although pike has a predominately sit-and-wait foraging mode, changes in resource densities can force pike to more actively forage in open water, similarly to that high water turbidity can induce more frequent use of open water (Andersen et al. 2008). However, Kobler et al. (2009) showed that only one part of the population responded by shifting to the open water habitat and suggested that density and interference competition could be important mechanisms involved in differential habitat use or creating different behavioural types of pike (see also Skov et al. 2007, Nilsson et al. 2012). The behavioural types related to swimming activity and habitat use of pike described in the Kobler et al. (2009) study could also correspond to individual specific personal traits (Sih et al. 2004, Réale et al. 2007). Such traits have been observed as ubiquitous in fish populations and suggested to be of high ecological and evolutionary importance since they, in turn, correlate with a whole suite of physiological and life-history traits (e.g. Bell 2007, Wolf and Weissing 2012, Mittelbach et al. 2014). In fact, within-population variation in behavioural attributes have been shown not only in swimming behaviour and habitat choice in pike, but also in behavioural characteristics including foraging behaviour, explorative behaviour and activity in larval, juvenile as well as adult pike (Kobler et al. 2009, Nyqvist et al. 2012, McGhee et al. 2013, Pintor et al. 2014, Pasquet et al. 2015, Laskowski et al. 2016). Needless to say, if such differences among individuals within populations prevail, optimal behaviours should also differ between individuals and cause variation in predation-cycle efficiency and performance, with consequences for the functional response and higher-order processes. If such variation causes shifts in optimality theory for pike or adds explanation to variation around predicted behaviours, it most certainly deserves the attention in future work pursuing mechanisms behind how pike find food while staying alive.

REFERENCES CITED

Abrams, P.A. and L.R. Ginzburg. 2000. The nature of predation: prey dependent, ratio dependent or neither? *Trends Ecol. Evol.* 15:337-341.

Andersen, M., L. Jacobsen, P. Grønkjaer and C. Skov. 2008. Turbidity increases behavioural diversity in northern pike, *Esox lucius* L., during early summer. *Fish. Manag. Ecol.* 15:377-383.

Andersen, N.G. 1999. The effects of predator size, temperature, and prey characteristics on gastric evacuation in whiting. *J. Fish. Biol.* 54:287-301.

Baktoft, H., K. Aarestrup, S. Berg, M. Boel, L. Jacobsen, N. Jepsen, A. Koed, J.C. Svendsen and C. Skov. 2012. Seasonal and diel effects on the activity of northern pike studied by high-resolution positional telemetry. *Ecol. Freshw. Fish.* 21:386-394.

Beaudoin, C.P., W.M. Tonn, E.E. Prepas and L.I. Wassenaar. 1999. Individual specialization and trophic adaptability of northern pike (*Esox lucius*): an isotope and dietary analysis. *Oecologia* 120(3):386-396.

Bell, A.M. 2007. Animal personalities. *Nature* 447:539-540.

Beyerle, G.B. and J.E. Williams. 1968. Some observations of food selectivity by northern pike in aquaria. *Trans. Am. Fish. Soc.* 97:28-31.

Billard, R. 1996. Reproduction of pike: gametogenesis, gamete biology and early development. pp 13-44. *In*: J.F. Craig (ed.). *Pike: Biology and exploitation*. London: Chapman & Hall.

Blake, R.W. 2004. Fish functional design and swimming performance. *J. Fish. Biol.* 65(5): 1193-1222.

Bleckmann, H. 1986. Role of the lateral line in fish behaviour. pp 177-202. *In*: T.J. Pitcher (ed.). *The behaviour of teleost fishes*. New York, NY: Springer.

Bolnick, D.I., R. Svanbäck, J.A. Fordyce, L.H. Yang, J.M. Davis, C.D. Hulsey and M.L. Forister. 2003. The ecology of individuals: Incidence and implications of individual specialization. *Am. Nat.* 161(1):1-28.

Bromley, P.J. 1994. The role of gastric evacuation experiments in quantifying the feeding rates of predatory fish. *Rev. Fish Biol. Fish.* 4:36-66.

Brown, G.E., D.P. Chivers and R.J.F. Smith. 1995. Localized defecation by pike: a response to labelling by cyprinid alarm pheromone? *Behav. Ecol. Sociobiol.* 36:105-110.

Brown, G.E. 2003. Learning about danger: chemical alarm cues and local risk assessment in prey fishes. *Fish and Fish.* 4(3):227-234.

Brown, R.J. and C. McIntyre. 2005. New prey species documented for northern pike (*Esox lucius*): bald eagle (*Haliaeetus leucocephalus*). *Arctic* 58:437-437.

Bry, C., E. Basset, X. Rognon and F. Bonamy. 1992. Analysis of sibling cannibalism among pike, *Esox lucius*, juveniles reared under semi-natural conditions. *Env. Biol. Fish.* 35:75-84.

Brönmark, C. and J.G. Miner. 1992. Predator-induced phenotypical change in body morphology in crucian carp. *Science* 258:1348-1350.

Brönmark, C., C. Skov, J. Brodersen, P.A. Nilsson and L.-A. Hansson. 2008. Seasonal migration determined by a trade-off between predator avoidance and growth. *PLoS ONE* 3(4):e1957.

Carpenter, S.R. and J.F. Kitchell. 1993. *The trophic cascade in lakes.* Cambridge, UK: Cambridge University Press.

Casselman, J.M. and C.A. Lewis. 1996. Habitat requirements of northern pike (*Esox lucius*). *Can. J. Fish. Aquat. Sci.* 53:161-174.

Chivers, D.P. and R.J.F. Smith. 1998. Chemical alarm signalling in aquatic predator-prey systems: a review and prospectus. *Écoscience* 5:338-352.

Chivers, D.P., R.S. Mirza, P.J. Bryer and J.M. Kiesecker. 2001. Threat-sensitive predator avoidance by slimy sculpins: understanding the importance of visual versus chemical information. *Can. J. Zool.* 79(5):867-873.

Christensen, B. 1996. Predator foraging capabilities and prey antipredator behaviours: pre- versus post capture constraints on size-dependent predator-prey interactions. *Oikos* 76:368-380.

Clemente, C.J. and R.S. Wilson. 2015. Balancing Biomechanical Constraints: Optimal escape speeds when there is a trade-off between speed and maneuverability. *Integr. Comp. Biol.* 55(6):1142-1154.

Clemente, C.J. and R.S. Wilson. 2016. Speed and maneuverability jointly determine escape success: exploring the functional bases of escape performance using simulated games. *Behav. Ecol.* 27(1):45-54.

Craig, J.F. (ed.). 1996. Pike: Biology and Exploitation. London: Chapman & Hall.

Diana, J.S. 1979. The feeding pattern and daily ration of a top carnivore, the northern pike (*Esox lucius*). *Can. J. Zool.* 57:2121-2127.

Domenici, P. and R.W. Blake. 1997. The kinematics and performance of fish fast-start swimming. *J. Exp. Biol.* 200(8):1165-1178.

Domenici, P., E.M. Standen and R.P. Levine. 2004. Escape manoeuvres in the spiny dogfish (*Squalus acanthilas*). *J. Exp. Biol.* 207(13):2339-2349.

Ebenman, B. and L. Persson. 1988. Size-structured Populations. Berlin: Springer-Verlag.

Eklöv, P. and S.F. Hamrin. 1989. Predatory efficiency and prey selection: interactions between pike *Esox lucius*, perch *Perca fluviatilis* and rudd *Scardinus erythrophthalmus*. *Oikos* 56:149-156.

Eklöv, P. and L. Persson. 1996. The response of prey to the risk of predation: proximate cues for refuging juvenile fish. *Anim. Behav.* 51:105-115.

Eklöv, P. 1997. Effects of habitat complexity and prey abundance on the spatial and temporal distributions of perch (*Perca fluviatilis*) and pike (*Esox lucius*). *Can. J. Fish. Aquat. Sci.* 54:1520-1531.

Eklöv, P. and R. Svanbäck. 2006. Predation risk influences adaptive morphological variation in fish populations. *Am. Nat.* 167(3):440-452.

Endler, J.A. 1991. Interactions between predators and prey. pp 169-196. *In*: J.R. Krebs and N.B. Davies (eds.). *Behavioural ecology: an evolutionary approach.* Oxford: Blackwell Scientific Publications.

Engström, H. 2001. Long term effects of cormoratn predation on fish communities and fishery in a freshwater lake. *Ecography* 24:127-138.

Engström-Öst, J., M. Lehtiniemi, S.H. Jonasdottir and M. Viitasalo. 2005. Growth of pike larvae (Esox lucius) under different conditions of food quality and salinity. *Ecol. Freshw. Fish* 14(4):385-393.

Foster, W.A. and J.E. Treherne. 1981. Evidence for the dilution effect in the selfish herd from fish predation on a marine insect. *Nature* 293:466-467.

Frith, H.R. and R.W. Blake. 1995. The Mechanical Power Output and Hydromechanical Efficiency of Northern Pike (*Esox lucius*) Fast-Starts. *J. Exp. Biol.* 198(9):1863-1873.

Fryxell, J.M. and P. Lundberg. 1998. *Individual Behavior and Community Dynamics.* London: Chapman & Hall.

Giles, N., M.R. Wright and M.E. Nord. 1986. Cannibalism in pike fry, *Esox lucius* L.: some experiments with fry densities. *J. Fish. Biol.* 29:107-113.

Gilliam, J.F. 1990. Hunting by the hunted: optimal prey selection by foragers under predation hazard. pp 797-819. *In*: R.N. Hughes (ed.). *Behavioural mechanisms of food selection,* NATO ASI Series Vol. G 20.

Giraldeau, L.-A. and T. Caraco. 2000. *Social foraging theory.* New Jersey: Princeton University Press.

Godin, J.-G.J. 1990. Diet selection under the risk of predation. pp 739-770. *In*: R.N. Hughes (ed.). *Behavioural mechanisms of food delection.* NATO ASI Series Vol. G 20.

Goss-Custard, J.D. 1996. *The Oystercatcher: From Individuals to Populations.* Oxford: Oxford University Press.

Grimm, M.P. 1981. Intraspecific predation as a principal factor controlling the biomass of northern pike (*Esox lucius* L.). *Fish. Manage.* 12:77-79.

Grimm, M.P. and M. Klinge. 1996. Pike and some aspects of its dependence on vegetation. pp 125-156. In: J.F. Craig (ed.). *Pike: Biology and exploitation.* London: Chapman & Hall.

Hambright, K.D. 1991. Experimental analysis of prey selection by largemouth bass: role of predator mouth width and prey body depth. *Trans. Am. Fish. Soc.* 120:500-508.

Hambright, K.D., R.W. Drenner, S.R. McComas and N.G.J. Hairston. 1991. Gape-limited piscivores: planktivore size refuges, and the trophic cascade hypothesis. *Arch. Hydrobiol.* 121:389-404.

Harper, D.G. and R.W. Blake. 1991. Prey capture and the fast-start performance of northern pike *Esox lucius*. *J. Exp. Biol.* 155:175-192.

Hart, P. and S.F. Hamrin. 1988. Pike as a selective predator. Effects of prey size, availability, cover and pike jaw dimensions. *Oikos* 51:220-226.

Hart, P.J.B. and B. Connellan. 1984. Cost of prey capture, growth rate and ration size in pike, *Esox lucius* L., as functions of prey weight. *J. Fish. Biol.* 25:279-292.

Hart, P.J.B. and S.F. Hamrin. 1990. The role of behaviour and morphology in the selection of prey by pike. pp 235-254. *In*: R.N. Hughes (ed.). *Behavioural mechanisms of food selection.* NATO ASI Series, Vol. G 20.

Harvell, C.D. 1990. The ecology and evolution of inducible defences. *Q. Rev. Biol.* 65:323-340.

Hassell, M.P. and G.C. Varley. 1969. New inductive population model for insect parasites and its bearing on biological control. *Nature* 223:1133-1137.

Hawkins, L.A., J.D. Armstrong and A.E. Magurran. 2005. Aggregation in juvenile pike (*Esox lucius*): interactions between habitat and density in early winter. *Funct. Ecol.* 19:794-799.

Holling, C.S. 1959. The components of predation as revealed by a study of small-mammal predation of the European pine sawfly. *Can. Entomol.* 91:293-320.
Holling, C.S. 1965. The functional response of predators to prey density and its role in mimicry and population regulation. *Mem. Entomol. Soc. Can.* 45:1-60.
Ince, B.W. and A. Thorpe. 1976. Effects of starvation and force-feeding on metabolism of northern pike, *Esox lucius* L. *J. Fish. Biol.* 8(1):79-88.
Jepsen, N., S. Beck, C. Skov and A. Koed. 2001. Behavior of pike (*Esox lucius* L.) >50 cm in a turbid reservoir and in a clearwater lake. *Ecol. Freshw. Fish* 10:26-34.
Jobling, M. 1987. Influences of food particle size and dietary energy content on patterns of gastric evacuation in fish: test of a physiological model of gastric emptying. *J. Fish. Biol.* 30:299-314.
Johnson, L. 1966. Experimental determination of food consumption of pike, *Esox lucius*, for growth and maintenance. *J. Fish. Res. Bd. Can.* 23(10):1495-1505.
Juanes, F. 1994. What determines prey size selectivity in piscivorous fishes? pp 79-100. *In*: D.J. Stouder, K.L. Fresh and R.J. Feller (eds.). *Theory and application in fish feeding ecology*. Columbia: Carolina University Press.
Jönsson, M., L. Ranåker, P.A. Nilsson and C. Brönmark. 2012. Prey-type-dependent foraging of young-of-the-year fish in turbid and humic environments. *Ecol. Freshw. Fish* 21(3):461-468.
Jönsson, M., L. Ranåker, P.A. Nilsson and C. Brönmark. 2013. Foraging efficiency and prey selectivity in a visual predator: differential effects of turbid and humic water. *Can. J. Fish. Aquat. Sci.* 70(12):1685-1690.
Keenleyside, M.H.A. 1979. *Diversity and Adaptation in Fish Behaviour*. Berlin: Springer-Verlag.
Kjernsmo, K. and S. Merilaita. 2013. Eyespots divert attacks by fish. *Proc. R. Soc. Lond. B* 280:20131458.
Knight, C.M., R.E. Gozlan and M.C. Lucas. 2008. Can seasonal home-range size in pike Esox lucius predict excursion distance? *J. Fish. Biol.* 73:1058-1064.
Kobler, A., T. Klefoth, C. Wolter, F. Fredrich and R. Arlinghaus. 2008. Contrasting pike (*Esox lucius* L.) movement and habitat choice between summer and winter in a small lake. *Hydrobiologia* 601:17-27.
Kobler, A., T. Klefoth, T. Mehner and R. Arlinghaus. 2009. Coexistence of behavioural types in an aquatic top predator: a response to resource limitation? *Oecologia* 161(4):837-847.
Krause, J. and G.D. Ruxton. 2002. *Living in Groups*. Oxford Series in Ecology and Evolution. Oxford: Oxford University Press.
Landeau, L. and J. Terborgh. 1986. Oddity and the 'confusion effect' in predation. *Anim. Behav.* 34:1372-1380.
Laskowski, K.L., C.T. Monk, G. Polverino, J. Alos, S. Nakayama, G. Staaks, T. Mehner and R. Arlinghaus. 2016. Behaviour in a standardized assay, but not metabolic or growth rate, predicts behavioural variation in an adult aquatic top predator *Esox lucius* in the wild. *J. Fish. Biol.* 88(4):1544-1563.
Lehtiniemi, M., J. Engström-Öst and M. Viitasalo. 2005. Turbidity decreases anti-predatory behaviour in pike larvae, *Esox lucius*. *Env. Biol. Fish.* 73:1-8.
Lönnstedt, O.M., M.I. McCormick and D.P. Chivers. 2013. Predator-induced changes in the growth of eyes and false eyespots. *Sci. Rep.* 3:2259.
Magnhagen, C. and E. Heibo. 2001. Gape size allometry in pike reflects variation between lakes in prey availability and relative body depth. *Funct. Ecol.* 15:754-762.
Magurran, A.E. 1990. The adaptive significance of schooling as an anti-predator defence in fish. *Ann. Zool. Fennici* 27:51-66.

Mathis, A. and R.J.F. Smith. 1993. Chemical alarm signals increase the survival-time of fathead minnows (*Pimephales promelas*) during encounters with northern pike (*Esox lucius*). *Behav. Ecol.* 4(3):260-265.
Mauck, W.L. and D.W. Coble. 1971. Vulnerability of some fishes to northen pike (*Esox lucius*) predation. *J. Fish. Res. Bd. Can.* 28:957-969.
McGhee, K.E., L.M. Pintor and A.M. Bell. 2013. Reciprocal Behavioral Plasticity and Behavioral Types during Predator-Prey Interactions. *Am. Nat.* 182(6):704-717.
Mittelbach, G.G. and L. Persson. 1998. The ontogeny of piscivory and its ecological consequences. *Can. J. Fish. Aquat. Sci.* 55:1454-1465.
Mittelbach, G.G., N.G. Ballew and M.K. Kjelvik. 2014. Fish behavioral types and their ecological consequences. *Can. J. Fish. Aquat. Sci.* 71(6):927-944.
New, J.G., L. Alborg Fewkes and A.N. Khan. 2001. Strike feeding behavior in the muskellunge, *Esox masquinongy*: contributions of the lateral line and visual sensory system. *J. Exp. Biol.* 204:1207-1221.
Nilsson, P.A., C. Brönmark and L.B. Pettersson. 1995. Benefits of a predator-induced morphology in crucian carp. *Oecologia* 104:291-296.
Nilsson, P.A. and C. Brönmark. 1999. Foraging among cannibals and kleptoparasites: effects of prey size on pike behavior. *Behav. Ecol.* 10:557-566.
Nilsson, P.A., K. Nilsson and P. Nyström. 2000. Does risk of intraspecific interactions induce shifts in prey-size preference in aquatic predators? *Behav. Ecol. Sociobiol.* 48:268-275.
Nilsson, P.A. and C. Brönmark. 2000a. Prey vulnerability to a gape-size limited predator: behavioural and morphological impacts on northern pike piscivory. *Oikos* 88:539-546.
Nilsson, P.A. and C. Brönmark. 2000b. The role of gastric evacuation rate in handling time of equal-mass rations of different prey sizes in northern pike. *J. Fish. Biol.* 57:516-524.
Nilsson, P.A. 2001. Predator behaviour and prey density: evaluating density-dependent intraspecific interactions on predator functional responses. *J. Anim. Ecol.* 70:14-19.
Nilsson, P.A., G.D. Ruxton and J.H. Nilsson. 2003. Temporally fluctuating prey and coexistence among unequal conspecific interferers. *Oikos* 101:411-415.
Nilsson, P.A., F.A. Huntingford and J.D. Armstrong. 2004. Using the functional response to determine the nature of unequal interference among foragers. *Proc. Roy. Soc. Lond. B.* (Suppl.) 271:S334-S337.
Nilsson, P.A. 2006. Avoid your neighbours: size-determined spatial distribution patterns among northern pike individuals. *Oikos* 113:251-258.
Nilsson, P.A., H. Turesson and C. Brönmark. 2006. Friends and foes in foraging: intraspecific interactions act on foraging-cycle stages. *Behaviour* 143:733-745.
Nilsson, P.A., P. Lundberg, C. Brönmark, A. Persson and H. Turesson. 2007. Behavioral interference and facilitation in the foraging cycle determine the functional response. *Behav. Ecol.* 18:354-357.
Nilsson, P.A., L. Jacobsen, S. Berg and C. Skov. 2009. Environmental conditions and intraspecific interference: unexpected effects of turbidity on pike (*Esox lucius*) foraging. *Ethology* 115:33-38.
Nilsson, P.A., H. Baktoft, M. Boel, K. Meier, L. Jacobsen, E.M. Rokkjær, T. Clausen and C. Skov. 2012. Visibility conditions and diel period affect small-scale spatio-temporal behaviour of pike *Esox lucius* in the absence of prey and conspecifics. *J. Fish. Biol.* 80:2384-2389.
Nyqvist, M.J., R.E. Gozlan, J. Cucherousset and J.R. Britton. 2012. Behavioural Syndrome in a Solitary Predator Is Independent of Body Size and Growth Rate. *PLoS ONE* 7(2):e31619.
O'Brien, W.J., H.I. Browman and B.I. Evans. 1990. Search strategies for foraging animals. *Amer. Sci.* 78:152-160.

Öhlund, G., P. Hedström, S. Norman, C.L. Hein and G. Englund. 2015. Temperature dependence of predation depends on the relative performance of predators and prey. *Proc. R. Soc. Lond. B.* 282(1799):20142254.

Pasquet, A., A. Sebastian, M.L. Begout, Y. LeDore, F. Teletchea and P. Fontaine. 2015. First insight into personality traits in Northern pike (*Esox lucius*) larvae: a basis for behavioural studies of early life stages. *Env. Biol. Fish.* 99(1):105-115.

Pedreschi, D., S. Mariani, J. Coughlan, C.C. Voigt, M. O'Grady, J. Caffrey and M. Kelly-Quinn. 2015. Trophic flexibility and opportunism in pike *Esox lucius*. *J. Fish. Biol.* 87(4):876-894.

Persson, L., J. Andersson, E. Wahlström and P. Eklöv. 1996. Size-specific interactions in lake systems: predator gape limitation and prey growth rate and mortality. *Ecology* 77:900-911.

Persson, L., P. Byström, E. Wahlström and E. Westman. 2004. Trophic dynamics in a whole lake experiment: size-structured interactions and recruitment variation. *Oikos* 106(2):263-274.

Pettersson, L.B. and C. Brönmark. 1993. Trading off safety against food: state dependent habitat choice and foraging in crucian carp. *Oecologia* 95:353-357.

Pettersson, L.B., P.A. Nilsson and C. Brönmark. 2000. Predator recognition and defence strategies in crucian carp. *Oikos* 88:200-212.

Pintor, L.M., K.E. McGhee, D.P. Roche and A.M. Bell. 2014. Individual variation in foraging behavior reveals a trade-off between flexibility and performance of a top predator. *Behav. Ecol. Sociobiol.* 68(10):1711-1722.

Pyke, G.H., H.R. Pulliam and E.L. Charnov. 1977. Optimal foraging: a selective review of theory and tests. *Q. Rev. Biol.* 50:137-154.

Raat, A.J.P. 1988. Synopsis of the biological data on the northern pike, *Esox lucius* Linnaeus, 1758. *FAO Fisheries Synopsis*: No. 30, Rev. 2, 178 p.

Ranta, E., P. Lundberg and V. Kaitala. 2006. *Ecology of populations*. Cambridge: Cambridge University Press.

Ranåker, L., M. Jönsson, P.A. Nilsson and C. Brönmark. 2012. Effects of brown and turbid water on piscivore-prey fish interactions along a visibility gradient. *Freshw. Biol.* 57(9):1761-1768.

Réale, D., S.M. Reader, D. Sol, P.T. McDougall and N.J. Dingemanse. 2007. Integrating animal temperament within ecology and evolution. *Biol. Rev.* 82(2):291-318.

Reimchen, T.E. 1991. Evolutionary attributes of headfirst prey manipulation and swallowing in piscivores. *Can. J. Zool.* 69(11):2912-2916.

Savino, J.F. and R.A. Stein. 1989. Behavioural interactions between fish predators and their prey: effects of plant density. *Anim. Behav.* 37:311-321.

Sih, A. and B. Christensen. 2001. Optimal diet theory: when does it work, and when and why does it fail? *Anim. Behav.* 61:379-390.

Sih, A., A. Bell and J.C. Johnson. 2004. Behavioral syndromes: an ecological and evolutionary overview. *Trends Ecol. Evol.* 19(7):372-378.

Sih, A., J. Cote, M. Evans, S. Fogarty and J. Pruitt. 2012. Ecological implications of behavioural syndromes. *Ecol. Lett.* 15(3):278-289.

Skalski, G.T. and J.F. Gilliam. 2001. Functional responses with predator interference: viable alternatives to the Holling type II model. *Ecology* 82:3083-3092.

Skov, C., P.A. Nilsson, L. Jacobsen and C. Brönmark. 2007. Habitat-choice interactions between pike predators and perch prey depend on water transparency. *J. Fish. Biol.* 70:298-302.

Stauffer, H.P. and R.D. Semlitsch. 1993. Effects of visual, chemical and tactile cues of fish on the behavioral responses of tadpoles. *Anim. Behav.* 46(2):355-364.

Stephens, D.W. and J.R. Krebs. 1986. *Foraging Theory*. Princeton: Princeton University Press.

Stephens, D.W., J.S. Brown and R.C. Ydenberg (eds.). 2007. *Foraging: Behavior and Ecology*. The University of Chicage Press.

Svanbäck, R., M. Quevedo, J. Olsson and P. Eklöv. 2015. Individuals in food webs: the relationships between trophic position, omnivory and among-individual diet variation. *Oecologia* 178(1):103-114.

Treasurer, J.W. 1988. Measurement of regurgitation in feeding studies of predatory fishes. *J. Fish. Biol.* 33:267-271.

Triplet, P., R.A. Stillman and J.D. Goss-Custard. 1999. Prey abundance and the strength of interference in a foraging shorebird. *J. Anim. Ecol.* 68:254-265.

Turesson, H. and C. Brönmark. 2004. Foraging behaviour and capture success in perch, pikeperch and pike and the effects of prey density. *J. Fish. Biol.* 65(2):363-375.

Vogel, S. 1994. *Life in moving fluids: the physical biology of flow*. 2nd ed. Princeton: Princeton University Press.

Wahl, D.H. and R.A. Stein. 1988. Selective predation by three esocids: the role of prey behavior and morphology. *Trans. Am. Fish. Soc.* 117:142-151.

Webb, P.W. and J.M. Skadsen. 1980. Strike tactics of Esox. *Can. J. Zool.* 58:1462-1469.

Webb, P.W. 1986. Locomotion and predator-prey relationships. pp 24-41. *In*: M.E. Feder and G.V. Lauder (eds.). *Predator-prey relationships*. London: The University of Chicago Press.

Webb, P.W. and A.G. Fairchild. 2001. Performance and maneuverability of three species of teleostean fishes. *Can. J. Zool.* 79(10):1866-1877.

Werner, E.E. 1974. The fish size, prey size, handling time relation in several sunfishes and some implications. *J. Fish. Res. Bd. Can.* 31:1531-1536.

Werner, E.E. and J.F. Gilliam. 1984. The ontogenetic niche and species interactions in size structured populations. *Ann. Rev. Ecol. Syst.* 15:393-425.

Wolf, M. and F.J. Weissing. 2012. Animal personalities: consequences for ecology and evolution. *Trends Ecol. Evol.* 27(8):452-461.

Wright, R.M. and N. Giles. 1987. The survival, growth and diet of pike fry, *Esox lucius* L, stocked at different densities in experimental ponds. *J. Fish. Biol.* 30(5):617-629.

Wright, R.M. and E.A. Shoesmith. 1988. The reproductive success of pike, *Esox lucius* - aspects of fecundity, egg density and survival. *J. Fish. Biol.* 33(4):623-636.

Chapter 3

Coping with Environments; Vegetation, Turbidity and Abiotics

Lene Jacobsen[1] *and Jonna Engström-Öst*[*2]

3.1 INTRODUCTION

In today's world, fish are exposed to multiple stressors and threats which can lead to collapse of predatory fish populations. Overfishing is probably the largest threat for pelagic fish (Myers and Worm 2003), but for littoral fish such as pike, biotic and abiotic environmental alterations are becoming more important (reviewed by Rypel 2012), including climate changes. Increased eutrophication caused by rising temperatures can lead to indirect effects on pike due to higher turbidity and a subsequent reduction in vegetation. Such habitat degradation, along with other human alterations in nature, can be a key factor in explaining pike declines as pike are highly dependent on (compared with pelagic fish) abundant vegetation, with density and structure governing individual success. More direct effects of climate change, e.g. temperature increases on pike growth, survival and behaviour have received surprisingly little attention as they should be particularly important for larvae and young pike that are considerably more sensitive to environmental change than adult fish. The reason for the low number of studies on this could, in part, be that the pike is mesothermic and can cope with a fairly large range of temperatures.

In the present chapter we discuss how the pike deals with the environment it lives in, both the biotic (vegetation, turbidity) and abiotic (temperature, oxygen, salinity, etc.) elements. We first discuss (section 3) the role of vegetation for individuals during their lifetime and the interactions with other environmental factors, acknowledging the growing literature showing new aspects and underlining

[1]DTU Aqua, Technical University of Denmark, Section for Inland Fisheries and Ecology, Vejlsøvej 39, 8600 Silkeborg, Denmark; Email: lj@aqua.dtu.dk

**Corresponding author*: [2]Novia University of Applied Sciences, Raseborgsvägen 9, 10600 Ekenäs, Finland; Email: jonna.engstrom-ost@novia.fi

the large versatility in pike behaviour (see for example chapter 5). We continue reviewing (section 3) how turbidity and water clarity affect the individual behaviour of pike and discuss the complexity of studies on this. In section 3, the influence of some of the main abiotic factors, e.g. temperature and oxygen, are reviewed followed by a section on how pike cope with other basic abiotic factors such as salinity and pH (section 3), recognizing that pike are not only found in freshwaters but also in coastal habitats like the Baltic Sea. Finally, in section 3, we report about existing knowledge on the effects of climate change on pike individuals and conclude with recommendations for future research.

3.2 THE ROLE OF VEGETATION FOR PIKE INDIVIDUAL BEHAVIOUR

The importance of vegetation for pike success has been well acknowledged in the literature (e.g. Raat 1988, Grimm 1989, Bry 1996, Casselman and Lewis 1996, Grimm and Klinge 1996, Craig 2008). The function and use of vegetation or structure vary between the different life stages of pike individuals and is less crucial for larger individuals. Vegetation is important, both as spawning grounds and nursery areas, for small pike. With growth and increased individual size, the role of vegetation changes, and acts in concert with many abiotic and biotic factors as, for instance, water depth and transparency, vegetation type, prey, predation and cannibalism, and is, hence, often more complex than simple. In the following section we review the importance of vegetation for spawning, nursery habitat, foraging, predator refuge and predation for larval, juvenile and adult pike (Fig. 3.1).

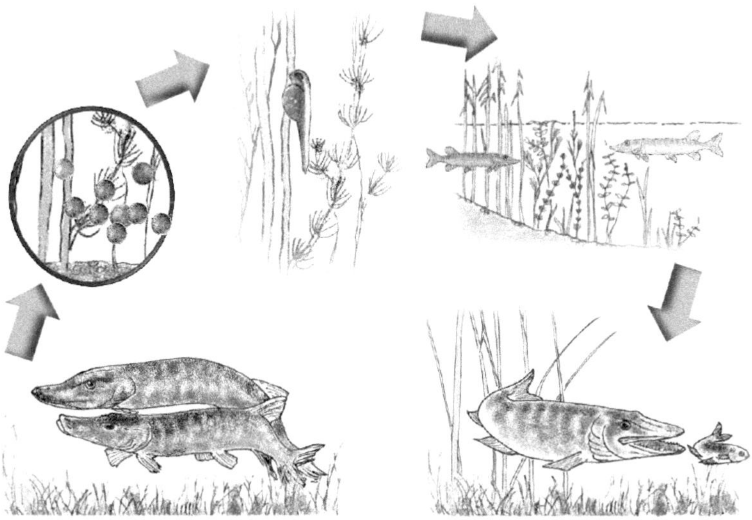

FIGURE 3.1 Pike depend on vegetation in different phases of their life cycle. The interaction with vegetation changes when pike grow larger and, for adult pike, vegetation can be important as a hideout for hunting, but not necessarily (Drawing by L. Jacobsen).

3.2.1 Vegetation as Spawning Substrate

One of the main roles of vegetation for pike success is its function as spawning substrate which has been extensively described and reviewed in both North American and European studies (Franklin and Smith 1963, McCarraher and Thomas 1972, Raat 1988, Bry 1996, Casselman and Lewis 1996, Nilsson et al. 2014). The pike is a phytophilic spawner, being dependent on plants. Vegetation is considered to be a visual stimulus triggering the start of the spawning act in combination with optimal temperature and light circumstances (Fabricius 1950). It has been observed that pike spawning does not start until the appropriate vegetation is available, i.e. after the water level has risen to a certain level and vegetated areas have been flooded, making vegetation a prerequisite for spawning and as important as e.g. water temperature. Fabricius and Gustafsson (1958) described, using large aquaria trials, how females maintained a position 5-15 cm above short vegetation during the spawning act and seemingly avoided areas without vegetation with bare sediment or areas with emergent reeds (*Phragmites* sp.). The fact that pike preferred to remain slightly above the vegetation was confirmed by observations in nature (Fabricius and Gustafsson 1958). Pike eggs are ovulated as single eggs in smaller or larger portions, depending on the size of the female, during the spawning session. The eggs have an adhesive surface which allows them to stick to plant structures in the water column. This prevents them from sinking to the bottom sediment where decaying mats of detritus or filamentous algae can cause egg mortality due to low oxygen levels or other environmental constraints such as high hydrogen sulphide concentrations (Siefert et al. 1973, Fago 1977, Dombeck et al. 1984, Casselman and Lewis 1996). In running waters, riverbeds or exposed coastal areas, adhesion to structure also prevents the eggs from being flushed away with water currents or movements.

To survive the first critical stages of life, pike are highly dependent on some kind of vegetation or structure, and spawning structures are chosen accordingly. There might be different success rates for different kinds of vegetation, though reported spawning structure encompasses almost all kinds of submerged macrophytes, emergent vegetation and inundated terrestrial plants like grasses, sedges or mosses. In general, pike seem to behave opportunistically. They select among available structures, but prefer short grass if given a choice (McCarraher and Thomas 1972, Alldridge and White 1980, Fortin et al. 1983). In river beds or impoundments, temporally flooded grassland and other terrestrial plants seem to be among the preferred substrates if available (Casselman and Lewis 1996). Fortin et al. (1983) described the highest numbers of eggs on flooded meadows, pastures and shrub grass areas in a Canadian river. In streams connected to the Baltic Sea, where pike have adapted to brackish water and perform anadromous spawning migration, pike prefer flooded meadows, particularly inundated grasses and sedges in heterogeneous patches (Nilsson et al. 2014, see also chapter 10). Very few eggs were deposited on emergent vegetation, and even fewer on submerged vegetation with bare sediment (Nilsson et al. 2014).

In lakes, pike spawning is successful in flooded meadows and on a variety of submerged macrophytes (see reviews in Raat 1988, Bry 1996). Macrophytes such as pondweed (*Potamogeton* sp.), duck weeds (*Lemna* sp.) and stonewort (*Chara* sp.) have been reported to be successful substrates (McCarraher and Thomas 1972, Farrell et al. 1996, Pierce et al. 2007). In comparison, egg density on plants such as *Myriophyllum* sp. or *Ceratophyllum demersum* Linnaeus, 1753. was relatively lower (0-5 eggs per 30 cm^2 sample compared to 26-30 eggs per 30 cm^2 sample for *Potamogeton* sp. and *Chara* sp.) (McCarraher and Thomas 1972). Casselman and Lewis (1996), however, classified *Potamogeton* sp. as a less suitable substrate for spawning. Pike have also been reported to spawn on emergent vegetation and weeds such as *Phragmites* sp., *Carex* sp. or *Equisitum* sp. (Fabricius 1950, Frost and Kipling 1967), while cattail (*Typha* sp.) and other thick types of weeds were avoided (Franklin and Smith 1967, Alldridge and White 1980, Casselman and Lewis 1996, Farrell 2001).

Pike also reproduce in brackish water, and use sheltered bays, archipelagoes, and estuaries for spawning in coastal areas of the Baltic Sea (Urho et al. 1990). In Kalmar Sound in Sweden, pike eggs were found on different kinds of vegetation with no particular preferences, including inundated grasses, *Phragmites* sp. and submerged macrophytes such as *Potamogeton* sp., *Ruppia* sp., *Myriophyllum* sp., *Chara* sp. and *Cladophora* sp., as well as loose floating bladderwrack *Fucus vesiculosus* Linnaeus, 1758 (Nilsson 2006). In areas of the Finnish archipelago, the majority of pike larvae were found in a habitat formed by the previous season's flattened reeds (*Phragmites* sp.), that had been cut by waves and ice during winter, pushed under the water surface into a horizontal position and forced against the coastline (Lappalainen et al. 2008, Kallasvuo et al. 2011). Here, pike eggs were particularly abundant among reed belts with mosses (Kallasvuo et al. 2011). Bladderwrack was previously considered as spawning substrate for brackish pike (Raat 1988), but more recent studies clarify that this is probably not the case as the bladderwrack has a more open structure, is coarse-leaved, and is often found in more exposed areas (Lehtonen 1986, Lappalainen et al. 2008). This misconception has probably arisen from pike juveniles often observed around bladderwrack. The density of vegetation seems to be important, and Farrell et al. (1996) showed that more eggs were found in denser cover and also in taller vegetation, especially in deeper water.

Natural fluctuations in water level are an important factor regulating the available areas of inundated grassland. Water levels are often high during springtime which coincides with the timing of pike spawning; however, the instability of temporarily flooded areas is a threat to successful pike spawning. If water levels are unstable after spawning and/or decline faster than normal and leave the flooded areas dry too soon after spawning, newly spawned pike eggs attached to the grasses will dessicate, and reproduction will consequently fail (Threinen 1969). Accordingly, fluctuations in water level have been linked to 0+ pike year-class strength (Johnson 1957, Smith et al. 2007). Johnson (1957) found, within a shallow two hectare lake in Minnesota, that during a seven year period, the lowest water level drop during egg incubation (9 cm) coincided with the highest year-class strength of pike (1.69 times the mean year class strength in the period).

3.2.2 Vegetation as Nursery Habitat for Larvae and Small Pike

The importance of vegetation for the success of pike larvae has been claimed to be even higher than for spawning (Casselman and Lewis 1996). This includes the first short period as yolk-sac larvae as well as when the larvae start feeding, hereafter defined as 'fry' (see also chapter 9 for definition of juvenile stages of pike). Once the eggs are fully developed after 6-10 days, pike hatch as yolk-sac larvae. The newly hatched larvae move to attach to vegetative structure by the use of adhesive papillae on their head (Frost and Kipling 1967, Raat 1988, Cooper et al. 2008). Once they attach, the larvae will hang from the vegetation for the next 5-6 days. This adhesion to structure prevents them from sinking to the bottom, thereby avoiding low oxygen conditions or other environmental stress factors. This adaptation is also likely to prevent the poorly mobile larvae from drifting into open water where they would be more vulnerable to predation. Being closely attached to vegetation might also provide the larvae with shelter from predators as it, to some degree, works as camouflage (Frost and Kipling 1967). During the following phase, when larvae are free-swimming and start to forage, vegetation plays a major role as nursery habitat in shallow water, and vegetated habitats are preferred to areas lacking vegetation. For instance, Holland and Huston (1984) found ten times more Young-of-the-Year (YOY) pike in emergent vegetation and three times more in submerged vegetation compared with open water in the Upper Mississippi River. This is probably linked to growth opportunity, as pike growth rates in unvegetated tanks were considerably lower than growth rates in tanks with vegetation (Johnson 1960, cited in Casselman and Lewis 1996).

The relationship between water depth and vegetation during the first months of a pike's life has been described by Casselman and Lewis (1996). They found that until a length of ca. 5 cm pike are mainly found in water shallower than 50 cm. As pike fry grow, they seek deeper waters. In general, it seems that preferred water depth increases with ca. 10 cm for each 1 cm of pike growth until the juvenile pike reaches a length of ca. 15 cm. Berg et al. (2012) found the same correlations, but pike fry were found in even shallower water. Pike fry of 4.5 cm were stocked in an experimental area of 10-70 cm depth with artificial vegetation made from spruce tree (Fig. 3.2). Here, pike initially occurred in water depths of 10-20 cm and gradually moved to ca. 50-70 cm depth at 14-15 cm body length. Cucherousset et al. (2009) studied YOY pike in a small and temporarily flooded nursery area and found that pike distribution correlated with depth but not with the presence of vegetation cover, indicating that water depth was the most important factor when the water level decreased. In contrast to these studies, Skov and Berg (1999) found no significant correlation between pike juvenile size and depth distribution during summer in a turbid lake with natural and artificial vegetation/ structure. Thus, well-established relationships between size, vegetation and depth preferences for YOY pike may be challenged, possibly due to interactions with other environmental factors such as turbidity or simply differences in available water depths or vegetation types.

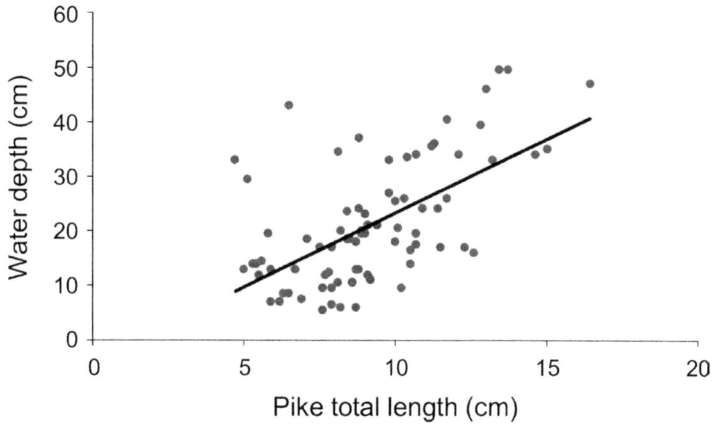

FIGURE 3.2 Pike juveniles inhabit increasing depths as they grow. Pike fry and fingerlings were caught by electrofishing in depths from 10-70 cm with artificial spruce trees during six electrofishing sessions from 1 June – 26 July 2004, $n = 74$. Linear regression line: $y = 2.7128 \times -3.7125$, $R^2 = 0.3593$. (Modified from Berg et al. 2012)

3.2.3 Importance of Vegetation Types and Densities

Different kinds of vegetation may be used during the various life stages of pike. Casselman and Lewis (1996) review that pike fry generally prefer emergent and flooded terrestrial vegetation; young pike prefer emergent, floating and submerged vegetation, while adults prefer submerged macrophytes such as *Potamogeton* sp., *Ceratophyllum demersum* Linnaeus, 1758 and *Elodea canadensis* Michx.. This change in vegetation preference during ontogeny is in correspondence with the preference for increasing water depth when pike grow larger. Nilsson et al. (2014) found more small pike in flooded terrestrial vegetation and *Phragmites* sp. belts, compared with submerged vegetation and open water.

A high density of vegetation is preferred by small pike (Grimm 1983, Casselman and Lewis 1996, Eklöv 1997, Skov and Berg 1999, Engström-Öst et al. 2007) while this preference decreases when pike grow larger (Grimm 1983). Juvenile and adult pike avoided dense vegetative mats and dense emergent vegetation (Grimm 1983) and were mainly caught in intermediate macrophyte densities ranging from 35 to 80% cover (Casselman and Lewis 1996). In an experimental approach, larger pike (30-35 cm) avoided the densest structure of 600 stems m^{-2}, but they inhabited lower densities of artificial stems (Bean and Winfield 1995). Adult pike do have a preference for vegetation but also use the open water, which has been confirmed in many field studies (Diana et al. 1977, Chapman and Mackay 1984, Cook and Bergersen 1988, Eklöv 1997, Kobler et al. 2008). Correspondingly, Grimm (1983) concludes that pike < 41 cm depend on vegetation, but when pike grow larger than 54 cm, vegetation is not indispensable for pike survival. Eklöv (1997) found that the largest pike in a lake inhabited more open structures such as submerged trees, and Chapman and Mackay (1984) found larger pike to be more connected to the interface between open water and vegetation, probably relating to feeding opportunities in the nearby vegetation.

3.2.4 Vegetation as Feeding Substrate

During their early life stages, pike are dependent on small organisms such as zooplankton and other invertebrates for food. In general, the diversity and abundance of different organisms increase with physical habitat complexity (Kohn and Leviten 1976, Eklöv 1997). The use of vegetation by pike can accordingly be linked to foraging opportunity. In combination with shallow water, vegetated areas provide warmer water in spring and more nutrients and substrate for a high zooplankton production (Jeppesen et al. 1997). Zooplankton are important prey for newly hatched fry (Franklin and Smith 1963, Wright and Giles 1987), and Nilsson et al. (2014) found the highest densities of pike fry in patches of vegetation in shallow flooded areas, co-occurring with the highest densities of zooplankton. In the margins of the nearby river there was less vegetation and lower zooplankton densities, and fry of the same age from this habitat were accordingly considerably smaller, indicating suboptimal foraging conditions (Nilsson et al. 2014). Skov and Koed (2004) also observed an increased use of vegetation by pike fry (20-31 mm) in experimental ponds in the presence of zooplankton compared to ponds without zooplankton. Juveniles up to 8-14 cm inhabiting dense emergent vegetation are able to feed on invertebrates which are abundant in this kind of vegetation (Grimm 1994). Feeding opportunity can also be a reason for large 0+ pike to associate with vegetation; Skov et al. (2002) recorded, in an experimental setup, that pike (9-17 cm) increased their use of macrophytes in the presence of prey fish compared with control ponds without prey fish. It has, furthermore, been shown that larger pike stay close to areas inhabited by prey fish (Eklöv and Diehl 1994, Eklöv 1997).

As pike grow larger, vegetation also plays an important role as cover for the ambush, sit-and-wait hunting behaviour of pike (Casselman and Lewis 1996). Larger pike inhabiting the edge of vegetation are camouflaged and able to conceal themselves prior to the ambush attack on prey fish passing in the open water or inhabiting the outer, less dense macrophyte belt. Despite this advantage of vegetation or structure for ambush hunting, Greenberg et al. (1995) reported similar foraging success of pike (38-40 cm) with and without artificial vegetation (plastic strings) in a field enclosure experiment. The importance of vegetation for pike foraging is lower in turbid water where pike more often inhabit the open water that provides equally high foraging success (Skov et al. 2003, Andersen et al. 2008).

3.2.5 Vegetation as Refuge against Predation

Structural complexity in vegetation can influence predator-prey interactions depending on size distribution and behaviour of both predator and prey (Crowder and Cooper 1982, Eklöv and Diehl 1994, Jacobsen and Berg 1998). Thus, an important feature of vegetation is to function as a refuge against predation. Small pike can be cannibalistic, and vegetation may protect the young pike against intra-cohort cannibalism. This can be attributed to lower detectability in the vegetation, but also to a higher density of alternative prey items in the vegetation that can reduce cannibalism (Casselman and Lewis 1996, Skov et al. 2003). A high level of cannibalism was correlated with a low use of vegetation, and there was a lower degree of cannibalism in more dense structural complexity in two experimental set-ups (Skov et al. 2003, Skov and Koed 2004). However, the importance of vegetation for these predator-prey interactions is complex, and one could argue that vegetation, besides providing shelter for the smaller individuals against cannibals, also provides cover for the cannibals to improve their attacks (Skov et al. 2003). In line with this idea, Skov and Koed (2004) demonstrated that the largest 0+ pike inhabited vegetation and forced the smaller pike fry into the open water which is likely to be a suboptimal habitat, thus inducing an indirect mortality effect on the smaller individuals. The paradox of vegetation being an ambush hideout for the predatory pike and at the same time functioning as a refuge for fish prey was reflected in an experimental study, where prey sought refuge in open water when the pike predator was inhabiting the macrophytes (Jacobsen and Perrow 1998).

A high density of vegetation in shallow habitats provides small YOY pike with shelter from predation from adult pike (Fig. 3.3) since larger pike prefer less dense vegetation (see section 3.2.3) and the manoeuvrability of adult pike is restricted by dense vegetation (Casselman and Lewis 1996). Likewise, it was demonstrated, in a field experiment, that water depth is important since shallow water with vegetation protected young pike fry from predation by large piscivorous perch inhabiting vegetation in deeper water (Berg et al. 2012). In an experiment

FIGURE 3.3. Pike fry use the vegetation both as shelter from predation and for foraging. Photo: S. Berg

with newly hatched pike larvae, the predation rate from three-spined stickleback (*Gasterosteus aculeatus* Linnaeus, 1758) was lower in dense mats of *Cladophora* sp. compared to in bladderwrack, supporting that dense structure is a good refuge for predation (Engström-Öst et al. 2007). The inverse relationship between vegetation density and size of pike could reduce cannibalism risk for smaller pike (Grimm 1983, Grimm 1994), but it might also interact with feeding and induce a trade-off between foraging and survival (Werner and Hall 1988). Hence, Grimm (1994) reports pike are able to sustain their energy maintenance on invertebrate food in dense emergent vegetation where they can avoid predation, but when they are more than 14 cm, they have to move to less dense vegetation in order to forage on fish. In a study by Eklöv and Diehl (1994), large pike had effects, not only on habitat use of smaller pike, but also on their growth rate. Large pike were always closer to prey fish, and the smaller pike had to move into denser vegetation with lower prey fish availability to avoid potential cannibals.

3.3 EFFECTS OF WATER VISIBILITY, LIGHT AND TURBIDITY ON INDIVIDUALS

Aquatic organisms that use vision as a major sense of perceiving their surroundings rely on water clarity to perform important tasks in life. Visibility conditions, affected by e.g. brown water or turbidity, are thus important for individual success. Increased eutrophication can lead to higher turbidity and subsequent habitat degradation. This is caused by increased biological productivity from excessive addition of nutrients to aquatic systems (Cooke et al. 1993), and as a result, microalgae densities rise and increase turbidity because of greater light attenuation in the water (Nielsen et al. 2002). Poor visibility can be of organic origin, i.e., algae-induced (Utne-Palm 2002) or appear as brownification caused by natural organic matter (humic substances, see sections 3.3.1 and 3.6). Turbidity can also consist of inorganic particles such as re-suspended sediments or different clays (Engström-Öst et al. 2013). In this section, we discuss the importance of water visibility and how it influences pike foraging, predator avoidance, habitat choice and activity.

3.3.1 Foraging Behaviour and Prey Choice

It is essential for young pike to optimize foraging, e.g. accurately time the ontogenetic transition from planktivory to piscivory, and simultaneously minimise the risk of size-dependent cannibalism and predation (Bry et al. 1995). The pike is a predator that mainly uses vision to detect prey and predators, and pike should, therefore, be negatively affected by increased turbidity and poor visibility conditions (Craig and Babaluk 1989). Pike is, however, also capable of using alternative sensory modes such as the lateral line (Raat 1988, see also chapter 2) which may, to some extent, compensate for impaired visibility in turbid water. There is also direct evidence that pike predation is not limited by low water visibility (Skov et al. 2007, Nilsson et al. 2009) as well as indirect evidence that pike can be abundant in turbid lakes (cited in Skov et al. 2002).

By affecting foraging, visually degraded environments expose pike fry to different conditions for early life success. Studies using young stages have demonstrated that high turbidity can have a negative effect on visually foraging fish (Vinyard and O'Brien 1976, Confer et al. 1978, Gregory and Northcote 1993, Engström-Öst et al. 2006, Salonen et al. 2009, Salonen and Engström-Öst 2010, 2013). Other studies have observed that intermediate levels of turbidity can lead to higher foraging rates (Boehlert and Morgan 1985, Miner and Stein 1993, Bristow et al. 1996, Utne-Palm 1999). This can either be due to (1) the physical effect hypothesis, where low turbidity can increase the contrast between the prey and its background (Hinshaw 1985), and (2) the motivation hypothesis suggesting that increased turbidity leads to higher feeding motivation caused by decreased risk of predation (Gregory and Northcote 1993).

Turbidity can affect foraging in a number of ways, such as reduced reactive distance to prey (Berg and Northcote 1985, Quesenberry et al. 2007), reduced visual range (i.e., reduced volume of water searched; Aksnes and Giske 1993, Turesson and Brönmark 2007), changed prey types (Rowe 1984, Gregory and Northcote 1993, Salonen and Engström-Öst 2010), decreased amount of consumed prey (Sweka and Hartman 2001, Engström-Öst et al. 2006, Salonen and Engström-Öst 2010), and higher attack rates on prey (Engström-Öst and Mattila 2008). Different methods and ways to colour the water may render different results; living algae change the water turbidity, whereas chlorophyll decreases the water clarity. Water visibility can also have no apparent effect on fry foraging (Skov et al. 2002). For example, YOY pike were unable to catch as many calanoid copepods *Eurytemora affinis* Poppe, 1880 in turbid water compared to clear water whereas the cladoceran *Daphnia longispina* Müller, 1776 was caught in similar abundances in clear and turbid waters (Salonen and Engström-Öst 2010). *Daphnia* are relatively slow-swimming and seem unable to escape predation as rapidly as the copepods. The movement pattern of copepods is more irregular than that of cladocerans (Viitasalo et al. 2001) which gives copepods an evasive advantage over cladocerans at high turbidity levels (Salonen and Engström-Öst 2010, Fig. 3.4).

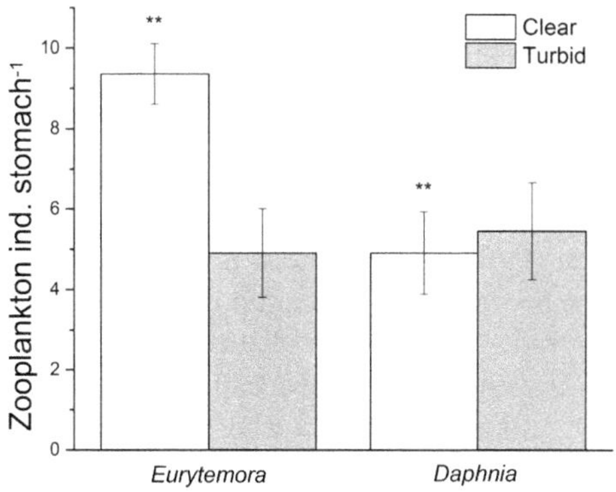

FIGURE 3.4 The predation success of pike fry (~ 1.8 cm) on crustacean zooplankton *Eurytemora affinis* and *Daphnia longispina* in clear (white columns) and turbid water (grey columns). Values are mean ± S.E.; $N = 10$. ** indicates $p < 0.01$. Modified from Salonen and Engström-Öst 2010.

Jönsson et al. (2012) reported significantly shorter reaction distance (i.e. distance between the snout of the pike and the nearest body part of the prey when pike turned its head or entire body towards the prey) by pike in turbid water when provided with roach *Rutilus rutilus* Linnaeus, 1758 larvae. Reaction distance did not change when pike were presented with *Daphnia*. Jönsson et al. (2012) concluded that pike did not swim more in order to compensate for low visibility; instead, when search efficiency decreased, more attacks towards prey were observed. Thus, it seems that turbidity can affect the predation on various prey types in different ways. However, Mauck and Coble (1973) found no difference in prey selection (for various North American prey fish species) by adult pike between turbid and clear-water experimental cages.

Hongve et al. (2004) suggested that 'brown' water is becoming more common due to climate-induced precipitation including more rain and snowfall. When rainfall increases on land, causing more flooding and storms, natural organic matter will drain to lakes and watersheds leading to 'browner' water. Changes in the amount and quality of dissolved natural organic matter will influence physical and chemical variables in the waters. 'Brown' water may actually become a greater problem for small pike in the future than algae-induced turbidity, as many nursery areas in lakes will get more 'brown' in the coming decades (Hongve et al. 2004 and references therein). Light is considered to be a major limiting factor for fish in lakes suffering from 'brownification' (Ask et al. 2009, Karlsson et al. 2009), and fish can, therefore, not grow at maximum rates due to low food availability and reduced foraging efficiency (Jönsson et al. 2011). Fish growth also varies depending on variations in day length. Day length affects primary production, which drives food availability, causing a synergistic effect for pike (Boeuf and Falcon 2001). Pike are also known to be good hunters in low light intensities due to their highly developed vision and lateral lines (Volkova 1973, Dobler 1977). In the study by Craig and Babaluk (1989), pike body condition was positively correlated with water clarity (but only 21% of the variation was explained by Secchi depth) in Canadian prairie lakes. The authors suggested that it could be due to foraging efficiency being improved in clear water. The use of open water in 0+ pike (length: 9-17 cm) did not change due to light intensity, indicating that the observed behaviour was similar during both day and night (Skov et al. 2002). However, light proved to be an important regulating factor for feeding of 0+ pike in the absence of predators demonstrated by fewer food attacks in 50% light level in both clear and turbid water (Lehtiniemi et al. 2005).

Zooplankton adjust their pigmentation in response to UV radiation, but colouration can increase the predation risk and visibility to visual hunters, indicating that zooplankton face a trade-off between radiation and predator avoidance (Hansson and Hylander 2009). Carotenoid-pigmented copepods in turbid water, induced by added micro-algae (here chlorophytes), allowed increased feeding for pike whereas pigmented prey in brown water, induced by humic substances, reduced pike foraging success (Jönsson et al. 2011). Background colour and brightness, therefore, seem to affect feeding success. Brown water absorbs a substantial part of the light and forms a dark background decreasing the detectability of pigmented copepods. In contrast, turbid water consisting either of algae or suspended matter backscatters light. The relative effects of absorbing and scattering light on fish vision are not well described, but it might affect pike foraging in different ways. Ranåker et al. (2012) found that the reaction distance of pike (ca. 22.5 cm) to roach prey decreased when visual range decreased in different optical treatments (algae, clay or brown humic water) in experimental tanks. The authors showed that the effect of turbidity (algae or clay) was stronger than the effect of brown water. Nevertheless, strike distance was affected by neither visual range, nor by optical treatment. In combination with a shorter escape distance for roach in brown water compared to turbid water, the authors concluded that brown water should increase the probability of successful attack for pike compared to algae or clay turbidity at the same visual range (Ranåker et al. 2012). Other studies have found little effect of turbidity on pike foraging on fish prey. Skov et al. (2002) found no difference in foraging success of 0+ pike (9-17 cm) preying on roach fry between clear water and chlorophyll-coloured water treatments. Likewise, young pike (mean 22 cm) were capable of eating just as many roach prey in mesocosm treatments with clay turbidity of medium and very high levels compared to clear water (Nilsson et al. 2009).

3.3.2 Anti-predator Effects

For a small planktivore, such as a pike fry that detects its planktonic prey at a short distance (Chesney 1989, Giske et al. 1994, Fiksen et al. 2002), turbidity may have a positive anti-predator effect, indicating that turbidity may function as shelter against predators (Utne-Palm 2002). The anti-predator behaviour was reduced in first-feeding pike in turbid water compared with clear water. The behaviour was observed as less pronounced when hiding in vegetation in turbid water when exposed to predator cues. The fry also displayed fewer attacks and fed less in clear water when predator cues were present (Lehtiniemi et al. 2005, Engström-Öst and Mattila 2008). This effect is further supported by significantly decreased foraging (~ 50%) in the presence of a visible predator (European perch *Perca fluviatilis* Linnaeus, 1758) compared with the non-predator control (Engström-Öst and Mattila 2008).

3.3.3 Habitat Choice and Swimming Activity

Few studies exist on the effects of turbidity on pike larval and fry habitat choice, while the relationship between high water turbidity and lack of appropriate vegetation for pike has been broadly acknowledged (Threinen 1969, Grimm 1989, Jude and Pappas 1992, Lehtonen et al. 2009). Leslie and Timmins (1992) showed how tightly connected turbidity and habitat (vegetation) are in a study conducted in a heavily polluted and degraded area, Hamilton Harbour in Lake Ontario, Great Lakes, USA. The area is largely devoid of suitable substrate and vegetation for spawning and refuge. Shortage of habitat induces low survival of fry (see vegetation section), and as turbidity negatively influences vegetation growth, distribution and dispersal of fish can be limited. Sandström et al. (2005) showed that there was a limited abundance of pike larvae in dredged (turbid) areas that were also heavily affected by boating. This result was strongly influenced by the lack of vegetation.

Experimental studies have confirmed that pike fry spend less time in the vegetation in turbid water than in clear waters (Snickars et al. 2004, Engström-Öst et al. 2007, Engström-Öst and Mattila 2008). Skov et al. (2002) found similar results by demonstrating that pike visited turbid open-water areas more frequently than clear water and especially when prey was present. Reduced water transparency is generally considered to provide shelter for larval stages against predators (Gregory 1993, Gregory and Northcote 1993, Maes et al. 1998), and prey can, hence redistribute more evenly between different habitats (Skov and Koed 2004). This includes active use of open-water areas and a reduced need for use of vegetative shelter (Utne-Palm 2002, Snickars et al. 2004).

Adult pike also seem to change their habitat use in turbid conditions from being mostly associated with the littoral zone and its vegetation in clear water to adopting a more pelagic behaviour in turbid water (Vøllestad et al. 1986, Andersen et al. 2008). Pike may associate with vegetation to avoid interactions with conspecifics or to hide in ambush to attack passing prey, both of which should be less important in turbid conditions. The reduction of anti-predator behaviour of prey fish in turbid water was demonstrated by Abrahams and Kattenfeld (1997). The authors also showed that pike spent more time in a less safe habitat when the water was turbid. A mesocosm experiment with pike (ca. 30 cm) further showed a higher use of vegetated areas in clear water compared to turbid water in which habitat use was more random (Nilsson et al. 2012). In this study, there were no behavioural interactions with prey or conspecifics indicating that the effect of water clarity on habitat choice was direct and not an adaptation to changes in e.g. prey fish distribution (Nilsson et al. 2012). On the contrary, Jepsen et al. (2001) demonstrated that pike in a clear-water lake were less associated with vegetation compared to a turbid reservoir, and this result is attributed to the great diversity of pike behaviour (see also chapter 5). In lakes with shorelines exposed to wind, turbidity may increase in the littoral zones during windy events, and pike will inhabit deeper areas farther from the shore (Chapman and Mackay 1984, Cook and Bergersen 1988). Cook and Bergersen (1988) found that the higher turbidity in the littoral rendered this habitat less suitable for pike, and an active avoidance of the impaired visual environment was suggested due to high levels of suspended matter.

Pike fry swim less in turbid water (Engström-Öst and Mattila 2008). Turbidity can also induce activities such as migration, reduced use of shelter (Vøllestad et al. 1986, Gregory 1993), and increased use of open water (reviewed by Utne-Palm 2002). Changed behaviour can be due to perceiving less threat in a turbid environment (Abrahams and Kattenfeld 1997). Andersen et al. (2008) showed a higher activity in pike inhabiting a turbid lake compared to pike in a clear lake, but these individuals were in the size range of ~ 60-80 cm. In the same study, pike also showed higher variability in activity level and habitat use between individuals in turbid water suggesting that diversity in individual behaviours can increase with turbidity. The combined more pelagic and active behaviour of pike in turbid water might be a response to a reduced need for shelter and a greater need to move around in order to encounter prey.

3.4 IMPORTANCE OF TEMPERATURE AND OXYGEN

3.4.1 Temperature

Temperature and dissolved oxygen are the most extensively studied and considered environmental abiotic factors affecting fish growth and survival (Fang et al. 2004). Swift (1965) stated in a study on water temperature affecting pike egg development and mortality that "*these facts are not recorded elsewhere; it was therefore felt that they were worth recording*". Since 1965, a large amount of temperature-related papers have appeared. Raat (1988) summarized in his extensive FAO review that pike embryonic development can take place between 4°C and 23°C with the most optimal temperature interval for reproduction between 8°C and 15°C. In the review by Souchon and Tissot (2012), temperature intervals tolerated by differently aged pike are covered in detail, and optimal temperature decreases with increasing size in pike. The authors indicate that the pike belongs to a mesothermic group and to a group of fish that can reproduce in temperatures < 10°C. It has been recognized from the work of Casselman (1978) on pike (28-43 cm) that optimum temperature for growth is 19°C for growth in biomass and 21°C for growth in length. The preferred temperatures, however, seem to be slightly higher (2-3 degrees). Moreover, testing for a continuum of temperatures showed that growth occurred down to 4°C, but only with 3.9% of the maximum growth, and the same was seen for the upper limit for growth at 27.5°C. Temperatures above 29.4°C were found to be lethal whereas, at the lower end of the scale, Casselman (1978) observed pike to cope with almost freezing water (0.1°C). Larval physiological optimum for growth is about 25.6°C (Hokanson et al. 1973), and, for the young of the year, 22-23°C was noted as the optimum for growth (Casselman and Lewis 1996).

Temperature preferences have been documented in nature as well. Pierce et al. (2013) found large (> 71 cm) pike in deep lakes to follow the thermocline during summer into deeper and cooler water of 16-21°C, despite the major part of the water masses being warmer (up to 28°C). Co-occurring medium-sized pike (48-55 cm) varied more in their thermal preferences during summer, but most preferred slightly warmer (21-22°C) and shallower waters. Although pike have median thermal requirements (meso-thermal) among the fish in the cool-water assemblage (Casselman 1978), it is not particularly well known how the species may adapt to warming waters (for more details, see part 6 on climate change). One reason may be that it is often impossible to isolate the single effect of temperature from other field conditions such as photoperiod and water turbidity (Baktoft et al. 2012). Neither water temperature nor ice cover, however, affected pike activity in lake conditions (Baktoft et al. 2012). Temperatures approaching thermal maxima are stressful (Beitinger et al. 2000). Temperature can also interact with other factors, which are not problematic for the fish in cold water, but can result in multiplicative effects in warm waters (Gingerich et al. 2007). Salinity and starvation belong to those factors (Jacobsen et al. 2007). Water temperature is also affecting species interactions; Hein et al. (2014) showed that pike and brown trout *Salmo trutta* Linnaeus, 1758 are able to co-occur in cold lakes, but not in small, warm lakes.

Their study showed that brown trout could not stand the increased predation by pike when the water gets warmer. Pike attacks on brown trout increase by two orders of magnitude between 6°C and 10°C. In large lakes, there are more refuges for both species allowing more possibilities for brown trout and pike co-existence (see chapter 14).

3.4.2 Oxygen and Hypoxia

The tolerance of pike to low oxygen was tested in a number of studies during the 1970s and 1980s by measuring growth, survival and embryonic development (Adelman and Smith 1970, Siefert et al. 1973). It was concluded that pike larvae are generally rather tolerant to low oxygen; Siefert et al. (1973) found that embryos and larvae survived well in low oxygen, as mortality did not start to increase until after 12 days in 12.5% oxygen saturation. Although pike are fairly tolerant to low oxygen concentrations, the behaviour of fry can change when oxygen levels are sub-optimal (Engström-Öst and Isaksson 2006). In general, young stages of fish tend to be more sensitive to hypoxia than adult individuals (Werner 2002). Concerning pike larvae, they performed fewer prey attacks in hypoxic (3.1 mg l^{-1}) compared to fully oxygenized water (normoxia 5.2 mg l^{-1}). In a field study by Öhman et al. (2006), pike distribution was negatively correlated with anoxic conditions in isolated lakes, and winter migration by pike between connected lakes was reduced due to oxygen deficiency (Todd and Magnusson 1982). Muscalu-Nagy and Muscalu-Nagy (2012) showed that an increase in temperature combined with foraging activities increased the consumption of dissolved oxygen for all sizes of pike studied. Pierce et al. (2013) reported that vertical movements by pike of all sizes were heavily constrained by hypoxia, with 3 mg l^{-1} as the lower oxygen threshold, collaborating the findings of Casselman (1996). Despite this general avoidance of oxygen levels below 3 mg l^{-1}, pike are capable of surviving lower oxygen for shorter or longer periods. For instance, Pierce at al. (2013) found a pike to stay 47 min in water depths (11 m) with only 0.5 mg O_2 l^{-1}, and Moyle and Clothier (1959) report that pike survived a winter in a shallow lake with oxygen concentrations that occasionally were as low as 0.9-2.7 mg l^{-1}. Casselman (1978) even caught live pike in gillnets at oxygen concentrations of 0.04 mg l^{-1}. Thus, adult pike can tolerate very low oxygen levels, but activity is reduced and feeding stops when concentrations are lower than 2 mg l^{-1} (Casselman 1996). Winterkills do happen in oxygen depleted, ice covered lakes and older and fast growing individuals are especially vulnerable (Casselman and Lewis 1996).

Pike actively avoid the lowest oxygen levels if possible, and, in deep lakes with summer thermoclines and low oxygen in the hypolimnion, pike may have to trade off avoidance of high temperatures and low oxygen concentrations which is why they stay close to the thermocline (Pierce et al. 2013). Casselman (1978) also found pike to congregate in water with slightly higher oxygen concentration in winter. There is an exponentially increasing relationship between temperature and lowest oxygen concentration tolerated (see Casselman 1978).

3.5 COPING WITH SALINITY AND pH

3.5.1 Salinity

Fish that experience changes in salinity have higher energetic costs of osmotic and ionic regulation, and, therefore, less energy to allocate to growth (Wootton 1990). Although the pike is primarily a freshwater fish, it occurs also in brackish waters (Ojaveer 1981) such as the Baltic Sea and Caspian Seas (Crossman 1996). Pike is a stenohaline fish, not being able to fully osmoregulate to cope with truly marine waters. Salinities of 6-8.5 ppt are considered to be the upper limit for females to produce viable eggs (Westin and Limburg 2002, Jørgensen et al. 2010). There is, however, only limited information available on the ecological effects of salinity and how it influences pike living in brackish waters such as the Baltic Sea, one of the largest brackish water basins on Earth. The salinity tolerance of pike is not well understood, but adult pike are able to live in areas of 10-12 ppt, and can probably cope with temporary increases in salinities up to 14-15 ppt as indicated by the pike distribution in southern coastal zones of Denmark (Jacobsen et al. 2017). The salinities in these areas normally vary between 8 and 14 ppt (Dahl 1961, Jacobsen et al. 2017), and southern Denmark and surrounding areas thus constitute the limit for pike survival and dispersal in the Baltic Sea. Massive mortality of pike has occasionally been detected after large saline water inflows from the North Sea into the Baltic Sea via the Danish Sounds (Dahl 1961). As an example, pike disappeared, and the pike fishery was poor for several years after saltwater inflows occurred in 1930 and 1951, which increased the salinity up to 18 ppt (Dahl 1961). There is a lack of experimental knowledge, though, on the exact salinity tolerance of adult pike, with the exception of Schlumpberger (1966, cited in Raat 1988) reporting pike to show stressful behaviour above 12.4 ppt. This experiment may have been carried out with pike from freshwater systems, while brackish water pike could be adapted to higher salinities (Jørgensen et al. 2010).

Young stages are, in general, considered much more sensitive to salinity than adults. Kallasvuo et al. (2011) found that the reed belt shore of inner bay areas in the Gulf of Finland having low salinity, high temperature and dense vegetation were especially favourable as nursery areas for pike fry (Fig. 3.5). Lappalainen et al. (2008) reported salinity to be the strongest (and the only) significant explanatory variable for pike larval field abundances in springtime in the western Gulf of Finland. Pike eggs and larvae from a Danish brackish water pike population hatched and developed at 8.5 ppt (Jørgensen et al. 2010); this is a higher salinity level than previously reported for successful pike reproduction. In the same hatching experiment, no eggs from brackish water pike hatched at 0 and 3 ppt; whether this is due to an inherent physiological adaptation to hatching in relatively high salinity remains to be answered (Jørgensen et al 2010). Westin and Limburg (2002) found pike eggs to be capable of hatching in 6.5 ppt; however, these individuals inhabited areas around Gotland in Sweden with 6-7 ppt at the highest. Fry of brackish water origin exhibited stress and decreased growth rates when subjected to salinities above 13 ppt (Fig. 3.6, Jørgensen et al. 2010). In contrast, pike fry of freshwater origin exhibited stress responses at 11 ppt and died at 13 ppt (Fig. 3.7, Jacobsen et al. 2007). These results indicate that brackish water living pike are adapted to local salinities. The mortality of pike fry in different salinities is affected by water temperature; pike fry can tolerate higher salinities at lower water temperatures (Jacobsen et al. 2007).

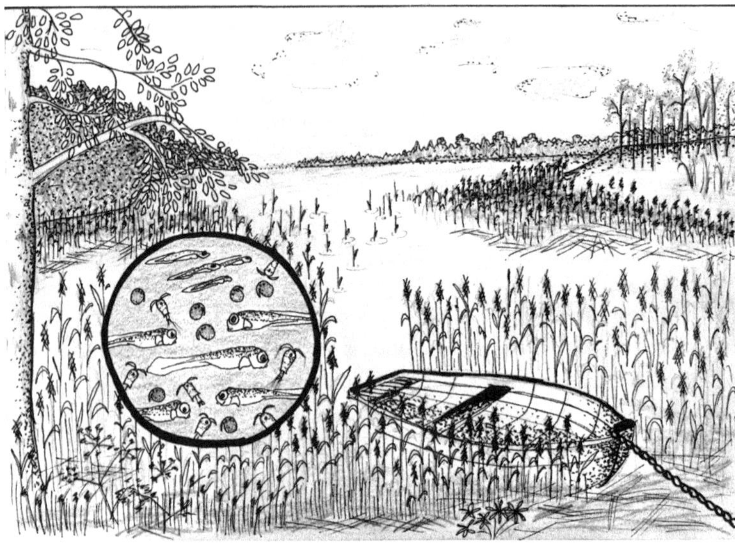

FIGURE 3.5 Many factors affect the success of fry such as water turbidity, zooplankton species and density, exposure level, shore vegetation, reed bed size and salinity. Spatial differences in intra- or interspecific competition and predation can also be important. The schematic is drawn from the inner archipelago zone based on a study with pike fry in the Gulf of Finland, Baltic Sea. Modified from Salonen (2012).

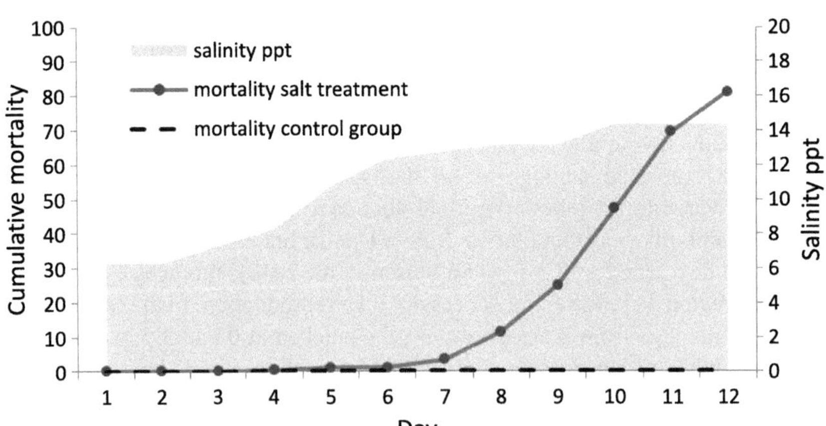

FIGURE 3.6 Salinity tolerances of brackish water pike fry. Cumulative mortality (%) of pike fry (mean length 16 mm ± 2 SE) hatched at 8.5 ppt, exposed to increasing salinities from 6-14 ppt, $n = 60$, temp. = 15.4°C. There was no mortality in the controls kept at 6 ppt. (Modified after Jørgensen et al. 2010).

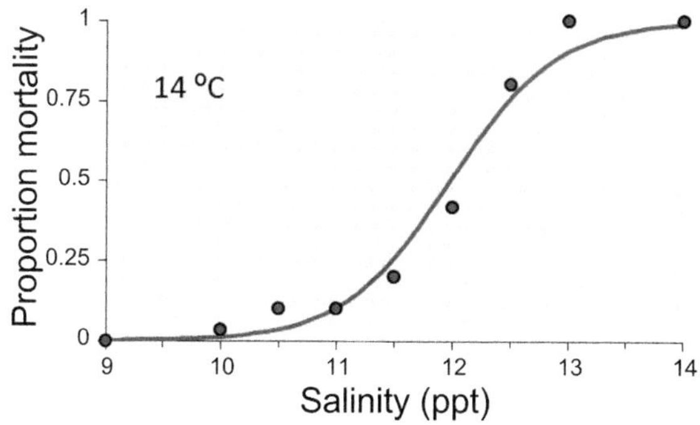

FIGURE 3.7 Salinity tolerances of freshwater pike fry. The proportion mortality of freshwater pike fry (21-41 mm total length) after 72 h in 14°C, introduced to varying salinities from 9-14 ppt. Number of fry in each salinity trial = 60. There was no mortality in control trials with freshwater. (Modified after Jacobsen et al. 2007).

Apart from affecting individuals directly, brackish water can have quite a large indirect influence on fry individuals via their food as salinity changes the size structure of potential prey (Flinkman et al. 1998, Rönkkönen et al. 2004, Salonen and Engström-Öst 2013). One example is from the Baltic Sea where pike live and move between the border of fresh and brackish water. Young stages of pike showed considerably higher growth in freshwater environments with a freshwater zooplankton community as food than in brackish-water conditions with associated brackish-water prey (Engström-Öst et al. 2005). The authors controlled for the effect of salinity on growth *per se*, by providing the fry with control food (*Artemia* shrimps) in different salinities. They found that pike fry had a lower growth in higher salinities when provided with similar food. Based on field data, however, salinity was not a significant predictor of body condition of pike fry (Salonen et al. 2009).

There are few examples of pike living in brackish water, such as the Baltic Sea and the Caspian Sea (Crossman 1996). In the brackish Baltic Sea along the coast of Gotland, two pike populations with different reproductive strategies have been identified; one is an obligate freshwater spawner (i.e. anadromous) and the other reproducing in saline water without entering freshwater at all (Westin and Limburg 2002). Accordingly, the salinity tolerance of the eggs of the two populations differ; the brackish spawners can reproduce successfully at 6.5 ppt, whereas the eggs of the freshwater spawners only were able to hatch at 6.0 ppt (Westin and Limburg 2002). Further populations from the Baltic area living in brackish water but spawning in freshwater have been described (Johnson 1982, Müller and Berg 1982, Müller 1986, Engstedt et al. 2010, Rothla et al. 2012), as well as other brackish spawning populations (Lappalainen et al. 2008, Engstedt et al. 2010, Jacobsen et al. 2017).

3.5.2 Acidification and pH

During the 1970s and 1980s, when lakes acidified due to industrial pollution with sulphur and nitrogen oxides, pH was extensively studied in relation to heavy metal uptake in both pike and other freshwater fish (reviewed by Spry and Wiener 1991). Recently, climatic changes also cause decreases in pH, but we have found no papers studying the effect of climate-induced pH declines on pike individuals; the reason could be that this phenomenon is considered more problematic in marine areas. Available literature reports how low pH levels affect pike larval stages. Growth was severely reduced in pH 5.5, and survival and development were impaired in pH 5 or lower (Duis and Oberemm 2000). Johansson and Kihlström (1975) report larval mortality being 17% at pH 6.8, 26% at pH 5 and 97% at pH 4. Le Louarn and Webb (1998) showed that pike eggs were less sensitive to low pH in comparison to young larvae. At pH levels below 6, larval survival and swimming performance decreased. On the other hand, Raitaniemi (1995) found no relationship between pike larval growth and pH in field data. He concluded that low pH conditions are not a central factor for larval growth in the field. The tolerance of adult pike seems to be greater. In low pH waters caused by acid rain, adult pike can live in pH levels between 4-4.4 and can reproduce in pH between 4.4 and 4.9 (Magnuson et al. 1984, Leuven and Oyen 1987). Öhman et al. (2006) found no relationship between the presence of pike and pH levels in neither isolated nor connected Scandinavian lakes, although pike presence was related to low methane (CH_4) concentrations. In general, the pike is considered an acidification-tolerant species and can be the last fish species left in acid lakes measuring pH levels 4.5-5.5 (Raitaniemi et al. 1988, Rask and Tuunainen 1990). Keinänen et al. (2000, 2004) confirm that pike are extremely tolerant to low pH.

3.6 CLIMATE CHANGES INCLUDING WARMING AND UV RADIATION

The changes in our climatic conditions due to global warming not only involve warming (IPCC 2013), but also changes in precipitation and water flows and levels (Jeppesen et al. 2015) as well as an increase in UV radiation (Taalas et al. 2000). Häkkinen et al. (2004) performed extensive experiments with pike larvae and showed that in three days only, larval mortality increased 10-20% after exposure to UV-B radiation. Even low UV-B doses resulted in serious neuro-behavioural disorders. The authors concluded that even a small increase in UV may affect larval survival negatively. Vehniäinen et al. (2007) stated that both fluence rate (i.e., irradiance from all angles onto a small region of space) and cumulative dose are important when monitoring UV effects, but fluence rate is of primary importance. Vehniäinen et al. (2012) have also investigated how UV-B radiation affects pike larval individuals at fluence rates that occur naturally in field conditions. UV-B radiation caused DNA damage and changes in protein expression as well as changed behaviour. These functional changes on molecular and behavioural levels were, according to the authors, most likely closely associated with each other and could have severe consequences for pike populations.

Cool-water fish species are thought to be differentially affected by recent climate change, and the information available concerning pike is still considered limited to some extent (Chu et al. 2005, Winfield et al. 2008, Lyons et al. 2010).

However, pike is especially sensitive to rapid warming during the spawning period in shallow habitats as well as receding water following deposition of eggs, as was demonstrated in a modeling study in the Lawrence estuary, Great Lakes, Canada (Mingelbier et al. 2008). Eggs in dewatered habitats run the risk of increasing mortality, and it has been shown that their mortality can be extensive (Mingelbier et al. 2008). Another potential climate-change induced factor that will increase in importance in boreal lakes due to higher run-off of dissolved organic carbon (DOC) is brownification (Stasko et al. 2012 and references therein). Ranåker et al. (2012) found that pike reaction distance decreased substantially as a function of reduced visual range in brown water. Effects of brown water on piscivores will probably increase in the future as changing climatic conditions may further deteriorate the visual conditions (Jönsson et al. 2012).

Many papers suggest that climate change effects on pike are complex and usually indirect and can take effect via changes in, for example, prey species composition or altered pike foraging success. Cyprinids are a group that is expected to benefit from higher temperatures in a future climate, (Kallasvuo et al. 2011) and this may impact cool-water fish such as pike. For instance, Winfield et al. (2008) showed that pike abundance and individual body condition increased in the more oligotrophic northern basin of Windermere, England compared to the more eutrophic south basin. The reason for lower success of pike in the eutrophic basin was suggested to be a reduction in prey abundance, mainly, Arctic char *Salvelinus alpinus* Linnaeus, 1758, even though an increase in roach density was detected. Hence, mechanisms behind effects of climate warming on pike can be complicated. An extensive meta-analysis by Rypel (2012) suggests that growth of pike in North America is sensitive to warming, whereas Eurasian pike are more resilient to such climate change.

Several authors have used a modelling approach to unravel effects of climatic changes on pike individuals. Fang et al. (2004) demonstrated that summer warming is forecasted to turn medium-depth lakes (Austin, Texas, USA) uninhabitable for pike and other cool-water fishes because of unsuitable temperatures. Vindenes et al. (2014) showed in their modelling exercise that the net effect of warming on population growth rate and fitness of pike in Lake Windermere (United Kingdom) was positive, but the sensitivity of pike to different temperatures seemed very dependent on both body size and vital rates (survival, fecundity and somatic growth). Modelling by Lyons et al. (2010) predicted that the pike belongs to a group of species that is forecasted to decline due to climate warming based on a large dataset from Wisconsin, USA. In conclusion, it seems like studies based on models and empirical data find very different results concerning how much pike will be affected by a future warmer climate. It is likely that the effects on pike will range according to geographical distribution and latitude (Jeppesen et al. 2012), but more studies are needed to investigate this matter.

3.7 CONCLUSION

To summarize, vegetation is essential for all life stages of pike, and anthropogenic habitat use and eutrophication can result in habitat destruction or even habitat loss, which can negatively impact pike (Nilsson et al. 2014). One important solution is the implementation of habitat restoration (see chapter 10). Restoration projects are far from always successful and they are also fairly expensive, but when they succeed, there are positive effects on pike. In terms of the environment, turbidity can eventually cause problems for pike as a visually foraging fish. The effects of turbidity on pike individuals are, however, complex and need to be addressed in future studies. This is not at least relevant since climate change is likely to increase the eutrophication of lakes and, thereby, increase turbidity. There are also gaps in our knowledge on the effects on pike from important environmental factors such as salinity tolerances in areas of high salinity in the western Baltic Sea and, not at least concerning, interactions between environmental factors and their effect on pike.

The direct and indirect effects of climate change on pike individuals and populations are still unclear. Although Winfield et al. (2008) draw the conclusion that direct negative impacts of climate change on pike are unlikely. One reason is the species' high tolerance to warming and its low use of hypoxic deep water. Different studies shown different predictions (see above) and it is likely that the effects on pike will differ between existing temperature regimes (Jeppesen et al. 2012). The increased understanding of warming effects on pike in its southern distribution range is particularly valuable, and future studies should focus on the role of habitat destruction combined with climate change effects.

REFERENCES CITED

Abrahams, M. and M. Kattenfield. 1997. The role of turbidity as a constraint on predator-prey interactions in aquatic environments. *Behav. Ecol. Sociobiol.* 40:169-174.

Adelman, I.R. and L.L. Smith. 1970. Effect of oxygen on growth and food conversion efficiency of northern pike. *Prog. Fish Cult.* 32:93-96.

Aksnes, D. and J. Giske. 1993. A theoretical model of aquatic visual feeding. *Ecol. Model.* 67:233-250.

Alldridge, N.A. and A. M. White. 1980. Spawning site preferences for northern pike (*Esox lucius*) in a New York marsh with widely fluctuating water levels. *Ohio J. Sci.* 80:57-73.

Andersen, M., L. Jacobsen, P. Grønkjær and C. Skov. 2008. Turbidity increases behavioural diversity in northern pike, *Esox lucius* L., during early summer. *Fish. Manag. Ecol.* 15:377-383.

Ask, J., J. Karlsson, L. Persson, P. Ask, P. Byström and M. Jansson. 2009. Terrestrial organic matter and light penetration: Effects on bacterial and primary production in lakes. *Limnol. Oceanogr.* 54:2034-2040.

Baktoft, H., K. Aarestrup, S. Berg, M. Boel, L. Jacobsen, N. Jepsen, A. Koed, J.C. Svendsen and C. Skov. 2012. Seasonal and diel effects on the activity of northern pike studied by high-resolution positional telemetry. *Ecol. Freshw. Fish* 21:386-394. 10.1111/j.1600-0633.2012.00558.x

Bean, C.W. and I.J. Winfield. 1995. Habitat use and activity patterns of roach (*Rutilus rutilus* (L.)), rudd (*Scardinius erythrophthalmus* (L.)), perch (*Perca fluviatilis* L.) and pike (*Esox lucius* L.) in the laboratory: the role of predation threat and structural complexity. *Ecol. Freshw. Fish* 4:37-46.

Beitinger, T.L., W.A. Bennett and R.W. McCauley. 2000. Temperature tolerances of North American freshwater fishes exposed to dynamic changes in temperature. *Environ. Biol. Fishes* 58:237–275.

Berg, L. and T. Northcote. 1985. Changes in territorial, gill-flaring, and feeding behaviour in juvenile coho salmon *Onchorhynchus kisutch* following short-term pulses of suspended sediment. *Can. J. Fish. Aquat. Sci.* 42:1410-1417.

Berg, S., C. Skov, J.S. Olsen and K. Michelsen. 2012. Lavt vand – en nødvendighed for geddeyngel. *Vand og Jord* 3:117-119. (in Danish)

Boehlert, G. and J. Morgan. 1985. Turbidity enhances feeding abilities of larval Pacific Herring, *Clupea harengus pallasi*. *Hydrobiologia* 123:161-170.

Boeuf, G. and J. Falcon 2001. Photoperiod and growth in fish. *Vie Milieu* 51:247-266.

Bristow, B., R. Summerfelt and R. Clayton. 1996. Comparative performance of intensively cultured larval walleye in clear, turbid, and colored water. *Prog. Fish Cult.* 58:1-10.

Bry, C., F. Bonamy, J. Manelphe and B. Duranthon.1995. Early-life characteristics of pike, *Esox lucius*, in rearing ponds – temporal survival pattern and ontogenic diet shifts. *J. Fish Biol.* 46:99-113.

Bry, C. 1996. Role of vegetation in the life cycle of pike. pp. 45-67. *In*: J.F. Craig (ed.). *Pike: Biology and exploitation.* London: Chapman & Hall.

Casselman, J.M. 1978. Effects of environmental factors on growth, survival, activity and exploitation of northern pike. *Am. Fish. Soc. Spec. Publ.* 11:114-128.

Casselman, J.M. 1996. Age, growth and environmental requirements of pike. pp. 68-101. *In*: J. Craig (ed.). *Pike: Biology and exploitation.* London: Chapman & Hall.

Casselman, J.M. and C.A. Lewis. 1996. Habitat requirements of northern pike (*Esox lucius*). *Can. J. Fish. Aquat. Sci.* 53:161-174.

Chapman, C.A. and W.C. Mackay. 1984. Versatility in habitat use by a top aquatic predator, *Esox lucius* L. *J. Fish Biol.* 25:109-115.

Chesney, E.J. 1989. Estimating the food requirement of striped bass larvae *Morone saxatilis*: effects of light, turbidity and turbulence. *Mar. Ecol. Prog. Ser.* 53:191-200.

Cucherousset, J., J.-M. Paillisson, A. Cuzol and J.-M. Roussel. 2009. Spatial behaviour of young-of –the-year northern pike (*Esox lucius* L.) in a temporarily flooded nursery area. *Ecol. Freshw. Fish* 18:314-322.

Chu, C., N.E. Mandrak and C.K. Minns. 2005. Potential impacts of climate change on the distributions of several common and rare freshwater fishes in Canada. *Divers. Distrib.* 11:299-310.

Confer, J., G. Howick, M. Corzette, L. Kramer, S. Fitzgibbon and R. Landesberg. 1978. Visual predation by planktivores. *Oikos* 31:27-37.

Cook, M.F. and E.P. Bergersen. 1988. Movements, habitat selection, and activity periods of northern pike in Eleven mile Reservoir, Colorado. *Trans. Am. Fish. Soc.* 117:495-502.

Cooke, G.D., E.B. Welch, S.A. Peterson and P.R. Newroth. 1993. *Restoration and Management of Lakes and Reservoirs*. Lewis Publishers, Boca Raton, FL, USA.

Cooper, J.E., J.V. Mead, J.M. Farrell and R.G. Werner. 2008. Potential effects of spawning habitat changes on the segregation of northern pike (*Esox lucius*) and muskellunge (*E. masquinongy*) in the Upper St. Lawrence River. *Hydrobiologia* 601:41-53.

Craig, J.F. and J.A. Babaluk. 1989. Relationship of condition of walleye (*Stizostedion vitreum*) and northern pike (*Esox lucius*) to water clarity, with special references to Dauphin Lake, Manitoba. *Can. J. Fish. Aquat. Sci.* 46:1581-1586.

Craig, J.F. 2008. A short review of pike ecology. *Hydrobiologia* 601:5-16.

Crossman, E.J. 1996. Taxonomy and distribution. pp. 1-11. *In*: J.F. Craig (ed.). *Pike: Biology and exploitation.* London: Chapman & Hall.

Crowder, L.B. and W.E. Cooper. 1982. Habitat structural complexity and the interaction between bluegill and their prey. *Ecology* 63:1802-1813.

Dahl, J. 1961. Alder og vækst hos danske og svenske brakvandsgedder. *Ferskvandsfiskeribladet* 59:34-38. (in Danish)

Diana J.S., W.C. Mackay and M. Ehrman. 1977. Movements and habitat preference of northern pike (*Esox lucius*) in Lac St. Anne, Alberta. *Trans. Am. Fish. Soc.* 10:560-565.

Dobler, E. 1977. Correlation between the feeding time of the pike (*Esox lucius*) and the dispersion of a school of *Leucaspius delineates*. *Oecologia* 27:93-96.

Dombeck, M.P., B.W. Menzel and P.N. Hinz. 1984. Muskellunge spawning habitat and reproductive success. *Trans. Am. Fish. Soc.*113:205-216.

Duis K. and A. Oberemm. 2000. Survival and sublethal responses of early life stages of pike exposed to low pH in artificial post-mining lake water. *J. Fish Biol.* 57:597-613.

Eklöv, P. 1997. Effects of habitat complexity and prey abundance on the spatial and temporal distributions of perch (*Perca fluviatilis*) and pike (*Esox lucius* L.). *Can. J. Fish. Aquat. Sci.* 54:1520-1531.

Eklöv P. and S. Diehl. 1994. Piscivore efficiency and refuging prey: the importance of predator search mode. *Oecologia* 98:344-353.

Engstedt, O., P. Stenroth, P. Larsson, L. Ljunggren and M. Elfman. 2010. Assessment of natal origin of pike (*Esox lucius*) in the Baltic Sea using Sr:Ca in otoliths. *Environ. Biol. Fishes* 89:547-555.

Engström-Öst, J., M. Lehtiniemi, S.H. Jónasdóttir and M. Viitasalo. 2005. Growth of pike larvae (*Esox lucius*) under different conditions of food quality and salinity. *Ecol. Freshw. Fish* 14:385-393.

Engström-Öst, J. and I. Isaksson. 2006. Effects of macroalgal exudates and oxygen deficiency on survival and behaviour of fish larvae. *J. Exp. Mar. Biol. Ecol.* 335:227-234.

Engström-Öst, J., M. Karjalainen and M. Viitasalo. 2006. Feeding and refuge use by small fish in the presence of cyanobacteria. *Environ. Biol. Fishes* 76:109-117.

Engström-Öst, J., Immonen, E., Candolin, U. and Mattila, J. 2007. The indirect effects of eutrophication on habitat choice and survival of fish larvae in the Baltic Sea. *Mar. Biol.* 151:393-400.

Engström-Öst, J. and J. Mattila. 2008. Foraging, growth and habitat choice in turbid water: an experimental study with fish larvae in the Baltic Sea. *Mar. Ecol. Prog. Ser.* 359:275-281.

Engström-Öst, J., S. Repka, A. Brutemark and A. Nieminen. 2013. Clay and algae-induced effects on biomass, cell size and toxin concentration in a brackish-water cyanobacterium. *Hydrobiologia* 714:85-92.

Fabricius, E. 1950. Heterogeneous stimulus in the release of spawning activities in fish. *Rept. Inst. Freshwater Res. Drottningholm.* 31:57-99.

Fabricius, E. and K.-J. Gustafson. 1958. Some new observations on the spawning behaviour of the pike, *Esox lucius* L. *Rept. Inst. Freshwater Res. Drottningholm.* 39:23-54.

Fago, D.M. 1977. Northern pike production in managed spawning and rearing marshes. *Tech. Bull. Wisc. Dep. Nat. Resour.* 96:1-30.

Fang, X., H.G. Stefan, J.G. Eaton, J.H. McCormick and S.R. Alam. 2004. Simulation of thermal/dissolved oxygen habitat for fishes under different climate scenarios Part 1. Cool-water fish in the contiguous US. *Ecol. Model.* 172:13-37.

Farrell, J.M., R.G. Werner, S.R. LaPan and K.A. Claypoole. 1996. Egg distribution and spawning habitat of northern pike and muskellunge in a St. Lawrence River marsh, New York. *Trans. Am. Fish. Soc.* 125:127-131.

Farrell, J.M. 2001. Reproductive success of sympatric northern pike and muskellunge in an Upper St. Lawrence river bay. *Trans. Am. Fish. Soc.* 130:796-808.

Fiksen, Ø., D.L. Aksnes, M.H. Flyum and J. Giske. 2002. The influence of turbidity on growth and survival of fish larvae: a numerical analysis. *Hydrobiologia* 484:49-59.
Flinkman, J., E. Aro, I. Vuorinen and M. Viitasalo. 1998. Changes in northern Baltic zooplankton and herring nutrition from 1980s to 1990s: top-down and bottom-up processes at work. *Mar. Ecol. Prog. Ser.* 165:127-136.
Fortin, R., P. Dumont and H. Fournier. 1983. La reproduction du grand brochet (*Esox lucius* L.) dans certains plains d'eau du sud de Québec (Canada). pp 39-51. *In*: R. Billard (ed.). *Le Brochet: gestion dans le milieu naturel et élevage*. Paris: INRA Publ. (in French)
Franklin, D.R. and L.L. Smith Jr. 1963. Early life history of northern pike (*Esox lucius* L.) with special reference to the factors influencing the numerical strength of year classes. *Trans. Am. Fish. Soc.* 92:91-110.
Frost, W.E. and C. Kipling. 1967. A study of reproduction, early life, weight-length relationship and growth of pike (*Esox lucius* L.) in Windermere. *J. Anim. Ecol.* 36:651-93.
Gingerich A.J., S.J. Cooke, K.C. Hanson, M.R. Donaldson, C.T. Hasler. C.D. Suski and R. Arlinghaus. 2007. Evaluation of the interactive effects of air exposure duration and water temperature on the condition and survival of angled and released fish. *Fish. Res.* 86:169-178.
Giske, J., D.L. Aksnes and Ø. Fiksen. 1994. Visual predators, environmental variables and zooplankton mortality risk. *Vie Milieu* 44:1-9.
Greenberg, L.A., C.A. Paszkowski and W.M. Tonn. 1995. Effects of prey species composition and habitat structure on foraging by two functionally distinct piscivores. *Oikos* 74: 552-232.
Gregory, R. 1993. Effect of turbidity on the predator avoidance behaviour of juvenile Chinook salmon (*Oncorhynchus tshawytscha*). *Can. J. Fish. Aquat. Sci.* 50:241-246.
Gregory, R. and T. Northcote. 1993. Surface, planktonic and benthic foraging of juvenile chinook salmon (*Oncorhynchus tshawytscha*) in turbid laboratory conditions. *Can. J. Fish. Aquat. Sci.* 50:233-240.
Grimm, M.P. 1983. Regulation of biomasses of small (< 41cm) northern pike (*Esox lucius* L.) with special reference to the contribution of individuals stocked as fingerlings (4-6 cm). *Aquacult. Res.* 14:115-134.
Grimm, M.P. 1989. Northern pike (*Esox lucius* L.) and aquatic vegetation, tools in the management of fisheries and water quality in shallow waters. *Hydrobiol. Bull.* 23:59-65.
Grimm, M.P. 1994. The influence of aquatic vegetation and population biomass on recruitment of 0+ and 1+ northern pike (*Esox lucius* L.). pp. 226-234. *In*: I. Cowx (ed.). *Rehabilitation of freshwater fisheries*. Oxford: Blackwell Scientific Publications.
Grimm, M.P. and M. Klinge. 1996. Pike and some aspects of its dependence on vegetation. pp 125-156. *In*: J.F. Craig (ed.). *Pike: Biology and exploitation*. London: Chapman & Hall.
Häkkinen, J., J. Vehniäinen and A. Oikari. 2004. High sensitivity of northern pike larvae to UV-B but no UV-photoinduced toxicity of retene. *Aquat. Toxicol.* 66:393-404.
Hansson, L.A. and Hylander, S. 2009. Effects of ultraviolet radiation on pigmentation, photoenzymatic repair, behavior, and community ecology of zooplankton. *Photochem. Photobiol. Sci.* 8:1266-1275.
Hein, C.L., G. Öhlund and G. Englund. 2014. Fish introductions reveal the temperature dependence of species interactions. *Proc. R. Soc.* B. 281:20132641.
Hinshaw, J.M. 1985. Effects of illumination and prey contrast on survival and growth of larval Yellow Perch *Perca flavescens*. *Trans. Am. Fish. Soc.* 114:540-545.
Hokanson, K.E.F., J.H. McCormick and B.R. Jones. 1973. Temperature requirements for embryos and larvae of the northern pike *Esox lucius* Linnaeus. *Trans. Am. Fish. Soc.* 102:89-100.

Holland, L.E. and M.L. Huston. 1984. Relationship of young-of-the-year northern pike to aquatic vegetation types in the backwater of the upper Mississippi River USA. *North Am. J. Fish. Manage.* 4:514-522.

Hongve, D., G. Riise and J.F. Kristiansen. 2004. Increased colour and organic acid concentrations in Norwegian forest lakes and drinking water - a result of increased precipitation? *Aquat. Sci.* 66:231-238.

IPCC. 2013. Climate Change 2013: *The Physical Science Basis. Contribution of Working Group I to the Fifth Assessment Report of the Intergovernmental Panel on Climate Change.* T.F. Stocker, D. Qin, G.-K. Plattner, M. Tignor, S.K. Allen, J. Boschung, A. Nauels, Y. Xia, V. Bex and P.M. Midgley (eds.). Cambridge, UK and New York, NY, USA: Cambridge University Press, 1535 pp.

Jacobsen, L. and S. Berg. 1998. Diel variation in habitat use by planktivores in field enclosure experiments: the effect of submerged macrophytes and predation. *J. Fish Biol.* 53:1207-1219.

Jacobsen, L. and M.R. Perrow. 1998. Predation risk from piscivorous fish influencing the diel use of macrophytes by planktivorous fish in experimental ponds. *Ecol. Freshw. Fish* 7:78-86.

Jacobsen, L., C. Skov, A. Koed and S. Berg. 2007. Short-term salinity tolerance of northern pike, *Esox lucius*, fry, related to temperature and size. *Fish. Manag. Ecol.* 14:303-308.

Jacobsen, L., D. Bekkevold, S. Berg, N. Jepsen, A. Koed, C. Skov, K. Aarestrup and H. Baktoft. 2017. Individual based migration patterns of pike (*Esox lucius*) in a brackish lagoon in the saltwater transition zone of the western Baltic. *Hydrobiologia* 784:143-154.

Jeppesen, E., J.P. Jensen, M. Søndergaard, T. Lauridsen, L.J. Pedersen and L. Jensen. 1997. Top-down control in freshwater lakes; the role of nutrient state, submerged macrophytes and water depth. *Hydrobiologia* 342:151-164.

Jeppesen, E., T. Mehner, I.J. Winfield, K. Kangur, J. Sarvala, D. Gerdeaux, M. Rask, H.J. Malmquist, K. Holmgren, P. Volta, S. Romo, R. Eckmann, A. Sandström, S. Blanco, A. Kangur, H.R. Stabo, M. Tarvainen, A.-M. Ventelä, M. Søndergaard, T.L. Lauridsen and M. Meerhoff. 2012. Impacts of climate warming on the long-term dynamics of key fish species in 24 European lakes. *Hydrobiologia* 694:1-39.

Jeppesen, E., S. Brucet, L. Naselli-Flores, E. Papastergiadou, K. Stefanidis, T. Nõges, P. Nõges, J.L. Attayde, T. Zohary, J. Coppens, T. Bucak, R.F. Menezes, F.R. Sousa Freitas, M. Kernan, M. Søndergaard and M. Beklioğlu. 2015. Ecological impacts of global warming and water abstraction on lakes and reservoirs due to changes in water level and related changes in salinity. *Hydrobiologia* 750:201-227.

Jepsen, N., S. Beck, C. Skov and A. Koed. 2001. Behavior of pike (*Esox lucius* L.) > 50 cm in a turbid reservoir and in a clearwater lake. *Ecol. Freshw. Fish* 10:26-34.

Johansson, N. and J.E. Kihlström. 1975. Pikes (*Esox lucius* L.) shown to be affected by low pH values during first weeks after hatching. *Environ. Res.* 9:12-17.

Johnson, F.H. 1957. Northern pike year-class strength and spring water levels. *Trans. Am. Fish. Soc.* 86:285-293.

Johnson, L. 1960. *Studies of the Behaviour and Nutrition of the Pike* (*Esox lucius L.*). Ph.D. thesis, University of Leeds, Leeds, UK.

Johnson, T. 1982. Seasonal migrations of anadromous fishes in a northern Swedish stream. *Mon. Biol.* 45:351-360.

Jönsson, M., S. Hylander, L. Ranåker, P.A. Nilsson and C. Brönmark. 2011. Foraging success of juvenile pike *Esox lucius* depends on visual conditions and prey pigmentation. *J. Fish. Biol.* 79:290-297.

Jönsson, M., L. Ranåker, P.A. Nilsson and C. Brönmark. 2012. Prey-type-dependent foraging of young-of-the-year fish in turbid and humic environments. *Ecol. Freshw. Fish* 21:461-468.

Jørgensen, A.T., B.W. Hansen, B. Vismann, L. Jacobsen, C. Skov, S. Berg and D. Bekkevold. 2010. High salinity tolerance in eggs and fry of a brackish *Esox lucius* population. *Fish. Manag. Ecol.* 17:554-560.

Jude, D.J. and J. Pappas. 1992. Fish utilization of Great Lakes coastal wetlands. *J. Great Lakes Res.* 18:651-672.

Kallasvuo, M., A. Lappalainen and L. Urho. 2011. Coastal reed beds as fish reproduction habitats. *Boreal Env. Res.* 16:1-14.

Karlsson, J., P. Byström, J. Ask, P. Ask, L. Persson and M. Jansson. 2009. Light limitation of nutrient-poor lake ecosystems. *Nature* 460:506-509.

Keinänen, M., S. Peuranen, M. Nikinmaa, C. Tigerstedt and P.J. Vuorinen. 2000. Comparison of the responses of the yolk-sac fry of pike (*Esox lucius*) and roach (*Rutilus rutilus*) to low pH and aluminium: sodium influx, development and activity. *Aquat. Toxicol.* 47:161-179.

Keinänen, M., C. Tigerstedt, S. Peuranen and P.J. Vuorinen. 2004. The susceptibility of early development phases of an acid-tolerant and acid-sensitive fish species to acidity and aluminum. *Ecotoxicol. Environ. Safety* 58:160-172.

Kobler, A., T. Klefoth, C. Wolter, F. Fredrich and R. Arlinghaus. 2008. Contrasting pike (*Esox lucius* L.) movement and habitat choice between summer and winter in a small lake. *Hydrobiologia* 601:17-27.

Kohn, A.J. and P.J. Leviten. 1976. Effect of habitat complexity on population density and species richness in tropical intertidal predatory gastropod assemblages. *Oecologia* 25:199-210.

Lappalainen, A., M. Härmä, S. Kuningas and L. Urho. 2008. Reproduction of pike (*Esox lucius*) in reed belt shores of the SW coast of Finland, Baltic Sea: a new survey approach. *Boreal Env. Res.* 13:370-380.

Lehtiniemi, M., J. Engström-Öst and M. Viitasalo. 2005. Turbidity decreases anti-predator behaviour in pike larvae, *Esox lucius*. *Environ. Biol. Fishes* 73:1-8.

Lehtonen, H. 1986. Fluctuations and long-term trends in the pike, *Esox lucius* (L.) population in Nothamn, western Gulf of Finland. *Aqua Fenn.* 16:3-9.

Lehtonen, H., E. Leskinen, R. Selén and M. Reinikainen. 2009. Potential reasons for the changes in the abundance of pike, *Esox lucius*, in the western Gulf of Finland, 1939-2007. *Fish. Manag. Ecol.* 16:484-491.

Leslie, J.K. and C.A. Timmins. 1992. Distribution and abundance of larval fish in Hamilton Harbor, a severely degraded embayment of Lake Ontario. *J. Great Lakes Res.* 18:700-708.

Leuven, R.S.E.W. and F.G.F. Oyen. 1987. Impact of acidification and eutrophication on the distribution of fish species in shallow and lentic soft waters of the Netherlands: an historical perspective. *J. Fish Biol.* 31:753-774.

Le Louarn, H. and D.J. Webb. 1998. Effets négatifs de pH extremes sur le développement embryonnaire et larvaire du brochet *Esox lucius* L. *Bull. Fr. Pêche Piscic.* 350-351:325-336.

Lyons, J., J.S. Stewart and M. Mitro. 2010. Predicted effects of climate warming on the distribution of 50 stream fishes in Wisconsin, U.S.A. *J. Fish Biol.* 77:1867-1898.

Maes, J., A. Taillieu, K. Van Damme, K. Cottenie and F. Ollevier. 1998. Seasonal patterns in the fish and crustacean community of a turbid temperate estuary (Zeeschelde Estuary, Belgium). *Estuar. Coast. Shelf Sci.* 47:143-151.

Magnuson, J.J., J.P. Baker and E.J. Rahel. 1984. A critical assessment of effects of acidification on fisheries in North America. *Phil Trans. R. Soc. Lond. B* 305:501-516.

Mauck, W.L. and D.W. Coble. 1973. Vulnerability of some fishes to northern pike (*Esox lucius*) predation. *J. Fish. Res. Bd. Can.* 28:957-969.

McCarraher, D.B. and R.E. Thomas. 1972. Ecological significance of vegetation to northern pike, *Esox lucius*, spawning. *Trans. Am. Fish. Soc.* 101:560-563.

Miner, G. and R. Stein. 1993. Interactive influence of turbidity and light on larval bluegill (*Lepomis macrochirus*) foraging. *Can. J. Fish. Aquat. Sci.* 50:781-788.
Mingelbier, M., P. Brodeur and J. Morin. 2008. Spatially explicit model predicting the spawning habitat and early stage mortality of Northern pike (*Esox lucius*) in a large system: the St. Lawrence River between 1960 and 2000. *Hydrobiologia* 601:55-69.
Moyle, J.B. and W.D. Clothier. 1959. Effects of management and winter oxygen levels on the fish population of a prairie lake. *Trans. Am. Fish. Soc.* 88:178-185.
Müller, K. and E. Berg. 1982. Spring migration of some anadromous freshwater fish species in the northern Bothnian Sea. *Hydrobiologia* 96:161-168.
Müller, K. 1986. Seasonal anadromous migration of the pike (*Esox lucius* L.) in coastal areas of the northern Bothnian Sea. *Arch. Hydrobiol.* 107:315-330.
Muscalu-Nagy, R. and C. Muscalu-Nagy. 2012. Variation of oxygen consumption for northern pike (*Esox lucius*). *J. Biotechnol.* 161S:25.
Myers R.A. and B. Worm. 2003. Rapid worldwide depletion of predatory fish communities. *Nature* 423:280-283.
Nielsen, S.L., K. Sand-Jensen, J. Borum and O. Geertz-Hansen. 2002. Depth colonization of eelgrass (*Zostera marina*) and macroalgae as determined by water transparency in Danish coastal waters. *Estuaries* 25:1025-1032.
Nilsson, J. 2006. Predation of northern pike (*Esox lucius* L.) eggs: a possible cause of regionally poor recruitment in the Baltic Sea. *Hydrobiologia* 553:161-169.
Nilsson J., O. Engstedt and P. Larsson. 2014. Wetlands for northern pike (*Esox lucius* L.) recruitment in the Baltic Sea. *Hydrobiologia* 721:145-154.
Nilsson, A.P., L. Jacobsen, S. Berg and C. Skov. 2009. Environmental conditions and intraspecific interference: unexpected effects of turbidity on pike (*Esox lucius*) foraging. *Ethology* 115:33-38.
Nilsson, P.A., H. Baktoft, M. Boel, K. Meier, L. Jacobsen, E. Rokkjær, T. Clausen and C. Skov, 2012. Visibility conditions and diel period affect small-scale spatio-temporal behaviour of pike *Esox lucius* in the absence of prey and conspecifics. *J. Fish Biol.* 80:2384-2389.
Öhman, J., I. Buffam, G. Englund, A. Blom, E. Lindgren and H. Laudon. 2006. Associations between water chemistry and fish community composition: a comparison between isolated and connected lakes in northern Sweden. *Fresh. Biol.* 51:510-522.
Ojaveer, E. 1981. Fish fauna of the Baltic Sea. pp. 275-292. *In*: A. Voipio (ed.). *The Baltic Sea*. Oceanogr. Ser. 30, Amsterdam: Elsevier.
Pierce, R.B., J.A. Younk and C.M. Tomcko. 2007. Expulsion of miniature radio transmitters along with eggs of muskellunge and northern pike – a new method for locating critical spawning habitat. *Environ. Biol. Fishes* 79:99-109.
Pierce, R.B., A.J. Carlson, B.M. Carlson, D. Hudson and D.F. Staples. 2013. Depths and thermal habitat used by large versus small northern pike in three Minnesota lakes. *Trans. Am. Fish. Soc.* 142:1629-1639.
Quesenberry, N., P. Allen and J. Cech Jr. 2007. The influence of turbidity on three-spined stickleback foraging. *J. Fish Biol.* 70:965-972.
Raat, A. 1988. *Synopsis of biological data on the northern pike*: Esox lucius Linnaeus, 1758. FAO Fish. Syn. 145:178 pp.
Raitaniemi, J., M. Rask and P.J. Vuorinen. 1988. The growth of perch, *Perca fluviatilis* L., in small Finnish lakes at different stages of acidification. *Ann. Zool. Fenn.* 25:209-219.
Raitaniemi, J. 1995. The growth of young pike in small Finnish lakes with different acidity-related properties and fish species composition. *J. Fish Biol.* 47:115-125.
Ranåker, L., M. Jönsson, A.P. Nilsson and C. Brönmark. 2012. Effects of brown and turbid water on piscivore-prey fish interactions along a visibility gradient. *Fresh. Biol.* 57:1761-1768.

Rask, M. and P. Tuunainen. 1990. Acid-induced changes in fish populations of small Finnish lakes. pp. 911-927. *In*: P. Kauppi, P. Anttila and K. Kenttämies (eds.). *Acidification in Finland*. Berlin: Springer.

Rönkkönen, S., E. Ojaveer, T. Raid and M. Viitasalo. 2004. Long-term changes in Baltic herring (*Clupea harengus membras*) growth in the Gulf of Finland. *Can. J. Fish. Aquat. Sci.* 61:219-229.

Rothla, M., M. Vetemaa, K. Urtson and A. Soesoo. 2012. Early life migration of Baltic Sea pike *Esox lucius*. *J. Fish. Biol.* 80:886-893.

Rowe, D. 1984. Factors affecting the foods and feeding patterns of lake-dwelling rainbow trout (*Salmo gairdnerii*) in the North Island of New Zealand. New Zeal. *J. Mar. Fresh. Res.* 18:129-141.

Rypel, A. L. 2012. Meta-analysis of growth rates for a circumpolar fish, the northern pike (*Esox lucius*), with emphasis on effects of continent, climate and latitude. *Ecol. Freshw. Fish* 21:521-532.

Salonen, M., L. Urho and J. Engström-Öst. 2009. Effects of turbidity and zooplankton availability on the condition and prey selection of pike larvae. *Bor. Environ. Res.* 14:981-989.

Salonen, M. and J. Engström-Öst. 2010. Prey capture of pike *Esox lucius* larvae in turbid water. *J. Fish Biol.* 76:2591-2596.

Salonen, M. 2012. *The Effect of Turbidity on the Ecology of Pike Larvae*. Ph.D. Thesis, University of Helsinki, Helsinki, Finland.

Salonen, M. and J. Engström-Öst. 2013. Growth of pike larvae: effects of prey, turbidity and food quality. *Hydrobiologia* 717:169-175.

Sandström, A. and P. Karås. 2005. Boating and navigation activities influence the recruitment of fish in a Baltic Sea archipelago area. *Ambio* 34:125-130.

Siefert, R.E., W.A. Spoor and R.F Syrett. 1973. Effects of reduced oxygen concentrations on northern pike (*Esox lucius*) embryos and larvae. *J. Fish. Res. Bd. Can.* 30:849-852.

Skov, C. and S. Berg. 1999. Utilization of natural and artificial habitats by YOY pike in a biomanipulated lake. *Hydrobiologia* 408-409:115-122.

Skov, C., S. Berg, L. Jacobsen and N. Jepsen. 2002. Habitat use and foraging success of 0+pike (*Esox lucius* L.) in experimental ponds related to prey fish, water transparency and light intensity. *Ecol. Freshw. Fish* 11:65-73.

Skov, C., L. Jacobsen and S. Berg. 2003. Post-stocking survival of 0+year pike in ponds as a function of water transparency, habitat complexity, prey availability and size heterogeneity. *J. Fish Biol.* 62:311-322.

Skov, C. and A. Koed. 2004. Habitat use of 0+ year pike in experimental ponds in relation to cannibalism, zooplankton, water transparency and habitat complexity. *J. Fish Biol.* 64:448-459.

Skov, C., P.A. Nilsson, L. Jacobsen, and C. Brönmark. 2007. Habitat-choice interactions between pike predators and perch prey depend on water transparency. *J. Fish Biol.* 70:298-302.

Smith, B.M., J.M. Farrell, H.B. Underwood and S.J. Smith. 2007. Year-class formation of Upper St. Lawrence River Northern pike. *N. Am. J. Fish. Manage.* 27:481-491.

Snickars, M., A. Sandström, and J. Mattila. 2004. Antipredator behaviour of 0+ year *Perca fluviatilis*: effect of vegetation density and turbidity. *J. Fish Biol.* 65:1604-1613.

Souchon, Y. and L. Tissot. 2012. Synthesis of thermal tolerances of the common freshwater fish species in large Western Europe rivers. *Knowl. Managt. Aquat. Ecosyst.* 405, 03.

Spry, D.J. and J.G. Wiener. 1991. Metal bioavailability and toxicity to fish in low-alkalinity lakes: a critical review. *Environ. Pollut.* 71:243-304.

Stasko, A.D., J.M. Gunn and T.A. Johnston. 2012. Role of ambient light in structuring north-temperate fish communities: potential effects of increasing dissolved organic carbon concentration with a changing climate. *Environ. Rev.* 20:173-190.

Sweka, J. and K. Hartman. 2001. Influence of turbidity on brook trout reactive distance and foraging success. *Trans. Am. Fish. Soc.* 130:138-146.

Swift, D.R. 1965. Effect of the temperature on mortality and rate of development of the eggs of the pike (*Esox lucius* L.) and the perch (*Perca fluviatilis* L.). *Nature* 206:528.

Taalas, P., J. Kaurola, A. Kylling, D. Shindell, R. Sausen, M. Dameris, V. Grewe, J. Herman, J. Damski and B. Steil. 2000. The impact of greenhouse gases and halogenated species on future solar UV radiation doses. *Geophys. Res. Lett.* 27:1127-1130.

Threinen, C.W. 1969. An evaluation of the effect and extent of habitat loss on northern pike populations and means of prevention of losses. *Wis. Dep. Nat. Resour. Bur. Fish Mgmt. Rep.* 28.

Turesson, H. and C. Brönmark. 2007. Predator-prey encounter rates in freshwater piscivores: effects of prey density and water transparency. *Oecologia* 153:281-290.

Urho, L., M. Hildén and R. Hudd. 1990. Fish reproduction and the impact of acidification in the Kyrönjoki River estuary in the Baltic Sea. *Environ. Biol. Fishes* 27:273-283.

Utne-Palm, A.C. 1999. The effect of prey mobility, prey contrast, turbidity and spectral composition on the reaction distance of *Gobiusculus flavescens* to its planktonic prey. *J. Fish Biol.* 54:1244-1258.

Utne-Palm, A.C. 2002. Visual feeding of fish in a turbid environment: physical and behavioural aspects. – *Mar. Freshw. Behav. Phys.* 35:111-128.

Vehniäinen, E.-R., J.M. Häkkinen and A.O.J. Oikari. 2007. Fluence rate or cumulative dose? Vulnerability of larval northern pike (*Esox lucius*) to ultraviolet radiation. *Photochem. Photobiol.* 83:444-449.

Vehniäinen, E.-R., K. Vähäkangas and A. Oikari. 2012. UV-B exposure causes DNA damage and changes in protein expression in northern pike (*Esox lucius*) posthatched embryos. *Photochem. Photobiol.* 88:363-370.

Viitasalo, M., J. Flinkman and M. Viherluoto. 2001. Zooplanktivory in the Baltic Sea: a comparison of prey selectivity by *Clupea harengus* and *Mysis mixta*, with reference to prey escape reactions. *Mar. Ecol. Prog. Ser.* 216:191-200.

Vindenes, Y., E. Edeline, J. Ohlberger, Ø. Langangen, I.J. Winfield, N.C. Stenseth and L.A. Vøllestad. 2014. Effects of climate change on trait-based dynamics of a top predator in freshwater ecosystems. *Am. Nat.* 183:243-256.

Vinyard, G. and W. O'Brien. 1976. Effects of light and turbidity on the reactive distance of bluegill (*Lepomis macrochirus*). *J. Fish. Res. Bd. Can.* 33:184-214.

Volkova, L. 1973. The effect of light intensity on the availability of food organisms to some fishes in Lake Baikal. *J. Ichthyol.* 13:591-602.

Vøllestad, L.A., J. Skurdal and T. Qvenild. 1986. Habitat use, growth, and feeding of pike (*Esox lucius* L) in 4 Norwegian lakes. *Arch. Hydrobiol.* 108:107-117.

Werner, E.E. and D.J. Hall. 1988. Ontogenetic habitat shift in bluegills: the foraging rate-predation risk trade-off. *Ecology* 69:1352-1366.

Werner, R.G. 2002. Habitat requirements. pp. 161-182. *In*: L.A. Fuiman and R.G. Werner (eds.). *Fishery Sciences – the unique contributions of early life stages*. Oxford: Blackwell Scientific Publications.

Westin, L. and K.E. Limburg. 2002. Newly discovered reproductive isolation reveals sympatric populations of *Esox lucius* in the Baltic. *J. Fish Biol.* 61:1647-1652.

Winfield, I.J., J.B. James and J.M. Fletcher. 2008. Northern pike (*Esox lucius*) in a warming lake: changes in population size and individual condition in relation to prey abundance. *Hydrobiologia* 601:29-40.

Wootton, R.J. 1990. *Ecology of Teleost Fishes*. Fish and Fisheries Series 1. Chapman & Hall, London, UK.

Wright, R.M. and N. Giles. 1987. The survival, growth and diet of pike fry, *Esox lucius* L., stocked at different densities in experimental ponds. *J. Fish Biol.* 30:617-629.

Žiliukienė, V. and V. Žiliukas. 2012. Spawning population characteristics of pike *Esox lucius* L. in Lake Rubikiai (Lithuania). *Centr. Eur. J. Biol.* 7:867-877.

Chapter 4

Bioenergetics and Individual Growth of Pike

John D. Armstrong[1]

4.1 INTRODUCTION

Bioenergetics concerns the flow of energy through animals and ecosystems and is fundamental in understanding the relationship between consumption (as prey) and retention (as growth) within pike *Esox Lucius* Linnaeus, 1758 populations and the individuals that comprise them. Pike are particularly interesting because they are the epitomy of the "lurking predator" and renowned for being able to reach a large body size (Fickling 2004) relative to most freshwater temperate fishes. This begs the question: is there anything unusual about the way that pike acquire and use energy that relates to their behaviour and growth potential? This chapter considers the question by bringing together a range of scientific information to develop conceptual frameworks and quantitative predictive models that summarise our current state of understanding. Pike acquire energy as food, metabolise energy for a range of metabolic and locomotory activities, use energy to produce gametes and may store energy in body tissues. Growth rate is determined by the balance between rates of energy intake and expenditure, together with rates of accumulation of elements such as calcium and phosphorus for body structures.

The numerous studies of growth of pike, in terms of annual increases in length, have recently been reviewed in relation to variation across spatial scales (Rypel 2012). Growth can be limited by a wide range of extrinsic environmental factors that affect metabolism and feeding success (chapters 2 and 3), and by intrinsic factors, which can be divided into two categories. The first of these is motivation to feed. In some situations, fish voluntarily constrain their food intake due to a trade-off

[1]Marine Scotland Science, Freshwater Fisheries Laboratory, Pitlochry, Perthshire, Scotland, UK; Email: john.armstrong@gov.scot

between advantages of fast growth and risk of mortality associated, for example, with exposure to predation during feeding (Mangel and Stamps 2001). Hence, there may be growth potential in reserve, such as is expressed in fish during a compensatory phase after restriction in food availability (Russell and Wootton 1992). The second category of intrinsic limitation is physiological and morphological constraints. These include factors that limit the rate at which food may be processed and assimilated even when it is readily available and motivation to feed is high.

This chapter focuses on intrinsic factors that affect growth, particularly physiological capacity, and considers information on how energy and power budgets relate to scope for growth. The first section considers relevant aspects of pike morphology and organ function to set out adaptations of the body tissue and the organic capacity that facilitates the growth process. The second section sets out principles of energetics that determine the quantity of energy acquired as food that can be retained as growth and the rate at which it can be transferred to body tissue. The third section considers what data are available to populate pike energy transfer models. The final section summarises and concludes on how metabolism of pike relates to their growth potential.

4.2 STRUCTURE AND FUNCTION

4.2.1 Body Shape and Acceleration

Pike have a relatively elongated body form with large dorsal and anal fins located posterially adjacent to the caudal peduncle and caudal fin. The main muscle myotomes of esocids comprise predominantly white fibres specialized for anaerobic burst activity. However, presence of a thin layer of red and pink musculature under the lateral line, particularly in the caudal region (Hoyle et al. 1986), provides tissue with primary role of powering aerobic swimming from body flex. Propulsion is also by skulling, particularly of pectoral fins. Red muscle represents approximately 5.2-6.9% of total muscle mass of pike (Schwalme and MacKay 1985).

Fast-start performance of pike corresponds with a life-style of capturing prey by bursts of activity (Webb and Skadsen 1980). A comparison of pike (mean length, 0.38 m) with rainbow trout, *Oncorhynchus mykiss* Walbaum, 1792, (mean length, 0.32 m) (Harper and Blake 1990) showed that only pike could start using an "S-shaped" body form that has higher stability and less yaw than the "C-shaped" form that is usual in teleosts. These pike accelerated at 120.2 ± 20.0 m s^{-2}, almost twice the rate of the trout. Therefore, it appears that the body-form, both in terms of muscle constitution and shape, is specifically adapted for fast-start performance and generating very high anaerobic power outputs.

4.2.2 Gills

Gill surface area is positively related to sustained activity levels (Gray 1954) and the associated respiratory gas exchange of fish. Hughes (1966) compared 14 temperate fish species, categorized as sluggish, intermediate or active, and found weight-specific gill area to range across 151-1241 mm^2 g bodyweight^{-1}. Pike gill area is towards the lower end of this range, having been measured as 125 mm^2 g bodymass^{-1} in a 650 g fish (Riess 1881), 95-179 mm^2 g bodymass^{-1} in 661-1398 g fish (n = 3) (De Jager et al. 1977) and 472-167 mm^2 g bodymass^{-1} across the range 87-1870 g (n = 9) (Jakubowski 1993). A scaling relationship is provided by Jakubowski (1993) whereby gill area (mm^2 g bodymass^{-1}) = 2199 $M^{-0.349}$ where M is mass (g).

4.2.3 The Pike Heart

4.2.3.1 Morphology

The teleost heart is supplied with blood via the sinus venosus, empties into the bulbous arteriosis and comprises two main muscular chambers – an atrium and a ventricle. Santer et al. (1983) examined 41 species of teleost and categorized the ventricle morphology as sac-like (saccular) in 30, tubular in four and pyramidal in seven species. They suggested that tubular and saccular ventricles differ only as a consequence of the effect of body shape on the position of the heart in relation to the gills. Pyramidal ventricles were found in those species, such as herring, *Clupea harengus* Linnaeus, 1758 and sprat, *Sprattus sprattus*, Linnaeus, 1758, with lifestyles that include periods of sustained fast aerobic swimming. The ventricles of all species of teleost have a trabeculated layer of myocardium (the spongy layer, or spongiosa) supplied with oxygen by venous blood passing through the heart. The hearts of some species also have an external layer of compact myocardium, or compacta. In the study of Santer et al. (1983), compacta was associated exclusively with pyramidal ventricles and hence highly active lifestyles. Santer and Greer Walker (1980) reported that of 93 species of teleost, 20 (21.5%) possessed compacta. A coronary vasculature, supplied by well oxygenated blood, was associated with compacta in all cases except two anguillids. They suggested that compacta is required by species that constantly or periodically expend large amounts of energy.

Cameron (1975) noted that the pike ventricle comprises both compacta, which constitutes about 29% of the tissue volume of the chamber, and spongiosa. Blood space inside the ventricle was estimated to be 35% of the total cross-sectional area. Examination of the ventricle reveals a well-developed coronary vasculature (Fig. 4.1) and a tubular structure (Armstrong 1987). Hence the structure of the pike heart is different to those of fish examined by Santer et al. (1983) and Santer and Greer Walker (1980) and seems to be typical of relatively sessile species in terms of shape but of more active species in terms of wall structure and coronary vasculature. However, perhaps the structure is important not for rapid sustained aerobic swimming but for sustaining growth at very low oxygen concentrations, for example under ice (Casselman 1978), when venous oxygen concentration may be insufficient alone to fuel the heart. Resting heart rate is < 5 beats min^{-1} under such conditions (Armstrong 1986) when blood circulation rate is evidently very low. A further alternative, or additional, possibility is that the rate-regulating heart of pike is associated with sustained metabolic demands during digestion and assimilation of large meals. Agnisola and Tota (1994) argue that whereas single-layer spongiosa ventricles are specialized for volume regulation, those with a layer of compacta have the potential to operate as high pressure, low volume pumps responding to variations in tissue oxygen demand by regulating rate.

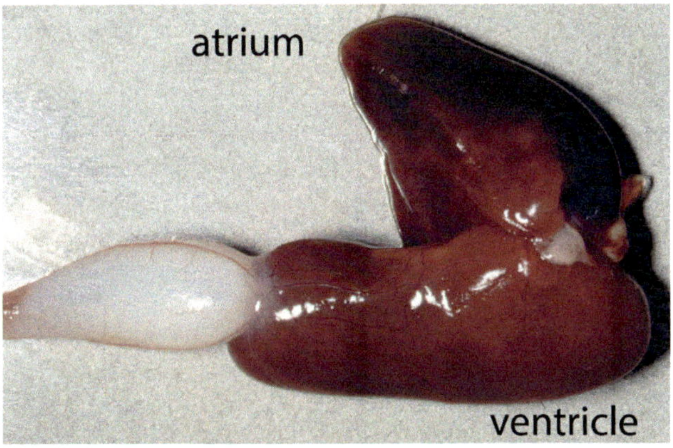

FIGURE 4.1 Photograph of the pike heart showing the positions of atrium and ventricle. Note the tubular structure of the ventricle and cardiac vasculature.

4.2.3.2 Cardiac Regulation

Variations in aerobic metabolic activity and tissue oxygen demand are accommodated in vertebrates by change in cardiac output (the volume of blood leaving the heart per unit time) together with variation in blood oxygen concentration (Kiceniuk and Jones 1977). Cardiac output is the product of heart rate and ventricular volume (stroke volume). Pike strongly regulate heart rate across much of the range of variation in metabolism, up to approximately 55 beats min^{-1} at 5 and 15°C (Armstrong 1986). As a consequence, averaged heart rate provides a good estimate of metabolic rate, within limits (Armstrong 1986, 1998). However, there is also evidence of substantial short-term variation in metabolic rate independent of heart rate, which is probably attributable largely to changes in stroke volume. Above a level of about 55 beats min^{-1} at 15°C, change in metabolic rate is mostly independent of change in heart rate. Although early studies led to a focus on predominantly regulation of stroke volume among teleost fishes, there is increasing recognition of the importance of rate regulation in some species (Clark et al. 2005).

Regulation of heart rate in pike is consistent with the presence of a layer of compacta according to the argument of Agnisola and Tota (1994). The pike heart structure clearly appears to be rather unusual among teleosts, and may be adapted for strong rate regulation.

4.2.3.3 Cardiac Electrical Activity as an Indicator of Function

Examination of the pike electrocardiogram (ECG) (Fig. 4.2) reveals functional adaptation to predominantly cardiac rate regulation. The ECG waveform records electrical activity associated with mechanical activities of the heart chambers (Labat 1966). The P wave corresponds to contraction of the atrium, which forces blood into the ventricle. The QRS complex signals the contraction of the ventricle. Finally, a T wave is associated with repolarization of the ventricle. Maximum heart rate is set by the time interval between the QRS and T wave, that is, the time taken until the ventricle repolarizes and is ready to contract again. The PT interval is constant in the cod *Gadus morhua* Linnaeus, 1758 (Wardle and Kanwisher 1973) and at maximum heart rates the T wave sits on the subsequent P wave so that a fixed ventricular systole sets a maximal heart rate. Similarly, Priede (1983) noticed no variation in the PT interval in plaice *Pleuronectes platessa* Linnaeus, 1758. Pike are different since the QRS-T interval decreases as heart rate increases and this process is essential for enabling the high heart rates that are observed (Armstrong 1987). Regulation of rate via several physiological feedback processes is suggested by the form of the heart rate variability signal (Armstrong et al. 1989a).

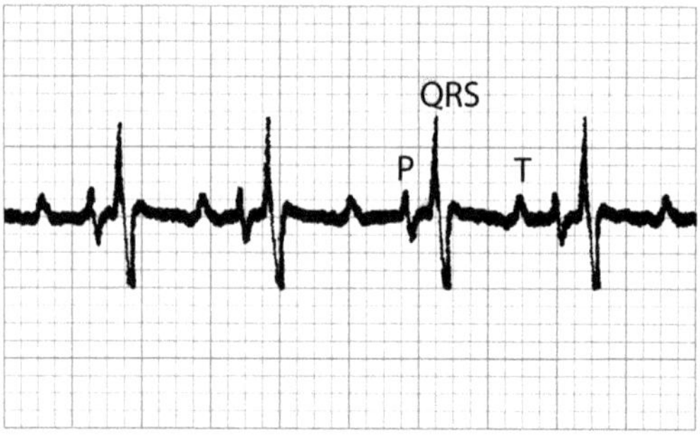

FIGURE 4.2 The pike ECG illustrating the P wave, corresponding with atrial contraction, QRS complex signalling ventricular contraction, and T wave associated with ventricular repolarization.

4.3 BIO-ENERGETICS

Bioenergetics encompasses the transfer of energy between animals and their environments and the storage and use of energy within the animal (Ney 1993). Here, consideration is given to energy budgets, which usually describe changes in the flow and distribution of energy over time, and power budgets, which determine peak aerobic levels of energy transfer and modulation of energy use at any given time. The energy budget is useful for capturing aspects of the characteristics of how pike function within a particular ecosystem and for comparing among such systems.

4.3.1 Form of the Energy Budget

Energy intake, I, (as prey) relates to growth (G) as

$$IA = G + R \tag{4.1}$$

where R is energy respired and A is assimilation efficiency, the proportion of the meal that is metabolised. This budget can be simplified further to

$$G = IC \tag{4.2}$$

where C is the conversion efficiency. C may vary due to a range of factors, some of which can be examined by expanding the budget into a wider set of sub-components, often of the form (Windell 1978)

$$G = I - (F + U + R_s + R_a + R_{sda}) \tag{4.3}$$

where F is faecal losses, U is ureic losses and R is subdivided here into standard metabolism of body maintenance (R_s), locomotory activity costs (R_a) and energy costs of handling, digesting and assimilating food (apparent specific dynamic action) (R_{sda}). G may be subdivided into somatic (G_s) and gonadal (G_g) growth and the equation may be further elaborated by separating R_s into components such as costs of transforming somatic to gonad tissue (Koch and Weiser 1983), which may vary seasonally.

4.3.2 The Power Budget and Metabolic Scope for Activity

The power budget describes how the respiratory system is apportioned at any instant to various simultaneous activities, such as aerobic locomotion and biochemical assimilation of food (Priede 1985). The metabolic scope for activity was originally defined as the difference between metabolism during resting (R_s) and maximum sustained aerobic swimming, which was taken to be the peak level (R_{max}) (Fry 1947). However, in principle, depending on the species and life stage, it is possible that R_{max} might be induced by any one or combination of the respiratory demands, including sustained swimming, R_{sda} and recovery from anaerobic activity. Experimentation is required to define the metabolic scope, the factor(s) that induce(s) R_{max} and the prioritization of metabolic demands where they exceed the metabolic scope (Priede 1985, Blaikie and Kerr 1996). Power budgeting is important because it determines the maximum rate at which the combined metabolic processes involved in surviving, feeding and growing can occur. It also determines whether there are conflicts between such processes. For example, assimilation of food may be delayed if too much of the metabolic scope is being used for activity, or peak swimming may be reduced during digestion.

The power budget includes only aerobic activities. As discussed earlier, pike can generate very high power outputs during acceleration. Such activity is largely anaerobic and builds up an oxygen debt that must subsequently be paid back more slowly within the metabolic scope.

4.3.2.1 *Standard or Resting Metabolic Rate, R_s*

Standard or resting metabolic rate (R_s), analogous to basal metabolic rate in mammals, is a basic cost of maintaining cells and organs in readiness for higher levels of biological activity, and includes protein turnover, osmoregulation, gill ventilation and pumping of the heart in a resting state (Calow 1985). To achieve an estimate of R_s, fish should be thermally acclimated (Duthie and Hoolihan 1982), activity should be discounted, biological oxygen demand of the water should be accounted for, R_{sda} from any meal should have fully subsided and the fish should not be unduly constrained (Hickman 1959). These criteria were fulfilled in the observations of Armstrong et al. (1992, 2004) compiled from pike of 40-1291 g. Resting heart rate of free-living pike (Lucas et al. 1991) was similar, when adjusted for temperature, to that of the same stock of fish within the respirometers used by Armstrong et al. (1992). This is an important observation because in some species, absence of a suitable shelter, as may be the case in some fish respirometry studies, can result in a substantial increase in *R* (Millidine et al. 2006). Hence, it would appear that the pike perceived their respirometer habitat to be sufficiently sheltered.

R can be expressed in the form aM^b, where *M* is mass and *b* is the scaling coefficient. R_s in units of $mgO_2\ h^{-1}$ across a pike mass (*M*, g) range 40-1291 g at 15°C was estimated to be of the form:

$$R_s = 0.162\ M^{0.80} \tag{4.4}$$

and across a relatively narrow mass range of 684-1282 g at 5°C, R_s was 16.9 $mgO_2\ kg^{-1}\ h^{-1}$.

From the review of Clarke & Johnstone (1999), R_s (mmol h^{-1} for a 50 g fish at 15°C) of 0.105 for pike is broadly similar to that of eels (0.086) and cyprinids (0.104) and lower than salmonids (0.231) and gadids (0.417). With reference to a more recent review of studies of R_s (Killen et al. 2007), both *a* and *b* in pike are similar to those of benthic teleosts.

Although the three studies of adult pike (Dolinin 1973, Diana 1982, Armstrong et al. 1992) generated closely similar estimates of R_s at temperatures of 2-5°C, there were substantial deviations at 14-15°C. This variation is likely to be due, at least in part, to elevation due to methodological factors (Armstrong and Hawkins 2008) and the lower estimate (Armstrong et al. 1992) is considered here. Nevertheless, it should be borne in mind that there may be differences among stocks of pike.

Wieser et al. (1992) found R_s of larval pike to be substantially lower than that of the fast-swimming predator, *Coregonus lavaretus* Linnaeus, 1758, at similar size. Extrapolation of equation (4.4) predicts 329 mg $kg^{-1}\ h^{-1}$ at a larval pike mass of 0.029 g, compared with Wieser et al.'s measured value of 364 mg $kg^{-1}\ h^{-1}$. However, the values for pike of about 1 g extrapolated from equation (4.4) (162 mg $kg^{-1}\ h^{-1}$) were much lower than those measured (258 mg $kg^{-1}\ h^{-1}$) by Wieser et al. (1992). It is possible, then, that *b* is not a constant across the full mass range. Simms (2000) measured R_s over the range 13-51 g to be about 19% lower than those predicted from equation (4.4) and to scale with a similar exponent. He found a substantial degree of plasticity in R_s in response to variations in feeding regime (in the order of 25% across a two-fold variation in ration). Hence, variation in R_s with size could reflect feeding regime to some extent.

4.3.2.2 *Maximum Metabolic Rate* (R_{max})

Data on sustained swimming speed (Jones et al. 1974) suggest that aerobic swimming activity does not use the full metabolic scope in pike and recovery from exhaustion probably provides a better indication of peak physiological capacity of the 40 g+ fish considered (Armstrong et al. 1992). In pike, R_{max} (mgO_2 $fish^{-1}\ h^{-1}$)

following exhaustive activity at 15°C (Armstrong et al. 1992) relates to wet mass(g) across the range 40-1291 g as

$$R_{max} = 0.162\ M^{0.99} \tag{4.5}$$

At 5°C, across a narrower mass range of 684-1282 g, R_{max} = 52.8 mg kg^{-1} h^{-1}. The scaling coefficient of R_{max} in pike is near unity, corresponding closely with observations of sockeye salmon, *Oncorhynchus nerka* Walbaum, 1792, (Brett and Glass 1973) and rainbow trout (Wieser 1985).

A comparison across species (Norin and Clark 2016) indicated that R_{max} of pike is relatively low, being higher only than that of the sessile flatfish *Solea solea* and almost 15-fold lower than the highly active salmonid *Oncorhynchus gorbuscha* Walbaum, 1792. Small R_{max} would hence appear to reflect a lifestyle involving low sustained activity. Simms (2000) found R_{max} of pike in the mass range 13-51 g to be 25% higher than those predicted from extrapolation of equation (4.5) and to scale with an exponent of 0.9. He also found substantial plasticity, in the order of 20%, such that R_{max} increases in response to sustained exercise. Wieser et al. (1992), measuring R_{max} as the peak value observed after feeding fish to satiation, found it to be substantially lower in larval pike than that of the more active *Coregonus lavaretus*.

4.3.2.3 Metabolic Scope and Power Budgeting

The consequence of differential scaling factors in equations (4.4) and (4.5) is that factorial metabolic scope (R_{max}/R_s) increases from 2.0 at 40 g to 3.7 at 1000 g. The larger factorial metabolic scope as pike get heavier may be due to need for greater aerobic capacity at certain times of the year, for example, during spawning (Lucas 1992), or to allow growth over winter by accommodating low oxygen concentrations under ice cover. Scaling of metabolism among species of fish varies with lifestyle and temperature (Killen et al. 2010) indicating a strong influence of ecological parameters.

Extrapolation of equations (4.4) and (4.5) to smaller pike would predict no metabolic scope at weights < 1 g and hence evidently scaling relationships of R_s and/or R_{max} must change with size. Simms (2000) measured factorial metabolic scopes ranging from 2.6-4.3 standardised to 35 g body mass depending on the feeding and activity regimes used to maintain the pike. Wieser et al. (1992) recorded factorial metabolic scopes, based on R_{max} at peak R_{sda} (R_{sda_max}), at 15°C of 2.36 for the weight range 0.008-0.05 g and 2.68 for the range 0.2-1.6 g.

How do these patterns of scaling compare with those of other species of teleost fishes? Decrease in factorial metabolic scope with increase in size has been recorded in sockeye salmon (Brett and Glass 1973) and rainbow trout (Wieser 1985). Killen et al. (2007) examined metabolic scaling and aerobic scope across the full range of body mass in three species of marine teleost. Contrary to the situation in pike, they found factorial scope to decrease to <= 1.5 at larval sizes. However, their estimates of R_{max} were from exhaustive swimming alone and it is possible that post-prandial metabolism may have caused higher elevations in *R*. Interestingly, Wieser et al (1992) found factorial scope, taking R_{max} to be that during peak R_{sda}, to be close to

2.0 for *Coregonus lavaretus* at larval sizes. This estimate is lower than in pike of similar size, but higher than the estimates of Killen et al. (2007).

4.3.2.4 R_s and Peak R_{sda}

Peak R during assimilation of food (that is $R_s + R_{sda\_max}$) is approximately 2-3 × R_s across a wide range of teleosts (Secor 2009), as high as 4 in a warm water catfish consuming meals of 24% bodyweight (Fu et al. 2005), up to 2.6 in small pike fed ad libitum (Wieser et al. 1992), and approximately double R_s in pike of 40-1291 g (Armstrong et al. 1992). In view of the close relationship between R_s and R_{sda_max} it is likely that a large proportion of R_s is associated with maintaining a capacity for assimilation of food. Indeed, at an intraspecific level, fish with relatively high R_s also had high R_{sda} and could process meals most quickly (Millidine et al. 2009). It is likely that the 21-28% elevation in R_s in pike associated with a high ration regime (Simms 2000) is a response to increased assimilation and hence growth capacities. Similarly, the high levels of R_s in pike larvae and small juveniles, could reflect high investment in capacity for very fast assimilation and growth at that life stage, as discussed later.

Overall, particularly in the case of small pike under normoxic conditions, assimilation of food can largely use the metabolic scope. At low oxygen concentrations, the metabolic scope can be expected to be reduced and then food assimilation rates in larger fish may be constrained (Jordan and Steffensen 2007). This scenario would explain low growth rates of pike under such conditions (Adelman and Smith 1970). The importance of metabolic scope in limiting growth of animals is becoming more widely recognized (Downs et al. 2016).

4.3.3 Parameterising Energy Budgets of Pike

Some components of the energy budget are functions of I. F has been estimated (Diana 1982) as $0.13I$, and excretion, U, has generally been taken to be $0.08I$ (Brett and Groves 1979). R_{sda} has been measured as $0.083I$ in larval and small juvenile pike (Wieser et al. 1992), as $0.08I$ using heart rate telemetry (Armstrong 1986, Lucas and Armstrong 1991) across a pike mass range 324-1930 g, and as $0.1I$ in relatively large pike using respirometry (Diana 1982). R_a is generally low and negligible in pike in aquaria and has been estimated from heart rate telemetry in the field to be 0.07-0.16 R_s (Lucas et al. 1991). Using these parameter estimates, with R_{sda} as $0.09I$ and R_a as zero, equation (4.3) can be re-written as

$$G = I - (0.13 + 0.08 + 0.1)\ I - R_s \tag{4.6}$$

or

$$G = 0.69I - R_s \tag{4.7}$$

This equation, with small adjustments as required to use different parameter values for R_{sda} and R_a, summarises our understanding of the energy budget of pike in terms of the general form laid out in equation (4.3). For application to field situations R_a can be incorporated with current best estimate of about $0.1R_s$ and hence

$$G = 0.69I - 1.1R_s \tag{4.8}$$

The energy budget is rearranged to allow the derivation of some key estimates of I and G as functions of R_s in following sections. Such estimates are further parameterized with R_s from equation (4.4) and compared with published empirical data.

4.3.4 Extending and Testing the Energy Budget

4.3.4.1 Maintenance Intake, I_{maint}

Maintenance ration (I_{maint}) is the level of I required such that a pike neither gains nor loses energy. Setting $G = 0$, using equation (4.7),

$$I_{maint} = 1.45R_s \tag{4.9}$$

I_{maint} has been estimated by ration to be 0.6 kcal day^{-1} for pike of 50 g at 14-15°C (Diana 1982). Using equation (4.9), I_{maint} is predicted to be 0.41 kcal day^{-1}, 32% less than the measured estimate. Ration experiments require substantial disturbance and handling of fish, which may well elevate estimates of I_{maint}. Using Diana's (1982) measured R_s yields I_{maint} of 1.08 kcal day^{-1}, 80% more than measured by his ration experiments. Hence, the lower values of R_s from equation (4.4) provide a better prediction and are on the lower side of measured I_{maint} as expected. Diana (1982) dismissed his winter ration estimate as being clearly flawed. In view of the evident methodological problems associated with maintenance ration experiments, estimation from the energy budget provides a useful alternative benchmark.

4.3.4.2 Predicted Maximum Intake, I_{max}, and Growth Rate, G_{max}

When food is readily available, young pike will re-feed before they have finished digesting a meal. Hence, they might maintain a continuous supply of food in the gut to be assimilated, and R_{sda} at R_{sda_max}, as in young cod (Soofiani and Hawkins 1982). R_{sda_max} determines maximum rate of assimilation and hence maximum rate at which I can be processed (I_{max}). I_{max} relates to the product of the proportion of I used in R_{sda} and the factorial increase over R_s at R_{sda_max} as

$$I_{max} = (I/R_{sda})\cdot(R_{sda\_max} - R_s/R_s)\cdot R_s \tag{4.10}$$

G at I_{max} (G_{max}) can be calculated from equations (4.7) and (4.8).

Estimates of I_{max} and G_{max} from equations (4.10) and (4.7) can be compared with published data from laboratory feeding experiments. For comparison of pike mass > 40 g, best estimates for necessary data can be compiled from the studies reviewed here. Taking (R_{sda}/I) as 0.09 (from study estimates of 0.08-0.1, see earlier) and best available estimate of (($R_{sda\_max} - R_s)/R_s$) as approximating to 1 (Armstrong et al. 1992),

$$I_{max} = 1/0.09.R_s \tag{4.11}$$

Using R_s from equation (4.4) for pike mass of 40 g-1000 g at 15°C and typical energy density of prey fish of 3600 J g^{-1} illustrates scaling, which follows that of R_s, and magnitude in units comparable with growth and ration experiments (g·g^{-1}·day^{-1}) (Fig. 4.3).

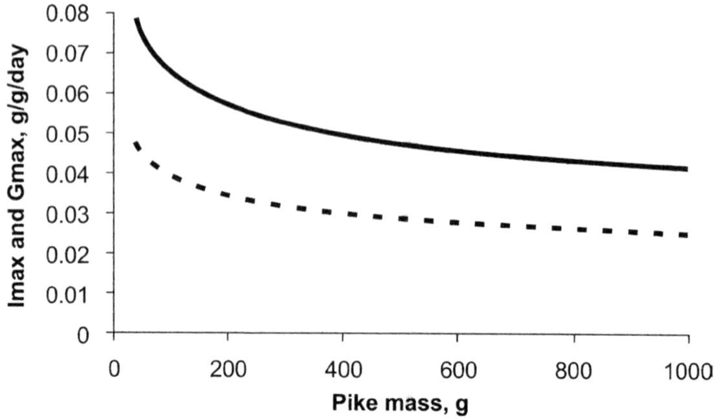

FIGURE 4.3 Estimates of theoretical maximum intake (I_{max}, solid line) and growth (G_{max}, dashed line) across a range of pike weights. See main text for explanation of derivation.

4.3.4.3 Maximum Conversion Efficiency, C_{max}

Derived from I_{max} and G_{max}, C_{max} is 60%, representing the predicted highest level across the mass range 40-1291 g at 15°C. A very wide range of *C* can be expected to occur in natural populations, depending on motivations of pike to feed, availability of food and constraints on rate of assimilation. Time spent below R_{sda_max} is time when costs of R_s are being accrued, full potential growth benefits are not being realized, resulting in a decrease in *C*. The value of *C* is 0 at I_{maint}.

4.3.4.4 Empirical Comparisons of I, G and C from ad libitum Feeding Tank Trials

Controlled, replicated trials feeding pike (7.2-43.5 g) at 15°C with *ad libitum* rations of live prey (estimated caloric content approximating 3600 J g^{-1}) were conducted by Bevelhimer et al. (1985). They measured *I* as 0.057 g g^{-1} day^{-1}, 73% of the predicted I_{max} for a 40 g pike. Measured *G* was 0.0185 g g^{-1} day^{-1} 38% of estimated G_{max}. Extrapolating from the allometry in Fig. 4.3, it is possible that the deviations would have been greater on average for their sample of fish (with mean < 40 g). Hence, *C* was 32% at most.

The only other corresponding empirical estimate of G_{max} for comparison (Diana 1996) appears to be the data of Casselman (1978). His estimate for *G* of 0.6% day^{-1} (0.006 g g^{-1} day^{-1}) for a mixture of "young-of-the-year and yearling" pike is low compared with G_{max} across the range of plausible sizes (Fig. 4.3) and much lower than the observations of Bevelhimer et al. (1985).

There are several reasons why pike held individually on *ad libitum* feeding may grow well below G_{max}. They may be constrained in how much food they can consume, for example, because of evasion by their prey (if a live diet is used). R_s may be heightened if there is insufficient shelter or there is perceived threat from other pike. Furthermore, as pike increase in size they seem to become more reluctant to feed in captivity (Armstrong 1986, Diana 1996), at least on dead fish.

Together, the theoretical estimates of I_{max} and G_{max} derived here and the empirical estimates of Bevelhimer et al. (1985) provide useful benchmarks for peak potential with which to compare field data. However, the question remains as to whether the deviations between the two sets of estimates (theoretical and empirical) could be narrowed under different experimental arrangements.

4.3.4.5 I_{max}, G_{max} and C in Early Life Stages

Unlike larger pike, fry and small juvenile stages often feed readily in captivity and hence provide perhaps the greatest potential for testing how I and G relate to theoretical physiological maxima. As discussed earlier, R_s in pike fry and small juveniles is higher than predicted based on scaling relationships derived from large juveniles and adults. Wieser et al. (1992) measured $R_{sda\_max} - R_s/R_s$ to be 1.36 in larvae and 1.68 in small juveniles, much higher than the best estimate of 1.0 for larger pike. Using budget parameters for these small fish yields I_{max} of 0.50 g g^{-1} day^{-1} in fry and 0.44 g g^{-1} day^{-1} in small juveniles. Estimates of G_{max} are 0.31 g g^{-1} day^{-1} in fry and 0.28 g g^{-1} day^{-1} in small juveniles. Evidently, very small pike have a disproportionately large capacity for assimilating food compared with older pike. Wieser et al. (1992) measured C to be 63% in larvae and 62% in small juveniles, higher than the maximum values predicted for larger pike, and similar to theoretical maximum of 0.62 and 0.64 for small fish. The pike grew at 22-24% day^{-1} compared with G_{max} of 28-31% day^{-1}, that is 78% of theoretical maximum values. If, as discussed earlier, peak metabolism coincides with peak SDA(sda) during digestion, then the implication is that these fish are working flat-out for most the time despite rarely exhibiting locomotory activity. The growth rate was similar to that observed in pike larvae in ponds (Bry et al. 1991).

4.3.5 Application of Energy Budgets

4.3.5.1 Juveniles in a Closed Environment

Wieser et al. (1992) could measure I, G and R accurately in a closed system. Young pike feed readily in captivity, growth is entirely somatic and activity is minimal. Such a system is ideal as a model of simple energy and power budgets under largely intrinsic constraints. At this scale, the budget is interesting as a comparative tool to understand the energetics among fishes with different life styles. In the Wieser et al. (1992) study, pike grew more quickly than *Coregonus lavaretus* of similar size despite lower I, largely due to a much lower R_s. It would be interesting to extend understanding further by feeding pike to the maximum levels possible using the approach of Soofiani and Hawkins (1982). This observation is consistent with growth advantage due to economy by pike in the fraction of R_s used for maintaining a potential for sustained swimming.

4.3.5.2 Pike in the Wild

4.3.5.2.1 Budget Imbalance Another application of the energy budget is to explore the ecology of pike in natural systems. In this case, multiple environmental variables, including temperature, presence of predators and availability of prey, fluctuate across a range of temporal and spatial scales. Furthermore, in adults energy may be allocated both to growth of gonads and soma and may vary between sexes. Estimating *I* is often of particular interest as part of a process of evaluating the impact of pike on prey populations (e.g. Muhlfeld et al. 2008).

Ideally, the energy budget parameters, R_s, R_{sda}, *F* and *U*, can be measured in the laboratory, *G*, R_a and *I* can be measured in the field, and the budget can be balanced (Diana 1982). If that was possible, then it might provide confidence in using components of budgets as a predictive tool, for example, to assess prey consumption rates. Diana (1983) combined budget components that he had measured from a population of pike and found a poor fit to field data. He suggested that the estimates of *I*, from stomach contents of netted pike and a digestion rate model from force-fed fish (Diana 1979), were probably the main sources of error. Nets may select for hungry, active fish (with less food in their stomachs than the average). Pike in nets may regurgitate food, although this can be identified by presence of a flaccid stomach (Treasurer 1988). Furthermore, there may be digestion while pike remain in the nets (Heikinheimo and Korhonen 1996). Also, force-feeding may affect digestion rates in some species of fish (Swenson and Smith 1973). Hence, there are potential errors due to the methodology for deriving *I*. Heikinheimo and Korhonen (1996) similarly found an underestimate of *I* using stomach analysis of gill-netted pike, compared with a general bioenergetics model for carnivorous fish.

Wahl and Stein (1991) applied derivations of several energy budget parameters from Bevelhimer et al. (1985) to a population of hatchery-selected stocked pike in a warm lake. They avoided potential errors due to netting by electrofishing pike from the lake margin. However, as with Diana (1982), their estimates of *I* greatly underestimated those predicted from their energy budget and measurements of *G*. They questioned the validity the R_s estimates of Bevelhimer et al. (1985), suggesting that these would need to be at least halved to balance budgets for pike and other esocids. Armstrong and Hawkins (2008) provided further support for this contention and demonstrated that use of the estimates for R_s from Armstrong et al. (1992) greatly improved fits for both the Diana (1982) and Wahl and Stein (1991) budgets.

4.3.5.2.2 Estimation using Heart Rate Telemetry In many studies of fish energy budgets, R_a is a large and variable component (Boisclair and Sirois 1993). Although pike are lurking predators, several studies have shown that they can be nomadic or move extensively over large home ranges (Diana et al. 1977, Diana 1980, Mackay and Craig 1983, Chapman and Mackay 1984, Cook and Bergersen 1988, Rosten et al. 2016, chapter 5). Nevertheless, such movements, resolved at a scale < 5 m, have been discounted in the energy budget (Diana 1980). Yet even very localized movements may invoke significant costs if a fish is stemming a current for example.

Monitoring of heart rate of free-ranging pike provides an option for determining both *I* and R_a at fine resolution since it mirrors changes in metabolism associated with aerobic activity, recovery from anaerobic activity and R_{sda} (Armstrong 1986). Indeed, the r^2 for R_{sda} as a function of *I* estimated from heart rate is very high

(0.988) (Lucas and Armstrong 1991) despite a range of pike weights, meal sizes, prey species and temperatures.

Heart rate can be telemetered from pike by using the *R* wave of the ECG to trigger the generation of pulses from an acoustic transmitter mounted on the fish (Armstrong et al. 1989b). It can then be recorded remotely via a standard hydrophone and acoustic receiver system that also permits locational tracking of the fish. This approach allows discrimination of short term localized locomotory activity, during which there is interference from electromyogram electrical activity. Also, moments of stimulating sensory input to the fish can be identified since they elicit missing heart beats, termed bradycardias (Labat 1966). Using such a system, Lucas et al. (1991) monitored three pike in a loch (Scottish lake) and determined detailed estimates of R_s, R_{sda}, R_a and I for two of the fish. Estimates of R_a using a conventional approach underestimated the level derived from heart rate telemetry by about ten-fold, because of localized but intense activity, such as that associated with prey capture. Nevertheless, R_a represented increases of only 6.6 and 15.5% of R_s. Estimated I was 4.7 and 2.3 fold higher than energy expended on R_s. Hence, on average, I was 23-47% of I_{max}. G was 36% and 10% of G_{max} and C was 50% and 29%. Lucas et al. (1991) believed that the pike were browsing on small perch, *Perca fluviatilis* Linnaeus, 1758, which were abundant in the loch margins at the time. The study demonstrated that such a strategy can maintain a positive energy balance, and as shown here, use up to 36% of the growth capacity.

4.3.5.2.3 Estimation using Otolith Analysis Otoliths are calcareous-proteinaceous structures that are part of the fish acoustico-lateralis system and located in the head. They grow throughout life and have been used extensively to estimate age, and in some cases, historic daily growth rates of fish (Panella 1971). Armstrong et al. (2004) established that growth of the pike otolith is closely related to energy intake via metabolism. Averaged over the course of digestion of a meal, daily otolith growth comprises the sum of two components – one relating to R_s and the other to R_{sda}. Further work is needed to confirm how closely post-prandial otolith growth correlates with a range of energy intakes. However, the basic physiological principles are now established and indicate that analysis of otoliths may be a useful tool for estimating food intake in pike.

Advantages of otolith analysis would be that a relatively large number of individuals could be examined over many days of records. However, it would be necessary to use sites where temperature is reasonably homogeneous so that effects of change in thermal habitat could be differentiated from those due to energy intake. Alternatively, otoliths might be used together with temperature sensing transmitters or data storage tags to measure thermal history to account for that effect in the energy budget.

4.4 SYNTHESIS AND CONCLUSIONS

Pike is a distinctive fish, recognizable in profile from the elongated shape, massive flattened head and locations of the fins. It is adapted for extremely powerful bursts of activity, due to the muscle constitution, the body form and the fast-start process. However, compared with the range of values for teleosts (Killen et al. 2016), the gill area is modest, and R_{max} and aerobic metabolic scope are relatively low. Hence, locomotory metabolism is probably largely anaerobic. However, pike do move slowly, nomadically or over large home ranges (chapter 5), and can adjust their physiology to adapt to flowing water in that R_s increases in response to exercise training. The heart is unusual, both in terms of morphology and mode of operation. The high contribution of heart rate in regulation of cardiac output is interesting from a functional point of view and also, when coupled with the lifestyle (large meals and sprint prey capture), in enabling relatively accurate telemetry and partitioning of metabolism of free-ranging pike to a degree that is probably not possible in many other species of teleost. The generally low R_a in pike allows simplification of the energy budget and reduces a major source of uncertainty, at least in some environments and seasons, such as summer in lentic waters (Lucas et al. 1991). In other environments (such as rivers) and seasons, activity costs may be higher. As an experimental animal, small pike are an excellent model for laboratory studies and large pike have the mass to carry substantial instrumentation, such as telemetry transmitters, for field studies. However, large pike do not generally adapt well in captivity to small-scale enclosed systems and dead food, when they may feed seldom, if ever.

Particularly in view of these experimental constraints, the application here of energy budgets for deriving I_{max} and G_{max} of captive pike is promising when the outputs are compared with the few empirical data available. Further tests of the concept using studies in which pike are fed unequivocally to satiation would be useful, preferably with continuous measurement of metabolic rate. It is evident that such predictions are very sensitive to estimations of R_s, as is the case also when applying field energy budgets to derive I (Heikinheimo and Korhonen 1996). In seeking to improve quantification of R_s, a focus has been on accounting for methodological flaws that are known to inflate estimates erroneously (Armstrong and Hawkins 2008). However, it is also important to acknowledge that there are factors that affect R in some species of fish, that are not routinely captured in energy budgets, and usually represented as variation in R_s. These include seasonal variations (Koch et al. 1992), effects of shelter (Millidine et al. 2006), social interactions (Millidine et al. 2009) and acute changes in temperature (Oligny-Hébert et al. 2015). It would be useful to establish how such factors influence R in pike and whether they might cause significant imbalances in budget estimations if not accounted for. Any such elevations will reduce I_{max} and G_{max} and tend to underestimate I using measured G in energy budgets.

There is a need for energy budget analysis that includes more robust estimates of I. Heart rate telemetry has potential to provide detailed insights into R and I. It is technically challenging to an extent and would benefit from further calibration of relationships between heart rate and metabolism under low oxygen conditions. However, application of heart rate telemetry to observe pike in a number of contrasting settings would be very informative. Otolith analysis has yet to be applied in a field setting, but has potential to provide a long-term record of key energy budget components.

This chapter set out to determine how metabolic factors might contribute to the high growth potential of pike. Aerobic metabolic scope and associated aerobic tissue masses (gills and red muscle) are small in pike relative to aerobically active fish species. Such tissues are expensive to maintain (Houlihan et al. 1986) and the low quantities in pike can be expected to economise on expenditure in R_s if it follows general interspecific relationships (Killen et al. 2010). However, R_s also includes costs of maintaining metabolic potential for assimilation (Boratynski and Koteja 2010) as reflected in fishes by close correlation with R_{sda_max}. Although R_s of pike is relatively low overall, it is not extremely so, being comparable with that of many other benthic teleosts reviewed by Killen et al. (2016). This observation is consistent with a combination of extreme economy in maintaining aerobic scope but investment in a good assimilation capacity to support growth potential. Very small pike seem to have an unusually high potential to assimilate food, reflected by high R_s and R_{sda_max} relative to body weight. High investment in growth, through elevated R_s, may be important for achieving an early size advantage over prey fishes but would depend on a high availability of prey to be a successful strategy since net energy drain would be very high between assimilation of meals.

Various studies have sought to distinguish whether R_s or R_{max} most closely correlate with growth of animals, and whether R_s competes with other biochemical activities for available energy (Downs et al. 2016, Auer et al. 2015). In the model developed here, magnitude of R_s is, in part, an investment for growth potential through supporting capacity to assimilate food. R_{max} may well directly measure that potential in small pike if it coincides with R_{sda_max}. In larger pike, additional scope beyond R_{sda_max} can be expected to sustain assimilation rates during conditions of moderately low environmental oxygen and when there are additional demands on the power budget such as from recovery from burst swimming. Hence, R_s, R_{max} and R_{sda_max} are part of a complex. The upper and lower limits of the metabolic scope are both important for sustaining fast growth and their relative importance depends on the context. It is predicted that investment in the component of R_s that supports R_{sda_max} should be at a level that optimizes the balance between energy costs and benefits in terms of increasing assimilation rate, as reflected in growth. Further investment in R_s to raise R_{max} can be expected to optimize the balance between those energy costs and the benefits of additional metabolic scope. These benefits may include evading predation, spawning successfully and allowing sufficient scope for R_{sda_max} to be minimally constrained. The optima in these components of R_s can be expected to vary depending on pike size, food regime and physico-chemical environment. In many ways the pike is an excellent model animal for exploring these concepts further both as juveniles in captivity, and adults in the field.

Acknowledgements

My thanks to Eef Cauwelier who translated reference material.

REFERENCES CITED

Adelman, I.R. and L.L. Smith. 1970. Effect of oxygen on growth and food conversion efficiency of northern pike. *Prog. Fish-Cult.* 32:93-96.

Agnisola, C. and B. Tota. 1994. Structure and function of the fish cardiac ventricle: flexibility and limitations. *Cardioscii* 5:145-153.

Armstrong, J.D. 1986. Heart rate as an indicator of activity, metabolic rate, food intake and digestion in pike, *Esox lucius*. *J. Fish Biol.* 29:207-221.

Armstrong, J.D. 1987. Metabolism, feeding and cardiac function in pike, *Esox Lucius*. Ph.D. Thesis, University of Aberdeen, Aberdeen, Scotland, UK.

Armstrong, J.D., L.De Vera and I.G. Priede. 1989a. Short-term oscillations in heart rate of teleost fishes: *Esox lucius* L. and *Salmo trutta* L. *Physiol. Lond.* 409:41.

Armstrong, J.D., M.C. Lucas, I.G. Priede and L.De Vera. 1989b. An acoustic telemetry system for monitoring the heart rate of pike, *Esox lucius* L., and other fish in their natural environment. *J. Exp. Biol* 143:549-552.

Armstrong, J.D., I.G. Priede and M.C. Lucas. 1992. The link between respiratory capacity and changing metabolic demands during growth of northern pike, *Esox lucius* L. *J. Fish Biol.* 41:65-75.

Armstrong, J.D. 1998. Relationships between heart rate and metabolic rate of pike: integration of existing data. *J. Fish Biol.* 52:362-368.

Armstrong, J.D., P.S. Fallon-Cousins and P.J. Wright. 2004. The relationship between specific dynamic action and otolith growth in pike. *J. Fish Biol.* 64:739-749.

Armstrong, J.D. and L.A. Hawkins. 2008. Standard metabolic rate of pike, *Esox lucius*: variation among studies and implications for energy flow modelling. *Hydrobiologia* 601:83-90.

Auer, S.K., K. Salin, G.J. Anderson and N.B. Metcalfe. 2015. Aerobic scope explains individual variation in feeding capacity. *Biol. Lett.* 11:20150793.

Bevelhimer, M.S., R.A. Stein and R.F. Carline. 1985. Assessing significance of physiological differences among three esocids with a bioenergetics model. *Can. J. Fish. Aquat. Sci.* 42:57-69.

Blaikie, H.B. and S.R. Kerr. 1996. Effect of activity level on apparent heat increment in Atlantic cod, *Gadus morhua*. *Can. J. Fish. Aquat. Sci.* 53:2093-2099.

Boisclair, D. and P. Sirois. 1993. Testing assumptions of fish bioenergetics models by direct estimation of growth, consumption, and activity rates. *Trans. Am. Fish. Soc.* 122:784-796.

Boratyński, Z. and P. Koteja. 2010. Sexual and natural selection on body mass and metabolic rates in free-living bank voles. *Funct. Ecol.* 24:1252-1261.

Brett, J.R. and N.R. Glass. 1973. Metabolic rates and critical swimming speeds of sockeye salmon (*Oncorhynchus nerka*) in relation to size and temperature. *J. Fish. Res. Bd. Can.* 30:379-387.

Brett, J.R. and Groves, T.D.D. (1979). Physiological energetics. *Fish physiol.* 8(6):280-352.

Bry, C., M.G. Hollebecq, V. Ginot, G. Israel and J. Manelphe. 1991. Growth patterns of pike (*Esox lucius* L.) larvae and juveniles in small ponds under various natural temperature regimes. *Aquaculture* 97:155-168.

Calow, P. 1985. Adaptive aspects of energy allocation. pp 13-32. *In*: P. Tytler and P. Calow (eds.). *Fish energetics: new perspectives*. London: Croom Helm.

Cameron, J.N. 1975. Morphometric and flow indicator studies of the teleost heart. *Can. J. Zool.* 53:691-698.

Casselman, J.M. 1978. Effects of environmental factors on growth, survival, activity, and exploitation of northern pike. *Am. Fish. Soc. Spec. Publ.* 11:114-128.

Chapman, C.A. and W.C. Mackay. 1984. Versatility in habitat use by a top aquatic predator, *Esox lucius* L. *J. Fish Biol.* 25:109-115.
Clarke, A. and N.M. Johnston. 1999. Scaling of metabolic rate with body mass and temperature in teleost fish. *J. Anim. Ecol.* 68:893-905.
Clark, T.D., T. Ryan, B.A. Ingram A.J. Woakes P.J. Butler and P.B. Frappell. 2005. Factorial aerobic scope is independent of temperature and primarily modulated by heart rate in exercising Murray cod (*Maccullochella peelii peelii*). *Physiol. Biochem. Zool.* 78:347-355.
Cook, M.F. and E.P. Bergersen. 1988. Movements, habitat selection, and activity periods of northern pike in Eleven Mile Reservoir, Colorado. *Trans. Am. Fish. Soc.* 117:495-502.
De Jager, S., M.E. Smit-Onel, J.J. Videler, B.J.M. Van Gils and E.M. Uffink. 1977. The respiratory area of the gills of some teleost fishes in relation to their mode of life. *Bijdr. Dierkd.* 46:199-205.
Diana, J.S., W.C. Mackay and M. Ehrman. 1977. Movements and habitat preference of northern pike (*Esox lucius*) in Lac Ste. Anne, Alberta. *Trans. Am. Fish. Soc.* 106:560-565.
Diana, J.S. 1979. The feeding pattern and daily ration of a top carnivore, the northern pike (*Esox lucius*). *Can. J. Zool.* 57:2121-2127.
Diana, J.S. 1980. Diel activity pattern and swimming speeds of northern pike (*Esox lucius*) in Lac Ste. Anne, Alberta. *Can. J. Fish. Aquat. Sci.* 37:1454-1458.
Diana, J.S. 1982. An experimental analysis of the metabolic rate and food utilization of northern pike. *Comp. Bioch. Physiol. A* 71:395-399.
Diana, J.S. 1983. An energy budget for northern pike (*Esox lucius*). *Can. J. Zool.* 61:1968-1975.
Diana, J.S. 1996. Energetics. pp 103-124. *In*: J.F. Craig (ed.). *Pike: Biology and exploitation.* London: Chapman & Hall.
Dolinin, V.A. 1973. The rate of basal metabolism in fish. *J. Ichthyol.* 13:430-438.
Downs, C.J., J.L. Brown, B.W. Wone, E.R. Donovan and J.P. Hayes. 2016. Speeding up growth: Selection for mass-independent maximal metabolic rate alters growth rates. *Am. Nat.* 187:295-307.
Duthie, G.G. and D.F. Houlihan. 1982. The effect of single step and fluctuating temperature changes on the oxygen consumption of flounders, *Platichthys flesus* (L.): Lack of temperature adaptation. *J. Fish Biol.* 21:215-226.
Fickling, N. 2004. Mammoth Pike. Lucebaits Publishing. Gainsborough, Lincolnshire, UK.
Fry, F.E.J. 1947. Effects of environment on animal activity. *Publs. Ont. Fish. Res. Lab.* 55:1-15.
Fu, S.J., X.J. Xie and Z.D. Cao. 2005. Effect of meal size on postprandial metabolic response in southern catfish (*Silurus meridionalis*). *Comp. Biochem. Physiol, A.* 140:445-451.
Gray, I.E. 1954. Comparative study of the gill area of marine fishes. *Biol. Bull.* 107:219-225.
Harper, D.G. and R.W. Blake. 1990. Fast-start performance of rainbow trout *Salmo gairdneri* and northern pike *Esox lucius. J. Exp. Biol.* 150:321-342.
Heikinheimo, O. and A.P. Korhonen. 1996. Food consumption of northern pike (*Esox lucius* L.), estimated with a bioenergetics model. *Ecol. Freshw. Fish* 5:37-47.
Hickman, C.P. 1959. The osmoregulatory role of the thyroid gland in the starry flounder, *Platichthys stellatus. Can. J. Zool.* 37:997-1060.
Hoyle, J., H.S. Gill and A.H. Weatherley. 1986. Histochemical characterization of myotomal muscle in the grass pickerel, *Esox americanus vermiculatus* (LeSeuer), and the muskellunge, *E. masquinongy* (Mitchell). *J. Fish Biol.* 28:393-401.
Houlihan, D.F., D.N. McMillan and P. Laurent. 1986. Growth rates, protein synthesis, and protein degradation rates in rainbow trout: effects of body size. *Physiol. Zool.* 59:482-493.

Hughes, G.M. 1966. The dimensions of fish gills in relation to their function. *J. Exp. Biol* 45:177-195.

Jakubowski, M. 1993. Re-examination of the gill respiratory surface area in the pike, *Esox lucius*, and remarks on the other fish species. *Acta Biol. Cracov. Ser. Zool.* 34:25-32.

Jones, D.R., Kiceniuk, J.W. and Bamford, O.S., 1974. Evaluation of the swimming performance of several fish species from the Mackenzie River. *J. Fish. Res. Board Can.* 31(10):1641-1647.

Jordan, A.D. and J.F. Steffensen. 2007. Effects of ration size and hypoxia on specific dynamic action in the cod. *Physiol. Biochem. Zool.* 80:178-185.

Kiceniuk, J. and D.R. Jones. 1977. The oxygen transport system in trout (*Salmo gairdneri*) during exercise. *J. Exp. Biol.* 69:247-260.

Killen, S.S., I. Costa, J.A. Brown and A.K. Gamperl. 2007. Little left in the tank: metabolic scaling in marine teleosts and its implications for aerobic scope. *Proc. Roy. Soc. Lond. B* 274:431-438.

Killen, S.S., D. Atkinson and D.S. Glazier. 2010. The intraspecific scaling of metabolic rate with body mass in fishes depends on lifestyle and temperature. *Ecol. Lett.* 13:184-193.

Killen, S.S., D.S. Glazier, E.L. Rezende, T.D. Clark, D. Atkinson, S. Astrid, T. Willener and L.G. Halsey. 2016. Ecological influences and morphological correlates of resting and maximal metabolic rates across telesot species. *Am. Nat.* 187:592-606.

Koch F. and W. Wieser. 1983. Partitioning of energy in fish: can reduction of swimming activity compensate for the cost of production? *J. Exp. Biol* 107:141-146.

Koch, F., W. Wieser and H. Niederstätter. 1992. Interactive effects of season and temperature on enzyme activities, tissue and whole animal respiration in roach, *Rutilus rutilus*. *Env. Biol. Fishes* 33:73-85.

Labat, R. 1966. Electrocardiologie chez les poissons téléostéens: influence de quelques facteurs écologique. *Ann Limnol-Int.* 2:1-175.

Lucas, M.C. and J.D. Armstrong. 1991. Estimation of meal energy intake from heart rate records of pike, *Esox lucius* L. *J. Fish Biol.* 38:317-319.

Lucas, M.C., I.G. Priede, J.D. Armstrong, A.N.Z. Gindy and L. Vera. 1991. Direct measurements of metabolism, activity and feeding behaviour of pike, *Esox lucius* L., in the wild, by the use of heart rate telemetry. *J. Fish Biol.* 39:325-345.

Lucas, M.C. 1992. Spawning activity of male and female pike, *Esox lucius* L., determined by acoustic tracking. *Can. J. Zool.* 70:191-196.

Mackay, W.C. and J.F. Craig. 1983. A comparison of four systems for studying the activity of pike, *Esox lucius* (L.), perch, *Perca fluviatilis* (L.) and *P. flavescens* (Mitchill). In *Proceedings of the Fourth International Conference of Wildlife Biotelemetry* (pp. 22-30). Applied Microelectronics Institute and Technical University of Nova Scotia.

Mangel, M. and J. Stamps. 2001. Trade-offs between growth and mortality and the maintenance of individual variation in growth. *Evol. Ecol. Res.* 3:611-632.

Millidine, K.J., J.D. Armstrong, and N.B. Metcalfe. 2006. Presence of shelter reduces maintenance metabolism of juvenile salmon. *Func. Ecol.* 20:839-845.

Millidine, K.J., J.D. Armstrong, and N.B. Metcalfe. 2009. Juvenile salmon with high standard metabolic rates have higher energy costs but can process meals faster. *Proc. R. Soc. Lon. B* 276:2103-2108.

Millidine, K.J., N.B. Metcalfe and J.D. Armstrong. 2009. Presence of a conspecific causes divergent changes in resting metabolism, depending on its relative size. *Proc. R. Soc. Lon. B* 276:3989-3993.

Muhlfeld, C.C., D.H. Bennett, R.K. Steinhorst, B. Marotz and M. Boyer. 2008. Using bioenergetics modeling to estimate consumption of native juvenile salmonids by nonnative northern pike in the upper Flathead River system, Montana. *N. Am. J. Fish. Manage*. 28:636-648.

Ney, J.J. 1993. Bioenergetics modeling today: growing pains on the cutting edge. *Trans. Am. Fish. Soc.* 122:736-748.

Norin, T., and T.D. Clark. 2016. Measurement and relevance of maximum metabolic rate in fishes. *J. Fish Biol.* 88:122-151.

Oligny-Hébert, H., C. Senay, E.C. Enders and D. Boisclair. 2015. Effects of diel temperature fluctuation on the standard metabolic rate of juvenile Atlantic salmon (*Salmo salar*): influence of acclimation temperature and provenence. *Can. J. Fish. Aquat. Sci.* 72:1306-1315.

Priede, I.G. 1985. Metabolic scope in fishes. pp 33-64. *In*: P. Tytler and P. Calow (eds.). *Fish energetics: new perspectives*. London: Croom Helm.

Priede, I.G. 1983. Heart rate telemetry from fish in the natural environment. *Comp. Biochem. Physiol. A* 76:515-524.

Riess, J.A. 1881. Der Bau der Kiemenblatter bei den Knochenfischen. *Archives fur Naturgesch.* 47:518-550.

Rosten, C.M., R.E. Gozlan and M.C. Lucas. 2016. Allometric scaling of intraspecific space use. *Biol. Lett.* 12:20150673.

Russell, N.R. and R.J. Wootton. 1992. Appetite and growth compensation in the European minnow, *Phoxinus phoxinus* (Cyprinidae), following short periods of food restriction. *Env. Biol. Fish.* 34:277-285.

Rypel, A.L. 2012. Meta-analysis of growth rates for a circumpolar fish, the northern pike (*Esox lucius*), with emphasis on effects of continent, climate and latitude. *Ecol. Freshw. Fish* 21:521-532.

Santer, R.M. and M.G.-Walker. 1980. Morphological studies on the ventricle of teleost and elasmobranch hearts. *J. Zool.* 190:259-272.

Santer, R.M., M.G.-Walker, L. Emerson and P.R. Witthames. 1983. On the morphology of the heart ventricle in marine teleost fish (Teleostei). *Comp. Biochem. Physiol.* A76:453-457.

Schwalme, K. and W.C. Mackay. 1985. The influence of angling-induced exercise on the carbohydrate metabolism of northern pike (*Esox lucius* L.). *J. Comp. Physiol. B* 156:67-75.

Secor, S.M. (2009). Specific dynamic action: a review of the postprandial metabolic response. *J. Comp. Physiol. B* 179:1-56.

Simms, L. 2000. *Intraspecific variation in the metabolism of juvenile Atlantic salmon Salmo salar and northern pike Esox lucius*. Ph. D. Thesis, University of Durham, Durham, UK.

Soofiani, N.M. and A.D. Hawkins. 1982. Energetic costs at different levels of feeding in juvenile cod, *Gadus morhua* L. *J. Fish Biol.* 21:577-592.

Swenson, W.A. and L.I. Smith Jr. 1973. Gastric digestion, food consumption, feeding periodicity, and food conversion efficiency in walleye (*Stizostedion vitreum*). *J. Fish. Res. Bd. Can.* 30:1327-1336.

Treasurer, J.W. 1988. Measurement of regurgitation in feeding studies of predatory fishes. *J. Fish Biol.* 33:267-271.

Wahl, D.H. and R.A. Stein. 1991. Food consumption and growth of three esocids: field tests of a bioenergetic model. *Trans. Am. Fish. Soc.* 120:230-246.

Wardle, C.S. and J.W. Kanwisher. 1973. The significance of heart rate in free swimming cod, *Gadus morhua*: Some observations with ultra-sonic tags. *Mar. Freshw. Behav. Physiol.* 2:311-324.

Webb, P.W. and J.M. Skadsen. 1980. Strike tactics of *Esox*. *Can. J. Zool.* 58:1462-1469.

Wieser, W. 1985. Developmental and metabolic constraints of the scope for activity in young rainbow trout (*Salmo gairdneri*). *J. Exp. Biol.* 118:133-142.

Wieser, W., A. Laich, and N. Medgyesy. 1992. Energy allocation and yield and cost of growth in young *Esox lucius* and *Coregonus lavaretus* (Teleostei): influence of species, prey type and body size. *J. Exp. Biol.* 169:165-179.

Windell, J.T. 1978. Estimating food consumption rates of fish populations. pp 227-254. *In*: T. Bagenal (ed.). *Methods for assessment of fish production in fresh water*. Oxford: Blackwell Scientific Publications.

Chapter 5

Spatial Ecology

Christian Skov[*1], Martyn C. Lucas*[2] *and Lene Jacobsen*[3]

5.1 INTRODUCTION

Most aquatic animals move during their lifetime, from tiny *Daphnia* which make vertical migrations between different layers of water to large whales (Cetacea) crossing the oceans. Not surprisingly, studies of spatial ecology are numerous for a wide spectrum of aquatic animals, and the pike *Esox lucius* Linnaeus, 1758 is no exception. These range from the first studies of pike movement where scientists tagged, released and recaptured pike to measure distances moved (e.g. Carbine and Applegate 1946, Miller 1948) until today where advanced telemetry methods allow mapping of the ranging behavior of wild pike with submeter precision 24 hours a day and 7 days a week for up to 3 years, resulting in an almost incomprehensible amount of data and a high level of detail about individual movement and habitat use (e.g. Baktoft 2012). Despite these recent advances in technology, researchers still debate whether pike should be considered as a sedentary fish that remains in a restricted area performing its ambush attacks (the traditional perspective) or if it is more mobile, frequently and swiftly moving between different habitats. This discussion reflects the fascinating complexity of the spatial behavior of pike as this species seems very versatile and able to adapt to different conditions by applying a wide range of space-use strategies.

**Corresponding author*: [1]DTU Aqua, Technical University of Denmark, Section for Inland Fisheries and Ecology, Vejlsøvej 39, 8600, Silkeborg, Denmark; Email: ck@aqua.dtu.dk

[2]Department of Biosciences, Durham University, South Road, Durham DH1 3LE, UK; Email: m.c.lucas@durham.ac.uk

[3]DTU Aqua, Technical University of Denmark, Section for Inland Fisheries and Ecology, Vejlsøvej 39, 8600, Silkeborg, Denmark; Email: lj@aqua.dtu.dk

5.1.1 Movement, Activity and Habitat Shifts; Background and Definitions

Why should a pike bother to move in the first place? From a broad perspective, four key factors can explain fish movements; feeding, reproduction, avoiding predators, and escaping unfavorable or even life threatening conditions. In terms of movements before, during and after spawning, these are often influenced by strict intrinsic factors, i.e., a biological clock driven by environmental conditions (e.g. advancing photoperiod, temperature), whereas predator avoidance and foraging movements often are influenced by a more complex matrix of intrinsic and extrinsic factors. A nice illustration of this is provided by Werner and Gilliam (1984) who hypothesized that an animal should choose its habitat as a result of a tradeoff between growth potential and predation risk and that at any given time the animal should seek the habitat where the net benefit is greatest. This has great relevance for pike which are highly susceptible to cannibalistic predation at small size, but are largely immune from it at large size.

Another model of animal movement in response to the availability of key resources, particularly in response to competition, is the "ideal free distribution" (Fretwell and Lucas 1970) which proposes that animals distribute in a way that optimizes their share of resources. As a very simplified example, consider two lakes (A and B) connected by a stream allowing movement between the lakes; if lake A contains twice as many prey fish as the neighboring lake B, then the ideal free distribution predicts that the number of predators in lake A will be proportionally higher than in lake B. In its simplest form the ideal free distribution has several assumptions which, for some groups of animals and especially those operating at large geographical scales, can be challenged. For example, it can be questioned whether animals are able to move unhindered between habitats, have similar competitive abilities as conspecifics and have simultaneous knowledge of resource availability in the different habitats. However, several attempts have been made to test these assumptions (e.g. Godin and Keenleyside 1984) including for pike in natural conditions (Haugen et al. 2006).

With regard to movements that are strictly related to foraging, the "giving up density" framework (GUD) (Brown 1988) is a third theory that can be used to predict movement. The GUD framework states that animals forage in patches and that the energy return from prey in these patches decreases with time as prey abundance declines. At some point the prey abundance is so low that energy return no longer counterbalances the costs associated with feeding in that patch, such as predation risk, missed opportunities elsewhere and the costs of processing and handling the prey. Hence, at this tipping point the GUD framework predicts that an animal should move to a new patch.

The examples above illustrate different concepts that may be used to explain or investigate why fish behave as they do and, as will become evident throughout this chapter, how this also applies to pike. Another challenge in spatial ecology is to understand the mechanisms behind a given movement, i.e., what intrinsic or extrinsic mechanisms trigger, for example, homing to a previously used locality rather than to some other area of similar habitat, as may occur for pike in the brackish Baltic Sea returning to their natal streams to reproduce (Engstedt et al. 2014).

Several terms are used in this chapter to describe space use by pike, so it is appropriate to provide working definitions (see Box 5.1). Key space-use decisions such as dispersal from a site of origin or the adoption of a particular size or structure of home range can be viewed in terms of the cost-benefit framework that is central to behavioral ecology (Krebs and Davies 1997). For example, use of a large home range potentially provides greater resource benefit, such as prey for pike, but also requires greater locomotion costs, especially if a sit-and-wait ambush tactic can minimize such costs. Migration, involving the movement of a large proportion of a population between differing localities or habitats, usually with a distinct periodicity (see Lucas and Baras 2001, Brönmark et al. 2014 for reviews) can comprise a wide variety of subtypes, including ontogenetic migrations and seasonal migration between habitats. Central to all migration subtypes is that the behavioral decision to migrate affords evolutionary fitness increase if the ultimate benefits of migration exceed the costs, again reflecting the central 'tradeoff' theme within behavioral ecology.

Box 5.1 Definitions of key space-use terminology used in this chapter

Movement: Pike may swim between different habitats and in this chapter we define movement as an active change in location, e.g. movements between or within habitats. This is usually measured on the individual level on a shorter (e.g. minutes) or longer (e.g. days) timescale.

Activity: There may be periods during the day or year where pike move more or move faster, i.e., have a higher level of activity. In this chapter we define activity as rate of movement, i.e., movement per time unit. This can be measured on the individual or population levels.

Home range: The area (or volume) traversed by an animal during its normal day-to-day activities, such as for food gathering, mating and caring for young (Burt 1943, Börger et al. 2008). For animals whose environment changes over the year (as is typical for pike), and/or for which their needs change during ontogeny (again typical for pike), the home range may shift in location or size. For example, areas of a home range used in summer may be unsuitable in winter, but may expand to incorporate new localities or habitat during reproduction. Some parts of a home range may be used more frequently (these are normally termed **core areas**) than others, and conspecifics' home ranges may overlap widely.

Site fidelity and homing: Philopatry (site fidelity) is the tendency for an organism to stay in one place or, in the more commonly applied meaning, the tendency to return to a previously used location (Greenwood 1980, Switzer 1993). The most common form is natal philopatry in which the animal homes to its birth place. Site fidelity can be expressed by long-range migrants, but has also been used to refer to restricted space use behavior, i.e. home ranges that are much smaller than could be expected from observed levels of mobility (Börger et al. 2008). This implies that animals that have a home range often show site fidelity, and if they are displaced from it, they usually return (Lucas and Baras 2001).

Migration: Although definitions of migration are numerous and variable (and contain numerous subtypes), we regard migration as a movement, involving a major part of the population, between two or more distinct habitats or localities that happens at regular intervals and involves a return movement at some stage (Northcote 1978, Dingle 1996).

Dispersal: Dispersal is a one-way movement, away from the site of origin (commonly the natal site) and without return (Greenwood 1980, Matthysen 2012). If dispersal is followed by successful reproduction, then gene flow occurs, so it is fundamental to colonization processes and in response to local competition and inbreeding risk. Dispersal is often considered to occur without strong directionality (radial dispersal from a central point), and particularly in younger individuals. In many aquatic environments dispersal can, however, be strongly directional and at least partly passive due to distinct currents (Lucas and Baras 2001).

5.1.2 Pike in Lentic, Lotic and Brackish Habitats

Pike can be encountered in aquatic habitats throughout the northern hemisphere. This species is common in lakes and ponds ('lentic' habitats, with little or a variable direction of water movement), in many slower-flowing rivers and streams ('lotic' habitats, with flowing water) and also in brackish habitats where salinity is below about 12 ppt. Pike are, perhaps, most familiar to those with an interest in lakes and ponds. Here pike is often the iconic top predator and the lack of water current and frequent abundance of vegetation provides the conventional notion of suitable habitat for this species. Regardless of the habitat, some pike move a lot, others do not (as will become evident later in this chapter), and it is a challenge to comprehend why this variation occurs and how it is influenced by biotic and abiotic factors. There are multiple explanatory factors, and although a few general rules seems to apply to pike spatial behavior, overall, strong variability exists in pike behavior within, as well as between, lentic, lotic and brackish habitats (Jepsen et al. 2001, Koed et al. 2006, Kobler et al. 2009, Pierce et al. 2013, Jacobsen et al. 2017).

Most studies on the spatial ecology of pike that have emerged during the last 40 years rely on telemetry techniques, with the exception of a few notable capture-mark-recapture studies (e.g. Miller 1948, Frost and Kipling 1967, Kipling and LeCren 1984). Over recent decades the longevity of acoustic and radio transmitters has increased, while their size has decreased, facilitating detailed studies in a range of habitats, varying environmental conditions and seasons and for differing sizes and sexes (Cooke et al. 2012).

This chapter includes the description and discussion of movement patterns and habitat use of pike in lentic, lotic and coastal brackish areas and, in some populations, between these distinct habitats. Much of the information provided is descriptive (though quantitative) in nature; this is deliberate, since multiple studies in recent decades have generated varying, sometimes conflicting, patterns in the spatial ecology of pike. The chapter divides into four main sections. Firstly, we deal with rates of movement reflecting activity levels and swimming speeds, how this varies temporally and is affected by biotic, abiotic and anthropogenic influences. Secondly, home range and site fidelity is considered. Thirdly, the use of different habitats and the movements between them is considered, as well as how this varies temporally and is affected by biotic, abiotic and anthropogenic influences. These three sections deal primarily with adult pike. This is because the majority of studies in pike spatial ecology so far have used telemetry methods, which most often have been constrained to relatively large, adult pike. In contrast, studies on fry, juveniles and small adults are underrepresented. Hence, the fourth section, deals solely with the spatial ecology of young pike and larvae.

5.2 ADULT PIKE; MOVEMENT AND ACTIVITY

Fish activity is usually measured as movement rate, i.e., distance covered per time unit. A few studies conclude that pike are generally sedentary (Diana 1980, Eklöv 1997) but many studies contradict this and show how pike may exhibit substantial movements which, on some occasions, can consist of several kilometers. In an acoustic telemetry study in a 5 km^2 brackish lagoon in the Danish western Baltic, pike moved all over the lagoon on a yearly basis (Jacobsen et al. 2017). Other examples focus on daily movements which can be extensive. A few examples are Carbine and Applegate (1946), in Houghton Lake and the Muskegon River, Michigan, 16 km in 22 h; Cook and Bergersen (1988), 26 km in three days in Eleven Mile Reservoir, Colorado; Moen and Henegar (1971), 240 km in 78 days in Lake Oahe, South and North Dakota; Ovidio and Philippart (2002), up to 4 km per day in the River Ourthe, Belgium; Diana et al. (1977), 4 km per day in Lac Ste. Anne, Alberta; Miller et al. (2001) up to 26 km per day in Kabetogama Lake, Minnesota.

5.2.1 Large versus Small Pike and Day to Day Individual Variation

Most studies on pike activity report great individual variability in the extent of activity by pike (e.g. Diana et al. 1977, Kobler et al. 2009, Baktoft et al. 2012). Because, across the animal kingdom, larger animals have greater absolute energy demands, metabolic theory indicates the extent and rate of movement of foraging animals can be expected to increase with body size. However, it is not expected to increase in direct proportion to mass (McNab 1963); i.e. it scales allometrically, not isometrically. Because of its large ultimate body size, the pike is a good model species in which to examine such predictions. River-dwelling pike, varying in mass from young-of-the-year (YOY) of under 10 g to large adults exceeding 10 kg have been shown to exhibit an allometric scaling relationship of mean activity, i.e. daily travel distance, with a mass exponent of about 0.4 (Rosten et al. 2016). This reflects a strong relative decline in activity with increasing size, even though the absolute rate of activity shows an increase with body size. The functional significance of allometric scaling in activity and area traversed, relative to the energetic demands of body maintenance and space use requirements are considered in section 5.3 below. Because of other causes of variability in behavior (e.g. due to differences in environmental conditions), such scaling is only obvious over large relative body size ranges; thus it is not surprising that individual studies have not always demonstrated such size effects on movement rate. An example is

Baktoft et al. (2012) who found no relationship between size and activity of pike despite a relatively large size range (36-99 cm) in a small Danish lake.

Similarly to Rosten et al. (2016), Kobler et al. (2008a) found an increase in activity, i.e., absolute movement rates, with size for pike between 45 and 76 cm in a German lake and discuss that this size-specific increase in activity could be explained by a higher demand for food among larger pike. However, it could also be related to reduced activity occurring among smaller pike as a strategy to avoid encounters with larger cannibalistic conspecifics. Also, Jepsen et al. (2001) found a weak but significant increase in activity with length (52-77 cm) in a turbid lake, but not so in a clearer lake. Similarly, Andersen et al. (2008) investigated pike activity (57-85 cm) in two lakes with differing turbidity and found a positive relationship between size and activity in the most turbid lake, but not in the clearer lake, suggesting that under some circumstances visibility may affect activity levels in a size-specific manner.

Large variation in activity occurs not only between individuals, but clearly also on the individual scale on a day to day basis. As an example, an 80 cm pike in a small 1ha lake made extensive movements all over the lake adding up to 5.6 km on one day, and moved only 1.3 km two days later, staying mainly at one side of the lake (H. Baktoft, unpublished data) (Fig. 5.1).

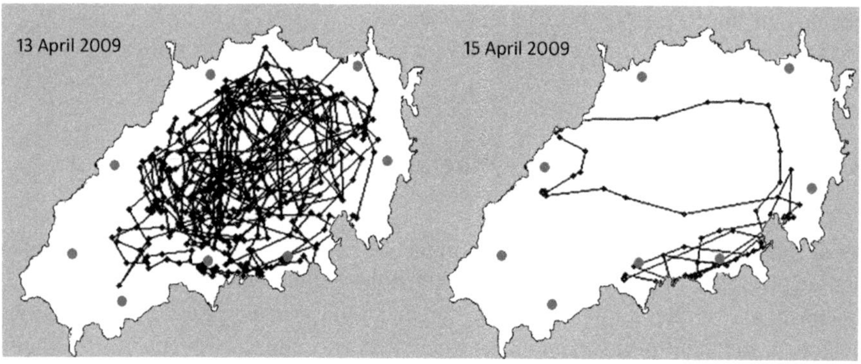

FIGURE 5.1 24 hr movements of a female pike (length 80 cm) in Lake Gosmer (1 ha), Denmark, on two different days: 13 April 2009, total distance moved 5642 m (left), and 15 April 2009, total distance moved 1319 m (right). Data were retrieved from an automatic acoustic telemetry system with 8 hydrophones covering the lake (grey dots) (See Baktoft 2012 for system details). The pike was tagged with a acoustic transmitter with a nominal burst rate interval of 45 seconds. (H. Baktoft, unpublished data).

5.2.2 Activity during the Spawning Period

Activity of mature pike often increases around spawning time. Lucas (1992) reported daily activity levels of small numbers of females and males, respectively of 1.13 km day^{-1} and 1.72 km day^{-1} at spawning time (24-30 April) compared to 0.29 and 0.24 km day^{-1} during the immediate post-spawning period (and similar values in the pre-spawning period) in Loch Davan (36 ha) in Scotland. Likewise, Cook and Bergersen (1988) reported increased activity of tracked pike in the spawning period, as did Casselman (1978) who used monthly catch per unit effort data from gill nets to show that highest activity of pike was found during the spawning period. Knight (2006) placed passive integrated transponder (PIT) antenna arrays at the entrances to a series of small side channels onto the floodplain of the River Frome, southern England, and showed that the activity of pike, measured as the frequency of detections at the PIT antennas, increased during the spawning period, with repeated entries and exits from the channels by different pike. In Danish Lake Gosmer, activity levels of male pike increased before and during the spawning period, whereas females appeared to have an activity peak shortly after the spawning period ended. Moreover, the activity of males during the spawning period was much higher than females (Fig. 5.2). The latter is in contrast to Koed et al. (2006) who in a lotic environment observed females to move more than males during the spawning period (15 March–15 May) and observed upstream prespawning migrations of 1.3-37 km.

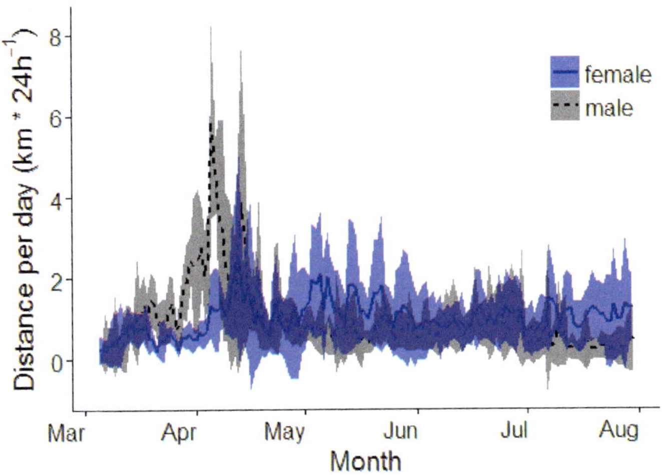

FIGURE 5.2 Daily distances moved of male (N = 5, length 53-76 cm, grey shade, dotted line) and female (N = 6, length 46-99 cm, light blue shade, solid line) pike from March to August 2009 in Lake Gosmer (1 ha), Denmark. Movements of male pike increased before and during the expected spawning time in April. Distance moved per day [median (line) ± standard deviation (shaded area)] was the total distance between recorded positions during 24 h (H. Baktoft, unpublished data). Dark blue shade is overlapping standard deviations. See Fig. 5.1 and Baktoft (2012) for more details about the telemetry system used.

5.2.3 Abiotic Factors

Variation in pike spatial ecology is to a large degree influenced by temporal variation in environmental, abiotic factors, resulting in distinct seasonal and even diel patterns. The most important factors are temperature, light (turbidity) and oxygen level. The way pike in general cope with these environmental factors is treated in chapter 3, but here we concentrate on how they influence pike movements and activity.

5.2.3.1 Temperature

It is not surprising that laboratory experiments have shown that swimming activity of pike which are, like most fishes, ectotherms, is positively correlated with water temperature within their range of thermal tolerance (Casselman 1978). However, under field conditions this pattern has rarely been shown. Only a few studies have reported lower activity levels in winter, such as Kobler et al. (2008a) who investigated movement rates in the 25 ha mesotrophic Lake Kleiner Döllnsee. By contrast, other studies have found no differences in seasonal activity. For example Baktoft et al. (2012) who used an array of hydrophones in a small Danish lake (1 ha) to continuously monitor pike activity for two years, found no difference in activity levels between late summer and winter. Even during periods of ice cover, activity levels remained similar to those at other times, i.e., average daily movements rates of between 621 and 1248 m. Likewise, in the much larger Eleven mile reservoir, Colorado (1362 ha) Cook and Bergersen (1988) found no difference in activity levels between January-March and July-October, suggesting that the marked drop in temperature between late summer and winter had no apparent impact on activity levels. Diana et al. (1977) (Lake Lac Ste. Anne, 5700 ha) and Jepsen et al. (2001) (Lake Ring, 58 ha and Lake Bygholm 22 ha) found no difference in activity between winter and summer. However, Jepsen et al. (2001) reported that an increase in temperature during winter (from around 0°C to 5°C) can boost activity. In two shallow lakes (Wickett Lake (71 ha) and Smoky Hollow Lake (30 ha)) Casselman (1978) found higher activity around 15-18°C (based on gill net CPUE) than at warmer and colder temperatures. In river-dwelling pike, daily activity appears to be stable and low in summer compared to in cooler months (Masters et al. 2005, Knight et al. 2009, Pauwels et al. 2014, Pankhurst et al. 2016), although elevated movement rates in spring are primarily associated with migratory behavior rather than local foraging (Masters et al. 2005, Knight et al. 2009, Pauwels et al. 2014).

5.2.3.2 Oxygen

Oxygen levels provide a strong controlling factor on fish activity through aerobic metabolism (Fry 1971, Claireaux and Chabot 2016) and activity can be expected to be reduced markedly under hypoxic conditions. So far, studies examining pike behavior in response to hypoxic conditions are few. However, such studies seem very relevant, since hypoxic conditions are frequent in lentic habitats, i.e., during prolonged ice cover in shallow eutrophic lakes or during warm, settled conditions in hypertrophic lakes with high biological oxygen demand. The pike has been shown to be tolerant to low ambient oxygen concentrations (Casselman 1978) and cold winter conditions (< 3°C). Casselman (1978) captured pike in static fishing gear in winter in Smoky Hollow Lake, Ontario, at local oxygen concentrations of 0.04 mg L^{-1} and for which the maximum concentration was 0.8 mg L^{-1}, indicating that pike were active even under these conditions. Armstrong (chapter 4) suggests the cardio-respiratory system of pike may be suited to cope with hypoxia, particularly at low temperatures, and chapter 3 contains further details about oxygen requirements.

5.2.3.3 *Turbidity*

Since pike primarily are visually oriented predators, water clarity might be expected to affect pike activity. For example, turbidity influences the visual interactions with prey (chapter 2), but could be compensated for by changes in hunting behavior and ultimately activity. However, turbidity has been related to activity in only a few studies. Andersen et al. (2008) compared activity levels in a turbid and a clear-water lake and found higher among-individual variation in the turbid lake and inferred that turbidity may induce increased diversity in individual behavioral strategies, especially among larger individuals. By contrast, Jepsen et al. (2001), who also compared activity of pike in a turbid and a clear-water lake, found no such difference.

Both Andersen et al. (2008) and Jepsen et al. (2001) included only two lakes in their respective studies, and naturally this lack of replication calls for care in the interpretation as additional factors to water clarity and turbidity may have influenced the observed patterns. For example, a reduction in water clarity often influences the characteristics of the habitats, e.g. the abundance and diversity of submerged vegetation. This could also influence predator-prey interactions and thereby activity levels. Likewise, temperature and turbidity, along with several other biological, environmental and physical variables are integrated factors in the interpretation of seasonal patterns of fish activity, and in field studies the effect of each of these as well as their interactions are often impossible to distinguish from each other. Hence, it may not be that surprising, as indicated in the examples presented in this chapter, that it is difficult to point out general seasonal patterns of pike activity other than ones that relate to distinct periods, such as those relating to spawning activity. In addition, several authors (Jepsen et al. 2001, Rossell and MacOscar 2002, Kobler et al. 2008a, Baktoft et al. 2012) identify differences in methods between telemetry studies as another mechanism to explain the variability between studies. The evidence to date is that pike typically display similar levels of activity between summer and winter (but see Pankhurst et al. 2016), whereas the role of turbidity on activity is not yet well-elucidated (see also chapters 2, 3 and 6).

5.2.4 Diel Activity Patterns

In general, pike activity displays a diel pattern with more activity at day and less activity at night (Fig. 5.3). Moreover, pike are often reported to show crepuscular activity, with peaks at dawn and dusk in lakes (Diana 1980, Cook and Bergersen 1988, Lucas 1992, Jepsen et al. 2001, Kobler et al. 2008a, Baktoft et al. 2012) and rivers (Beaumont et al. 2005, Hodder et al. 2007). The crepuscular behavior, may reflect foraging profitability, i.e., that foraging success may be higher at low light intensities where prey may be more susceptible to predation (Volkova 1973, Dobler 1977, Pitcher and Turner 1986). Likewise, higher daytime activity was related to foraging by Lucas et al. (1991), who used heart rate telemetry, which also enabled detection of muscle electromyogram (EMG) interference, and observed an order of magnitude greater foraging activity (generating EMG interference and resulting in tachycardia) by day than at night in a Scottish lake. Diel patterns in activity can also be found in pike migrating from brackish water into rivers in the spring.

Müller (1986) reported a peak in activity (trap catches) in the early morning and a smaller peak in the afternoon, whereas Engstedt (2011) found comparable peaks in spring migratory activity during morning and evening.

The above patterns of pike activity are contrasted by studies that present little diel variation in activity (e.g. Andersen et al. 2008). Moreover, there are examples where individual behavioral patterns stand out and do not follow the diel patterns displayed by the majority of the fish in the population. For example, in Lake Gosmer, Denmark, some individual pike were active at night (at least in periods) while most individuals in the population were active during day (H. Baktoft, unpublished results). Kobler et al. (2009) even found diel activity differences between different behavioral types of pike; a group of pike that preferred submerged macrophytes as habitat were less active at night than during the day, while another group, displaying a habitat opportunistic behavior, generally had higher activity and moved more at night than at day and finally a third group (reed selectors) were relatively inactive and showed no diurnal differences. Once again this underlines the versatility that exists among individuals within and between lakes.

Diel patterns may also vary with season (Fig. 5.3), and whereas Baktoft et al. (2012) found that the diel pattern was consistent from summer to winter, Jepsen et al. (2001) showed clear seasonal variability, with higher nocturnal activity during summer at least in one of two study lakes, and Kobler et al. (2008a) found a more uniform diel activity pattern during winter whereas crepuscular activity was obvious only during summer. Casselman (1978) used gill net catches to report seasonal activity patterns and found that during periods without ice cover pike were least active at high light intensities and most active at twilight and nocturnal periods. During winter with ice cover they were most active during the day. The variation between studies could relate to varying light intensities with latitudes, e.g. during summer, nights are less dark at northern latitudes. Nevertheless, the examples show diversity in pike activity patterns between localities.

5.2.5 Activity and Anthropogenic Factors

Pike activity can be affected by anthropogenic factors such as fishing or boat noise (see also chapter 12). Several studies suggest that short-term changes in pike behavior, in terms of activity, can be expected after capture and release by anglers. Klefoth et al. (2008) found that radiotracked caught and released pike reduced activity for up to a week post release. Likewise, Baktoft et al. (2013), who used a continuous tracking system, found that angled and released pike exhibited reduced activity for up to 48 h post-release, in comparison to a control group. Arlinghaus et al. (2008) studied the behavior of radio-tagged pike that were released post-angling with a lure retained in the mouth, as an attempt to simulate the situation where a lure breaks off the line during pike capture. The post-release behavior of these pike was monitored using small surface floats (for the first hours) and by externally attached radio transmitters (for the following 3 weeks). Pike with retained lures showed reduced activity for the first hours, increased activity compared to controls for the following 20 h, but thereafter exhibited no difference from controls.

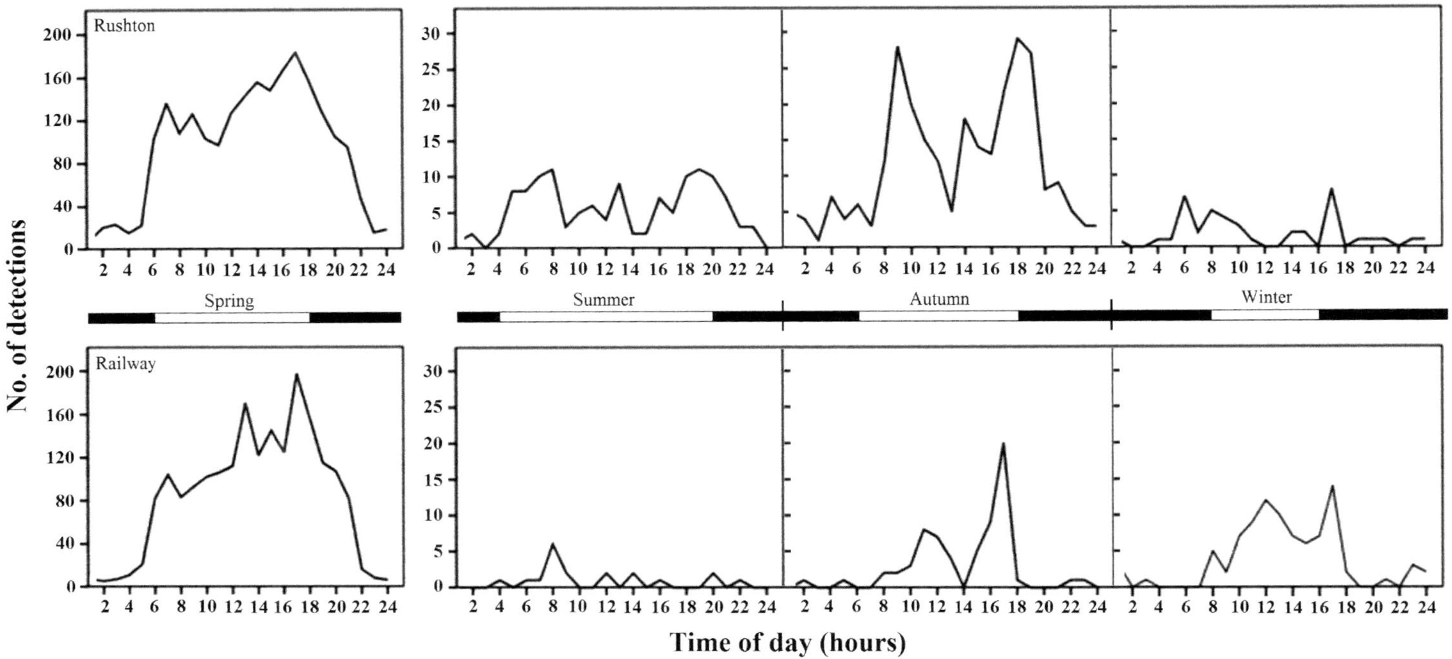

FIGURE 5.3 Diel and seasonal pattern of PIT tagged juvenile and adult pike activity in two side channels of the River Frome, southern England, presented as the number of detections of unique fish (one-hour filter). Spring movements are shown on a different scale due to the much higher levels of activity associated with spawners entering and leaving. Rushton and Railway are blind-ending, heavily vegetated drainage ditches used for spawning and nursery habitat by pike, but experience low water levels and periodic low oxygen levels in summer. Data are for March 2004 – Feb 2005 (spring Mar-May, summer Jun-Aug, autumn Sep-Nov, winter Dec-Feb). Average day-night bars are given for each season. The side channel entrances are directly accessible from the same unobstructed 4 km river reach. Pike (12-101 cm) were tagged with 23 mm HDX tags and monitored with PIT readers interrogating pairs of antennae 20 m upstream of the confluence of each channel. Further information is given in Knight (2006).

Interestingly, this pattern could not be confirmed by Pullen et al. (2017) who, in laboratory settings, used a combination of blood-based physiological metrics and metabolic rate measurements to explore the physiological consequences of lure retention in pike along with video observations to quantify activity patterns. Their study concluded that a retained lure had little effect on either physiological recovery to exhaustive exercise or resting metabolic rate in the laboratory setting, and had no effect on activity in the short term.

Indirect effects of angling on pike behavior have also been examined. For example, the effect of the disturbance by repeated throws with a fishing lure in the water has been studied, but no clear effects of this on pike activity have so far been established (e.g. Klefoth et al. 2011, Jacobsen et al. 2014). In one case, Klefoth et al. (2011) found similar behavior of radiotagged pike in angling areas compared to non-angling areas. Another example is from the Danish Lake Gosmer (1 ha), where Jacobsen et al. (2014) explored how repeated angling with artificial lures affected pike; neither here was there evidence of any disturbance effect. The same study found that although boating with a small outboard engine affected the behavior of tagged roach *Rutilis rutilus* Linnaeus, 1758, pike activity did not change significantly during or after boating events, but there was a tendency for a decrease in swimming speed after a few hours of boating disturbance. Although pike may not alter their behavior substantially in response to disturbance, as indicated above, this does not mean that they are unaware of such disturbances. Heart rate telemetry in field and laboratory environments has shown that pike respond to novel or disturbing stimuli (e.g. an anchor weight hitting the lake bed) by exhibiting a distinct bradycardia (dramatic slowing of the heart rate) (Lucas 1989). Hence, based on existing knowledge, pike seem quite resilient to indirect anthropogenic disturbances, at least when evaluated as changes in behavior such as activity, but they are likely very aware of such disturbances around them.

5.3 ADULT PIKE; HOME RANGE AND DISPERSAL

As already mentioned, pike exhibit huge individual variation in activity and habitat use. It is hence not surprising that studies show varying home range sizes (see Box 5.1 for definitions). For example, based on repeated electrofishing (May-September) in two consecutive years in a 27 ha Swedish lake, Eklöv (1997) found that, within a given year, the majority of a group of previously tagged pike were recaptured in the same or neighboring areas (each being 100-440 m wide) of littoral habitat, which suggests use of a restricted home range. Also, Grimm and Klinge (1996) report restricted movement of pike in a 4.5 ha Dutch lake. In contrast to these studies there are numerous examples of pike having considerable home ranges (Cook and Bergersen 1988, Lucas 1992, Jepsen et al. 2001, Rosell and MacOscar 2002, Midwood and Chow-Fraser 2015, Sandlund et al. 2016). Capture-mark-recapture data of adult pike by Moen and Henegar (1971) for over 2000 recaptured pike in Lake Oahe, a Missouri reservoir, showed that the furthest distance travelled was 322 km, but 70-89% of tagged pike were recovered within 32 km of the tagging locations.

There are various possible reasons for variation in home range patterns in pike. They could for instance relate to lake size and the spatio-temporal distribution of key resources, as suggested by Kobler et al. (2008b), as well as individual specific variation in absolute energy demands. Absolute energy demands, met through food consumption or depletion of body energy reserves (chapter 4) are greater for large pike than smaller ones, but the relative energy use (metabolic rate per unit mass) decreases with body mass. Although the exact body mass scaling of energy demand varies somewhat across species it has frequently been estimated to be about 0.75 in vertebrates (McNab 1963), and has been measured as 0.8 in pike (Armstrong et al. 1992). It was originally suggested that the home range size required by animals of a given species should be directly proportional to maintenance metabolic rate (McNab 1963), and so, for pike, should scale close to 0.8. However, for river-dwelling pike, home range has been found to scale with body mass (for pike of 7-12060 g) at a significantly greater scaling coefficient of 1.08 (Rosten et al. 2016) and this is also greater than the scaling coefficient of maximum metabolic rate for pike of 0.99 (Armstrong et al. 1992). Thus, in pike, home range size increases more rapidly with mass than metabolic scaling theories predict. Since daily travel distance for the same pike scaled as $mass^{0.4}$ (Rosten et al. 2016), much less than the home range scaling coefficient (1.08), this supports a theory by which larger fish cover the full extent of their home range less often, leading to lesser expulsion of competitors, greater resultant overlap of home range, increased shared resource use, and a requirement for larger relative home range for a given body mass (Jetz et al. 2004).

River-dwelling pike generally adopt restricted home ranges centered around pool-type habitats (Langford 1981, Hodder et al. 2007), concentrating their activity in one or multiple core areas (see Box 5.1 for definitions) and occasionally making excursions to the periphery or outside the main home range (Koed et al. 2006, Hodder et al. 2007, Knight et al. 2008). In addition, Hodder et al. (2007) found a high degree of home range overlap of adult pike (average length 69 cm), with up to five tagged pike exhibiting overlap in core areas of use. Based upon a capture-mark-recapture study with low temporal resolution in the River Frome, a chalkstream in southern England, Mann (1980) proposed that the pike population comprised static (74% of fish) and mobile components. However, telemetry of the same population has shown a continuum of space-use strategies, from resident individuals with small home ranges of a few hundred metres river length, to indviduals making repeated movements of several km, broadly but not exclusively, on a seasonal basis (Masters et al. 2005). This pattern of variation has been found in other studies too (Ovidio and Philippart 2005, Vehanen et al. 2006, Pankhurst et al. 2016). Maximum linear home range of adult River Frome pike observed over 8-25 months, varying between 0.16-5.92 km (Masters et al. 2005), was positively related to fish length, but with no effect of sex. The location and size of river-dwelling pike home ranges was found to vary by season in the Frome (Knight et al. 2009) and the magnitude of excursions outside the home range was positively correlated with home range area (Knight et al. 2008). Hodder et al. (2007) found that, for the same river, home range size was not correlated with discharge during periods when the water remained within the river channel, but during extensive floods home ranges were significantly greater.

5.3.1 Homing and Site Fidelity

Several studies have shown clear indications of spawning fidelity (reproductive philopatry) through homing in pike (DosSantos 1991, Miller et al. 2001, Engstedt et al. 2010, Engstedt 2011, Sandlund et al. 2016) including natal spawning site fidelity (Miller et al. 2001, Engstedt et al. 2014) (see Box 5.1 for definitions). In DosSantos' (1991) study, several radio-tagged pike were recorded homing between alternating spawning and feeding areas over 2 years (three spawning seasons), exhibiting upstream and downstream migrations of 16 km between these areas, each taking 2-3 weeks. In anadromous Baltic pike, repeated reproductive homing to the same streams was evident from more than 3400 PIT tagged pike that ascended into four different brooks. All pike that returned to spawn in the streams in subsequent years returned to the same river in which they were tagged, and no incidents of tagged pike appearing in the other rivers occurred (Engstedt 2011, Larsson et al. 2015). Natal homing to the stream where the pike was spawned was already suggested by Müller (1986) and was recently indicated by otolith composition of juveniles and adult pike in small streams (Engstedt et al. 2014, but also see Rohtla et al. 2014). Often reproductive homing is observed only in parts of the population (e.g. Vehanen et al. 2006). In the western Baltic Sea, distinct genetic differences are evident for anadromous stream-spawning populations and genetically differentiated populations can be close together (Larsson et al. 2015). Evidence for local phenotypic differentiation and adaptation to spawning habitat in sympatric subpopulations (Tibblin et al. 2015, Berggren et al. 2016) suggests that the degree of natal homing is of strong functional significance. Further consideration of pike population genetics is given in chapter 7.

Homing has also been shown to occur outside of the spawning period, i.e., pike showing site fidelity to a particular area during summer, fall or winter (see Box 5.1 for definitions). Kobler et al. (2008b) translocated eight radiotagged fish between 377 and 550 m away from their original activity center in a small (25 ha) lake and found that all pike returned to their original home range between 4 and 143 h later. However, the wide and varying home ranges observed in other studies (Diana et al. 1977, Cook and Bergersen 1988, Jepsen et al. 2001) suggest that strong site fidelity may be the exception rather than the rule in lentic pike. In the Baltic coastal areas of Sweden and Finland, pike were caught and tagged with conventional tagging in one area and then translocated various distances before release. If these pike were moved to areas up to 10 km away they were found to home back to the area of capture, if moved more than 60 km away, they did not return (Karås and Lehtonen 1993). Other experiments in Finnish coastal areas showed that pike that were tagged outside of the spawning season and translocated 0.5-2.6 km away were repeatedly recaptured at their original tagging sites (Halme and Korhonen 1960, cited in Karås and Lehtonen 1993). Hence, overall, there appears to be a tendency to site fidelity and homing to feeding areas for coastal pike.

5.3.2 Behavioral Types

The existence of behavioral and physiological differences between individuals which are stable over time, i.e., animal personalities, has received much attention in recent years. Often these behavioral and physiological traits are intercorrelated with each other and can be referred to as temperament, behavioral syndromes, coping styles or simply predispositions (Carere and Maestripieri 2013). Pike, and especially young pike, have been the focus of studies on personalities and behavioral syndromes (e.g. Nyquist et al. 2012, McGhee et al. 2013, Laskowski et al. 2016). In relation to this, several authors have suggested that pike populations consist of different behavioral types that reflect both activity and home range size. Mann (1980) proposed that in the River Frome, UK, two behavioral groups of pike exist, one group being sedentary and occupying a restricted area, and another group that move extensively. This reflected a widespread opinion at around the same time for similar behavior (stayers, movers) of a range of stream fish species studied by capture-mark-recapture (e.g. Stott 1967 who suggested static and mobile components in cyprinid populations), with its rather coarse temporal resolution of movement data (few locations, widely spaced in time). However, working on the same river, with telemetry, Masters et al. (2005) refute this view and argue that the extent of movement by different pike comprises a continuum (see the introduction to section 5.3). In contrast, Vehanen et al. (2006) and Sandlund et al. (2016) who used telemetry, though with quite intermittent location, also considered their tagged pike, in regulated river and reservoir environments to comprise sedentary and much more mobile components. Similarly, but for lentic pike, Jepsen et al. (2001) grouped pike into those that stay in one restricted area, another that utilize larger areas and, in addition, a third intermediate group that shift between two or three "favorite areas" each having a relatively restricted range. Kobler et al. (2009) also report three lentic behavioral types, two types that only selected vegetated littoral habitats being either reed or submerged vegetation, and a third more opportunistic group that used all habitats and in addition had higher activity. No differences in size or age could explain the variation in behavioral strategies, and Kobler et al. (2009) further suggests that despite the higher rate of movements and consequently higher energy cost among the group of active opportunistic pike, some compensatory mechanisms, i.e. elevated food intake, enable them to balance the cost of increased movement rates. This is supported by the fact that life time growth was similar between the three groups. Further, Kobler et al. (2009) proposed that the diversification of behavior among individual pike could reduce intraspecific competition and therefore play a role in the emergence of an ideal free distribution among the pike. Jepsen et al. (2001) found more females than males in one of their behavioral groups, but as shown by Kobler et al. (2009) who studied only females, behavioral types are not strictly related to sex. In a brackish lagoon two thirds of tagged pike moved outside the lagoon outside spawning time, while the rest stayed in the lagoon. This behavior seemed consistent between individuals, since the same individuals moved out of the lagoon after spawning again the year after (Jacobsen et al. 2017). There was no sex or size difference in the two behavioral groups, nor was there any genetic component to explain these two phenotypic traits. Different spawning behavioral groups occurs in other populations of Baltic pike, where some pike stay in the coastal brackish bays to spawn and the others migrate up small streams to spawn in freshwater (Westin and Limburg 2002, Engstedt et al. 2014). In conclusion, there seems to be broad agreement that individuals within pike populations group into different behavioral types.

5.4 ADULT PIKE; HABITAT USE AND HABITAT SHIFTS

So far, we have examined rates of activity, i.e., factors that influence how much and how fast pike move around within a given habitat, and patterns of homing and site fidelity. Further, we have also touched upon sizes of home range and how this may be influenced by body size, sex and behavioral type. In this section we describe the habitat use by pike, i.e., which habitats are used and when. This will also consider migration patterns related to spawning (see Box 5.1 for definitions). We also explore potential biotic and abiotic factors that could influence the movements between habitats. However, we consider habitat preferences only briefly as this is covered in chapter 3.

5.4.1 The Use of Vegetation and Other Structured Habitat

Lentic pike are often reported to stay in areas with vegetation, though the dependence on vegetation varies according to life stages and function (see chapter 3). However, there is huge variation in observed habitat use between studies, probably reflecting both intra- and inter-lake variation in vegetation abundance and types. As discussed more thoroughly in chapter 3, the importance of vegetation as a habitat characteristic may relate to the size of the pike, i.e., as pike grow, the use of unvegetated and more offshore areas may increase. Specifically, it has been suggested that the spatial distribution of pike > 53 cm are less restricted to vegetated areas compared to smaller pike (Grimm and Klinge 1996). Indeed, large pike (> 65 cm) may be free ranging between littoral and offshore areas as reported by Chapman and Mackay (1984a). Likewise, gillnet catches of pike in open water and at depths down to 10 m suggest less dependence of adult pike on vegetation as the preferred habitat (Vøllestad et al. 1986). Nevertheless, other studies indicate that vegetated areas are often preferred habitats even for larger lentic adult pike. For example, Cook and Bergersen (1988) found in Eleven Mile reservoir (Colorado) that, in summer, 60-89 cm pike occupied vegetated areas far more often than could be accounted for by random association with available habitats. In Lake Ste Anne, pike were associated with vegetation 95% of the time during summer (Diana et al. 1977) whereas Kobler et al. (2009) found half of the study fish > 50 cm to stay in or near vegetated areas in Kleiner Dölnsee, Germany (25 ha, mean depth 4.1 m). Chapman and Mackay (1984b) observed pike in Twin Lake (Alberta, 23 ha, mean depth 40 m) through Scuba diving in July-September and found that 73% of pike > 25 cm were found in vegetated areas, principally at the interface between open water and vegetation. A similar observation was made by Brosse et al. (2007) based upon electrofishing surveys in the vegetated areas of a French reservoir.

In lotic environments, due to the poor sustained swimming performance of pike (Jones et al. 1974), slack water seems the most important habitat determinant at a meso scale (DosSantos 1991, Lamouroux et al. 1999). However, pike in rivers tend also to be associated with structural habitat features, often (Mann 1980) but not always (Langford 1981) associated with vegetation, not least because some rivers do not have extensive in-channel vegetation or it may have died back substantially in winter. Also water depth seems to influence habitat use. In the Frome, South England, core areas of habitat use were found in significantly deeper water than occurred throughout the linear home range of tracked pike (Masters et al. 2005). These core areas averaged just 0.16 m deeper than the surrounding areas, perhaps reflecting the availability of local refuges from faster flow in these stream bed depressions. In regulated or impounded rivers, pike use only the lower gradient reaches characterized by slow flow (Vehanen et al. 2006, Sandlund et al. 2016). In summer, Vehanen et al. (2006) found that pike in the River Kajanninjoki, Finland utilized a wide range of depths (0.2-19.5 m) and vegetation coverage (0-90%).

In Brackish coastal systems most studies on habitat use have focused on spawning and nursery habitats (e.g. Lappaleinen et al. 2008), and there is not much knowledge on habitat use for adult pike outside spawning time. Likewise, most of the documented shifts between habitats are during spawning migrations from coastal waters into small streams, described in the section on migration below. In coastal areas pike may stay relatively offshore outside the spawning season. In a Finnish tagging experiment, pike were recaptured offshore (5 km) around small archipelagos outside the spawning season, whereas they were caught close to the mainland shore around inlets and estuaries during spawning (Karås and Lehtonen 1993). Likewise, some of the acoustic tagged pike in a Danish brackish lagoon moved out of the more sheltered lagoon during summer with the possibility of foraging out in the bay in a more open water environment (Jacobsen et al. 2017). However, mesohabitat use of brackish water pike outside the spawning period remains unknown.

5.4.2 Abiotic Factors

Temperature, oxygen, vegetative cover, water transparency and light intensity are primary factors that interact to affect life history patterns of pike (Casselman 1996), though water flow and salinity are crucial factors also in lotic and brackish environments respectively, and of course these factors also act on mechanisms mediating both the horizontal as well as the vertical space use of pike. Especially, changes in environmental conditions outside the normal range of function may be expected to cause local movements of pike. Likewise, when environmental conditions stabilize pike may reinvade the areas. In a network of Swedish lakes Spens et al. (2007) reported that in lakes where pike had been eradicated by rotenone treatments, recolonization of pike occurred in most of the lakes as a result of immigration from nearby lakes. However, the authors also found that pike did not recolonize upstream positioned lakes if the slope in the stream connecting the lakes were too high. More specifically, if the maximum stream channel slope was > 7%, pike dispersal and recolonization did not occur

5.4.2.1 *Oxygen, pH and Temperature*

Numerous studies have shown temperature and oxygen to be primary drivers in pike habitat shifts. During summer, habitats may become unsuitable for pike if oxygen levels drop below 2-3 mg l^{-1} (Casselman 1978, Pierce et al. 2013). Headrick and Carline (1993) found 95% of the locations of radio-tagged fish to be at dissolved oxygen levels above 3 mg/l (average 5.5 mg/l). Not surprisingly, pike may therefore move between habitats to meet their environmental requirements. For example, as summer hypoxia occurs in the deeper and cooler water of deep productive lakes, there are examples of pike seeking a summer habitat at the 3 mg/l dissolved oxygen isopleth (Headrick and Carline 1993, Pierce at al. 2013) and thereby balancing the use of foraging habitat with access to as cool water as possible but sufficient oxygen levels. Based on gill net catches in two small shallow lakes Casselman (1978) observed a change in depth use as oxygen levels declined during summer, suggesting an important role of oxygen and temperature on vertical distribution. Likewise, Headrick and Carline (1993) found that pike moved from inshore to offshore habitats in summer to find cooler water in deeper parts of two Ohio impoundments. In opposition, Cook and Bergersen (1988) found pike closer to the shore in Eleven Mile Reservoir during summer but consider ice cover and vegetation die back as the major causes of forcing the pike into deeper water in winter. In a more recent study, pike > 71 cm in length selected areas with temperatures of 16-21°C in summer when temperatures up to 28°C were available, which implied vertical movements to water volumes with relatively restricted distribution (Pierce et al. 2013).

In eutrophic lakes, changes in water quality may result in stressful conditions, including night-time oxygen sags and daytime high pH as a result of high photosynthetic plant biomass, or as a result of longer-term depletion of oxygen, including due to ice cover in shallow 'winterkill' lakes. Although, as indicated above, pike exhibit thermal and oxygen refuging responses in summer, for high pH episodes (approaching pH 10) in shallow, eutrophic, normoxic lakes, pike appear to choose a sit-and-tolerate strategy rather than to move to more environmentally benign areas (Scott et al. 2005), even though this incurs significant ion regulation costs.

Rivers exhibit greater mixing of oxygen than lakes and tend to exhibit less temperature variability and so pike in lotic environments may be less likely to exhibit behavioral responses to local thermal/oxygen conditions. Pike tracked in the River Thames, England, spent short periods (1-3 days) in the warm water plume of a power station which raised the water temperature 3-6°C above ambient, but such fish showed no greater residence to that locality than elsewhere (Langford 1981). However, pike in the River Tryggevælde Å in South eastern Denmark, tracked by acoustic telemetry, reacted to severe environmental conditions with low oxygen in the lower part of the river following very warm temperatures in August by moving downstream close to the river outlet, and a few individuals even moved out into in the brackish bay for a few days (Højrup 2015).

Apart from Headrick and Carline (1993), few studies have explored individual patterns of vertical movements in pike as a mechanism to optimize habitat use in relation to abiotic factors. A nice exception to this is Pierce et al. (2013) who used acoustic telemetry in combination with archival tags to explore the use of depth and found huge variation between years, individual fish and lakes. Large pike (> 71 cm) exhibited strong individual variation in depth use and, the pattern between lakes was similar, i.e., pike generally followed the thermocline. Among smaller adult pike (47-61 cm), there were clear inter lake differences. In one lake these pike preferred the cooler, deeper water during summer while in two other lakes the warmer, shallow water was preferred. Pierce et al. (2013) relate this to the fact that in the latter two lakes there was abundant near shore vegetative cover which the pike could benefit from in terms of shading, habitat for forage and shelter from predation. In the lake where the pike used deeper water, vegetative cover was poor and the benefits of choosing the shallower near shore habitats were therefore few.

5.4.2.2 Turbidity and Light

Changes in light regimes may influence pike behavior as illustrated by the diel changes in pike activity (section 5.2.4.3). As the pike is a visual predator (though chemical and tactile cues may be used) it can be argued that water clarity may affect habitat use by pike as a result of changes in foraging profitability (Dobler 1977). In line with this, Vøllestad et al. (1986) observed with gill net catches in four Norwegian lakes that pike extended the use of the open water in turbid waters (down to 10 meters depth), whereas Jepsen et al. (2001) saw an increased use of vegetated areas in turbid Lake Bygholm, compared to clear water Lake Ring (Denmark). Andersen et al. (2008) found that pike had higher individual variation in habitat use (distance to shore) in a turbid lake compared to a clear water lake in Denmark. Since neither Jepsen et al. (2001) nor Andersen at al. (2008) found differences in body condition of pike between the turbid and clear water lakes it is possible that alternating behavior compensates for any adverse impact that the changing water clarity may have caused. Also light intensity may influence habitat use. Chapman and Mackay (1984a) found pike closer to the shore on sunny days than on overcast days while on windy days pike used habitats which were further from shore than on calm days supposedly because the wind made the shore areas unfavorable due to increased turbidity. In opposition, Cook and Bergersen (1988) found pike to be in deeper water, albeit not further from shore. The same authors also found diel patterns in preferred water depths with pike occupying shallow water in late afternoon and early evening and deeper water at night, while DosSantos (1991) observed the opposite for river-dwelling pike.

5.4.2.3 Flow

In streams and rivers, flow has a strong effect on the habitat use of pike, since although they have excellent burst swimming capabilities (Webb and Skadsen 1980), their sustained swimming performance is low, with a critical swimming speed (*Ucrit*) (a standard measurement, to assess swimming capabilities of fish, based upon swimming at incrementally increased water velocities) of about 1.3 body lengths (BL) s^{-1} (Jones et al. 1974). This reduces their ability to remain in faster flowing areas. Instead they tend to be restricted to slower-flowing glides, pools, backwaters and side channels (DosSantos 1991, Lamouroux et al. 1999,

Masters et al. 2002, Knight 2006). Simms (2000) was able to experimentally swim juvenile pike of 18-20 cm length continuously at 0.5 BL s^{-1} (0.1 ms^{-1}) for 11 weeks, this being close to the estimated maximum sustained swimming speed of ~ 0.8 BL s^{-1} estimated as 60% of *Ucrit*. Adult pike in the Flathead River, Montana, were reported to prefer habitat with water velocities under 0.18 ms^{-1} and averaging 0.13 ms^{-1} (Dos Santos 1991). Fast swimming performance over shorter periods by pike enables their upstream movement in many unobstructed river environments, including swiftly-flowing chalkstreams, with pike tracked moving through swift glides and riffles to access slower-flowing habitat patches (Knight 2006). Masters et al. (2002) demonstrated that adult pike moved out of the main channel habitat of the River Frome, England, into flooded areas in winter, and because movements between flooded land, the river and side channels were frequent, they interpreted this as being due to exploiting foraging opportunities (following prey fish out of the channel and/or eating earthworms, amphibians etc), rather than for refuge from flow. During floods in the regulated River Thames, England, some tracked pike moved or were displaced downstream up to 1.5 km and usually homed upstream to their pre-flood location afterwards (Langford 1981).

5.4.3 Biotic Factors

Optimizing foraging and reproductive opportunities against fitness impacts, such as predation risk, are central biotic factors that influence the propensity of pike to move between habitats. Below we discuss potential roles of prey distribution and prey behavior on the habitat shifts of pike. Moreover, since the pike is both a predator and prey, e.g. to birds and to conspecifics, and sometimes has to compete for limited resources, antipredator behavior and mechanisms that relate to competition are also highlighted. A further section considers habitat use and habitat shifts for spawning.

5.4.3.1 Prey

Pike should maximize foraging profitability by aggregating in areas of high prey density, and hence prey abundance and distribution are likely to influence habitat choice and habitat shifts. However, relatively few studies have demonstrated this. Kobler et al. (2008a) found pike to move further from the shore at night (from vegetated littoral areas to more open littoral areas/pelagic areas) and argue that this behavior could result from pike following prey performing nightly horizontal migrations. Kobler et al. (2008b) found that average home ranges of female pike (580-2287 g) in 25 ha German Lake Kleiner Dôlnsee increased threefold from summer to winter and attributed this to altered seasonal availability of structured habitats and prey distribution. Using quantitative, giant traps, Hladík and Kubečka (2003) report that about 12% of the estimated stock of pike in the Římov Reservoir, Czech Republic, migrated between the reservoir and its major tributary the Malŝe River in spring, and they interpret entry into the tributary in spring as a way to optimize foraging on the large spawning aggregations of cyprinids arriving to spawn in the tributary. Sandlund et al. (2016) argue that migration of older larger pike from Lake Løpsjøen to the River Rena, Norway during summer relates to higher prey opportunities in the river. Other studies find the opposite pattern. For example, only a few pike followed potential prey such as roach and white bream *Blicca bjoerkna* Linnaeus, 1758 into small connecting streams in Swedish Lake Krankesjön during fall, winter and spring. Instead they overwintered in the lake (Skov et al. 2008).

Eklöv (1997) found in Lake Degersjön, Sweden, that the areas pike occupied had higher prey densities than other areas suggesting that prey distribution and prey movements play a role in pike habitat use. Haugen et al. (2006) used 40 years of gill net catches of pike to explore pike densities in two neighboring lake basins of Windermere and found indications of quite high levels of between-basin dispersal from one spring to the next spring. There was overall a net movement to the more productive southern basin containing more prey. However, during a manipulation period where pike were heavily fished and therefore removed from the northern, less productive basin, the net movement was reversed. This suggests that pike distributed between the two lake basins in an ideal free manner and distributed themselves in a way that equalized fitness across habitats (Haugen et al. 2006). Subsequently, Haugen et al. (2007) showed that prey abundance, and for larger pike also competition from conspecifics, played a major role for the dispersal. Recently, Sandlund et al. (2016) discuss that similar density dependent mechanisms could explain the distribution of pike between Løpsjøen Reservoir and the adjacent River Rena.

Between lake movements, e.g. in order to optimize foraging, of course relate to the physical conditions of the connecting streams. Spens et al. (2007) who examined recolonization of pike following rotenone eradication found that dispersal probably is more common when lakes are linked by wide and short, rather than long and narrow, connections. Likewise, eight years of PIT monitoring a short but narrow stream (3 km long, 2-3 m wide, average gradient 0.02%) between two adjacent lakes in which hundreds of pike were PIT tagged, revealed virtually no movements between the two lakes, despite several inter-lake movements of potential prey (C. Skov unpublished data).

5.4.3.2 Predators and Conspecifics

As suggested in the studies from Windermere described in section 5.4.3.1, the distribution and density of conspecific pike are likely to affect habitat use and habitat shifts of pike. Like all animals pike may experience intraspecific competition for food and other resources, which can affect growth rate and the use of space. In addition their cannibalistic behavior results in extreme asymmetry for individuals that differ markedly in size. The result is that small pike, principally juveniles, occupy shallow, heavily vegetated habitat to minimize exposure to predation risk from larger conspecifics (Grimm and Klinge 1996). However, pike density and size have more complex effects on local patterns of space use, since even in similar habitat of shallow reed margin, Nilsson (2006) demonstrated that at higher densities of pike, smaller pike tended to be found more distant from larger pike (almost all pike in the study were < 50 cm) and this generated a clumped distribution, while at lower densities distances between individuals increases resulting in more even distribution. Based upon multiple studies in semi-controlled and natural conditions, it seems likely that structural complexity in the littoral has the capacity to alter behavior, space use and demography of pike and conspecific and heterospecific prey (Eklöv 1997). Even among groups of adult pike, size differences might lead to choice of different habitats in order to avoid cannibalism from larger conspecifics. In the small Danish Lake Gosmer, acoustic telemetry (Baktoft et al. 2012) revealed that groups of smaller adults (47-57 cm) and larger adults (67-80 cm), respectively, distributed themselves along the lake littoral zone, but the smaller adults were found in different parts of the littoral than the larger adults. At the same time, both groups seems to avoid the offshore habitat occupied by a single large apex pike (99 cm), which inhabited the whole lake and cruised in the offshore habitat on a regular basis in depths of 3-4 m (Schnedler-Meyer 2014).

5.4.3.3 Spawning

Habitat shifts in relation to spawning are well documented. Pike have specific habitats for spawning – shallow areas rich in vegetation or other complex structures on which eggs may be laid and develop (McCarraher and Thomas 1972, Bry 1996, chapters 3 and 10). In some systems, usually small, shallow lakes and slow-moving rivers, such habitat may be a large part of the system, and so access to and use of that habitat may fall within the same area used outside the spawning season (e.g. Lucas 1992, Pankhurst et al. 2016). However, for pike inhabiting larger, deeper lakes, rivers and brackish environments and faster flowing rivers, suitable spawning habitat may not be locally available and distinct migrations into seasonally inundated marshes, shallow bays, or vegetated side channels are common (Carbine 1942, Miller 1948, Clark 1950, Müller 1986, Masters et al. 2005, Knight 2006, Larsson et al. 2015, chapter 10).

Pike in many lakes undertake local migrations to discrete spawning areas, often in shallow bays with flooded grass, permanently submerged macrophytes (e.g. *Elodea*), stands of *Phragmites* or into marshes (Frost and Kipling 1967, Rosell and MacOscar 2002). However, migrations from lakes into afferent, or sometimes efferent streams, and drainage ditches are common (Clark 1950, Franklin and Smith 1963). Clark's (1950) study of Lake Erie pike showed that they migrated in spring into or adjacent to feeder streams soon after ice melt, to penetrate shallow, marshy areas for spawning, with males entering earlier than females, a similar pattern found by Carbine (1942) who observed that most movement occurred at night. Pike also entered an afferent stream in Lake George, Minnesota, but Franklin and Smith (1963) found no difference in sex ratio or size distribution during the migration. Sixty two percent of pike stayed in the stream for 40-60 days after spawning. Hladík and Kubečka (2003) report migrations around the time of spawning between the Římov Reservoir and its major tributary the Malŝe River, Czech Republic, in spring, and they concluded that pike spawn in suitable habitat within the reservoir and tributary. Similar patterns were seen by Sandlund et al. (2016) who report pike to spawn in both the River Rena and Løpsjøens Reservoir as well as spawning migrations between the two.

After spawning, pike normally disperse from their spawning grounds to feeding grounds. Such movements are quite distinct when from streams, ditches and marshes to lake areas, but may also be distinct within lakes; for example in Lower Lough Erne pike dispersed to areas of medium depth water between 5 and 10 m (Rosell and MacOscar 2002).

River-dwelling pike may exhibit a protracted period of side channel entry and residence for spawning, with males tending to enter before females (Fig. 5.4). Knight (2006) observed a substantial proportion of spermiating males in side channels already in December, about 3 months before the start of the spawning season. Most of these males were small (25-40 cm) and it is suggested that these adopted a strategy of early arrival to maximize chances of spawning, despite low food availability, but also to limit their risk of cannibalism since most larger pike (including many females) remained in the main river channel until close to spawning. In main river channels where shallow or side channel backwaters with abundant vegetation for spawning are not nearby, pike may exhibit migratory behavior along the main channel (Gerlier and Luquet 2000), either in search of suitable habitat or following cues towards likely areas. Even where spawning habitat is locally available, river pike may migrate upstream (DosSantos 1991, Ovidio and Philippart 2005, Pauwels et al. 2014, Table 5.1). Since main channel movements close to the spawning season often appear to be in an upstream direction (e.g. Ovidio and Philippart 2005, Pauwels et al. 2014, Table 5.1), it is possible that rheotactic cues are involved and Pauwels et al. (2014) found a

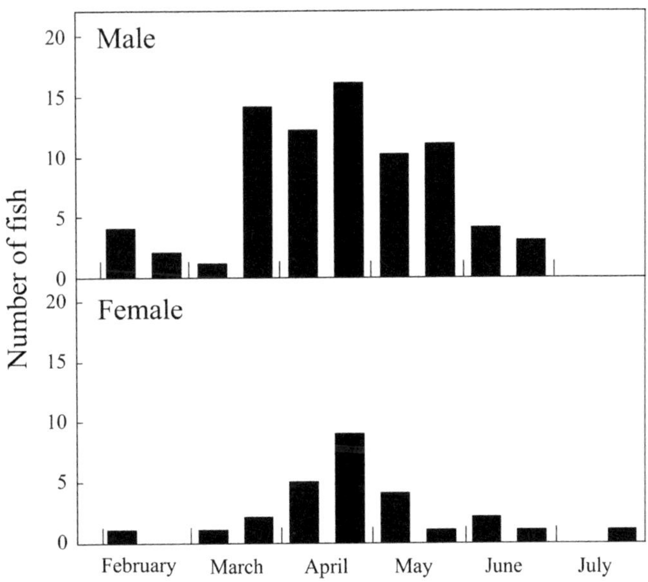

FIGURE 5.4 Visits by different PIT tagged male and female pike to PIT (cross-channel) instrumented side channels of the River Frome, England, before, during and after the 2005 spawning season. Total number of pike detected by PIT readers in each fortnight is presented. Courtship and spawning was observed from early April to early May. Further information is given in Knight (2006).

TABLE 5.1 Comparison of pike spawning migrations between several river systems. U = upstream migration; D = downstream migration.

River, country	***Timing***	***Distance of migration* (km)**	***Notes***	***Reference***
Ourthe and Amblève, Belgium	U 8 Feb to 30 Mar, D mid Mar to mid May	Average, 7.7; Range, 0.8-15.7	*n* = 6, Period at potential spawning sites ~ 11 days	Ovidio and Philippart, (2002, 2005)
Yser, Belgium	Feb to Mar (prespawning)	Variable, but 75% of movements > 2 km/day in Feb-Mar	*n* = 15, no obstacles, but artificial embankments in study reach. Pike in some zones moved more than others, suggesting local habitat affects distance moved	Pauwels et al. (2014)
Ill, France	Winter to spring	0.45-12.3	*n* = 10, five of these ascended four obstacles, 0.8-1.5 m high fitted with Denil passes	Gerlier and Luquet (2000)
Frome, England	Feb to Jun	Less than linear range measures of 0.16-5.92	*n* = 15 (Masters et al. 2005), *n* = 51 (Knight 2006), some, not all, moved upstream in spring, individual movement not consistent across years; slackwater and drainage ditches utilized	Masters et al. (2005) Knight (2006)
Rideau, Canada	No clear spawning migration. Spawning Apr to May	None observed	*n* = 18 Spring spawning areas contained in small areas of larger winter range, resulting in no clear migration. Average linear core range in spring (46 m), summer (58 m) vs. winter (318 m)	Pankhurst et al. (2016)
Flathead, Montana, US	Apr to Jun	Up to 56 km from wintering to spawning areas	*n* = 11, some exhibited repeat homing. Males spent up to 3 months in spawning area, females up to 6 weeks	DosSantos (1991)

significant effect of daily discharge over December to May on the mean daily distance moved. Pike are periodically captured in upstream-migration fishways and traps at low-head obstacles (M. Lucas, unpublished data) although most technical fishways throughout the pike's geographical range are designed for salmonids and not well-suited to pike passage. Thus the records of very low incidence in fishway traps in some rivers such as the Meuse (Prignon et al. 1998) are not surprising. PIT tagged pike of 45-70 cm have been recorded passing upstream through a 20% gradient 10-m long Denil fishway (Lucas et al. 2000) and through a 15% gradient, 11 m long superactive baffle Larinier fishway between spring and autumn (M. Lucas, unpublished data) (but see Spens et al. 2007). At both sites average water velocity exceeded 1 m s^{-1} and PIT telemetry showed that the time between last detection at the entrance and first detection at the exit was 10-35 seconds; it is evident that burst swimming was used to ascend the pass, since the sustained swimming speed of pike is only a little more than 1 body length per second (Jones et al. 1974).

The extent of movement observed during spawning migration is likely to be a function of the size of the system, and the distance between wintering, spawning and summer habitats, and the occurrence of partial or complete obstacles to migration. Pike movements of up to 100 km have been recorded for river-lake systems in Alaska (Burkholder and Bernard 1994), and similar scale movements could occur in the Baltic to spawning streams (Tibblin et al. 2016), although as yet the extent of migration from spawning streams into the Baltic (as distinct from mark-recapture of fish between sites in the Baltic) has not been well documented. Such large scale movements for spawning may be exceptional. More commonly, documented movements, in small to medium sized rivers, vary from those where the range of movement over the prespawning to postspawning period is less than at other times of the year (Pankhurst et al. 2016) to those with linear ranges, over the pre- to postspawning period of 30 km or more (DosSantos et al. 1991), summarized in Table 5.1.

Pike are able to survive, forage and grow successfully in brackish habitats up to about 12-15 ppt (Jacobsen et al. 2017) while successful spawning has been recorded at salinities of up to about 8 ppt, but potentially may occur even at higher salinities (Engström-Öst et al. 2005, Jørgensen et al. 2010, chapter 3). Pike in the Baltic Sea move and forage extensively around archipelagos and more open areas, and a large proportion move to sheltered bays with reed fringes to spawn (Lappalainen et al. 2008), but anadromous (*sensu* Lucas and Baras 2001) migration into coastal streams also occurs widely (Müller and Berg 1982, Müller 1986, Johnsson and Müller 1978). Müller (1986) described migrations of Baltic pike into the River Ängerån in spring, when ice broke, in order to access spawning grounds in areas where the water was warming up earlier than in the sea. Adult pike migrate downstream again from the Ängerån after spawning. Besides this migrating subpopulation, there was also a freshwater population in the river. Through otolith microchemistry studies these alternative life history strategies have been shown to occur among pike stocks living sympatrically within the Baltic (Westin and Limburg 2002). More recently it has been shown that 47% of adult Baltic pike (n = 175) caught on the coast around Kalmar in Southern Sweden were spawned in freshwater, the rest in brackish water (Engstedt et al. 2010). The anadromous pike ascend into small brooks and streams in the area in spring to spawn in inundated grasslands or other wetlands (chapter 10) and soon after spawning they migrate out of the streams again. A very thorough study by Tibblin et al. (2016) recently gave new information on spawning migration behavior of Baltic sea pike as well as rare evidence of how between- and within- individual variation in migratory timing across breeding events are correlated with phenotypic and fitness traits. They sampled > 2000 migrating pike (individually tagged) over six years by fyke netting in the streams when pike migrated from the Baltic Sea to spawn. Environmental conditions when the migration was initiated (average water temperature during the first week of migration) were surprisingly similar between years (4.1°C (1.1 SD)). Moreover, during the spawning migration, which on average lasted 50 days (11.5 SD), males arrived on average earlier than females (6 days). In addition, the arrival time for males was length dependent, i.e., larger males arrived later than smaller males. Furthermore there was a consistency in arrival time between individuals; individuals that arrived early in one year also arrived early the following years, and there was also indication that older individuals benefit from experience as they seem to fine tune arrival time the older they get. Finally, and perhaps most interestingly, Tibblin et al. (2016) show that individuals that are more flexible in their timing of arrival during the first reproductive years survive longer compared with less flexible individuals. The study by Tibblin et al. (2016) clearly inspires further exploration of spawning behavior and ecology of Baltic Sea pike, both those that spawn in freshwater and those that spawn at the coast and in sheltered saline water.

During the spawning period, aquatic vegetation and flooded marshes are important habitats for pike (chapters 3 and 10) but water depth can also be a strong predictor of habitat use. The appropriate water depth is of particular importance for successful reproduction and can be even more influential for spawning than the specific type of substrate (Alldridge and White 1980, Fortin et al. 1983). Shallow water is an optimal habitat in spring due to several reasons. First of all, the water warms up at a faster rate, and early spawners are correspondingly found in such habitats. Also, the shallow vegetated areas afford the newly hatched larvae with appropriate food resources and protection from predators. Preferred water depths for spawning can cover various depth intervals ranging from 5 cm down to 100 cm (e.g. Alldridge and White 1980, Fortin et al. 1983, Nilsson et al. 2014). In the review by Casselman and Lewis (1996) preferred spawning depth was found to be 10-70 cm, with the most used depths to be 20-40 cm. Interestingly, there are also studies that demonstrate that pike may frequent deeper water to spawn. Frost and Kipling (1967) reported that spawning occurred 2-3.5 m above the vegetation in Windermere, northwest England. Likewise, in the St. Lawrence River and some North-American lakes, part of the pike population spawns in deeper water down to 6 m depth above submerged vegetation (Farrell et al. 1996, Farrell et al. 2006, Pierce et al. 2007). The spawning in deeper parts takes place later than the shallow-water spawning, due to slower warming of the water. Pierce et al. (2007) demonstrated that mostly smaller individuals spawn in deeper water between 1.5 and 1.8 m. In some areas the deep water spawning seems to be a result of the disappearance of the habitat in shallow areas; hence a suboptimal choice in several ways. In line with this, Farrell et al. (2006) state that production might be limited in deep-water spawning, i.e., that a deep-water spawning strategy is an ecological sink.

5.5 YOUNG PIKE

5.5.1 First Movements and Activity

So far, this chapter has mostly dealt with adult pike; the remainder focuses on young pike, especially young of the year (YOY). Soon after hatching, young pike become mobile, especially after yolk sac depletion when the pike fry start to disperse (Carbine 1942). However, regular movements occur even before that. Bry and Davigny (2010) showed that pike embryos are mobile already 5-6 days after hatching. Although at this stage they spend most of their time attached to physical structures, they also make vertical movements between attachment locations. The same authors recorded high mobility levels 14–15 days after hatching which probably was related to the repeated visits to the surface where the swim bladder was filled.

As pike fry leave the spawning grounds, lentic pike disperse among the shallow vegetated areas of the lakes. Carbine (1942) observed that pike fry, with the yolk sac almost absorbed, emigrated from the marshes and ditches in which they were spawned to Houghton Lake, Michigan, by day. The first migrated 10 days after hatching, at an average length of 19 mm, and within 30 days (by end of May) 85% had migrated. Most of the fry drifted close to the surface and Carbine (1942) reports that there was greater activity on bright days than cloudy ones.

For pike spawned in shallow wetlands by migrating Baltic pike, emigration of fry from spawning areas back to the sea may start when larvae are only 13-17 mm long (Nilsson et al. 2014, Larsson et al. 2015) and most larvae emigrated at a size < 60 mm. In a study from the Ängerån in the northern Baltic, juvenile pike downstream migration started in June and ended in August at a length of 80 mm. Here, migration took place during daylight hours (Johnson and Müller 1978).

Using hatchery reared pike, Skov et al. (2011) showed that pike fry (30 mm) stocked in restricted areas of a small eutrophic lake (31 ha) dispersed throughout the lake within a few days, illustrating the potential for movement and dispersal even in small juveniles. Within shallow lentic nursery habitat, stocked PIT tagged fry (44-63 mm) tracked with hand-held PIT antennas were found to explore a relatively limited area of nursery habitat, moving on average of about 8 m per day (Cucherousset et al. 2009).

As for larger pike, juveniles may display diel patterns in behavior such as activity. In a pond experiment, Nilsson et al. (2012) showed higher activity of pike (average length 29.3 cm) at dusk than at other times of the day. Also the presence of cannibalistic conspecifics can affect activity rates. One of the mechanisms prey can use to avoid predators is to reduce activity, which reduces encounter rates (Lima and Dill 1990), hence it could be expected that pike at risk of predation would reduce activity rates. Nyqvist et al. (2012) demonstrated that YOY pike exhibited higher activity rates and prey capture rates under control (alone, visual contact to a neighboring empty tank) or competition conditions (visual contact with YOY conspecifics in a neighboring tank) than when in the presence of a potential cannibal (Age 1 pike in a neighboring tank), indicating a vigilance effect and a very localized reduction of activity.

5.5.2 Habitat Use and Habitat Shifts

Normally, pike fry and juveniles distribute themselves mainly in shallow, vegetated lentic areas as a means to optimize growth and survival (Grimm 1981). However, where adult pike migrate into streams and marshes to spawn, the young usually emigrate to avoid desiccation as water levels drop, or in response to increased competition and reduced availability of suitable food as they grow. Hence, water depth may influence the spatial distribution of young pike. Casselman and Lewis (1996) report a relationship between the length of juvenile pike (3-22 cm) and the depth at which they were caught by electrofishing, and suggest that the water depth frequented by pike up to 15 cm increases 10 cm for every 12 mm pike grow. Of course, this depth relationship may have an indirect link to vegetation and predation risk. Exceptions to this pattern have been reported (e.g. Skov and Berg 1999) and Cucherousset et al. (2009) found that PIT tagged pike (average 5.1 cm) aggregated in the deepest part of a nursery area (depths up to 35 cm), especially during periods of decreasing water levels, potentially illustrating how environmental conditions may affect habitat use. The role of water visibility and vegetation and especially the role of different types of vegetation on the life cycle of young pike are thoroughly described and discussed in chapter 3.

Also biotic factors influence habitat use of small pike. Young pike face predation risk from larger cannibalistic conspecifics, and a few studies have illustrated how this may affect their spatial distribution. Nilsson (2006) showed that pike in the vegetated area of a shallow Swedish lake distributed in such a way that they avoided larger potential cannibalistic conspecifics, i.e., the distance of individuals to the nearest cannibal-sized pike was on average over six times greater than the distance to nearest similar-sized pike. In another Swedish lake, pike below 16 cm avoided areas with a high density of perch *Perca fluviatilis*, Linnaeus, 1758 and also a higher than expected density of older pike (Eklöv 1997). The inverse distribution between differently sized conspecifics has also been observed in a pond experiment including YOY pike < 5 cm (Skov and Koed 2004). That study showed that larger cannibalistic individuals took position in the vegetated and sheltered part of the pond, whereas the smaller individuals were constrained to the unsheltered and potentially riskier habitats. Clumped distributions of young pike have been observed in several studies (Nilsson 2006, Cucherousset et al. 2009, Hawkins et al. 2003), but these pike may be of similar size (i.e., as reported in Hawkins et al. 2003) and so perceived risk of the neighboring pike is therefore smaller. As suggested by Hawkins et al. (2003) the occurrence of clumped distributions (rather than a more evenly spaced distribution) indicates that territoriality does not occur for these pike.

Seasonal patterns may also affect the spatial distribution of juvenile pike. Chapman and Mackay (1984b) found that the spatial distribution of young pike differed before and after autumn turnover. Also, in autumn and winter a lot of vegetation often dies and disappears and the pike may be forced to shift habitat as inferred by Grimm and Klinge (1996) and Hawkins et al. (2003).

As for many other fish species, behavioral syndromes (Nyqvist et al. 2012) and dominance hierarchy effects (Hawkins et al. 2005) occur in juvenile pike. Both of these have the capacity for a strong influence on the behavior and microhabitat use of young pike. Hawkins et al.'s (2005) mesocosm experiments on similar-sized (~ 14 cm) YOY pike demonstrated that single fish preferred shallow water habitat to a deeper pool, whereas in pairs or groups of four, a dominant fish used primarily the shallow habitat, while subordinate(s) used the pool. McGhee et al. (2013) explored the foraging behavior of individual juvenile pike (18.2-22.3 cm) in small tanks and also found distinct behavioral types, i.e., some pike consistently attacked their prey faster than others. Interestingly, time to first attack of prey was also positively related to the metabolic rate and eye size of individual pike. Clearly such coupling between foraging behavior, metabolic rate and organ size could have implications for both hunting strategy and habitat use of individual pike (McGhee et al. 2013).

5.5.3 Home Range

Few studies have attempted to establish the home ranges of small pike, and the ones that have make different conclusions. The only study to have telemetered locations of Age 0-1 pike to derive home ranges is that of Rosten et al. (2016) in which standardized summer home ranges (95% Kernel, 39 locations) of river-dwelling pike weighing 7-100 g (n = 9, 10-23 cm), averaged 99 m^2. These fish had free access to many kilometres of river channel, tributaries and ditches, but their home range over several weeks was typified by a 10-30 m strip of slack-water, vegetated river edge habitat. During repeated electrofishing sessions between June and September, Nilsson (2006) tagged and released > 200 smaller pike (the great majority < 20 cm) in the littoral zone of a 3 km^2 Swedish lake. Based on the spatial distribution of recaptures, the study suggests that young pike did not have small stable home ranges, but moved extensively between the vegetated areas in the lake. In contrast, in a 4.5 ha lake in Holland, 0+ pike were sampled and individually tagged in 1977, and 21 individuals (below 32 cm at tagging) were recaptured the following year. Of these fish 5% (one individual) had moved farther than 200 m, 20% (4 individuals) had moved 100-200 m, and 75% had moved less than 100 m (Grimm and Klinge 1996). As the scope for extensive movement of young pike could be more restricted in smaller lakes, the size differences between the two lakes mentioned could play a role in the different results.

5.6 CONCLUSIONS AND PERSPECTIVE

This chapter reviews a large number of studies that have generated a range of frequently conflicting conclusions on patterns in the spatial ecology of pike. Some patterns are clear, such as the strong habitat association of small pike to heavily vegetated areas compared to, for example, the much more variable habitat association of larger pike. It is also clear from this chapter, that a great deal of study of the spatial ecology of pike has been carried out in the last 25 years, especially concentrating on telemetry methods across the range of habitats that pike occupy, to build on the earlier classical observation, trapping and capture-mark-recapture studies. These have been combined alongside dispersal modelling approaches, laboratory mesocosm studies of space use and behavioral syndromes, as well as detailed field studies of phenotypic variation in life history and migratory behavior, supported by population genetic and otolith microchemistry approaches. Although pike species, particularly *E. lucius*, undoubtedly can be highly sedentary and forage in a sit-and-wait style, these studies have shown that they are very flexible in their spatial ecology between and within populations across the diverse range of habitats which they inhabit.

Knowledge of spatial ecology of pike could still be furthered, and we would like to conclude this chapter by highlighting some of these knowledge gaps, such as, for example, the behavior of adult pike in large lakes and their potential use of the pelagic zone as habitat. It is also under-explored how other predatory species like perch, pikeperch *Sander lucioperca,* Linnaeus, 1758 and catfish *Silurus glanis* Linnaeus, 1758 as well as avian predators affect the spatial ecology of pike. Due to the variety in observed behavior in different studies there is also a need to move towards more mechanistic studies of pike behavior in the natural environment. Many studies to date have been carried out from an applied ecological perspective of needing to understand the habitat requirements and spatial behavior of pike with regards to water management (e.g. biomanipulation), fisheries management or habitat conservation needs. However, the pike is frequently used as a model species in ecology and behavior (Forsman et al. 2015), so there are opportunities for mechanistic, controlled studies in field, laboratory and mesocosm environments from a 'basic' research point of view. There are difficulties too. Large, adult pike are highly suited for long-term telemetry studies, but are difficult to study under controlled laboratory or mesocosm conditions (but see Laskowski et al. 2016). Young pike can be tagged with PIT tags as YOY and recorded remotely at fixed PIT stations or tracked manually in restricted habitats, but cannot be tracked for more than a few weeks currently with battery-powered telemetry tags due to short tag life, whereas they can be used effectively in carefully controlled laboratory and mesocosm studies.

There is a need for a greater understanding of the spatial ecology of juvenile pike, especially in environments where spawning and nursery habitats are patchy or sparse, including in anthropogenically modified systems. There is also an obvious gap in our knowledge of the detailed spatial ecology of young and adult pike in brackish water ecosystems, despite the excellent utilization of otolith microchemistry and population genetics in recent years. Understanding the migration behavior and access to freshwater spawning grounds can be important for conservation of these stocks and thereby for management of waterways and artificial barriers to optimize migrations routes. Given that there are often long time series of mark-recapture and fishing effort data for pike, there are opportunities for greater use of spatial modelling approaches of the type used by Haugen et al. (2006) to test for changes in response to, for example, species invasion and changes in key habitat availability over time, although ecosystems in which pike occur are often subject to multiple changes simultaneously. Greater use could also be made of combining multiple methods, such as telemetry and underwater video to generate better insight of young pike as prey both for cannibals and also for other top predator species. Combination of stable isotope analysis of biopsied and telemetry tagged pike also offers an approach towards understanding alternative foraging tactics including littoral/limnetic, fish/non-fish components.

REFERENCES CITED

Alldridge, N.A. and A.M. White. 1980. Spawning site preferences for northern pike (*Esox lucius*) in a New York marsh with widely fluctuating water levels. *Ohio J. Sci.* 80:57-73.

Andersen, M., L. Jacobsen, P. Grønkjær and C. Skov. 2008. Turbidity increases behavioural diversity in northern pike, *Esox lucius* L., during early summer. *Fish. Manag. Ecol.* 15:377-383. 10.1111/j.1365-2400.2008.00635.x

Arlinghaus, R., T. Klefoth, A.J. Gingerich, M.R. Donaldson, K.C. Hanson and S.J. Cooke. 2008. Behaviour and survival of pike, *Esox lucius*, with a retained lure in the lower jaw. *Fish. Manag. Ecol.* 15:459–466, http://dx.doi.org/10.1111/j. 1365-2400.2008.00625.x.

Armstrong, J.D., I.G. Priede and M.C. Lucas 1992. The link between respiratory capacity and changing metabolic demands during growth of northern pike, *Esox lucius* L. *J. Fish Biol. (Supplement B)* 41:65-75.

Baktoft, H. 2012. Aspects of lentic fish behaviour studied with high resolution positional telemetry. PhD Thesis. Technical University of Denmark, Kgs. Lyngby, Denmark.

Baktoft, H., K. Aarestrup, S. Berg, M. Boel, L. Jacobsen, N. Jepsen, A. Koed, J.C. Svendsen and C. Skov. 2012. Seasonal and diel effects on the activity of northern pike studied by high-resolution positional telemetry. *Ecol. Freshw. Fish* 21:386-394. 10.1111/j.1600-0633.2012.00558.x

Baktoft, H., K. Aarestrup, S. Berg, M. Boel, L. Jacobsen, A. Koed, M.W. Pedersen, J.C. Svendsen and C. Skov. 2013. Effects of angling and manual handling on pike behaviour investigated by high-resolution positional telemetry. *Fish. Manag. Ecol.* 20:518-525. 10.1111/fme.12040

Beaumont, W.R.C., K.H. Hodder, J.E.G. Masters, L.J. Scott and J.S. Welton. 2005. Activity patterns in pike (*Esox lucius*), as determined by motion sensing telemetry. pp 231-243. *In*: M.T. Spedicato, G. Lembo and G. Marmulla (eds.). *Aquatic telemetry: advances and applications*. Rome: FAO/COISPA.

Berggren, H., O. Nordahl, P. Tibblin, P. Larsson, and A. Forsman. 2016. Testing for local adaptation to spawning habitat in sympatric subpopulations of pike by reciprocal translocation of embryos. *PLOS ONE* DOI: 10.1371/journal.pone.0154488.

Börger, L., B.D. Dalziel and J.M. Fryxell. 2008. Are there general mechanisms of animal home range behaviour? A review and prospects for future research. *Ecol. Lett.* 11:637-650.

Brown, J.S. 1988. Patch use as an indicator of habitat preference, predation risk and competition. *Behav. Ecol. Sociobiol.* 23:27-43.

Brönmark, C., K. Hulthén, P.A. Nilsson, C. Skov, L.-A. Hansson, J. Brodersen and B. Chapman. 2014. There and back again: migration in freshwater fishes. *Can. J. Zool.* 92:467-479. doi:10.1139/cjz-2012-0277

Brosse, S., G.D. Grossman and S. Lek. 2007. Fish assemblage patterns in the littoral zone of a European reservoir. *Freshwat. Biol.* 52:448-458. doi:10.1111/j.1365-2427.2006.01704.x

Bry, C. 1996. Role of vegetation in the life cycle of pike. pp 45-67. *In*: J.F. Craig (ed.). *Pike: Biology and exploitation*. London: Chapman & Hall.

Bry, C. and S. Davigny. 2010. Temporal changes in vertical distribution, body orientation, and mobility of pike (*Esox lucius* L.) free embryos: individual observations under controlled conditions. *Knowl. Manag. Aquat. Ec.* 399:03.

Burkholder, A. and D.R. Bernard. 1994. Movements and distributions of radio-tagged northern pike in Minto Flats. Alaska Department of Fish and Game, Division of Sport Fish. Achorage, Alaska.

Burt, W.H. 1943. Territoriality and home range concepts as applied to mammals. *J. Mammal.* 24:346-352.

Carbine, W.F. 1942. Observations on the life history of the northern pike, *Esox lucius* L., in Houghton Lake, Michigan. *Trans. Am. Fish. Soc.* 71:149-164.

Carbine, W.F. and V.C. Applegate. 1946. The movement and growth of marked northern pike (*Esox lucius* L.) in Houghton Lake and the Muskegon River. *Pap. Mich. Acad. Sci. Arts Lett.* 32:215–238.

Carere, C. and D. Maestripieri. 2013. Introduction: Animal Personalities: Who Cares and Why? pp 1-12. *In*: C. Carere and D. Maestripieri (eds.). *Animal personalities: Behaviour, physiology and evolution.* Chicago: The University of Chicago Press.

Casselman, J.M. 1978. Effects of environmental factors on growth, survival, activity and exploitation of northern pike. *Am. Fish. Soc. spec. Publ.* 11:114-128.

Casselman, J.M. 1996. Age, growth and environmental requirements of pike. pp 69-102. *In*: J.F. Craig (ed.). *Pike: Biology and exploitation.* London: Chapman & Hall.

Casselman, J.M. and C.A. Lewis. 1996. Habitat requirements of northern pike (*Esox lucius* L.). *Can. J. Fish. Aquat. Sci.* 53:161-174.

Chapman, C.A. and W.C. Mackay. 1984a. Versatility in habitat use by a top aquatic predator *Esox lucius* L.. *J. Fish Biol.* 25:109-115.

Chapman, C.A. and W.C. Mackay. 1984b. Direct observations of habitat utilization by northern pike. *Copeia* 1984:225-258.

Claireaux, G. and D. Chabot. 2016. Responses by fishes to environmental hypoxia: integration through Fry's concept of aerobic metabolic scope. *J. Fish Biol.* 88:232-251.

Clark, C.F. 1950. Observations on the spawning habits of the northern pike, *Esox lucius*, in north-western Ohio. *Copeia* 4:285-288.

Cook, M.F. and E.P. Bergersen. 1988. Movements, habitat selection, and activity periods of northern pike in Eleven-mile Reservoir, Colorado. *Trans. Am. Fish. Soc.* 117:495-502.

Cooke, S.J., S. Hinch, M.C. Lucas and M. Lutcavage. 2012. Biotelemetry and biologging. pp. 819-881. *In*: A. Zale, D. Parrish and T. Sutton (eds.). *Fisheries techniques 3rd ed.* Bethesda, Maryland: American Fisheries Society.

Cucherousset, J., J.-M. Paillisson, A. Cuzol and J.-M. Roussel. 2009. Spatial behaviour of young-of-the-year northern pike (*Esox lucius*) in a temporarily flooded nursery area. *Ecol. Freshw. Fish* 18:314-322.

Diana J.S., W.C. Mackay and M.Ehrman. 1977. Movements and habitat preferences of northern pike *Esox lucius* in Lac St Anne, Alberta. *Trans. Am. Fish. Soc.* 106:560-565.

Diana, J.S. 1980. Diel activity patterns and swimming speeds of northern pike (*Esox lucius*) in Lac Ste. Anne, Alberta. *Can. J. Fish. Aquat. Sci.* 37:1454-1458.

Dingle, H. 1996. *Migration: The Biology of Life on the Move.* Oxford University Press, New York, USA.

Dobler, E. 1977. Correlation between the feeding time of the pike (*Esox lucius*) and the dispersion of a school of *Leucaspius delineatus*. *Oecologia* 27:93-96.

DosSantos J.M. 1991. Ecology of a riverine pike population. pp. 155-159. *In*: J.L. Cooper and R.H. Hamre (eds). *Proceedings of warmwater fisheries symposium I.* Arizona: US Forest Service General Technical Report RM-207.

Eklöv, P. 1997. Effects of habitat complexity and prey abundance on the spatial and temporal distributions of perch (*Perca fluviatilis*) and pike (*Esox lucius*). *Can. J. Fish. Aquat. Sci.* 54:1520-1531.

Engstedt, O., P. Stenroth, P. Larson, L. Ljunggren and M. Elfman. 2010. Assessment of natal origin of pike (*Esox lucius*) in the Baltic Sea using Sr:Ca in otoliths. *Environ. Biol. Fishes* 89:547-555.

Engstedt, O. 2011. *Anadromous pike in the Baltic Sea*. PhD Thesis, Linnaeus University, Kalmar, Sweden.

Engstedt, O., R. Engkvist and P. Larsson. 2014. Elemental fingerprinting in otoliths reveals natal homing of anadromous Baltic Sea pike (*Esox lucius*). *Ecol. Freshw. Fish* 23:313-321.

Engström-Öst, J., M. Lehtiniemi, S.H. Jónasdóttir and M. Viitasalo. 2005. Growth of pike larvae (*Esox lucius*) under different conditions of food quality and salinity. *Ecol. Freshw. Fish* 14:385-393.

Farrell, J.M., R.G. Werner, S.R. LaPan and K.A. Claypoole. 1996. Egg distribution and spawning habitat of northern pike and muskellunge in a St. Lawrence River marsh, New York. *Trans. Am. Fish. Soc.* 125:127-131.

Farrell, J.M., J.V. Mead and B.A. Murry. 2006. Protracted spawning of St. Lawrence River northern pike (*Esox lucius*): simulated effects on survival, growth, and production. *Ecol. Freshw. Fish* 15:169-179.

Forsman, A., P. Tibblin, H Berggren, O. Nordahl, P. Koch-Schmidt and P. Larsson. 2015. Pike *Esox lucius* as an emerging model organism for studies in ecology and evolutionary biology: a review. *J. Fish Biol.* 87:472-479.

Fortin, R., P. Dumont and H. Fournier. 1983. La reproduction du grand brochet (*Esox lucius* L.) dans certains plains d'eau du sud de Québec (Canada). pp 39-51. *In*: R. Billard (ed.). *Le Brochet: gestion dans le milieu naturel et élevage*. France: INRA Publ. Paris. (in French)

Franklin, D.R. and L.L.Jr. Smith. 1963. Early life history of the northern pike, *Esox lucius* L., with special reference to the factors influencing the numerical strength of year classes. *Trans. Am. Fish. Soc.* 92:91-110.

Fretwell, D.S. and H.L. Lucas. 1970. On territorial behavior and other factors influencing habitat distribution in birds. *Acta Biotheor.* 19:16-32.

Frost, W.E. and C. Kipling. 1967. A study of reproduction, early life, weight-length relationship and growth of pike, *Esox lucius*, in Windermere. *J. Anim. Ecol.* 36:651-693.

Fry, F.E. 1971. The effect of environmental factors on the physiology of fish. pp 1-98. *In*: W.S. Hoar and D.J. Randall (eds.). *Fish Physiology Vol. 6*. New York: Academic Press.

Gerlier, M. and J.-F. Luquet. 2000. Preliminary study on the spawning migration of pike (*Esox lucius*) in the River Ill, a tributary of the Rhine. pp. 129-136. *In*: A. Moore and I. Russell (eds.). *Advances in Fish Telemetry, Proceedings of the 3rd Conference on Fish Telemetry in Europe*. Lowestoft, UK: CEFAS.

Godin, J.-G.J. and M.H.A. Keenleyside. 1984. Foraging on patchily distributed prey by a cichlid fish (Teleostei, Cichlidae): a test of the ideal free distribution theory. *Anim. Behav.* 32:1201-1213.

Greenwood, P.J. 1980. Mating systems, philopatry and dispersal in birds and mammals. *Anim. Behav.* 28:1140-1162.

Grimm, M.P. 1981. The composition of northern pike (*Esox lucius* L.) populations in four shallow waters in the Netherlands, with special reference to factors influencing 0+ pike biomass. *Fish. Manage.* 12:61-76.

Grimm, M.P. and M. Klinge. 1996. Pike and some aspects of its dependence on vegetation. pp 125-156. *In*: J.F. Craig (ed.). *Pike: Biology and exploitation*. London: Chapman & Hall.

Halme, E. and M. Korhonen. 1960. On the migrations of pike in our coastal areas. *Kalamies* 4:1-12. (in Finnish).

Haugen, T.O., I.J. Winfield, L.A. Vøllestad, J.M. Fletcher, J.B. James and N.C. Stenseth. 2006. The ideal free pike: 50 years of fitness-maximizing dispersal in Windermere. *Proc. R. Soc. B-Biol. Sci.* 273:2917-2924.

Haugen, T.O., I.J. Winfield, L.A. Vøllestad, J.M. Fletcher, J.B. James and N.C. Stenseth. 2007. Density dependence and density independence in the demography and dispersal of pike over four decades. *Ecol. Monogr.* 77:483-502.

Hawkins, L.A., J.D. Armstrong and A.E. Magurran. 2003. Settlement and habitat use by juvenile pike in early winter. *J. Fish Biol.* 63:174-186.

Hawkins, L.A., A.E. Magurran and J.D. Armstrong. 2005. Aggregation in juvenile pike (*Esox lucius*): interactions between habitat and density in early winter. *Funct. Ecol.* 19:794-799.

Headrick M.R. and R.F. Carline. 1993. Restricted summer habitat and growth of northern pike in two southern Ohio impoundments. *Trans. Am. Fish. Soc.* 122:228-36.

Hladík, M. and J. Kubečka. 2003. Fish migration between a temperate reservoir and its main tributary. *Hydrobiologia* 504:251-266.

Hodder, K.H., J.E.G. Masters, W.R.C. Beaumont, R.E. Gozlan, A.C. Pinder, C.M. Knight and R.E. Kenward. 2007. Techniques for evaluating the spatial behaviour of river fish. *Hydrobiologia* 582:257-269.

Højrup, L.B. 2015. *Pike (Esox lucius) in River Tryggevælde - focusing on population structure, habitat choice and movements.* Master Thesis, University of Copenhagen, Copenhagen, Denmark.

Jones, D.R., J.W. Kicenuik, and O.S. Bamford. 1974. Evaluation of the swimming performance of several species from the Mackenzie River. *J. Fish. Res. Board Can.* 31:1641-1647.

Jacobsen, L., H. Baktoft, N. Jepsen, K. Aarestrup, S. Berg and C. Skov. 2014. Effect of boat noise and angling on lake fish behaviour. *J. Fish Biol.* 84:1768-1780. 10.1111/jfb.12395.

Jacobsen, L., D. Bekkevold, S. Berg, N. Jepsen, A. Koed, Aarestrup, K., H. Baktoft and C. Skov, 2017. Pike (*Esox lucius* L.) on the edge – consistent individual movement patterns in transitional waters of the western Baltic. *Hydrobiologia* 784, 143-154, DOI 10.1007/s10750-016-2863-y.

Jepsen, N., S. Beck, C. Skov and A. Koed. 2001. Behaviour of pike (*Esox lucius* L.) >50 cm in a turbid reservoir and in a clearwater lake. *Ecol. Freshw. Fish* 10:26-34.

Jetz, W., C. Carbone, J.H. Fulford and J. Brown. 2004. The scaling of animal space use. *Science* 306:266-268.

Johnson, T. and K. Müller. 1978. Migration of juvenile pike, *Esox lucius* L., from a coastal stream to the Northern part of the Bothnian Sea. Aquilo, *Ser. Zool* 18:57-61.

Jones, D.R., J.W. Kiceniuk and O.S. Bamford. 1974. Evaluation of the swimming performance of several fish species from the Mackenzie River. *J. Fish. Res. Board Can.* 31:1641-1647.

Jørgensen, A.T., B.W. Hansen, B. Visman, L. Jacobsen, C. Skov, S. Berg and D. Bekkevold. 2010. High salinity tolerance in eggs and fry of a brackish *Esox lucius* population. *Fish. Manag. Ecol.* 17:554-560.

Karås P. and H. Lehtonen. 1993. Patterns of movement and migration of pike (*Esox lucius* L.) in the Baltic Sea. Nord. *J. Freshw. Res.* 68:72-79.

Klefoth, T., A. Kobler and R. Arlinghaus. 2008. The impact of catch-and-release angling on short-term behavoiur and habitat choice of northern pike (*Esox lucius* L.). *Hydrobiologia* 601:99-110.

Klefoth, T., A. Kobler and R. Arlinghaus. 2011. Behavioural and fitness consequences of direct and indirect non-lethal disturbances in a catch-and-release northern pike (*Esox lucius* L.) fishery. *Knowl. Manag. Aquat. Ec.* 401:11. DOI: 10.1051/kmae/2011072.

Kipling, C. and E.D. LeCren. 1984. Mark recapture experiments on pike in Windermere, 1943-1982. *J. Fish Biol.* 24:395-414.

Knight, C.M. 2006. *Utilisation of off-river habitats by lowland river fishes.* PhD thesis, University of Durham, Durham, UK.

Knight, C.M., R.E. Gozlan and M.C. Lucas. 2008. Can seasonal home-range size in pike *Esox lucius* predict excursion distance? *J. Fish Biol.* 73:1058-1064.

Knight, C.M., R.E. Kenward, R.E. Gozlan, K.H. Hodder, S.S. Walls and M.C. Lucas 2009. Home range estimation within complex restricted environments: importance of method selection in detecting seasonal change. *Wildlife Res.* 36:213-224.

Kobler A., T. Klefoth, C. Wolter, F. Fredrich and R. Arlinghaus. 2008a. Contrasting pike (*Esox lucius* L.) movement and habitat choice between summer and winter in a small lake. *Hydrobiologia* 601:17-27.

Kobler A., T. Klefoth and R. Arlinghaus. 2008b. Side-fidelity and seasonal changes in activity centre size of female pike *Esox lucius* L. in a small lake. *J. Fish Biol.* 73:1-13.

Kobler, A., T. Klefoth, T. Mehner and R. Arlinghaus. 2009. Coexistence of behavioural types in an aquatic top predator: a response to resource limitation? *Oecologia* 161:837-847

Koed A., K. Balleby, P. Mejlhede and K. Aarestrup. 2006. Annual movement of adult pike (*Esox lucius* L.) in a lowland river. *Ecol. Fresh. Fish* 15:191-199.

Krebs, J.R. and N.B. Davies (eds.). 1997. *Behavioural Ecology: An Evolutionary Approach.* Blackwell Science Ltd, Oxford, UK.

Lamouroux, R.E., H. Capra, M. Pouilly and Y. Souchon. 1999. Fish habitat preferences in large streams of southern France. *Freshw. Biol.* 42:673-687.

Langford, T.E. 1981. The movement and distribution of sonic-tagged coarse fish in two British rivers in relation to power station cooling-water outfalls. pp. 197-232. *In:* F.M. Long (ed.). *Proceedings of the 3rd International Conference on Wildlife Biotelemetry.* Wyoming, USA: University of Wyoming.

Larsson P., P. Tibblin, P. Koch-Schmidt , O. Engstedt, J. Nilsson, O. Nordahl, and A. Forsman. 2015. Ecology, evolution, and management strategies of northern pike populations in the Baltic Sea. *AMBIO* 44:451-461.

Lappalainen, A., M. Härmä, S. Kuningas and L. Urho. 2008. Reproduction of pike (*Esox lucius*) in reed belt shores of the SW coast of Finland, Baltic Sea: a new survey approach. *Boreal Environ. Res.* 13:370-380.

Laskowski, K.L., C.T. Monk, G. Polverino, J. Alos, S. Nakayama, G. Staaks, T. Mehner and R. Arlinghaus. 2016. Behaviour in a standardized assay, but not metabolic or growth rate, predicts behavioural variation in an adult aquatic top predator *Esox lucius* in the wild. *J. Fish. Biol.* 88 (4):1544-1563.

Lima, S.L. and L.M. Dill. 1990. Behavioural decisions made under the risk of predation: a review and prospectus. *Can. J. Zool.* 68:619-640.

Lucas, M.C. 1989. *Metabolic power budgeting in fishes: laboratory studies in zebra fish, Brachydanio rerio and heart rate telemetry in pike, Esox lucius.* PhD thesis, University of Aberdeen, Aberdeen, UK.

Lucas, M.C., I.G. Priede, J.D. Armstrong, A.N.Z. Gindy and L. De Vera. 1991. Direct measurements of metabolism, feeding and activity by heart rate telemetry in wild pike, *Esox lucius* L. *J. Fish Biol.* 39:325-345.

Lucas, M.C. 1992. Spawning activity of male and female pike, *Esox lucius* L., determined by acoustic tracking. *Can. J. Zool.* 70:191-196.

Lucas, M.C., T. Mercer, S. McGinty and J.D. Armstrong. 2000. Development and evaluation of a flat-bed passive integrated transponder detector system for recording movement of lowland-river fishes through a baffled fish pass. pp. 117-127. *In:* A. Moore and I. Russell (eds.). *Advances in fish telemetry.* Lowestoft: CEFAS.

Lucas, M.C. and E. Baras 2001. *Migration of Freshwater Fishes.* Blackwell Science Ltd., Oxford, UK.

Mann, R.H.K. 1980. The numbers and production of pike (*Esox lucius*) in two Dorset rivers. *J. Anim. Ecol.* 49:889-915.

Masters, J.E.G., J.S. Welton, W.R.C. Beaumont, K.H. Hodder, A.C. Pinder, R.E. Gozlan and M. Ladle. 2002. Habitat utilization by pike *Esox lucius* L. during winter floods in a southern English chalk river. *Hydrobiologia* 483, 185-191.

Masters, J.E.G., K.H. Hodder, W.R.C. Beaumont, R.E. Gozlan, A.C. Pinder, R.E. Kenward and J.S. Welton. 2005. Spatial behaviour of pike *Esox lucius* L. in the River Frome, UK. pp 179-190. *In*: M.T. Spedicato, G. Lembo and G. Marmulla (eds.). *Aquatic telemetry: Advances and applications*. Rome: FAO/COISPA.

Matthysen, E. 2012. Multicausality of dispersal: a review. pp 3-18. *In:* J. Colbert, M. Baguette, T. Benton and J.M. Bullock (eds.). *Dispersal ecology and evolution.* Oxford: Oxford University Press.

McCarraher, D.B. and R.E. Thomas. 1972. Ecological significance of vegetation to northern pike (*Esox lucius*) spawning. *Trans. Am. Fish. Soc.* 101:81-95.

McGhee, K.E., L.M. Pintor, and A.M. Bell. 2013. Reciprocal behavioral plasticity and behavioral types during predator-prey interactions. *Am. Nat.* 182:704-717.

McNab, B.K. 1963. Bioenergetics and the determination of home range size. *Am. Nat.* 97:133-140.

Midwood, J.D. and P. Chow-Fraser. 2015. Connecting coastal marshes using movements of resident and migratory fishes. *Wetlands* 35:69-79. doi:10.1007/s13157-014-0593-3

Miller R.B. 1948. A note on the movement of the pike, *Esox lucius. Copeia* 1948 (1):62.

Miller, L.M., L. Kallemeyn and W. Senanan. 2001. Spawning-site and natal site fidelity by northern pike in a large lake: mark recapture and genetic evidence. *Trans. Am. Fish. Soc.* 130:307-316.

Moen T. and D. Henegar. 1971. Movement and recovery of tagged Northern pike in Lake Oahe, South and North Dakota, 1964–1968. *Am. Fish. Soc. Spec. Pub.* 8:85-93.

Müller, K. and E. Berg. 1982. Spring migration of some anadromous freshwater fish species in the northern Bothnian Sea. *Hydrobiologia* 96:161-168.

Müller, K. 1986. Seasonal anadromous migration of the pike (*Esox lucius* L.) in coastal areas of the northern Bothnian Sea. *Arch. Hydrobiologia* 107:315-330.

Nilsson, J., O. Engstedt and P. Larsson. 2014. Wetlands for northern pike (*Esox lucius* L.) recruitment in the Baltic Sea. *Hydrobiologia* 721:145-154.

Nilsson, P.A. 2006. Avoid your neighbours: size-determined spatial distribution patterns among northern pike individuals. *Oikos* 113:251-258.

Nilsson, P.A., H. Baktoft, M. Boel, K. Meier, L. Jacobsen, E.M. Rokkjær, T. Clausen, and C. Skov. 2012. Visibility conditions and diel period affect small-scale spatio-temporal behaviour of pike *Esox lucius* in the absence of prey and conspecifics. *J. Fish Biol.* 80:2384-2389. 10.1111/j.1095-8649.2012.03284.x

Northcote, T.G. 1978. Migratory strategies and production in freshwater fishes. pp 326-359. *In:* S.D. Gerking (ed.). *Ecology of freshwater production.* Oxford U.K.: Blackwell Scientific Publications.

Nyqvist, M.J., R.E. Gozlan, J. Cucherousset and R. Britton. 2012. Behavioural syndrome in a solitary predator is independent of body size and growth rate. *PLoS ONE* 7:e31619.

Ovidio, M. and J.C. Philippart. 2002. The impact of small physical obstacles on upstream movements of six species of fish. *Hydrobiologia* 483:55-69.

Ovidio, M. and J.C. Philippart. 2005. Long range seasonal movements of northern pike (*Esox lucius* L.) in the barbel zone of the River Ourthe (River Meuse basin, Belgium). pp 191-202. *In*: M.T. Spedicato, G. Lembo and G. Marmulla (eds.). *Aquatic telemetry: Advances and applications*. Rome: FAO/COISPA..

Pankhurst, K., J.D. Midwood, H. Wachelka and S.J. Cooke. 2016. Comparative spatial ecology of sympatric adult muskellunge and northern pike during a one year period in an urban reach of the River Rideau, Canada. *Environ. Biol. Fish.* 99:409-421.

Pauwels, I.S., P.L.M. Goethals, J. Coeck and A.M. Mouton. 2014. Movement patterns of adult pike (*Esox lucius* L.) in a Belgian lowland river. *Ecol. Freshw. Fish* 23:373-382.

Pierce, R.B., J.A. Younk and C.M. Tomcko. 2007. Expulsion of miniature radio transmitters along with eggs of muskellunge and northern pike – a new method for locating critical spawning habitat. *Environ. Biol. Fishes.* 79:99-109.

Pierce R.B., A.J. Carlson, B.M. Carlson, D. Hudson and D.F. Staples. 2013. Depths and thermal habitat used by large versus small northern pike in three Minnesota Lakes. *Trans. Am. Fish. Soc.* 142:1629-1639.

Pitcher, T.J. and J.R. Turner. 1986. Danger at dawn: experimental support for the twilight hypothesis in shoaling minnows. *J. Fish Biol.* 29(Suppl. A):59-70.

Prignon, C., J. Micha and A. Gillet. 1998. Biological and environmental characteristics of fish passage at the Tailfer dam on the Meuse River, Belgium. pp 69-84. *In*: M. Jungwirth, S. Schmutz, and S. Weiss (eds.). *Fish migration and fish bypasses*, Oxford: Fishing News Books, Blackwell Science Publications.

Pullen, C.E., K. Hayes, C.M. O'Connor, R. Arlinghaus, C.D. Suski, J.D. Midwood and S.J. Cooke. 2017. Consequences of oral lure retention on the physiology and behaviour of adult northern pike (*Esox lucius* L.). *Fish. Res.* 186:601-611. http://dx.doi.org/10.1016/j.fishres. 2016.03.026

Rohtla, M., M. Vetemaa, R. Svirgsden, I. Taal, L. Saks, M. Kesler, A. Verliin and T. Saat. 2014. Using otolith 87Sr:86Sr as a natal chemical tag in the progeny of anadromous Baltic Sea pike (*Esox lucius*) – a pilot study. *Boreal Environ. Res.* 19:379-386.

Rosell, R. S. and K. C. MacOscar. 2002. Movements of pike, *Esox lucius*, in Lower Lough Erne, determined by mark recapture between 1994 and 2000. *Fish. Manag. Ecol.* 9:189-196.

Rosten, C.M., R.E. Gozlan and M.C. Lucas. 2016. Allometric scaling of intraspecific space use. *Biol. Lett.* 12:20150673.

Sandlund, O.T., J. Museth, and S. Øistad. 2016. Migration, growth patterns, and diet of pike (*Esox lucius*) in a river reservoir and its inflowing river. *Fish. Res.* 173: 53-60.

Schnedler-Meyer, N.A. 2014. *Impacts of intraspecific predation risk on behavior across adult pike size classes*. Master Thesis, DTU Aqua, Charlottenlund, Denmark.

Scott, D.M., M.C. Lucas, and R.W. Wilson. 2005. The effect of high pH on ion balance, nitrogen excretion and behaviour in freshwater fish from an eutrophic lake: a laboratory and field study. *Aquat. Toxicol.* 73:31-43.

Simms, L.D. 2000. *Intraspecific variation in the metabolism of juvenile Atlantic salmon Salmo salar and northern pike Esox lucius*. PhD thesis, University of Durham, Durham, UK.

Skov, C. and S. Berg. 1999. Utilization of natural and artificial habitats by YOY pike in a biomanipulated lake. *Hydrobiologia* 408/409:15-122.

Skov, C. and A. Koed. 2004. Habitat use of 0+ year pike in experimental ponds in relation to cannibalism, zooplankton, water transparency and habitat complexity. *J. Fish. Biol.* 64:448-459.

Skov, C., J. Brodersen, P.A. Nilsson, L.-A. Hansson and C. Brönmark. 2008. Inter- and size- specific patterns of fish seasonal migration between a shallow lake and its streams. *Ecol. Freshw. Fish* 17:406-415.

Skov, C., A. Koed, L. Baastrup-Spohr and R. Arlinghaus. 2011. dispersal, growth, and diet of stocked and wild northern pike fry in a shallow natural lake, with implications for the management of stocking programs. *N. Am. J. Fish. Mgmt.* 31(6):1177-1186.

Spens, J., G. Englund and H. Lundqvist. 2007. Network connectivity and dispersal barriers: using geographical information system (GIS) tools to predict landscape scale distribution of a key predator (*Esox lucius*) among lakes. *J. Appl. Ecol.* 44:1127-1137.

Stott, B. 1967. The movements and population densities of roach (*Rutilis rutilus* [L.]) and gudgeon (*Gobio gobio* [L.]) in the River Mole. *J. Anim. Ecol.* 36:407-423.

Switzer, P.V. 1993. Site fidelity in predictable and unpredictable habitats. *Evol. Ecol.* 7:553-555.

Tibblin, P., A. Forsman, P. Koch-Schmidt, O. Nordahl, P. Johannessen, J. Nilsson, and P. Larsson. 2015. Evolutionary divergence of adult body size and juvenile growth in sympatric populations of a top predator in aquatic ecosystems. *Am. Nat.* 186:98-110.

Tibblin, P., A. Forsman, T. Borger and P. Larsson. 2016. Causes and consequenses of repeatability, flexibility and individual fine-tuning of migratory timing in pike. *J. Anim. Ecol.* 85:136-145.

Vehanen, T., P. Hyvärinen, K. Johansson, and T. Laaksonen. 2006. Patterns of movement of adult northern pike (*Esox lucius* L.) in a regulated river. *Ecol. Freshw. Fish* 15:154-160.

Vøllestad, L.A., J. Skurdal and T. Qvenild. 1986. Habitat use, growth, and feeding of pike (*Esox lucius* L.) in four Norwegian lakes. *Arch. Hydrobiol.* 108:107-117.

Volkova, L.A. 1973. The effect of light intensity on the availability of food organisms to some fishes in Lake Baikal. *J. Ichthyol.* 13:591-602.

Webb, P.W. and J.M. Skadsen. 1980. Strike tactics of *Esox*. *Can J. Zool.* 58:1462-1469.

Werner, E.E. and J.F. Gilliam. 1984. The ontogenetic niche and species interactions in size structured populations. *Ann. Rev. Ecol. Syst.* 15:393-425.

Westin, L. and K.E. Limburg. 2002. Newly discovered reproductive isolation reveals sympatric populations of *Esox lucius* in the Baltic. *J. Fish Biol.* 61:1647-1652.

Section II

Populations and Communities

Chapter 6

Pike Population Size and Structure: Influence of Density-Dependent and Density-Independent Factors

Thrond O. Haugen[*1] *and L. Asbjørn Vøllestad*[2]

6.1 INTRODUCTION

Demographic population structure and composition in northern pike (*Esox lucius* Linnaeus, 1758, hereafter pike) intimately link to individual-level processes (Section I of this book). Because individual performance links to both abiotic factors, such as temperature, turbidity and habitat structures (vegetation), and to biotic factors such as zooplankton, fish prey species and predator density, demographic population structure is often affected by complex interactions among these factors. In addition to a range of natural factors acting on pike populations, anthropogenic factors may also substantially modify the population structure by for instance imposing size-biased harvesting or changing food availability (Edeline et al. 2007, Arlinghaus et al. 2010, Crane et al. 2015). Other chapters will deal with fisheries-induced effects on pike (chapters 12 and 13), but in this chapter, we will utilize information from manipulations of environmental factors for understanding environmental effects on pike population processes. Because pike may inhabit diverse freshwater systems ranging from small, homogenous ponds to large and spatially complex habitats, population sub-structuring into genetically divergent and locally adapted subpopulations may occur even at small spatial scales (e.g. Bekkevold et al. 2015, Larsson et al. 2015, Tibblin et al. 2015). These genetic structuring processes are

**Corresponding author*: [1]Faculty of Environmental Sciences and Natural Resource Management, Norwegian University of Life Sciences, PO Box 5003, NO-1432 Ås, Norway; Email: thrond.haugen@nmbu.no

[2]Centre for Ecological and Evolutionary Synthesis (CEES), Department of Biosciences, University of Oslo, P.O. Box 1066 Blindern, 0316 Oslo, Norway; Email: asbjorn.vollestad@mn.uio.no

somewhat in contrast to previous findings (e.g. Laikre et al. 2005) and will be covered in chapter 7, and are therefore not addressed further in this chapter.

Here, we address the roles of abiotic and biotic factors, and their interactions, in shaping pike population structure. We start by exploring how environmental factors affect the recruitment process, i.e., the annual supply of new individuals to the population, and move on to exploring how these effects propagate to later post-recruit life stages. In particular, we discuss how environmental factors affect mortality and individual growth.

6.2 Pike Recruitment

In pike, recruitment, i.e., reproduction and thus production of a new cohort, takes place annually during spring (Craig 2008). Recruitment is measured and used differently in different studies and contexts, and has to be defined in each instance. In this section, we will mostly explore the pike recruitment processes using young-of-year (YOY) data. However, when YOY data/studies are not available, studies using up to age-3 individuals as recruits, may be used instead to elucidate important mechanisms. Details on the spawning habitat and behavior are found in chapter 3 where the critical role of vegetation as spawning habitat is underscored as is the importance of access to safe shallow areas that warm up rapidly favoring rapid larval development (e.g. Raat 1988). Loss and degradation of pike spawning and nursery habitats has since long been acknowledged as the main reason for population declines within the pike's natural distribution area (Casselman and Lewis 1996, Minns et al. 1996, and chapter 10). During the larval stage, the strength of a given year class is determined by a large list of biotic and abiotic factors that may act alone, in concert, or they may interact (Fig. 6.1). Therefore, making predictions of year-class strength becomes a challenging, yet important research task within the field of pike population biology. For example, because YOY survival of pike is size-dependent (e.g. Grønkjær et al. 2004), all environmental factors affecting growth rate and absolute size attained in the first year of life can affect recruitment (Pagel et al. 2015). Hence, when exploring drivers of the recruitment process in pike, external environmental factors in addition to aspects of total population fecundity (stock-recruitment relationships) seems like a good place to start.

Inter-annual variation in pike recruitment has repeatedly been shown to vary less than in most other sympatric populations of freshwater fish species (e.g. Mills and Hurley 1990, van Kooten et al. 2010), suggesting that density-dependent mechanisms (in particular density-dependent juvenile survival) largely control pike recruitment (Mills and Mann 1985, Walrath et al. 2015). Example systems for this low inter-annual variation in pike recruitment can be found in the north-western USA lake Coeur d'Alene (Walrath et al. 2015) and in the UK lake Windermere (Mills and Hurley 1990, Paxton et al. 2009). Despite strong indications of density-dependent mechanisms controlling pike recruitment, an evident stock-recruitment (SR) relationship has proven hard to disentangle. Using catch per unit of effort (CPUE) data as an index of recruitment, it was early concluded that recruitment in Windermere pike is chiefly determined by ambient first growth-season temperature

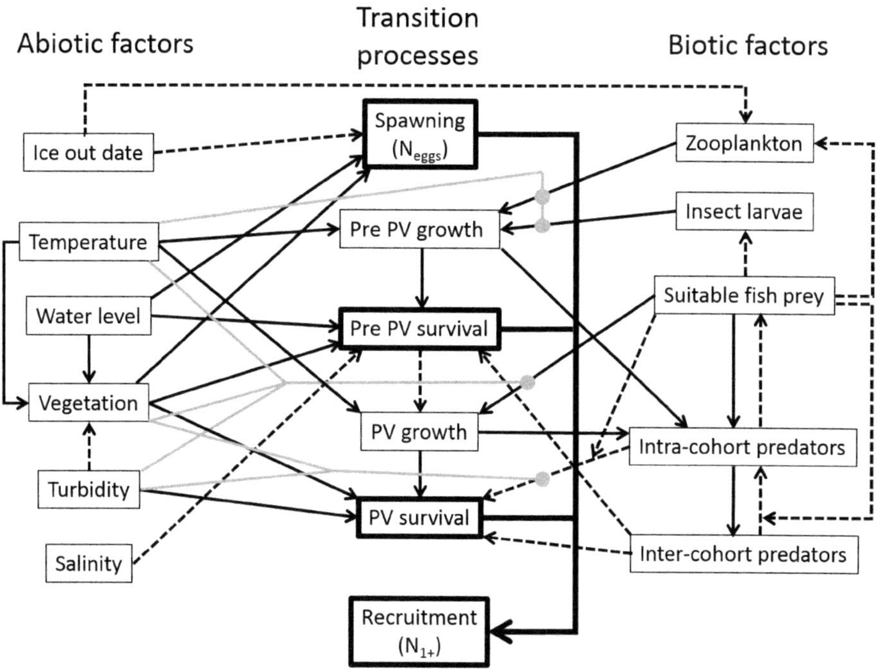

FIGURE 6.1 Process diagram of northern pike recruitment and the influence from abiotic and biotic factors. Positive effects are drawn as solid black arrows and negative effects as dashed arrows. Grey arrows illustrate modifying effects (interactions). PV = Piscivory.

and not influenced by stock size (Kipling and Frost 1970, Paxton et al. 2009). Paxton et al. (2009) used data from 1944-1991 and found higher inter-annual variation (27-fold variation) in age-3 recruit CPUE compared to earlier studies (10-fold variation, Mills and Hurley 1990). The analyses were performed on lake level. Edeline et al. (2008), however, used data on *estimated* (i.e., corrected for age-specific catchability and mortality) pike numbers for the 1945-1995 period and adjusted for various environmental drivers such as prey density, pathogen effect and temperature. Motivated by the differential productivity and morphology of the two lake basins of Windermere, they modeled the SR process on basin level using age-2 as recruits. They did find a significant compensatory Ricker SR relationship (Fig. 6.2A). In the most recent analysis by Langangen et al. (2011), they used new methods to estimate annual population sizes (also taking age-specific catchability and mortality into account) for each age class independently and used data from 1945-2002. As Paxton et al. (2009) they modeled the SR process on whole-lake level and used age-3 as recruits. Using a generalized additive model, they found a weak, but significant, compensatory SR relationship when accounting for the same environmental variables as Edeline et al. (2008) (see Fig. 6.2B for a reconstruction and a fitted Ricker curve). Hence, the somewhat confusing SR results reported by different papers based on Windermere gillnet time-series data seem largely to be due to use of different data types (CPUE *vs* estimated population sizes),

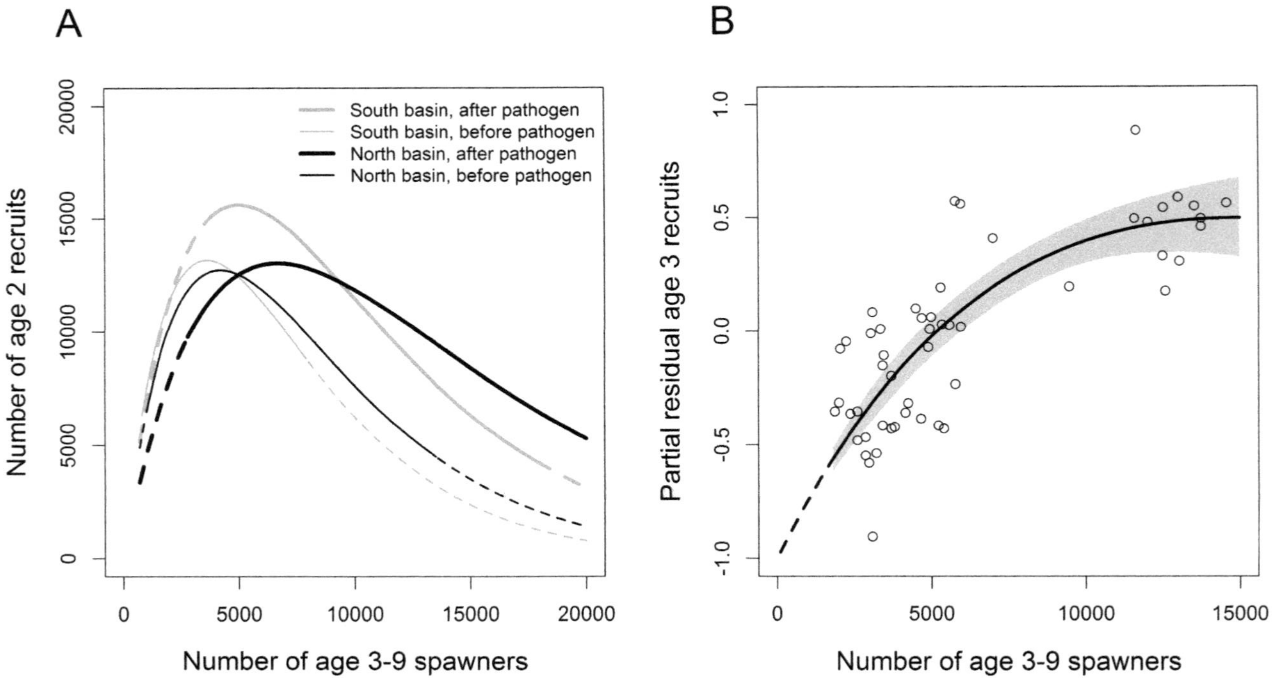

FIGURE 6.2. A. Windermere pike Ricker stock-recruitment curves fitted to estimated basin- and period-wise annual age-2 population sizes (recruits) and age 3-9 population sizes (stock) for the 1945-1995 period. The curves were reconstructed from Edeline et al. (2008) and are adjusted to average temperature of 15°C and mean period- and basin-specific perch population sizes. Periods cover pre-pathogen period (1945-1975) and pathogen period (1976-1995). Dashed parts of curves display parts not covered by stock-size data (i.e., extrapolations). **B.** Ricker curve with corresponding 95% confidence bounds (shaded area) fitted to partial residuals of whole-lake stock-recruitment data provided in Langangen et al. (2011). Partial residuals represent residual values corrected for temperature, perch abundance and effect from pathogen.

differential recruitment ages (age-2 vs age-3) and differential spatial scales (basin *vs* whole lake). The strongest SR relationship was found when using age-2 as recruits and basin level resolution on catchability-adjusted data (i.e., the Edeline et al. (2008) study). This could indicate, as also found in other species (e.g. Elliott 1985), density-dependent interaction processes to be most prominent during first year of life and, thus, signals from these early-life interaction processes vanish over time as other mortality and dispersal processes comes into play. Analyzing on lake level is likely to involve dilution of spatial differences masked by density-dependent dispersal of recruits. To conclude, the Windermere SR analyses, despite being contradictory, show that pike in this lake most likely is regulated by an overcompensatory SR relationship, *sensu* Ricker, but because YOY recruitment data is not available, we cannot provide the most correct SR relationship for this lake. Despite the obvious complexity of environmental mechanisms affecting pike recruitment, we will address these by first exploring effects on recruitment from abiotic and biotic factors separately, before exploring interaction effects.

6.2.1 Abiotic Effects on Pike Recruitment

In a spring-spawning, circumpolar fish like pike one might expect spring and late winter conditions to have a large influence on recruitment success. Early warming may lead to early spawning, and potentially to a longer first summer. A long first summer may, provided sufficient food and feeding opportunities, produce juveniles with a larger individual size at onset of first winter (Pagel et al. 2015). It is generally assumed that first-year survival in fishes is size-dependent (Perez and Munch 2010). This is possibly also the case for pike as it has been demonstrated in stocking experiments (Skov et al. 2003a, Grønkjær et al. 2004, chapter 9). However, strong evidence about size-dependent mortality from natural systems is scarce for YOY pike, but Kipling and Frost (1970) found YOY smaller than 180 mm in December to have lower over-winter survival prospects than for larger YOY individuals. Pagel et al. (2015) studied otoliths of surviving YOY pike in a small and unexploited German lake at carrying capacity revealing evidence that early-hatching juveniles maintained a size advantage over late-hatching juveniles in two of three study years. In laboratory studies, Murry et al. (2008) found that pike larvae hatched from late-spawned eggs caught up the head start of larvae from early-spawned eggs by developing faster and being larger from the very start. These latter findings may just have relevance under artificial lab conditions. An interesting mechanism may have been discovered by Trabelsi et al. (2013) who demonstrated that early hatched pike larvae are smaller at hatch and have faster growth and higher yolk-use efficiency than late-hatching ones. However, early-hatched larvae seem to be premature and hatch at a precocious developmental stage whereas late hatched individuals continue their growth within the egg and hatch at larger size but with lower yolk reserves. A compensatory growth phase was observed for the early hatching pike larvae, particularly during the first five post hatch days. In a more recent study, some of the same researchers documented the existence of two development strategies in pike, where early-season hatchers are smaller at hatch than late-season individuals, but survived longer under starvation (Trabelsi et al. 2015). If these laboratory-derived findings can be transferred to wild conditions, the contribution of recruits from late-spawning *vs* early-spawning females may differ among years as a function of spring-condition development (Pagel et al. 2015). This may be due to variation in the spring phenology of prey, competitors and predators, as shown for the Eurasian perch *Perca fluviatilis* Linnaeus, 1758 (Ohlberger et al. 2014).

6.2.1.1 Vegetation as Recruitment Constraint

Vegetation plays a key role as spawning habitat for pike (chapter 3), but the role of vegetation in pike recruitment extends far beyond serving as spawning- and incubation substrate/structure (e.g. Minns et al. 1996). The number of arrows attached to the vegetation box in Fig. 6.1 clearly demonstrates the extent to which this variable is involved as a determinant of pike recruitment, mainly through the refuge effect provided by structure to small YOY. In a series of population-monitoring and manipulation experiments in ponds and shallow lakes, Grimm and coworkers demonstrated a tight connection between vegetation and biomass of pike YOY recruits (Grimm 1981a, 1989, Grimm and Klinge 1996). Similarly, Pierce and Tomcko (2003) revealed tight positive correlations between stock sizes in pike and the percent littoral area (holding submerged vegetation) of Minnesota natural lakes. Also, Wright (1990) found pike November YOY abundance to be more than three times higher (23 vs. 6.6 fish ha^{-1}) in a gravel pit lake with high density of aquatic vegetation compared to a contrasting gravel pit lake with little aquatic vegetation. Based on stocking experiments in shallow Dutch lakes, a maximum of YOY biomass in Dutch shallow lakes was achieved at 75 kg ha^{-1} – provided full coverage of submerged vegetation (Grimm 1983). Vegetation effects on pike recruitment and YOY abundance are usually attributed to the importance of vegetation for pike juvenile survival and connects directly to the role of vegetation as refuge habitat against predation and cannibalism (Giles et al. 1986, Bry 1996). However, such vegetation-YOY production relationships are not always found. Wright and Giles (1987) failed to find a correlation between amount (measured as dry weight m^{-2}) of emergent and submerged vegetation in June and pike YOY density in December. It is unclear, though, if vegetation was still present in December. If the submerged vegetation had collapsed by the time of YOY density measurements, the lack of correlation may simply be due to reduced differences in structural refuges among the experimental units during the last period of the experiments when mortality was shown to be density-dependent and cannibalism shown to occur. The vegetation development and extent for example relates to variation in water level. In an early study covering seven years (1945-1952) in Ball Club Lake of north-central Minnesota, Johnson (1957) found that strong year-classes were associated with years with high spring water levels and slow water level decline during the incubation period. Rapid water-level descent during incubation is detrimental to egg and larvae that are attached to the vegetation and get exposed to air. A long-term study in eastern Lake Ontario showed a negative correlation between water level during late summer-early autumn and relative year-class strength (Casselman and Lewis 1996).

The importance of vegetation as a generator of habitat complexity clearly differs depending on biotic conditions (presence of food, competitors and predators) (Skov and Koed 2004) and on the size of pike (Eklöv 1997). Vegetation may serve as hiding structure for juvenile pike when ambushing prey such as juvenile fish larvae and invertebrates, and serve as refuge from predation when larger conspecifics or other predators are around (Casselman and Lewis 1996). The use of vegetation types in YOY pike changes during ontogeny. In a field experiment where artificial vegetation was used to increase vegetation diversity, Skov and Berg (1999) found that small YOY during early summer preferred highly complex habitats, whereas larger individual in late summer use less complex vegetation habitats. Potentially, therefore, the availability of different vegetation types (and their complexity) may serve as bottleneck factors in pike recruitment. Lack of complex vegetation types may serve as bottleneck in some lakes (affecting early summer YOY survival/growth) whereas constraints on less complex vegetation reduce YOY performance in other lakes. Because most submerged vegetation collapse during late autumn and winter, mortality may increase over winter due to the lost refuge effect (Grimm and Klinge 1996). Reaching large size before this collapse is critically affecting over-winter survival in YOY pike (Kipling and Frost 1970). Together this may lead to population regulation through density-dependent mechanisms moderated by habitat structure, and will therefore receive more attention in section 6.2.

6.2.1.2 Water Temperature

As in all fishes, temperature has a profound effect on the early development and growth of pike (e.g. Bry et al. 1991, Bondarenko et al. 2015). Fertilization success, egg and embryo development and hatching success all depend on temperature. Since pike is commonly reared in hatcheries, there are numerous studies on the effect of temperature for these developmental processes (see also chapter 13). In a recent study, Bondarenko et al. (2015) found that the optimal temperatures for production of high quality larvae was between 6-10°C. The lowest temperature for egg development was in this study estimated to be 3.3°C. Other studies show mainly similar, but also slightly different optimal, minimum and maximum temperatures (Hokanson et al. 1973, Hassler 1982). Such differences in temperature requirements are probably due to local thermal adaptation in the pike populations used in the experiments (e.g. Tibblin 2015).

Survival in pike recruits is very likely to be size-dependent, i.e., larger individuals have better prospects for surviving the first year (Grønkjær et al. 2004), also including first winter (Kipling and Frost 1970). This positive size effect can be related to both prey accessibility and predation risk, and owing to the pervasive effect of temperature on growth, it is not surprising that temperature often has been appointed as a key factor affecting year-class strength/recruitment (Kipling and Frost 1970, Le Cren 1987, Casselman 2002, Edeline et al. 2008, Paxton et al. 2009). The important role of temperature as a driver in pike recruitment is due to both direct and indirect effects on individual growth. In an extensive meta-analysis of 119 pike populations, Rypel (2012) found that temperature, expressed as growing degree days (GDD*), explained a large fraction of variation in size-at-age

(age ≤ 3) among populations. Similar results were reported by Pagel et al. (2015) for pike YOY growth within a lake system across three years. Overall, through its joint influence on egg and larval development as well as growth of YOY pike, water temperature strongly drives pike recruitment (Edeline et al. 2008, Paxton et al. 2009). Because temperature affects growth it will also modulate how biotic resources such as prey become available, as well as how fast the individual pike grow through the predation window. Spring and summer thermal development will therefore affect both individual growth and timing of when critical prey categories become accessible. Paxton et al. (2009) found that water temperature during late summer (August-September) of the first growth season was the most important factor for age-3 recruitment (CPUE-based) in Windermere. This factor alone accounted for 45% of the age-3 CPUE variation. Similarly, Casselman and Lewis (1996) found year-class strength of pike from Bay of Quinte and eastern Lake Ontario during 1971 to 1992 (sampled using multimesh gillnets and bottom trawls) to be highly associated with mean midsummer (July–August) water temperature, but there was no association with water temperatures earlier in the growth season. They argue that due to highest influence of water temperature relatively late in first growth season, early development stages (spawning and incubation conditions) seem less critical than later juvenile stages. Mid and late summer correspond to the period where pike YOY start exploring less complex vegetation areas and move towards more open habitats for utilizing larger prey (e.g. fish prey). High temperatures during this ontogenetic diet shift (invertivory to piscivory) will facilitate individual growth (provided sufficient food is available) and thus increase first-year survival. Casselman and Lewis (1996) found an optimum surface temperature for year-class strength at 23.5°C, which is far beyond the highest summer temperatures covered by the Paxton et al. (2009) study.

In a recent analysis of the long-term data series from Windermere where demographic data were used to parameterize a female-based integral projection model (IPM), Vindenes et al. (2016) found that population growth rate (λ) was sensitive to early temperature conditions through its effects on early growth and survival. This study operated with annual mean temperatures and can therefore not be used for assessing how within-year temperature variation impacts the sensitivity of early vital rates. The annual temperature effect, however, was much larger than the effect of size-dependent maternal quality (through egg size) on early growth and survival, which has been supported empirically by Pagel et al. (2015). This indicates that density-independent processes driven by temperature during the first growth season is important for pike recruitment, and other studies clearly suggest mid-to-late summer temperatures as particularly important.

6.2.1.3 Water Level

In most natural pike systems, water level and variation in water level during the spawning period, and especially during incubation and early larval development

---

*GDD $= \sum_t \frac{T_{max} - T_{min}}{2} T_{base}$, T_{max} = maximum daily temperature, T_{min} = minimum daily temperature, T_{base} = 0 or 10, t = day of year (starting at 1 March).

periods, are critical by affecting the amount and quality of spawning and nursery habitats and the availability of warm, shallow water. Pike spawning is very often left on hold awaiting a rise in water level if the water level is very low and suitable spawning substrates are not available (Fortin et al. 1982, Hill 2004). If available, spawning takes place in very shallow flooded land areas (with some exceptions, chapter 3) and so does early development (Casselman and Lewis 1996). This is a vulnerable spawning system as, obviously, subsequent reductions in water level may lead to loss of this habitat. This may be detrimental to the newly-hatched juveniles who may be unable to track the change in water level and end in dewatered habitat (Braum et al. 1996). Using telemetry based on passive integrated transponders (PIT) of 51.0 ± 5.3 mm YOY pike in a shallow, flooded nursery area, Cucherousset et al. (2007) found that only 19.3% and 6.9% of the PIT-tagged and untagged control fish, respectively, emigrated from the nursery before drying out. Small individuals (i.e., < 50 mm) experienced higher mortalities, and this size-dependent mortality was attributed to cannibalism rather than tagging procedure (same size-dependent mortality in control group). In addition, some 30% of the tagged individuals disappeared from the study site, most likely due to avian predation. A 90% loss of individuals over the 3-4 week nursery period covered by this PIT study clearly demonstrates how critical this nursery period is for pike recruitment. The water level regime was, to our knowledge, following a natural decrease pattern in this study. If the water level had dropped faster, stranding would probably have been even more pronounced. Indeed, water-level management has been identified as a key factor threatening pike spawning and recruitment (Casselman and Lewis 1996, Farrell 2001, Crane et al. 2015, chapter 10). In a seven-year study in Ball Club Lake of north-central Minnesota, USA, Johnson (1957) found that high water level during spawning combined with a small decline in water levels during egg incubation resulted in strong pike year classes. Similar findings have been reported from other North American lake systems (e.g. Franklin and Smith 1963, Hassler 1970). Motivated by this insight, Mingelbier et al. (2008) constructed a spatial model over the entire fluvial St. Lawrence River, Canada, aiming at constructing tools for sustainable water management with emphasize on pike recruitment. River discharge, wetland type and water temperature were included as drivers in the model that predicted, at a high spatial resolution, habitat suitability for spawning and early development in pike. The model predictions revealed that discharge had a substantial effect on both availability of suitable habitat for egg deposition and potential mortality following dewatering. As mentioned earlier, other modeling approaches, also motivated by water management objectives, found that water level should be managed so as to maximize nursery and juvenile habitat supply, as these habitats were shown to limit recruitment to a larger extent than spawning habitat (Minns et al. 1996).

6.2.1.4 Turbidity

Turbidity may influence pike recruitment in different ways. First, because pike recruits largely use vision for finding and catching prey, high turbidity levels may have a negative impact on their feeding efficiency (chapters 2 and 3). Potentially, the pike recruits can use the lateral line organ for locating prey even under both dark and high-turbidity conditions (Volkova 1973). However, development of the lateral line occurs when pike are between 22 and 56 mm (Raat 1988). Therefore, turbidity is likely to have more negative impact on YOY performance for individuals smaller than this developmental stage. Indeed, in a series of field studies and experiments, Salonen and coworkers (Salonen et al. 2009, Salonen and Engström-Öst 2010, Salonen and Engström-Öst 2013) showed that increased turbidity leads to reduced < 25 mm pike larvae condition, probably due to reduced ability to capture prey such as calanoid and cyclopoid copepods. At the same time, turbidity could also provide pike larvae visual refuge from other predators, as for instance found in some estuarine fish communities (Maes et al. 1998). We have, however, found no literature documenting this refuge hypothesis as being relevant for pike. High turbidity levels have a negative effect on subsurface vegetation, mainly due to light limitation (e.g. Penning et al. 2008b). On the other hand, high turbidity is often correlated with high trophic level and thus availability of zooplankton food for pike larvae. This may lead to better feeding conditions for those that survive. Because production of pike recruits is highly affected by vegetation (see above and chapters 3 and 10), high turbidity levels, but depending on the source of turbidity, may have a strong influence on recruitment in this species. In a series of experiments Skov & Nilsson and coworkers tested combined effects from water transparency, habitat complexity and density of conspecifics (including cannibals) and prey (Skov et al. 2002, Skov and Koed 2004, Nilsson et al. 2009). Even though visual range decreases with increased turbidity, Nilsson et al. (2009) suggested that overlapping habitat use with potential prey will increase prey encounter probability for pike under higher turbidities. Under experimentally controlled highly turbid conditions > 90 mm YOY pike were found to utilize the open-water habitat to a larger extent than under clear conditions, and they foraged also under low light intensities, possibly largely by using the lateral line (Skov et al. 2002). Surprisingly, these authors found even 20-31 mm YOY to have faster growth under high-turbidity conditions than in clear water conditions, and attribute this to reduced costs from a lowered alertness against cannibalism facilitated by the sheltering effect from reduced visibility (Skov et al. 2003a). The smaller YOY increased their use of complex habitat structures with decreasing water transparency, but were often forced into more open habitats in the presence of larger YOY individuals – exposing them to cannibalism (Skov and Koed 2004). The larger YOY individuals (> 90 mm) displayed a prey-density-dependent habitat use, with increased use of the open habitat under high-turbidity conditions, utilizing the increased chance of staying unnoticed in face of potential prey. In conclusion, there is consensus that pike recruitment is generally negatively affected by high turbidity conditions at levels where vegetation is heavily reduced, but the outcome from less severe turbidity levels is likely to depend upon a pile of additional environmental conditions, such as availability to important habitat structures and also density of suitable prey.

6.2.2 Biotic Effects – Different Threats and Opportunities through Ontogeny

As mention earlier, long-term studies report low inter-annual variation in year-class strength in pike (e.g. Franklin and Smith 1963, Treasurer et al. 1992, Langangen et al. 2011), suggesting density-dependent regulation of the pike recruitment process. One should therefore expect to find a somewhat compensatory stock-recruitment relationship, as for instance documented in Grimm and Klinge (1996) using data from Fago (1977). The study by Fago (1977) took place in small marsh ponds in Wisconsin, USA, and extrapolating these findings to natural lakes is not straightforward. As shown later for larger lake and river systems, complicated interaction effects between abiotic and biotic environmental variables make it hard to unravel the underlying density-regulating pattern. Before looking into this, we will explore biotic factors with important unique effects on pike recruitment.

Predation of pike juveniles can start even before eggs are fertilized or shortly after. Pike roe has high energy content and is likely to be eaten by both invertebrates and vertebrates (Bry 1996). For instance, in the brackish-water spawning pike populations living in Blekinge Archipelago in South-Eastern Sweden, egg predation from three-spined stickleback *Gasterosteus aculeatus* Linnaeus, 1758 has been suggested as an important mortality factor contributing to the general population decline observed in the area (Nilsson 2006a). Recently it was suggested that a negative association between the abundance of pike and perch juveniles was due to strong negative interactions such as stickleback predation on egg and larvae (Bergström et al. 2015, Byström et al. 2015).

Access to zooplankton before the yolk sac is resorbed is critical for pike larval survival (Billard 1996). Zooplankton plays a key role as food for pike up to sizes 15-20 mm. Some studies have documented zooplankton to be important for pike reaching sizes up to 50 mm and to be included in the diet all the way up to sizes at 100 mm (Skov et al. 2003b, Skov et al. 2011). The zooplanktivorous period is followed by a period where young pike feed predominantly on chironomids and other macroinvertebrates, followed by piscivory – including intracohort cannibalism. Intra-cohort cannibalism in YOY pike constitute a key regulation factor, but there are conflicting reports on when (i.e. at what size) this cannibalism starts, and the magnitude seems to vary greatly from system to system, and probably also from year to year. Bry et al. (1992) documented density-dependent YOY survival where cannibalism was prominent under experimental conditions in ponds even after just 12 days of exogenous feeding. Other studies do not report cannibalism at sizes below 45-100 mm (Bry et al. 1995, Mittelbach and Persson 1998, Skov et al. 2003b), but there are exceptions (Kucharczyk et al. 1997). Zaikov et al. (2006) found victim sizes to range between 42 and 73% of predator size and cannibalism started at predator sizes larger than 6 cm. According to Giles et al. (1986), who conducted tank experiments, intracohort cannibalism started when within-cohort ratio of predator size : prey size was larger than 2 : 1. The daily per capita mortality rates in this experiment suggested no density effects before cannibalism, but significant density-dependent mortality after onset of cannibalism. Consumption rates of cannibals varied between 0.63 and 6.0 fry per cannibal and day, accounting for 54-96% of daily mortality in the experimental tanks. These numbers clearly demonstrate the potential for intra-cohort cannibalism regulation in this species, but is likely not transferable to natural conditions.

It is likely that drivers of intra-cohort cannibalism are dependent on availability of other food resources such as zooplankton, insects and YOY of other fish species as well as amount of refuges. Lessons from YOY stocking programs have provided some insight into these dynamics (see also chapters 9 and 11). Hühn et al. (2014) showed that stocking of YOY pike under controlled and replicated experimental conditions in ponds resulted in no additional production of YOY when stocked into ponds holding natural recruits, and yielded same level of YOY production when stocked into ponds without natural recruits (see also Fig. 9.1). In their experiment, intracohort cannibalism started right after juveniles attained TLs of 30 mm, which is in line with other experiments and observations (Kucharczyk et al. 1997). Intracohort cannibalism seems postponed as long as there is a sufficient supply of other prey items (Skov et al. 2003a), but as just described vegetation coverage may play an important role here, as good access to refuge may reduce the likelihood of cannibal encounters and also provide favorable conditions for zooplankton prey (Grimm 1989).

Inter-cohort cannibalism from mostly yearlings on YOY victims has been predicted, from individual-based models as well as empirically, to constitute a stabilizing factor of pike recruitment with a large impact on individual growth trajectories (Persson et al. 2004b). The key mechanism behind intercohort cannibalism is the broad cannibalism window (i.e., the size range of cannibal and victim sizes allowing for cannibalism to occur) that characterize pike (Claessen et al. 2002). Typically, minimum and maximum victim:cannibal size ratio varies between 0.03 and 0.8 in this species (Bry et al. 1992, Persson et al. 2004b). Because YOY and yearlings of pike largely live in the same habitat, the size difference between the two allows for intercohort cannibalism from a very early stage. This cannibalism reduces competition among recruits, with a dampening effect on inter-annual variation in initial recruitment (i.e., numbers of eggs). This dampening effect of inter-cohort cannibalism on initial recruitment stabilizes density-dependent interactions and results in both a stable population age structure and also size structure, as most cohorts will experience comparable growth conditions. In a seven-year follow-up study of three small Italian stream-dwelling pike populations, Persson et al. (2006) attained population structure data and individual growth trajectory data strongly supporting these qualitative predictions of inter-cohort cannibalism-induced stabilization of the pike population dynamics. They also showed these cannibal-driven dynamics to be robust towards abiotic environmental variation. Hence, the role of inter-cohort cannibalism as a regulating factor of pike recruitment seems pervasive in small stream populations. Moreover, as stable inter-annual recruitment has been found in many other, and larger, natural populations as well (e.g. Franklin and Smith 1963, Treasurer et al. 1992, Langangen et al. 2011), intercohort cannibalism seems to play a key regulation role in most pike populations.

Pike recruits are not only subjected to intraspecific predation and competition, but interspecific predation and competition may have just as high influence on the recruitment process (Persson et al. 2004b). Pike juveniles are subjected to predation from a large range of both invertebrate as well as vertebrate predators (e.g. Bry 1996). Eurasian perch is a key prey species for the Windermere pike (Frost 1954, Winfield et al. 2012), but during juvenile stages, this species also affects the pike negatively through predation and competition. Interestingly, the effect of perch on recruitment of pike age-3 depends strongly on individual growth and demographic structure of the perch population. In 1974, the Windermere perch got infected by a pathogen (Bucke et al. 1979, Mills and Hurley 1990) that reduced individual growth and adult survival, and hence changed the demographic structure (Le Cren 1987). For the periods prior to the pathogen, perch density had a negative impact on pike age-3 recruitment, whereas after the pathogen perch density had a positive impact (Edeline et al. 2008, Langangen et al. 2011). Due to the changed growth and mortality pattern imposed by the pathogen, the combined predation and competition effects of perch on juvenile pike, pertinent to the pre-pathogen period, changed to a predator-prey relationship dominated by pike following the pathogen period.

6.2.3 Maternal Effects

Maternal effects, such as the effect of female size on fecundity, egg size, egg quality, and timing of spawning (Murry et al. 2008, Arlinghaus et al. 2010, Kotakorpi et al. 2013, Trabelsi et al. 2015) may have population-level consequences. As mentioned earlier, egg size may have a positive effect on offspring survival and starvation capacity (Kotakorpi et al. 2013, Trabelsi et al. 2015). Because both egg size and fecundity vary with female size and thus respond according to environmental factors affecting female size, inter-annual variations in both density-dependent and density-independent factors affecting female growth are likely to affect both recruitment potential (total initial fecundity) and offspring quality (egg size). Consequently, favorable growth conditions may yield both high initial population fecundity and high-quality offspring. Under natural conditions, this high recruitment potential will often be countered by intra-cohort agonistic and antagonistic interactions (according to SR-curves, Fig. 6.1A, see also chapter 2) or by unfavorable abiotic incubation conditions (e.g. low temperatures). Hence, evaluating population-level consequences from maternal effects requires detailed long-term time-series data or realistic population simulations. In a recent modeling study, Vindenes et al. (2016) parameterized their models using estimates from the long-term gillnet survey data from the Windermere lake system in Northwest England, and aimed at assessing recruitment effects on population growth resulting from maternal effects. Their results revealed that population growth in this particular population is sensitive to environmental conditions, including both density-dependent and density-independent factors, experienced in early life, whereas maternal effects through positive egg-size effect on offspring survival had little impact on population growth. Similarly, a comprehensive simulation study by Arlinghaus et al. (2010) showed that size-dependent maternal effects had limited impact on population dynamics and the way exploited populations rebounded after harvesting. Results from studies exploring effects from fecundity responses to female size show that pike population dynamics is far more sensitive towards this maternal effect. For instance, a follow-up simulation study exploring population-level consequences from impaired individual growth imposed by social stress in female pike showed that the population intrinsic rate of increase to potentially be decreased by > 50% (Edeline et al. 2010). Hence, even though studies like Pagel et al. (2015) show 2-3 year lasting benefit effects for early-hatching pike juveniles, the population level consequences from such maternal effects remains enigmatic and seem less important than fecundity-related maternal effects. Further studies within this field should be conducted, particularly under natural conditions.

6.2.4 Interaction Effects

Due to complex density-dependent and -independent interaction processes involved in pike recruitment, disentangling eventual underlying compensational, depensational or overcompensational stock-recruitment relationships (*sensu* Ricker (1954), and Myers (2001)) requires long-term time series. Even when such data are available, "-pensational" patterns are hard to disentangle as the many data analyses from Windermere show.

As described earlier, Windermere pike recruitment (to age 2 and age 3) is affected by water temperature, perch density and abundance of pike spawners (Edeline et al. 2008, Langangen et al. 2011). The effect from perch density on pike recruitment depended heavily upon the demography and size structure of perch, which in this study system changed dramatically before and after the perch pathogen infection. Before the pathogen, the perch-pike interaction dynamic constituted a mixture of competition and predation dominated by perch, whereas after the pathogen it changed to a simpler predator-prey relationship dominated by pike. Due to this, the SR-dynamics also changed resulting in a higher per capita recruitment rate in both basins (Edeline et al. 2008) (Fig. 6.2). Generalization from one study system is generally a bad idea. However, the lesson learned from the Windermere system is that pike stock-recruitment in a large lake system is complex and highly dynamic. A similar experience of complexity was concluded from a Finnish study. Kallasvuo et al. (2010) showed that in a brackish water system, pike YOY recruitment was affected by an interplay between habitat type (reed in this system), temperature, Secchi depth and availability of zooplankton (cyclopid copepods and cladocerans), but due to substantial collinearities among these variables the authors could not separate their relative roles in pike recruitment. Maybe just as important was the lack of information about spawning stock size in this particular study. In a small Swedish lake, Eklöv (1997) found strong indications that YOY size distribution was affected by cannibalism, but cannibalism was largely modified by habitat type (vegetation density) and interspecific interactions among > 160 mm pike (potential cannibals) and perch individuals. Interaction effects between habitat structures, and vegetation in particular, and predation/cannibalism on YOY production and performance is a returning conclusion from many pike studies. It can be inferred that availability of habitat structures constrain access to preferred habitat and therefore largely defines the system's carrying capacity for pike YOY production as the habitat structure will affect both risk of antagonistic interactions with predators/cannibals and agonistic interactions with YOY peers (Grimm and Klinge 1996, Nilsson 2006a, Edeline et al. 2010). As pointed out by Skov et al. (2011, see also chapters 9 and 11), stocking YOY individuals into systems already saturated with natural recruits should lead to rapid removal of these surplus individuals from predators and cannibals as the less fit individuals will be forced into suboptimal, often pelagic/open habitats with little shelter and low densities of appropriate food (Skov and Koed 2004, Skov et al. 2011). Indeed, this has been the general finding in most stocking programs carried out in systems with natural recruitment (Gres et al. 1996, Skov et al. 2003a, Skov et al. 2011, Hühn et al. 2014, chapters 9 and 11). The findings in these stocking studies is highly supportive of an underlying (over) compensational SR relationship where the strength of the (over) compensation is influenced by interaction effects from predator-prey dynamics which is largely constrained by access to structural habitats. As a consequence, altering the structural habitat in the littoral zone (e.g. Skov and Koed 2004) may potentially alter the shape of the SR relationship. Likewise, changing environmental conditions affecting demography of potential prey and predators, like the pathogen effect on perch in Windermere, changes the SR relationship as well (Edeline et al. 2008, Langangen et al. 2011). Finally, the existence of an (over) compensational SR relationship has management consequences where, for instance, the effect of a culling program depends strongly on what side of the SR-curve's maximum recruitment point you are.

6.3 ENVIRONMENTS AND POST-RECRUIT POPULATION COMPOSITION

As we have seen thus far, the recruitment process is largely, but not exclusively, affected by characteristics related to the post-recruit share of the population (i.e., pike older than age 1), both via intraspecific competition and cannibalism. Abiotic environmental factors (temperature and vegetation structures) constrain and modulate the effect from these biotic factors via complex interaction processes. Because intraspecific interaction processes are involved, recruits (i.e., pike < age 1) may also affect features of the post-recruit population, both as prey and as competitors. In the following, we will explore how processes like intraspecific competition, but also other factors, influence processes pertinent to the post-recruit population structure. The key post-recruit population structuring processes addressed will be largely connected to individual growth and survival processes (chapters 2, 3 and 4). Some important quantitative contributions to processes affecting survival- and growth are displayed in Table 6.1.

6.3.1 Size and Survival

Cannibalism is a key regulating mechanism in pike, with profound implications for the population structure. This was evident for the recruitment process reviewed above, but cannibalism poses a constant threat to pike individuals almost throughout

TABLE 6.1 Effects of density and density-independent factors and interactions between these on survival, growth and dispersal in post-recruit (> 30 cm) pike. "+" = positive effect; "–" negative effect; *X* = density of small pike individuals (post-recruits); *Y* = density of large pike individuals; *Z* = density of suitable fish prey species; *T* = growth season water temperature; Veg = vegetation coverage; *Turb* = Turbidity; TP = total phosphorus; Hab = habitat type; TS = Time series; B-A = before-after; CMR = capture-mark-recapture; REV = Review; MA = Meta analysis; REG = Regression; NLME = non-linear mixed effects models; LM = linear models; COR = correlation analysis; GS = gillnet survey data; EF = Electrofishing data. Studies presented cover post Raat (1988) period.

Vital rate	*Study systems*	*Size range* cm	*Biotic (Bi)*				*Abiotic (Ab)*			*Bi×Ab*	*Data/analysis*	*Reference*
			X	*Y*	*Z*	*XYZ*	*T*	*Turb*	*Other*			
Growth	Windermere, UK	30–80	–	+	+	+/–	+		Bas	*XZT*	TS/CMR	Haugen et al. (2007)
	Årungen, NOR	30–100		–							B-A/CMR	Sharma and Borgstrøm (2008a)
	19 small WI&MN lakes	10–90	–	–	–			–			GS/REG	Margenau et al. (1998)
	Windermere, UK	30–80	+	+	–	+/–	+		Bas	*XYZT*	TS/NLME	Edeline et al. (2007)
	119 US & Eurasia lakes	30–90					+		Hab		REV/MA	Rypel (2012)
	4 NOR lakes	20–90			+			–			GS/LM	Vøllestad et al. (1986)
	29 WI & MN lakes	NA	–	–				–			GS&EF/REG	Pierce et al. (2003)
	112 MN lakes	20–80	–	–	+/–		+	–	Veg (+/–) TP (+)		GS/COR	Jacobson (1993)
	37 CAN lakes	> 20						–			GS/REG	Craig and Babaluk (1989)
Survival	Windermere, UK	30–80	–	–	+	+/–	+			*XZT*	TS/CMR	Haugen et al. (2007)
	Årungen, NOR			–							B-A/CMR	Sharma and Borgstrøm (2008a)

XZT = interaction effects between X, Z and T, Bas=basin effect, NA = Not available.

life. Due to the elongated body shape, even large individuals are theoretically at risk for falling victims of cannibals. As other piscivorous predators, the pike is gape limited (Hart and Hamrin 1988, chapter 2). Both experimental and field studies have quantified the linear relationship between pike body length (total length, *TL*) and maximum prey body depth (BD_{max}) and arrived at the following equation for roach (*Rutilus rutilus* Linnaeus, 1758) as prey: $BD_{max} = 0.401 + 0{\cdot}131 TL$ (Nilsson and Brönmark 2000). This relationship differs for different prey species, but even for a deep-bodied species like bream (*Abramis brama* Linnaeus, 1758) the *BL* slope is surprisingly similar as in roach (0.127), but with higher intercept (1.427). A similar linear function for the cannibal *TL vs* victim BD_{max} relationship has, to our knowledge, never been fitted for pike. However, owing to the elongated body shape there is, from a gape-limitation point of view, likely to be a slow separation between the not-at-risk size and cannibal size in this species. Despite being at risk almost throughout life, field studies rarely find pike maximum victim:cannibal size ratios to exceed 0.5 (e.g. Persson et al. 2006). Mittelbach and Persson (1998, and see chapter 2) also document that pike generally feed on smaller prey than they potentially can eat, and prey > 40-50 cm are rarely eaten. This finding also receives support from reports on, what is considered, largely cannibalism-driven size-related pike mortalities retrieved from natural populations where natural mortality levels off at sizes above ca. 50 cm (e.g. Carlson et al. 2007, Haugen et al. 2007). For larger pike individuals, cannibalism therefore seems not constrained by gape size, but this propensity is more likely controlled by other ecological factors like for instance availability of small individuals relative to other prey species and amount of vegetation (Smith and Reay 1991, Nilsson and Brönmark 1999). Importantly, the pike has been found to have a low minimum victim:cannibal *TL* ratio and pike populations are therefore expected to exhibit sustained low-amplitude cannibal-driven dynamics where the survival of YOY pike is controlled by cannibalism (Persson et al. 2004b, see earlier). Therefore, YOY cannibalism is probably one of the most important mechanisms for survival and thus recruitment in pike, but this does not rule out later-life cannibalism as being important as well. The role of cannibalism in small- and medium-sized pike survival has been demonstrated in a number of removal studies (Le Cren 1965, Kipling and Frost 1970, Kipling 1983). For instance, removing a large proportion (55-65%) of large (i.e., > 65 cm), and largely cannibalistic, individuals from the Norwegian lake Årungen resulted in a significant proportional increase of medium-sized, largely age 3, individuals (Sharma and Borgstrøm 2008a). Due to the instant increase in abundance of medium-sized individuals, this effect could largely be attributed to increased survival of younger individuals (age 2 individuals in particular) both from reduced cannibalism, but also due to changed behavior flexibility of smaller individuals that could explore more optimal habitats under a reduced cannibalism threat (Sharma and Borgstrøm 2008a, 2008b). The large *vs* smaller pike interaction dynamics has also been documented in multi-lake studies. In the 29 lake study of Pierce et al. (2003) from Minnesota and Wisconsin USA, they documented a strong negative association between proportion of large (i.e., > 53 cm) individuals (proportional stock density, PSD) and total population density. Interestingly, cannibalism was not suggested as a potential mechanism explaining this pattern, even though cannibalism was documented to occur in 80% of the Wisconsin lakes (Margenau et al. 1998). Instead, the authors argue that high PSDs can only prevail under low-density conditions (i.e., < 14 individuals ha^{-1}), suggesting density dependent growth to be the structuring mechanism. Under high-density conditions both intra- and inter-cohort competition will result in low individual growth for most individuals and few individuals will reach sizes required for cannibalism (Diana 1987). Indeed, Pierce (2010) found no relationship between the relative abundance of very large pike and CPUE as index of abundance of smaller size classes, suggesting that the inverse relationship of density and size structure is more about growth depression. Density-dependent growth is also known from Windermere (Edeline et al. 2007, Haugen et al. 2007). In addition, under warm conditions, smaller individuals are competitively superior in systems where small and large individuals compete over common prey categories as the energy gain of small individuals increases faster with temperature than for large individuals (Claessen et al. 2000, Ohlberger et al. 2011). Indeed, the studied Minnesota and Wisconsin lakes have more than 2.5 months of water temperatures above 21°C during the growth season (Margenau et al. 1998). Hence, Pierce et al. (2003) may have good reasons for suggesting intraspecific competition to be the main driver behind the observed inter-lake PSD-density pattern in their Minnesota and Wisconsin pike populations, possibly reinforced by high temperature conditions. However, unraveling the relative role of cannibalism and intraspecific competition for pike growth and population structuring is not easy, and require either removal experiments or long-term time-series data within the same system for proper testing. We will explore this further in the interaction section 6.3.5.

The importance of cannibalism in pike population regulation cannot be inferred solely from stomach content analysis. Most pike diet studies reveal low percentages of pike prey, typically << 10% (e.g. Frost 1954, Margenau et al. 1998, Nilsson 2006b, Persson et al. 2006, Sharma and Borgstrøm 2008a, Winfield et al. 2012). However, since pike densities (hence availability) is often more than an order of magnitude lower than other potential prey species, cannibalism may still have a significant absolute impact (Craig 1996). In addition, the threat of cannibalism may also constrain the activity level and habitat use of small- and medium-sized individuals with potential negative impact on growth and survival (Nilsson 2006b, Sharma and Borgstrøm 2008a, chapter 2).

Cannibalism is not the only mechanism by which intraspecific interactions affect pike demography and population structure. As mentioned previously, prey-size preference remains surprisingly similar in both large and smaller pike even though larger individuals have the capacity to eat larger prey (Nilsson and Brönmark 2000). Consequently, intraspecific interactions over the same prey resources impose further negative effects on smaller pike individuals. Furthermore, smaller individuals face yet another intraspecific interaction burden as kleptoparasitism (i.e., stealing of prey) will impose both less efficient habitat use and selection of less optimal prey for smaller individuals living in situations with larger individuals in close vicinity (Nilsson and Brönmark 1999, Nilsson 2006b, chapter 2).

As mentioned, cannibalism plays a great role in shaping population structure in pike. Therefore, processes affecting individual growth become candidate drivers of the entire population dynamics in this species (Persson et al. 2004a, Persson et al. 2004b). Pike is sexually dimorphic, where females generally grow faster and attain larger sizes than males (Raat 1988, and references therein). This growth dimorphism may therefore influence the population sex ratio via mechanisms of cannibalism.

6.3.2 Sex Ratio

In a heterogametic (XY male chromosomes) species like pike, the expected sex ratio is 1:1 (West 2009). However, in many systems the population sex ratio deviates from this expectation (Casselman 1975, Paxton et al. 1999). Owing to the sexual dimorphism in growth, size-biased selection will potentially alter the sex ratio both in the short term and also on evolutionary time scales. Clearly, human-induced size-biased mortality from fisheries is a candidate for inducing a male-biased population structure (Crane et al. 2015, chapter 12), but because natural size-specific mortality generally is higher in males (Casselman 1975, Carlson et al. 2007, Haugen et al. 2007) sex-ratio as such cannot be used as an index of harvest pressure. However, having access to sex ratio time-series data covering periods of changed harvesting may allow for disentangling the true sex-biased effect from harvesting. In Windermere, the male:female ratio in gillnet catches increased over a 50-year period of decreasing fishing intensity (Paxton et al. 1999). The overall female bias documented in this study was interpreted as reflecting the larger size and faster growth of females and greater susceptibility to gillnet capture (Frost and Kipling 1967, Haugen et al. 2007).

Sexually biased populations structures may also arise from size-biased mortality resulting from winterkill episodes in eutrophic shallow lakes (see later). This takes place in ice-covered lakes, where oxygen depletion builds up from bottom to the ice-covered surface during winter (Crane et al. 2015). Critically low oxygen levels influence large, old and fast-growing individuals to a greater extent than other individuals. Consequently, females are at highest risk under such episodes and following dramatic winterkill episodes the sex-ratio may be dramatically altered (Casselman and Harvey 1975). Pagel (2009) showed that females preferred to reproduce with larger males and reported a selection pressure on male size, such that it is well possible that altered sex structures might strongly affect population dynamics. Research in this area is duly needed.

6.3.3 Abiotic Factors Affecting Growth and Survival

6.3.3.1 Water Temperature

Owing to the pervasive effect of temperature on individual growth, it comes as little surprise that this environmental variable vastly affect pike population structure and dynamics (Vindenes et al. 2014) – also beyond the recruitment phase. As temperature may differentially effect swimming performance of pike and its potential prey species (Öhlund et al. 2015), the prey-species composition is also likely to affect population dynamics of pike. The prey (and competitor) species composition is also likely to vary according to lake thermal regimes and this temperature-related effect may also affect pike population structure (Winfield et al. 2008, Ohlberger 2013). In the Windermere pike study of Haugen et al. (2007), it was shown that water temperature modified density effects from prey and conspecifics on both post-recruit individual growth and survival. Following from this, these effects were shown to have implications on fecundity (Edeline et al. 2007), pertinent to population growth potential. Hence, the effect of temperature is both a direct one, acting through metabolic pathways and indirectly acting through effects on individual growth with consequences for size-dependent survival and life-history traits like fecundity and maturation.

In the previously mentioned meta-analysis by Rypel (2012) it was revealed that individual growth rates of pike in North America were primarily driven by water temperature and therefore decreased with increasing latitude. When accounting for variation in the thermal opportunity for growth (i.e., growing degree days), a highly significant countergradient growth pattern was found in northern American populations. The same thermal and latitudinal patterns of growth were more vague in Eurasian populations. The same study also documented differences in longevity between populations inhabiting the two continents, but for individuals from both continents there was a clear negative correlation between growth rate and longevity – following well-documented life-history trade-off patterns (Roff 1984, Stearns 1992, Partridge and Mangel 1999). Rypel suggests that the vague signature of climate/thermal forcing in Eurasian pike population size- and life-history structure is due to a combination of less pronounced longitudinal climate gradients than in North America, and that local drivers such as lake productivity, population density and prey density (for unmentioned reasons), play a greater role in Eurasian populations. In the following, we will explore effects from these potential drivers on post-recruit individual growth and survival.

6.3.3.2 Lake Productivity

Total phosphorus (TP) is generally considered a key constraining factor for freshwater ecosystem production (Schindler 1974), and strongly influences the carrying capacity of most organisms living in a given freshwater ecosystem. However, in an analysis of 29 Wisconsin and Minnesota pike populations, Pierce et al. (2003) found only weak support for lake productivity (using Secchi depth as proxy) having an effect on population size structure and individual growth. The authors did neither report direction nor strength of this effect, but underscore that the effect only appeared when Wisconsin populations were included in the analysis. Abundance of 248 northern Europe pike populations, measured as catch per unit efforts (CPUE) from standardized Nordic gillnet** surveys, increase with increasing TP levels at low concentrations (Fig. 6.3A). The reason for this positive response is probably linked to the positive effect of TP on both physical habitat (i.e., submerged macrophytes) and availability of prey. The TP effect levels off and

**Multi-mesh 1.5 × 30 m gillnets comprised of 12 individual 2.5 m net panels with mesh sizes ranging from 5 mm to 55 mm.

reaches a CPUE peak at 15–20 μg TP L^{-1} where the general tendency is decreasing population densities beyond this maximum level (Fig. 6.3A). A similar maximum pike abundance at intermediate trophic index values (an index including spring and summer TP, Chl-a and Secchi depth, see e.g. Mischke et al. 2008) was found in a recent study from 57 lakes in northeastern Germany (Lewin et al. 2014). Here, pike abundance was estimated from electrofishing data. A maximum abundance at intermediate trophic index levels was evident when all lakes were analyzed in one model. However, when separating the analysis into shallow (max depth < 5.22 m) and deep lakes the maximum trophic index shifted to slightly lower values in shallow lakes, whereas pike abundance remained low at lower trophic index values and increased markedly at higher values in deep lakes. When fitting a generalized additive model (GAM, Hastie and Tibshirani 1990) with lake deepness ("shallow" *vs* "deep", mean depth < 5 m or ≥ 5 m, respectively) as a conditional effect to the Fig. 6.3 data, the same differential TP effect on pike abundance pattern between deep and shallow lakes as found in Lewin et al. (2014) appeared. In shallow lakes, a maximum pike CPUE was found at 15.1 μg TP L^{-1} and in deep lakes pike CPUE increased slowly along the TP-gradient (Fig. 6.3B). These converging results may indicate pike to be favored from increased eutrophication in deeper lakes possibly due to being in a better position of utilizing the increased abundance of omnivorous fish (cyprinids) that generally increase with lake productivity (e.g. Jeppesen et al. 2000).

This can arise from either higher prey abundance in the relatively narrower littoral zone, typically found in deeper lakes, and/or more efficient pelagic foraging in eutrophic lakes (e.g. Andersen et al. 2008). In the 71 lakes study of Jeppesen et al. (2000), pike CPUE (kg net^{-1} $night^{-1}$) showed an almost flat response to TP, but with lowest mean CPUE at mean TP > 400 μg L^{-1}. The highest 90% percentile CPUE values (assessed from boxplot) were found in lakes with TP values ranging from 100-400 μg L^{-1}, which deviates substantially from the maximum at 15-20 μg TP L^{-1} in Fig. 6.3. Interestingly, mean individual pike weight was highest at high TP levels (i.e., > 400 μg L^{-1}), a result the authors attributed to increased cannibalism due to reduced structural complexity (less/lack of macrophytes) and, consequently, loss of cannibalism refuges (sensu Grimm and Backx 1990). The negative effect of high TP concentrations on pike abundance found for shallow lakes in Lewin et al. (2014) and in Fig. 6.3B is most likely an indirect one, where TP affects distribution and density of submerged macrophytes (Penning et al. 2008a), and water turbidity via the direct effect of TP on phytoplankton production. This mechanism cannot explain the deviating effect of high TP on pike abundance in the Jeppesen et al. (2000) study. This latter study was based on mainly shallow (average depth < 3 m) Danish lakes that often do not freeze or just freeze for a couple of weeks during winter (Jackson et al. 2007), whereas most lakes involved in Fig. 6.3 freeze for months during winter. The combination of ice cover during winter and higher TP values in shallow lakes has potential effects on both oxygen levels and submerged macrophytes that do not to the same extent apply to deeper lakes (Hargeby et al. 2004, Kosten et al. 2009, Ejankowski and Lenard 2015). In shallow ice-covered lakes with high TP levels, anoxic winter conditions are common, and this may lead to winterkill episodes (Casselman and Harvey 1975,

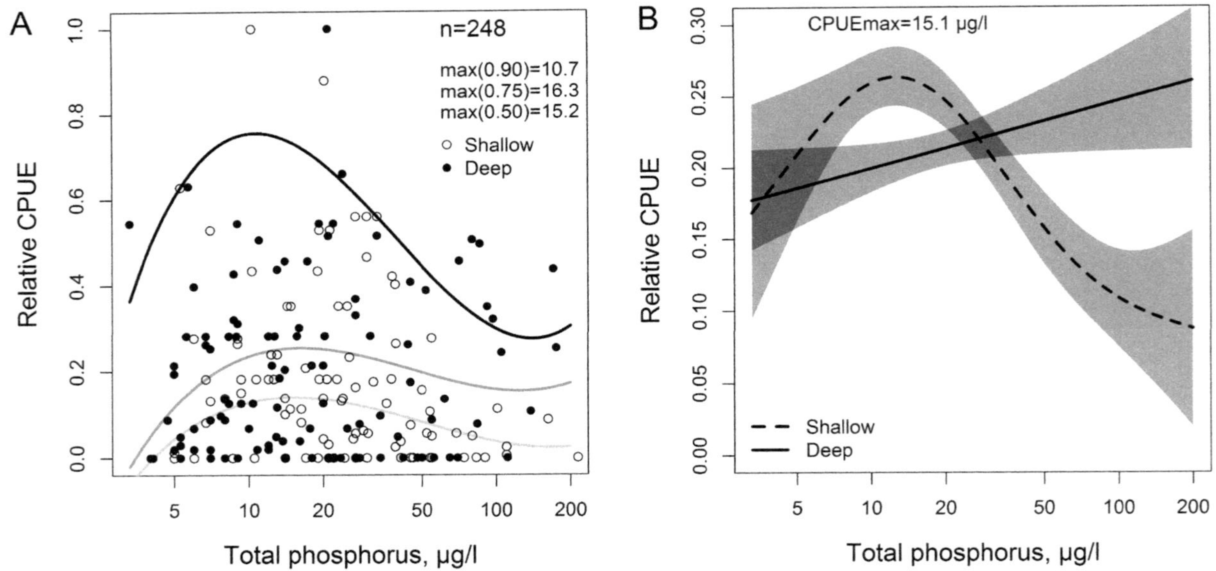

FIGURE 6.3 **A.** Relative catch per unit effort (number of individuals per Nordic gillnet per night compared to maximum CPUE in data set (1.6 inds net^{-1} $night^{-1}$)) of northern pike Nordic gillnet catches plotted as function of total phosphorus concentrations from Finnish (n = 104), Swedish (n = 77) and northern Germany (n = 67) lowland lakes. Lines represent quantile spline regressions for 90% (black), 75% (dark grey) and 50% (light grey) relative CPUE quantiles. Shallow lakes constitute lakes with mean depth < 5 m and deep lakes ≥ 5 m. Total phosphorus values of corresponding maximum quantile values are provided. **B.** Model predictions from generalized additive model (GAM) with corresponding 95% confidence bounds. The fitted GAM ($\arcsin(\sqrt{(y/\max(y))}) = s(TP \mid DepthGroup, df = 4)$, where $y = CPUE_{pike}$, $s(x \mid X, df)$ = thin plate spline function of predictor variable x conditional on group effect X with maximum degrees of freedom of 4) explained 7% of the variance. All data were sampled following procedures in (Appelberg et al. 1995), where gillnets were distributed in a randomized and stratified fashion with effort systematically increasing with lake area and depth. Data has been provided by Thomas Mehner (Leibniz-Institute of Freshwater Ecology and Inland Fisheries, Berlin), Martti Rask (Finnish Game and Fisheries Research Institute, Helsinki) and Anders Kinnerbäck (Swedish University of Agricultural Sciences, Department of Aquatic Resources, Drottningholm).

Jackson et al. 2007, Hilt et al. 2015), but even reduced oxygen (hypoxic) conditions may constrain pike performance and production. Tonn and Magnuson (1982) found in an 18 lakes northern Wisconsin study that pike were present in shallow lakes with an inlet or outlet, but were present only in summer and not in winter, suggesting that in winter, these individuals emigrate from the lakes into stream refuges. These observations are consistent with experimental findings showing that pike movements under hypoxia take the fish to the highest oxygen levels available in the immediate vicinity (Petrosky and Magnuson 1973, chapter 5). In a more recent northern Wisconsin small-lake study with histories of periodic winterkill, Margenau et al. (1998) found that larger and older individuals are most affected by such episodes (Casselman and Harvey 1975, Crane et al. 2015). However, this is not always the case (Hilt et al. 2015). Casselman and Harvey (1975) found low oxygen levels to select against larger, faster-growing pike individuals, and they noted that female fish were more vulnerable to low oxygen concentrations than males. Therefore, a male-biased population structure most likely will result (see section 6.3.2) because male pike are more likely to survive such episodes. It is also likely that the surviving females will have reduced quality, with possible effects on offspring viability. The reason for this relates to female gonadal investments taking place during winter (as opposed to their male counterparts, e.g. Diana 1983) and poor winter conditions will likely impair these investments (Neumann and Willis 1995). Jacobson (1993) found that pike populations from Minnesota lakes with a history of winterkill episodes produced fewer large-sized individuals compared with lakes without low winter oxygen episodes. Hence, it is evident that lake productivity, especially in shallow ice-covered lakes, has profound effects on pike population structure where smaller, slower-growing individuals may be better able to tolerate environmental stress associated with low oxygen concentrations most frequently imposed by high TP levels.

In temperate lakes, submerged macrophytes generally collapse and decay during late fall and winter (Ejankowski and Lenard 2015) providing less shelter during this part of year. Due to reduced metabolism, cannibalism is likely less pronounced in cold-water conditions (e.g. Edeline et al. 2007) and therefore not likely a contributing factor behind the differing TP effects on pike abundance between shallow ice-covered and non-ice-covered lakes. In springtime, the increased temperature and day length initiate re-growth of submerged macrophytes (Ejankowski and Lenard 2015). However, in ice-covered lakes this re-vegetation of submerged macrophytes is postponed mainly due to the light-inhibiting effect of the ice cover prior to ice out. Owing to the important role of submerged macrophytes as shelter habitats for smaller pike individuals, the inhibiting effect by ice cover on spring re-vegetation may, together with oxygen limitation, play a significant role as a limiting factor for pike production in shallow, productive ice-covered lakes.

In total, deeper lakes and shallow lakes with no ice cover during winter, pike abundance seems to increase with lake productivity at least up to TP levels of 400 $\mu g\ L^{-1}$. Pike populations living in shallow lakes with an ice cover during winter seem to respond positively at mesotrophic levels ($< 20\ \mu g\ L^{-1}$) but negatively above this level, most likely due to reduced winter oxygen levels and negative effects on aquatic vegetation.

6.3.3.3 Turbidity

As mentioned, high TP levels may result in high turbidity due to high phytoplankton production. High-turbidity conditions may have other sources than TP. For instance, lakes in clay-affected watercourses can experience high turbidities after heavy rainfalls. As for recruits also post recruit pike are visually oriented predators and therefore may perform more poorly under high-turbidity conditions – like in ultra-eutrophic lakes or in highly clay-affected lakes. The source of turbidity may affect pike performance differently. In the study of Ranåker et al. (2012) it was shown that reaction distance in TL 22.5 ± 1.4 cm pike (mean ± SD), decreased with decreasing visual range caused by increasing levels of algae, clay or humic matter. However, the effect on reaction distance was stronger in turbid water (clay, algae) than in the brown-water treatment, and the escape distance for roach was longer in turbid than in brown water treatments. The authors concluded that the optical environment is likely to have consequences for the strength of predator–prey interactions through changes in piscivore foraging efficiency and prey escape behavior (chapters 2 and 3).

In contrast to the previously mentioned Wisconsin/Minnesota pike study of Pierce et al. (2003), a multi-lake study of Margenau et al. (1998), found individual growth and population density to be negatively influenced by water transparency (Secchi depth), but by density-dependent factors as well. However, because other abiotic factors such as thermal habitat distribution and winterkill episodes also largely correlated with turbidity and had negative effects on pike growth and size structure, the authors concluded that it is the mixture of conditions, rather than just turbidity, that combine to limit growth. Craig and Babaluk (1989) found, in a 37-lake study from the Central Region of Canada, Secchi depth to be negatively correlated with post-recruit pike condition factor. A linear regression model predicted weight for a 50 cm pike to decrease from 940 g at 3 m Secchi depth conditions to 830 g under 1 m Secchi depth conditions. The authors attributed this effect to impaired food consumption under low water transparency.

In a four-lake study from south-eastern Norway, Vøllestad et al. (1986) found population density to be highest in the most turbid lakes, and individual growth to be highest in the least turbid lake. Interestingly, the fraction of pike using a pelagic habitat increased with turbidity. This increased use of pelagic habitat with increasing turbidity is in line with findings by Chapman and Mackay (1984b) and Andersen et al. (2008). The latter of these studies found, using radio telemetry, > 55 cm pike to be increasingly more active with increasing body size in a turbid-water lake, but no size-dependent activity effect in a clear-water lake. The authors argue that pelagic habitat use and active hunting (as opposed to ambush hunting), may be more favorable for pike under high-turbidity conditions as detection distance for prey decreases and the pike may utilize the lateral line for prey localization. In the same study, condition factor was found to be similar for pike in both lakes. The conflicting results from the Norwegian lakes study on individual growth performance were attributed by the authors to higher search costs induced by reduced visibility in high-turbidity lakes and density-dependent mechanisms, such as access to suitable prey sizes for maintenance growth being highest in the least turbid lake. Again, a situation where collinearity among environmental variables makes it difficult to separate unique effects from turbidity, but the general message is that this pike group seems to perform well also under high-turbidity conditions.

6.3.3.4 Habitat Composition

Pike, throughout its distribution range, live in heterogeneous habitats varying at many spatial scales and across seasons. Therefore, pike population densities may vary accordingly, but since potential prey resources are likely to reflect the pike population density, density-dependent survival and dispersal (and growth) may vary in both time and space – even within the same water body. The Windermere lake system consists of two basins where the northern basin is less productive than the southern basin (Talling 1993). Multi-state mark-recapture analyses applied to long-term mark-recapture data from this system revealed highly differentiated density-dependent survival, growth and dispersal patterns between pike individuals from the two basins (Haugen et al. 2007). Age 3-9 pike living in the more productive southern basin experienced a more pronounced interspecific density-dependent survival pattern, but the survivors grew better and therefore had higher fecundities than their north basin counterparts. Interestingly, north basin individuals were far more prone to migrate into the south basin when the between-basin density gradient was in favor of higher densities in the north basin, than south basin individuals were under a flipped density gradient. As a result, the high-density conditions in the north basin during spring were largely reduced during summer and autumn by emigration, whereas high-density conditions in the south basin were largely regulated by density-dependent mortality. When estimating the annual mean intrinsic fitness (the spring situation) for a 50-year period, Haugen et al. (2006) showed that apart from a couple of years (including a three-year period of experimental removal of pike in the northern basin) intrinsic fitness was higher for south basin females. As predicted from the ideal-free-distribution theory (IFD, Fretwell and Lucas 1970, Fretwell 1972), north basin females migrated to the south basin and thereby reduced the between-basin density gradient and increased south basin mortality. The realized fitness was found to be equal between basins, again according to the IFD theory. The between-basin population density pattern was shown to follow a highly significant positive linear relationship ("isodar", Morris 1988) that could be predicted from the IFD theory. Interestingly, adding basin-specific perch density effects to the picture did not change the predicted isodar slope or intercept, but increased the precision. In a spatio-temporally more detailed telemetry study, a fine-scaled IFD pattern was found in the German Kleiner Döllnsee (Kobler et al. 2009). In their study, about 50% of the pike individuals regularly used the pelagic habitat, whereas the other half of the individuals stayed almost exclusively in littoral areas. Despite the higher activity level related to a pelagic habitat use, all individuals exhibited equal lifetime growth rates and were similarly sized. The authors argue that growth can be considered a reasonable proxy for fitness in pike, and suggest the different habitat uses to be driven by spatial gradients in prey availability. In Kleiner Döllnsee, there was a drop in littoral prey resource availability through the course of the season. The costs of movement from the littoral to the pelagic areas and back were considered lower than or equal to the energy gain obtained by feeding in the pelagic area. Hence, the spatio-temporal density distribution pattern in pike seems largely driven by relative resource gradients. Consequently, the spatial population density and structure will change across seasons.

At between-lake level, habitat differences may have large effects on pike populations. Pierce and Tomcko (2005) examined 16 Minnesota lakes with diverse morphometric and biotic characteristics. Lake morphometry was identified as a key factor in determining pike density, which has important effects on growth rates, production, and population size structure. Greater numbers and mass of post-recruit pike larger than 35 cm were found in lakes with more littoral habitat (< 5 m depth) and higher optimal thermal habitat. Percent littoral area was the most important variable explaining density differences, exceeding the effects of other ecological factors such as water productivity, exploitation, or prey fish abundance. Densities of large pike (i.e., > 50 cm) were higher in lakes with larger total area and shoreline length. Similar results were found in the previously mentioned electrofishing study of 57 lakes from north-eastern Germany (Lewin et al. 2014). Also in this study, the most influential lake characteristic, which was positively correlated with pike abundance, was shoreline length in both deep and shallow lakes. Mean depth also had a high impact on pike abundance in shallow lakes (< 5.22 m), suggesting access to pelagic areas may be important, particularly at high turbidity levels (see section 6.3.3.3).

In total, pike abundance seems sensitive to habitat composition, and is capable of utilizing spatial variation in resource distribution. In particular, distribution of littoral habitat plays a key role in pike post-recruit growth, structure and production, which has been suggested to be due to provisioning of optimal thermal habitat, but also provisioning of vegetation/habitat structure.

6.3.3.5 Vegetation

The role of vegetation in pike post-recruit survival and growth is not as clear as for the recruitment process (e.g. Grimm 1981b, Casselman and Lewis 1996) (Fig. 6.2). According to Grimm (1989), the biomass of pike larger than 54 cm is indirectly related to vegetation because they recruit from this habitat. He documented that it is not just the amount of vegetation that affects biomass, but the distribution pattern and the spacing between plants is crucial for a habitat characterized by the interface between open water and vegetation. Hence, availability of open water-vegetated water transition zones seems to play an important role in post-recruit population density and structure (Chapman and Mackay 1984a). The use of these transition zones is largely dependent on individual size and risk of agonistic interactions (Nilsson and Brönmark 1999, Nilsson 2006b). As the post-recruit grows beyond 50-60 cm, inter- and intraspecific predation risk drops markedly (e.g. Carlson et al. 2007, Haugen et al. 2007) and the sheltering function from vegetation gets less critical, allowing for increased habitat-use flexibility. Field observations by Eklöv (1997) support this as pike individual size in the small Swedish lake Degersjön was negatively correlated with aquatic vegetation density. The author links this finding to larger individuals constraining habitat use of smaller conspecifics that have to seek deeper into the less favorable dense aquatic vegetation to seek shelter from their cannibalistic conspecifics (Eklöv and Diehl 1994). Typically, realization of the increased habitat-use flexibility arising from growing may be linked to turbidity, as exemplified in a telemetry study by Andersen et al. (2008). They found probability of using the pelagic habitat increased with size (> 50 cm) in a turbid lake, but not in a clear-water lake. However, Jepsen et al. (2001) found no such effect of turbidity on > 50 cm pike pelagic habitat use, indicating that the letting-go of the vegetation association in pike habitat use not only is a question of cannibalism-risk relaxation, but also has a resource side to it. If key resources are located within or in the proximity to the aquatic vegetation, there is no reason to seek other habitats. See chapter 5 for further discussions of habitat use and turbidity.

The existence of behavioral types in post-recruit pike opens new perspectives on vegetation constraints on pike growth and survival. The Kobler et al. (2009) radio telemetry study shows that two out of three pike behavioral types were tightly linked to aquatic vegetation habitats where fish-prey resources were slightly less abundant than in the open habitat of the mesotrophic to slightly eutrophic German study lake. The third behavior group, comprising 50% of the pike individuals, opportunistically used the open habitat and performed equally well as the two others in terms of growth and survival (see also chapter 5). The authors present two alternative hypotheses, where one proposes the opportunistic group to comprise competitively superior individuals that use the open habitat under conditions with reduced prey availability in the vegetated habitat. Alternatively, the authors suggest these individuals to be subordinately displaced from their preferred vegetated habitat into open habitat due to intraspecific agonistic density-dependent interactions (Nilsson and Brönmark 1999, Nilsson et al. 2006). Because all three behavioral types performed equally well, this behavioral flexibility, irrespective of mechanisms behind, will result in a higher pike production than what would have been the case under a less flexible vegetation-dependent habitat use. Behavioral types as the ones found by Kobler et al. (2009) is discussed further in chapter 5, but clearly the generality of such behavioral patterns remains to be explored in future studies.

6.3.4 Biotic Factors Affecting Post-Recruit Growth and Survival

Intraspecific density-dependent effects in post-recruit pike range from social stress, with implication for growth and fitness, to cannibalism (Nilsson and Brönmark 1999). Social stress may sound like a tiny detail when assessing the role of potential population-structuring factors in a top predator like the pike. However, an experimental study performed to quantify potential population-level consequences from social stress showed otherwise (Edeline et al. 2010). Using mature pike (TL 306-534 mm) they controlled pike densities at two levels, high and low, in experimental ponds under controlled prey (roach and crucian carp *Carassius carasssius* Linnaeus, 1758) availability. During the 3.5-5 months experimental period no cannibalism occurred and just two individuals (out of 120) died. The results showed no change or difference in average prey intake per pike (i.e., no change in interference or exploitative competition), but the high-density treatment pike experienced neuroendocrine stress (measured from blood stress hormones) implying a size-dependent dominance hierarchy. Further, the high-density group had depressed energetic status and, importantly, somatic growth rate was reduced by 23%. Applying the stressed-induced growth response in a Windermere pike life-table simulation model, revealed that the experienced stressed-induced growth deprivation potentially could transfer to a 37-56% decrease in population intrinsic growth rate (largely due to reduced fecundity), making population persistence more dependent on old individuals. The stress level measured in the Edeline et al. (2010) experiment are probably out of scope compared to what will be the case under natural conditions because the experimental pike densities (i.e., 0.5 and 1 pike m^{-2}) were unnaturally high over such a long time span and no suitable habitat structures were made available. However, the experiment demonstrates that social stress is a factor with potentially substantial population-level consequences in pike.

The fraction of large post-recruit individuals will through cannibalism and other agonistic interference behaviors determine the abundance of medium- and small-sized individuals – with potential cascading ecosystem effects (e.g. Magnhagen and Heibo 2001, Persson et al. 2004b, Langangen et al. 2011, chapter 8). The densities and size distributions of suitable prey species should affect the rate of cannibalism. From this, we can infer that pike population structure is largely controlled by the dynamics between the fraction of large individuals and availability of suitable prey species through cannibalism and interspecific predation. Hence, important determinants of post-recruit pike population dynamics and structuring are linked to factors facilitating individuals to become large and thus potential cannibals.

The ontogenetic diet shift from invertivory to piscivory has important effects on individual growth in pike (e.g. Venturelli and Tonn 2006) and constitutes a critical transition with huge implications for population structure and density. By exploring pike populations holding individuals that never get past the invertivorous stage, and get stunted (low juvenile growth rate and a near cessation of adult growth), researchers have gained further knowledge on the importance of this transition process as a determinant of pike population structure. North American pike studies frequently report stunted populations, but this is rarely reported in Eurasian pike populations. The reason for this difference is unclear. A meta-analysis of pike individual growth by Rypel (2012) revealed differential countergradient and climate-gradient response patterns between North American and Eurasian pike, and generally lower climate-adjusted growth potential in North American pike. However, the differences found were generally too small to attribute them to the higher stunting tendency in North American populations, and the author suggested phylogeny-related genetic differences to play a role in this.

Strong density-dependent competition, lack of appropriately sized prey, and lack of thermal refuges in summer are mechanisms that may impose stunting in pike (Diana 1987). Using simulations of female growth, Diana (1987) found size at age-3 to be highly sensitive to even slight (i.e., 5-10%) reductions in daily rations, and even more sensitive to reductions in availability of appropriately sized prey. Finally, inappropriate thermal regimes could reduce pike growth by up to 58%, but only under extremely high-temperature conditions. Hence, under most circumstances, the availability of appropriately sized prey seems like the most important driver of stunting in pike. After the larval and juvenile period of invertebrate feeding, pike shift to larger prey species, and often the size and abundance of available prey species may limit their growth (Diana 1979). The characteristics of lake systems that contain many large pike include abundant deep-water habitat and intermediate-size prey species, such as small coregonids and other appropriately sized fishes like e.g. roach or whitefish *Coregonus* sp. (Chapman and Mackay 1984b, Vøllestad et al. 1986, Jacobson 1993). Such lakes are quite different from lakes maintaining large abundances of smaller pike where abundant spawning and rearing habitat allow for high levels of reproduction (Jacobson 1993). These lake systems are often shallow and weedy and not thermally stratified. Often pike populations become stunted under such conditions (Diana 1987).

A density-manipulation experiment carried out in lakes with stunted pike populations by Margenau (1995) showed that lack of access to sufficient and appropriately sized fish prey constrained individual growth. Studies conducted in small boreal Alberta lakes where pike were introduced (35 kg ha^{-1}) to small lakes where they had access to nothing but invertebrate prey (due to a winterkill episode that eliminated all fish) yielded poor individual growth and stunting (Venturelli and Tonn 2005, 2006). Thus, stunting in pike is a response to limited access to appropriately sized prey in adequate numbers. Strong intraspecific competition for food may also strengthen these effects. The role of appropriate prey availability in pike population stunting may hold the key to understanding the difference in frequency of stunting between North American and Eurasian populations. Most Eurasian pike populations have access to cyprinids, and roach in particular, known to be ideal prey species for pike. Most North American small-lake systems are bluegill-rich, a less suitable prey species for pike, especially for smaller pike individuals (e.g. Paukert et al. 2003).

6.3.5 Interaction Effects

After going through both biotic and abiotic effects on pike growth and mortality, it is evident that none of these factors act alone, but rather in concert and/or conditionally on each other in shaping pike population density and structure. Quantifying such interaction effects rests heavily on access to data from either long-term time series from real populations, extensive multi-population data from comparable systems, or on experimental studies performed under ecologically relevant conditions. For instance, the multi-lake study of Minnesotan pike by Pierce and Tomcko (2005) concluded: "*Density is a predominant driving force in pike population dynamics and seems to be the key link between population dynamics and the suitability of the habitat to support pike. Basin morphometry has strong influences on density, which in turn has important effects on growth rates, production, and size structure of pike populations.*" Using data from both Minnesota and Wisconsin lakes in an earlier study, interaction effects between population density and lake productivity on both individual growth and population size structure was revealed (Pierce et al. 2003). Hence, abiotic conditions like lake morphology and its effect on habitat distribution (e.g. distribution of vegetation) frame how density-dependent processes act to shape pike population dynamics.

The effect of turbidity on pike population structure and individual growth is another factor that cannot be inferred without taking other factors into account. For instance, lake morphology and access to suitable prey may modify the negative effect from increased turbidity (e.g. Vøllestad et al. 1986). Turbidity may also potentially affect density-dependent interaction effects in pike as reduced visual contact may decrease among-individual interference. Supporting this, Nilsson et al. (2006) showed that conspecific presence decreased pike attack frequency and per capita consumption rates in clear-water conditions, and attributed this decrease to intimidation interference among individuals. In another experimental set-up, Nilsson et al. (2009) surprisingly found that the contrary was the case: per capita foraging by pike, when kept in groups, remained relatively constant through a range of turbidity levels. These authors also found that individuals foraging alone foraged more successfully at the highest level of turbidity, indicating high-turbidity conditions to shelter against visual contact with potential prey, and due to absence of interference, feeding was enhanced. Turbidity- and density-dependent interference may stabilize consumer-resource dynamics (Fryxell and Lundberg 1998, chapters 2 and 8) due to reduced predation intensity, and may therefore affect availability and size structure of prey. In turn, this will affect the degree of cannibalism, but also contribute to the mechanisms pertinent to maintenance of turbid alternative ecosystem states in lakes (Scheffer et al. 1993, chapter 8). The pike population-level consequences from this density-turbidity interaction is unclear, but increasing interference effects on feeding efficiency is likely to result in reduced growth and thus potentially alter size structure.

Temperature is crucially affecting performance of pike as predators – both related to attack success and swimming speed. According to Öhlund et al. (2015), climate warming may affect the performance of pike as a predator as temperature may have differential effects on pike attack speeds and prey (here: salmonids) escape speeds. Temperature may also affect pike attack rates. In an encounter experiment Öhlund et al. (2015) found a threshold temperature for these rates at 11°C. Clearly, changes in temperatures may therefore profoundly affect the population structure and dynamics. Performance is strongly temperature dependent, but the performance curves (e.g. swimming speed as a function of temperature, Videler 1993) will vary strongly among species indicating that the predator-prey relationships will differ among prey communities. Because prey communities differ over various geographic scales, we can expect pike performance, and thus individual growth, to differ at similar scales.

The interactions between temperature and density of predators and prey were analyzed by Haugen et al. (2007) in their long-term mark-recapture data of Windermere pike. Profound interaction effects between temperature and density-related factors were found with different size-specific mortalities for mature males and females. Males < 50 cm had far higher mortality rates than females of equal size, whereas larger females tended to have higher mortality rates. In addition, intraspecific density dependence was found to be acting on this size-specific mortality. The authors suggested size-dependent mortality in males to be largely due to agonistic encounters with larger conspecifics (both sexes) during the spawning season, but also due to cannibalism throughout the year. The analysis

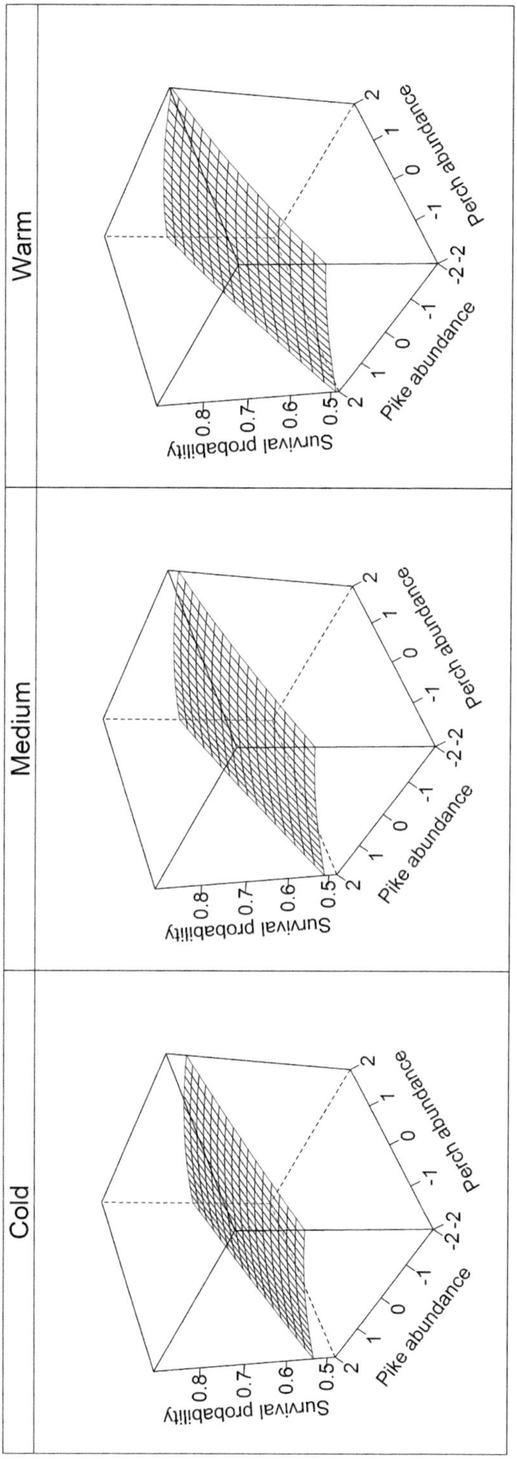

FIGURE 6.4 Interaction prediction plot for the joint effect of perch (prey) abundance, pike (competitor) abundance and water temperature on > 30 cm pike survival. Three temperature regimes are displayed corresponding to −1SD, mean and +1SD temperature conditions for Windermere during the 1944-1995 period. The figure has been reconstructed from Haugen et al. (2007).

also demonstrated that the density-dependent effect from conspecifics on size-dependent survival increased under conditions with low Eurasian perch availability (i.e., the main prey species). However, this perch effect was modified by summer water temperature, where pike survival increased sharply with increasing perch densities under warm summer conditions, but was less affected by perch densities under cold conditions (Fig. 6.4). Furthermore, there was strong evidence for spatial variation in these density-size interaction effects as the interaction effects were very different between individuals living in the two lake basins. Based on analysis of back-calculated individual growth trajectories (from operculum bones) from pike individuals sampled over the same 1949-1995 time-span as covered by the mark-recapture analyses, the same interaction effects between temperature, prey abundance and pike abundance was found for age-2 and age-3 growth as for survival (Haugen et al. 2007). The authors thus concluded individual post-recruit size-at-age and survival to be largely affected by the same density-dependent and density-independent mechanisms. These complex interaction effects of basin, sex, abiotic factors and density thus directly and indirectly impact on growth, survival and dispersal and also have profound population-level consequences.

6.4 CONCLUSIONS AND FUTURE STUDY TOPICS

Pike population size and structure is driven by a multitude of interacting density-dependent and density-independent factors. Clearly, the most important population-regulating processes take place during early stage of recruitment (i.e., first year) and both intra- and inter-cohort interactions, including cannibalism, play key regulating roles. However, strengths of these agonistic and antagonistic interactions may be heavily modified, and even overruled, by both biotic factors (e.g. alternative prey abundance) and abiotic factors (e.g. habitat structure/availability and temperature). All these interacting processes makes it challenging to unravel the most likely underlying Ricker-like (overcompensational) stock-recruitment relationship. Intra-cohort cannibalism is expected to impose size-biased survival in YOY pike, but this remains to be documented under natural conditions. This, most likely, size-dependence of YOY survival makes factors affecting individual growth important drivers of the pike recruitment process, including density-dependent as well density-independent mechanisms – and interactions among these. In particular, distribution and phenology of aquatic vegetation has been shown to play a key constraining role for density-dependent effects (access to prey and cannibalism in particular) to act on YOY growth and survival. The post-recruit part of the population is also largely regulated by processes affecting individual growth, as individual size affects survival, maturation and fecundity. As for the recruitment process, many biotic and abiotic factors interact to affect growth but also survival directly. Spatial patterns of vegetation along with lake morphology, and thus thermal conditions, largely define the growth potential to which density-dependent factors contribute. Availability of suitable fish prey categories has a large impact on the realization of the growth potential, and the roles of cannibalism and intraspecific interference vary strongly among water bodies, spanning from being dominating size-structuring factors to having but minor effects. Finally, lake productivity and turbidity seem to play differential constraining roles on pike abundance in shallow and deep lakes, where abundance generally increases with increasing productivity in deep lakes as well as in shallow lakes with no or short ice-cover during winter. In shallow ice-covered lakes, pike abundance reaches a maximum under mesotrophic condition, where the decreasing pike abundance in eutrophic ice-covered shallow lakes may be linked to oxygen-depletion and/or collapse of vegetation during winter/early spring.

The role of size-dependent mortality in both pike recruits and post-recruits is important for pike population biology, but the mechanisms behind are still not fully understood. Future studies on size-dependent mortality under contrasting natural environmental contexts (e.g. small/shallow lakes *vs* large/deep lakes, turbid *vs* non-turbid water conditions, low *vs* high productivity, little *vs* plenty vegetation, contrasting prey communities and size structures) is likely to elucidate remaining enigmatic pike population processes like the role of sex-related biology (sex ratio, sexual selection, maternal effects, sex-related mortality) and bottleneck factors for pike recruitment.

REFERENCES CITED

Andersen, M., L. Jacobsen, P. Grønkjær and C. Skov. 2008. Turbidity increases behavioural diversity in northern pike, *Esox lucius* L., during early summer. *Fish. Manag. Ecol.* 15(5-6):377-383.

Appelberg, M., H.M. Berger, T. Hesthagen, E. Kleiven, M. Kurkilahti, J. Raitaniemi and M. Rask. 1995. Development and intercalibration of methods in Nordic freshwater fish monitoring. *Water Air Soil Pollut.* 85(2):401-406.

Arlinghaus, R., S. Matsumura and U. Dieckmann. 2010. The conservation and fishery benefits of protecting large pike (*Esox lucius* L.) by harvest regulations in recreational fishing. *Biol. Conserv.* 143(6):1444-1459.

Bekkevold, D., L. Jacobsen, J. Hemmer-Hansen, S. Berg and C. Skov. 2015. From regionally predictable to locally complex population structure in a freshwater top predator: river systems are not always the unit of connectivity in northern pike *Esox lucius*. *Ecol. Fresw. Fish* 24(2):305-316.

Bergström, U., J. Olsson, M. Casini, B.K. Eriksson, R. Fredriksson, H. Wennhage and M. Appelberg. 2015. Stickleback increase in the Baltic Sea – A thorny issue for coastal predatory fish. *Estuar. Coast. Shelf Sci.* 163, Part B:134-142.

Billard, R. 1996. Reproduction of pike: gametogenesis, gamete biology and early development. pp 13-44. *In*: J.F. Craig (ed.). *Pike: Biology and exploitation.* London: Chapman & Hall.

Bondarenko, V., B. Drozd and T. Policar. 2015. Effect of water temperature on egg incubation time and quality of newly hatched larvae of northern pike (*Esox lucius* L., 1758). *J. Appl. Ichthyol.* 31:45-50.

Braum, E., N. Peters and M. Stolz. 1996. The adhesive organ of larval pike *Esox lucius* L., (Pisces). *Int. Rev. Ges. Hydrobiologia* 81:101-108.

Bry, C., M.G. Hollebecq, V. Ginot, G. Israel and J. Manelphe. 1991. Growth-patterns of pike (*Esox lucius* L) larvae and juveniles in small ponds under various natural temperature regimes. *Aquaculture* 97(2-3):155-168.

Bry, C., E. Basset, X. Rognon and F. Bonamy. 1992. Analysis of sibling cannibalism among pike, *Esox lucius*, juveniles reared under seminatural conditions. *Env. Biol. Fish.* 35(1):75-84.

Bry, C., F. Bonamy, J. Manelphe and B. Duranthon. 1995. Early-life characteristics of pike, *Esox lucius*, in rearing ponds - temporal survival pattern and ontogenic diet shifts. *J. Fish Biol.* 46(1):99-113.

Bry, C. 1996. Role of vegetation in the life cycle of pike. pp 45-67. *In*: J.F. Craig (ed.). *Pike: Biology and exploitation*. London: Chapman & Hall.

Bucke, D., G.D. Cawley, J.F. Craig, A.D. Pickering and L.G. Willoughby. 1979. Further studies of an epizooic of perch, *Perca fluviatilis* L., of uncertain aetiology. *J. Fish Diseases* 2:297-311.

Byström, P., U. Bergström, A. Hjalten, S. Stahl, D. Jonsson and J. Olsson. 2015. Declining coastal piscivore populations in the Baltic Sea: Where and when do sticklebacks matter? *Ambio* 44:S462-S471.

Carlson, S.M., E. Edeline, L.A. Vøllestad, T.O. Haugen, I.J. Winfield, J.M. Fletcher, J.B. James and N.C. Stenseth. 2007. Four decades of opposing natural and human-induced artificial selection acting on Windermere pike (*Esox lucius*). *Ecol. Letters* 10(6):512-521.

Casselman, J.M. 1975. Sex-ratios of northern pike, *Esox lucius* Linnaeus. *Trans. Am. Fish. Soc.* 104(1):60-63.

Casselman, J.M. and H.H. Harvey. 1975. Selective fish mortality resulting from low winter oxygen. *Verh. Internat. Verein. Limnol.* 19:2418-2429.

Casselman, J.M. and C.A. Lewis. 1996. Habitat requirements of northern pike (*Esox lucius*). *Can. J. Fish. Aquat. Sci.* 53:161-174.

Casselman JM (2002) Effects of temperature, global extremes, and climate change on year-class production of warmwater, coolwater, and coldwater fishes in the Great Lakes Basin. pp 39-59. *In*: N.A. McGinn (ed.). *Fisheries in a changing climate*. Bethesda, Maryland: American Fisheries Society.

Chapman, C.A. and W.C. Mackay. 1984a. Direct observation of habitat utilization by northern pike. *Copeia* 1(1):255-258.

Chapman, C.A. and W.C. Mackay. 1984b. Versatility in habitat use by a top aquatic predator, *Esox lucius* L. *J. Fish Biol.* 25(1):109-115.

Claessen, D., A.M. de Roos and L. Persson. 2000. Dwarfs and giants: Cannibalism and competition in size-structured populations. *Am. Nat.* 155(2):219-237.

Claessen, D., C. Van Oss, A.M. de Roos and L. Persson. 2002. The impact of size-dependent predation on population dynamics and individual life history. *Ecology* 83(6):1660-1675.

Craig, J.F. and J.A. Babaluk. 1989. Relationship of condition of walleye (*Stizostedion vitreum*) and Northern pike (*Esox lucius*) to water clarity, with special reference to Dauphin lake, Manitoba. *Can. J. Fish. Aquat. Sci.* 46(9):1581-1586.

Craig, J.F. 1996. Population dynamics, predation and role in the community. pp 201-217. *In*: J.F. Craig (ed.). *Pike: Biology and exploitation*. London: Chapman & Hall.

Craig, J.F. 2008. A short review of pike ecology. *Hydrobiologia* 601:5-16.

Crane, D.P., L.M. Miller, J.S. Diana, J.M. Casselman, J.M. Farrell, K.L. Kapuscinski and J.K. Nohner. 2015. Muskellunge and Northern Pike Ecology and Management: Important Issues and Research Needs. *Fisheries* 40(6):258-267.

Cucherousset, J., J.-M. Paillisson and J.-M. Roussel. 2007. Using PIT technology to study the fate of hatchery-reared YOY northern pike released into shallow vegetated areas. *Fish. Res.* 85(1-2):159-164.

Diana, J.S. 1979. Feeding pattern and daily ration of a top carnivore, the northern pike (*Esox lucius*). *Can. J. Zool.* 57(11):2121-2127.

Diana, J.S. 1983. An energy budget for northern pike (*Esox lucius*). *Can. J. Zool.* 61(9):1968-1975.

Diana, J.S. 1987. Simulation of mechanisms causing stunting in northern pike populations. *Trans. Am. Fish. Soc.* 116(4):612-617.

Edeline, E., S.M. Carlson, L.C. Stige, I.J. Winfield, J.M. Fletcher, J. Ben James, T.O. Haugen, L.A. Vollestad and N.C. Stenseth. 2007. Trait changes in a harvested population are driven by a dynamic tug-of-war between natural and harvest selection. *Proc. Natl. Acad. Sci. USA* 104(40):15799-15804.

Edeline, E., T. Ben Ari, L.A. Vollestad, I.J. Winfield, J.M. Fletcher, J. Ben James and N.C. Stenseth. 2008. Antagonistic selection from predators and pathogens alters food-web structure. *Proc. Natl. Acad. Sci. USA* 105(50):19792-19796.

Edeline, E., T.O. Haugen, F.A. Weltzien, D. Claessen, I.J. Winfield, N.C. Stenseth and L.A. Vollestad. 2010. Body downsizing caused by non-consumptive social stress severely depresses population growth rate. *Proc. Roy. Soc. B-Biol. Sci.* 277(1683):843-851.

Ejankowski, W. and T. Lenard. 2015. The effect of ice phenology exerted on submerged macrophytes through physicochemical parameters and the phytoplankton abundance. *J. Limnol.* 75(1).

Eklöv, P. and S. Diehl. 1994. Piscivor efficiency and refuging prey: the importance of predator seach mode. *Oecologia* 98:344-353.

Eklöv, P. 1997. Effects of habitat complexity and prey abundance on the spatial and temporal distributions of perch (*Perca fluviatilis*) and pike (*Esox lucius*). *Can. J. Fish. Aquat. Sci.* 54(7):1520-1531.

Elliott, J.M. 1985. Population regulation for different life-stages of migratory trout *Salmo trutta* in a Lake District stream, 1966-83. *J. Anim. Ecol.* 54(2):617-638.

Fago, D.M. 1977. Northern pike production in managed spawning and rearing marshes. *Tech. Bull. Wisc. Dept. Nat. Res.* 96:1-30.

Farrell, J.M. 2001. Reproductive success of sympatric northern pike and muskellunge in an upper St. Lawrence River Bay. *Trans. Am. Fish. Soc.* 130(5):796-808.

Fortin, R., P. Dumont, H. Fournier, C. Cadieux and D. Villeneuve. 1982. Reproduction et force des classes d'âge du Grand Brochet (*Esox lucius* L.) dans le Haut-Richelieu et la baie Missisquoi. *Can. J. Zool.* 60(2):227-240.

Franklin, D.R. and L.L. Smith. 1963. Early life history of the northern pike, *Esox lucius* l., with special reference to the factors influencing the numerical strength of year classes. *Trans. Am. Fish. Soc.* 92(2):91-110.

Fretwell, S.D. and H.J. Lucas. 1970. On territorial behavior and other factors influencing habitat distributions in birds. *Acta Biotheor.* 19:16-36.

Fretwell, S.D. 1972. *Populations in a seasonal environment.* Vol. 5. Princeton, N.J.: Princeton University Press.

Frost, W.E. 1954. The food of pike, *Esox lucius* L, in Windermere. *J. Anim. Ecol.* 23(2):339-360.

Frost, W.E. and C. Kipling. 1967. A study of reproduction, early life weight-length relationship and growth of pike *Esox lucius* L in Windermere. *J. Anim. Ecol.* 36(3): 651-693.

Fryxell, J.M. and P. Lundberg. 1998. *Individual Behavior and Community Dynamics.* London: Chapman & Hall.

Giles, N., R.M. Wright and M.E. Nord. 1986. Cannibalism in pike fry, *Esox lucius* L.: some experiments with fry densities. *J. Fish Biol.* 29(1):107-113.

Gres, P., P. Lim and A. Belaud. 1996. Effects of the initial stocking density of larval pikes (*Esox lucius* L, 1758) on survival, growth and daily food consumption (zooplankton, Chaoboridae) in intensive culture. *Bull. Franc. Peche Piscicult.* (343):153-174.

Grimm, M.P. 1981a. The composition of northern pike (*Esox lucius* L) populations in 4 shallow waters in the Netherlands, with special reference to factors Influencing 0+ pike biomass. *Fish. Manage.* 12(2):61-76.

Grimm, M.P. 1981b. Intraspecific predation as a principal factor controlling the biomass of northern pike (*Esox lucius* L). *Fish. Manage.* 12(2):77-79.

Grimm, M.P. 1983. Regulation of biomasses of small (< 41 cm) northern pike (*Esox lucius* L), with special reference to the contribution of individuals stocked as fingerlings (4-6 cm). *Fish. Manage.* 14(3):115-134.
Grimm, M.P. 1989. Northern pike (*Esox lucius* L.) and aquatic vegetation, tools in the management of fisheries and water quality in shallow waters. *Hydrobiol. Bull.* 23(1):59-65.
Grimm, M.P. and J.J.G.M. Backx. 1990. The restoration of shallow eutrophic lakes, and the role of northern pike, aquatic vegetation and nutrient concentration. *Hydrobiol.* 200(1):557-566.
Grimm, M.P. and M. Klinge. 1996. Pike and some aspects of its dependence on vegetation. pp 125-156. *In*: J.F. Craig (ed.). *Pike: Biology and exploitation*. London: Chapman & Hall.
Grønkjær, P., C. Skov and S. Berg. 2004. Otolith-based analysis of survival and size-selective mortality of stocked 0+ year pike related to time of stocking. *J. Fish. Biol.* 64(6):1625-1637.
Hargeby, A., I. Blindow and L.-A. Hansson. 2004. Shifts between clear and turbid states in a shallow lake: multi-causal stress from climate, nutrients and biotic interactions. *Arch. Hydrobiol.* 161(4):433-454.
Hart, P. and S.F. Hamrin. 1988. Pike as a selective predator: effects of prey size, availability, cover and pike jaw dimensions. *Oikos* 51(2):220-226.
Hassler, T.J. 1970. Environmental influences on early development and year-class strength of northern pike in lakes Oahe and Sharpe, South Dakota. *Trans. Am. Fish. Soc.* 99(2):369-375.
Hassler, T.J. 1982. Effect of temperature on survival of northern pike embryos and yolk-sac larvae. *Prog. Fish-Cult.* 44(4):174-178.
Hastie, T.J. and R.J. Tibshirani. 1990. *Generalized Additive Models*. London: Chapman & Hall.
Haugen, T.O., I.J. Winfield, L.A. Vøllestad, J.M. Fletcher, J.B. James and N.C. Stenseth. 2006. The ideal free pike: 50 years of fitness-maximizing dispersal in Windermere. *Proc. R. Soc. Lond. B* 273(1604):2917-2924.
Haugen, T.O., I.J. Winfield, L.A. Vøllestad, J.M. Fletcher, J.B. James and N.C. Stenseth. 2007. Density dependence and density independence in the demography and dispersal of pike over four decades. *Ecol. Monogr.* 77(4):483-502.
Hill, C.G. 2004. *Dynamics of northern pike recruitment into the Yampa River*. MSc-thesis, Dep of Fishery and Wildlife biology, Colorado State University.
Hilt, S., T. Wanke, K. Scharnweber, M. Brauns, J. Syväranta, S. Brothers, U. Gaedke, J. Köhler, B. Lischke and T. Mehner. 2015. Contrasting response of two shallow eutrophic cold temperate lakes to a partial winterkill of fish. *Hydrobiologia* 749(1):31-42.
Hokanson, K.E.F., J.H. McCormick and B.R. Jones. 1973. Temperature requirements for embryos and larvae of the northern pike, *Esox lucius* (Linnaeus). *Trans. Am. Fish. Soc.* 102(1):89-100.
Hühn, D., K. Luebke, C. Skov and R. Arlinghaus. 2014. Natural recruitment, density-dependent juvenile survival, and the potential for additive effects of stock enhancement: an experimental evaluation of stocking northern pike (*Esox lucius*) fry. *Can. J. Fish. Aquat. Sci.* 71(10):1508-1519.
Jackson, L.J., T.L. Lauridsen, M. Søndergaard and E. Jeppesen. 2007. A comparison of shallow Danish and Canadian lakes and implications of climate change. *Fresw. Biol.* 52(9):1782-1792.
Jacobson, P.C. 1993. Analysis of factors affecting growth of northern pike in Minnesota. In *Investigational Report*. St. Paul, Minnesota.: Minnesota Department of Natural Resources.
Jeppesen, E., J.P. Jensen, M. Søndergaard, T. Lauridsen and F. Landkildehus. 2000. Trophic structure, species richness and biodiversity in Danish lakes: changes along a phosphorus gradient. *Fresw. Biol.* 45(2):201-218.

Jepsen, N., S. Beck, C. Skov and A. Koed. 2001. Behavior of pike (*Esox lucius* L.) > 50 cm in a turbid reservoir and in a clearwater lake. *Ecol. Fresw. Fish* 10(1):26-34.
Johnson, F.H. 1957. Northern pike year-class strength and spring water levels. *Trans. Am. Fish. Soc.* 86(1):285-293.
Kallasvuo, M., M. Salonen and A. Lappalainen. 2010. Does the zooplankton prey availability limit the larval habitats of pike in the Baltic Sea? *Estuar. Coast. Shelf Sci.* 86(1): 148-156.
Kipling, C. and W.E. Frost. 1970. A study of mortality, population numbers, year class strengths, production and food consumption of pike, *Esox lucius* L, in Windermere from 1944 to 1962. *J. Anim. Ecol.* 39(1):115-157.
Kipling, C. 1983. Changes in the population of pike (*Esox lucius*) in Windermere from 1944 to 1981. *J. Anim. Ecol.* 52(3):989-999.
Kobler, A., T. Klefoth, T. Mehner and R. Arlinghaus. 2009. Coexistence of behavioural types in an aquatic top predator: a response to resource limitation? *Oecologia* 161(4):837-847.
Kosten, S., A. Kamarainen, E. Jeppesen, E.H.v. Nes, E.T.H.M. Peeters, G. Lacerot and M. Scheffer. 2009. Climate-related differences in the dominance of submerged macrophytes in shallow lakes. *Global Change Biol.* 15(10):2503-2517.
Kotakorpi, M., J. Tiainen, M. Olin, H. Lehtonen, K. Nyberg, J. Ruuhijarvi and A. Kuparinen. 2013. Intensive fishing can mediate stronger size-dependent maternal effect in pike (*Esox lucius*). *Hydrobiologia* 718(1):109-118.
Kucharczyk, D., A. Mamcarz, R. Kujawa and A. Skrzypczak. 1997. Development of cannibalism in larval northern pike, *Esox lucius* (Esocidae). *Ital. J. Zool.* 65(Suppl.): 261-263.
Laikre, L., L.M. Miller, A. Palme, S. Palm, A.R. Kapuscinski, G. Thoresson and N. Ryman. 2005. Spatial genetic structure of northern pike (*Esox lucius*) in the Baltic Sea. *Mol. Ecol.* 14(7):1955-1964.
Langangen, O., E. Edeline, J. Ohlberger, I.J. Winfield, J.M. Fletcher, J. Ben James, N.C. Stenseth and L.A. Vøllestad. 2011. Six decades of pike and perch population dynamics in Windermere. *Fish. Res.* 109(1):131-139.
Larsson, P., P. Tibblin, P. Koch-Schmidt, O. Engstedt, J. Nilsson, O. Nordahl and A. Forsman. 2015. Ecology, evolution, and management strategies of northern pike populations in the Baltic Sea. *Ambio* 44:S451-S461.
Le Cren, E.D. 1965. Some factors regulating the size of populations of freshwater fish. *Mitt. Int. Ver. Theor. Ange. Limnol.* 13:88-105.
Le Cren, E.D. 1987. Perch (*Perca fluviatilis*) and pike (*Esox lucius*) in Windermere from 1940 to 1985 - Studies in population-dynamics. *Can. J. Fish. Aquat. Sci.* 44(Suppl. 2):216-228.
Lewin, W.C., T. Mehner, D. Ritterbusch and U. Bramick. 2014. The influence of anthropogenic shoreline changes on the littoral abundance of fish species in German lowland lakes varying in depth as determined by boosted regression trees. *Hydrobiologia* 724(1):293-306.
Maes, J., A. Taillieu, P.A. Van Damme, K. Cottenie and F. Ollevier. 1998. Seasonal Patterns in the Fish and Crustacean Community of a Turbid Temperate Estuary (Zeeschelde Estuary, Belgium). *Estuar. Coast. Shelf Sci.* 47(2):143-151.
Magnhagen, C. and E. Heibo. 2001. Gape size allometry in pike reflects variation between lakes in prey availability and relative body depth. *Func. Ecol.* 15(6):754-762.
Margenau, T.L. 1995. Stunted northern pike: a case history of community manipulations and field transfer In *Research report*: Wisconsin. Dept. of Natural Resources.
Margenau, T.L., P.W. Rasmussen and J.M. Kampa. 1998. Factors affecting growth of northern pike in small northern Wisconsin lakes. *N. Am. J. Fish. Manage.* 18(3):625-639.
Mills, C.A. and M.A. Hurley. 1990. Long-term studies on the Windermere populations of perch (*Perca fluviatilis*), pike (*Esox lucius*) and Arctic charr (*Salvelinus alpinus*). *Freshw. Biol.* 23(1):119-136.

Mingelbier, M., P. Brodeur and J. Morin. 2008. Spatially explicit model predicting the spawning habitat and early stage mortality of Northern pike (*Esox lucius*) in a large system: the St. Lawrence River between 1960 and 2000. *Hydrobiologia* 601:55-69.

Minns, C.K., R.G. Randall, J.E. Moore and V.W. Cairns. 1996. A model simulating the impact of habitat supply limits on northern pike, *Esox lucius*, in Hamilton Harbour, Lake Ontario. *Can. J. Fish. Aquat. Sci.* 53(S1):20-34.

Mischke, U., U. Riedmüller, E. Hoehn, I. Schönfelder and B. Nixdorf. 2008. Description of the German system for phytoplankton-based assessment of lakes for implementation of the EU Water Framework Directive (WFD). In *Gewässerreport* Bad Saarow, Freiburg, Berlin: Univ. Cottbus, Lehrstuhl Gewässerschutz.

Mittelbach, G.G. and L. Persson. 1998. The ontogeny of piscivory and its ecological consequences. *Can. J. Fish. Aquat. Sci.* 55(6):1454-1465.

Morris, D.W. 1988. Habitat-dependent population regulation and community structure. *Evol. Ecol.* 2:253-269.

Murry, B.A., J.M. Farrell, K.L. Schulz and M.A. Teece. 2008. The effect of egg size and nutrient content on larval performance: implications to protracted spawning in northern pike (*Esox lucius* Linnaeus). *Hydrobiologia* 601:71-82.

Myers, R.A. 2001. Stock and recruitment: generalizations about maximum reproductive rate, density dependence, and variability using meta-analytic approaches. *ICES J. Mar. Sci.* 58(5):937-951.

Neumann, R.M. and D.W. Willis. 1995. Seasonal variation in gill-net sample indexes for northern pike collected from a glacial prairie lake. *N. Am. J. Fish. Manage.* 15:838-844.

Nilsson, J. 2006a. Predation of northern pike (*Esox lucius* L.) eggs: A possible cause of regionally poor recruitment in the Baltic Sea. *Hydrobiologia* 553(1):161-169.

Nilsson, P.A. and C. Brönmark. 1999. Foraging among cannibals and kleptoparasites: effects of prey size on pike behavior. *Behav. Ecol.* 10(5):557-566.

Nilsson, P.A. and C. Brönmark. 2000. Prey vulnerability to a gape-size limited predator: behavioural and morphological impacts on northern pike piscivory. *Oikos* 88(3): 539-546.

Nilsson, P.A., H. Turesson and C. Bronmark. 2006. Friends and foes in foraging: intraspecific interactions act on foraging-cycle stages. *Behaviour* 143:733-745.

Nilsson, P.A. 2006b. Avoid your neighbours: size-determined spatial distribution patterns among northern pike individuals. *Oikos* 113(2):251-258.

Nilsson, P.A., L. Jacobsen, S. Berg and C. Skov. 2009. Environmental conditions and intraspecific interference: unexpected effects of turbidity on pike (*Esox lucius*) foraging. *Ethology* 115(1):33-38.

Ohlberger, J., E. Edeline, L.A. Vøllestad, N.C. Stenseth and D. Claessen. 2011. Temperature-driven regime shifts in the dynamics of size-structured populations. *Am. Nat.* 177(2): 211-223.

Ohlberger, J. 2013. Climate warming and ectotherm body size – from individual physiology to community ecology. *Func. Ecol.* 27(4):991-1001.

Ohlberger, J., S.J. Thackeray, I.J. Winfield, S.C. Maberly and L.A. Vøllestad. 2014. When phenology matters: age–size truncation alters population response to trophic mismatch. *Proc. Roy. Soc. B-Biol. Sci.* 281(1793).

Öhlund, G., P. Hedström, S. Norman, C.L. Hein and G. Englund. 2015. Temperature dependence of predation depends on the relative performance of predators and prey. *Proc. R. Soc. Lond. B* 282(1799).

Pagel, T. 2009. *Determinants of individual reproductive success in a natural pike* (*Esox lucius* L.) *population: a DNA-based parentage assignment approach.* MSc-thesis, Faculty of Agriculture and Horticulture, Humboldt Universität zu Berlin, Berlin.

Pagel, T., D. Bekkevold, S. Pohlmeier, C. Wolter and R. Arlinghaus. 2015. Thermal and maternal environments shape the value of early hatching in a natural population of a strongly cannibalistic freshwater fish. *Oecologia* 178(4):951-965.

Partridge, L. and M. Mangel. 1999. Messages from mortality: the evolution of death rates in the old. *Trends Ecol. Evol.* 14(11):438-442.

Paukert, C.P., W. Stancill, T.J. DeBates and D.W. Willis. 2003. Predatory effects of northern pike and largemouth bass: bioenergetic modeling and ten years of fish community sampling. *J. Freshw. Ecol.* 18(1):13-24.

Paxton, C.G.M., J.M. Fletcher, D.P. Hewitt and I.J. Winfield. 1999. Sex ratio changes in the long-term Windermere pike and perch sampling program. *Ecol. Fresw. Fish* 8(2):78-84.

Paxton, C.G.M., I.J. Winfield, J.M. Fletcher, D.G. George and D.P. Hewitt. 2009. Investigation of first year biotic and abiotic influences on the recruitment of pike *Esox lucius* over 48 years in Windermere, UK. *J. Fish Biol.* 74(10):2279-2298.

Penning, W.E., B. Dudley, M. Mjelde, S. Hellsten, J. Hanganu, A. Kolada, M. van den Berg, S. Poikane, G. Phillips, N. Willby and F. Ecke. 2008a. Using aquatic macrophyte community indices to define the ecological status of European lakes. *Aquat. Ecol.* 42(2):253-264.

Penning, W.E., M. Mjelde, B. Dudley, S. Hellsten, J. Hanganu, A. Kolada, M. van den Berg, S. Poikane, G. Phillips, N. Willby and F. Ecke. 2008b. Classifying aquatic macrophytes as indicators of eutrophication in European lakes. *Aquat. Ecol.* 42(2):237-251.

Perez, K.O. and S.B. Munch. 2010. Extreme selection on size in the early lives of fish. *Evolution* 64(8):2450-2457.

Persson, L., D. Claessen, A.M. De Roos, P. Byström, S. Sjögren, R. Svanbäck, E. Wahlström and E. Westman. 2004a. Cannibalism in a size-structured population: Energy extraction and control. *Ecol. Monogr.* 74(1):135-157.

Persson, L., A.M. de Roos and A. Bertolo. 2004b. Predicting shifts in dynamics of cannibalistic field populations using individual-based models. *Proc. R. Soc. Lond. B* 271(1556):2489-2493.

Persson, L., A. Bertolo and A.M. De Roos. 2006. Temporal stability in size distributions and growth rates of three *Esox lucius* L. populations. A result of cannibalism? *J. Fish Biol.* 69(2):461-472.

Petrosky, B.R. and J.J. Magnuson. 1973. Behavioral responses of northern pike, yellow perch and bluegill to oxygen concentrations under simulated winterkill conditions. *Copeia* 1973(1):124-133.

Pierce, R.B. and C.M. Tomcko. 2003. Interrelationships among production, density, growth, and mortality of northern pike in seven north-central Minnesota lakes. *Trans. Am. Fish. Soc.* 132(1):143-153.

Pierce, R.B., C.M. Tomcko and T.L. Margenau. 2003. Density dependence in growth and size structure of northern pike populations. *N. Am. J. Fish. Manage.* 23(1):331-339.

Pierce, R.B. and C.M. Tomcko. 2005. Density and biomass of native northern pike populations in relation to basin-scale characteristics of north-central Minnesota lakes. *Trans. Am. Fish. Soc.* 134(1):231-241.

Pierce, R.B. 2010. Long-term evaluations of length limit regulations for northern pike in Minnesota. *N. Am. J. Fish. Manage.* 30(2):412-432.

Raat, A.J.P. 1988. *Synopsis of biological data on the northern pike Esox lucius Linnaeus, 1758*. Rome: FAO.

Ranåker, L., M. Jönsson, P.A. Nilsson and C. Brönmark. 2012. Effects of brown and turbid water on piscivore–prey fish interactions along a visibility gradient. *Freshw. Biol.* 57(9):1761-1768.

Ricker, W.E. 1954. Stock and Recruitment. *J. Fish. Res. Board Can.* 11(5):559-623.

Roff, D.A. 1984. The evolution of life history parameters in teleosts. *Can. J. Fish. Aquat. Sci.* 41:989-1000.

Rypel, A.L. 2012. Meta-analysis of growth rates for a circumpolar fish, the northern pike (*Esox lucius*), with emphasis on effects of continent, climate and latitude. *Ecol. Fresw. Fish* 21(4):521-532.

Salonen, M., L. Urho and J. Engström-Öst. 2009. Effects of turbidity and zooplankton availability on the condition and prey selection of pike larvae. *Boreal Env. Res.* 14(6):981-989.

Salonen, M. and J. Engström-Öst. 2010. Prey capture of pike *Esox lucius* larvae in turbid water. *J. Fish Biol.* 76(10):2591-2596.

Salonen, M. and J. Engström-Öst. 2013. Growth of pike larvae: effects of prey, turbidity and food quality. *Hydrobiologia* 717(1):169-175.

Scheffer, M., S.H. Hosper, M.L. Meijer, B. Moss and E. Jeppesen. 1993. Alternative equilibria in shallow lakes. *Trends Ecol. Evol.* 8(8):275-279.

Schindler, D.W. 1974. Eutrophication and recovery in experimental lakes: implications for lake management. *Science* 184(4139):897-899.

Sharma, C.M. and R. Borgstrøm. 2008a. Increased population density of pike *Esox lucius* - a result of selective harvest of large individuals. *Ecol. Fresw. Fish* 17(4):590-596.

Sharma, C.M. and R. Borgstrøm. 2008b. Shift in density, habitat use, and diet of perch and roach: An effect of changed predation pressure after manipulation of pike. *Fish. Res.* 91(1):98-106.

Skov, C. and S. Berg. 1999. Utilization of natural and artificial habitats by YOY pike in a biomanipulated lake. pp 115-122. *In*: N. Walz and B. Nixdorf (eds.). *Shallow Lakes '98: Trophic interactions in shallow freshwater and brackish waterbodies*. Dordrecht, Netherlands: Springer.

Skov, C., S. Berg, L. Jacobsen and N. Jepsen. 2002. Habitat use and foraging success of 0+ pike (*Esox lucius* L.) in experimental ponds related to prey fish, water transparency and light intensity. *Ecol. Fresw. Fish* 11(2):65-73.

Skov, C., L. Jacobsen and S. Berg. 2003a. Post-stocking survival of 0+year pike in ponds as a function of water transparency, habitat complexity, prey availability and size heterogeneity. *J. Fish Biol.* 62(2):311-322.

Skov, C., O. Lousdal, P.H. Johansen and S. Berg. 2003b. Piscivory of 0+pike (*Esox lucius* L.) in a small eutrophic lake and its implication for biomanipulation. *Hydrobiologia* 506(1-3):481-487.

Skov, C. and A. Koed. 2004. Habitat use of 0+year pike in experimental ponds in relation to cannibalism, zooplankton, water transparency and habitat complexity. *J. Fish Biol.* 64(2):448-459.

Skov, C., A. Koed, L. Baastrup-Spohr and R. Arlinghaus. 2011. Dispersal, growth, and diet of stocked and wild northern pike fry in a shallow natural lake, with implications for the management of stocking programs. *N. Am. J. Fish. Manage.* 31(6):1177-1186.

Smith, C. and P. Reay. 1991. Cannibalism in teleost fish. *Rev. Fish Biol.* 1:41-64.

Stearns, S.C. 1992. *The Evolution of Life Histories*. Oxford: Oxford University Press.

Talling, J.F. 1993. Comparative seasonal changes, and inter-annual variability and stability, in a 26-year record of total phytoplankton biomass in four English lake basins. *Hydrobiologia* 268(2):65-98.

Tibblin, P. 2015. *Migratory behaviour and adaptive divergence in life-history traits of pike* (*Esox lucius*). PhD-thesis, Department of Biology and Environmental Sciences, Linnaeus University, Kalmar.

Tibblin, P., A. Forsman, P. Koch-Schmidt, O. Nordahl, P. Johannessen, J. Nilsson and P. Larsson. 2015. Evolutionary divergence of adult body size and juvenile growth in sympatric subpopulations of a top predator in aquatic ecosystems. *Am. Nat.* 186(1):98-110.

Tonn, W.M. and J.J. Magnuson. 1982. Patterns in the species composition and richness of fish assemblages in northern Wisconsin lakes. *Ecology* 63(4):1149-1166.

Trabelsi, A., J.-N. Gardeur, F. Teletchea, J. Brun-Bellut and P. Fontaine. 2013. Hatching time effect on the intra-spawning larval morphology and growth in Northern pike (*Esox lucius* L.). *Aquacult. Res.* 44(4):657-666.

Trabelsi, A., J.-N. Gardeur, H. Ayadi and P. Fontaine. 2015. Effect of spawning time on egg quality, larval morphometrics and survival of Northern pike *Esox lucius*. *Cybium* 39(2):91-98.

Treasurer, J.W., R. Owen and E. Bowers. 1992. The population-dynamics of pike, *Esox lucius*, and perch, *Perca fluviatilis*, in a simple predator-prey system. *Env. Biol. Fish.* 34(1):65-78.

van Kooten, T., J. Andersson, P. Byström, L. Persson and A.M. de Roos. 2010. Size at hatching determines population dynamics and response to harvesting in cannibalistic fish. *Can. J. Fish. Aquat. Sci.* 67(2):401-416.

Venturelli, P.A. and W.M. Tonn. 2005. Invertivory by northern pike (*Esox lucius*) structures communities of littoral macroinvertebrates in small boreal lakes. *J. N. Am. Benthol. Soc.* 24(4):904-918.

Venturelli, P.A. and W.M. Tonn. 2006. Diet and growth of northern pike in the absence of prey fishes: initial consequences for persisting in disturbance-prone lakes. *Trans. Am. Fish. Soc.* 135(6):1512-1522.

Videler, J.J. 1993. *Fish Swimming*. T.J. Pitcher (ed.). Fish and Fisheries Series. London: Chapman & Hall.

Vindenes, Y., E. Edeline, J. Ohlberger, Ø. Langangen, I.J. Winfield, N.C. Stenseth and L.A. Vøllestad. 2014. Effects of climate change on trait-based dynamics of a top predator in freshwater ecosystems. *Am. Nat.* 183(2):243-256.

Vindenes, Y., Ø. Langangen, I.J. Winfield and L.A. Vøllestad. 2016. Fitness consequences of early life conditions and maternal size effects in a freshwater top predator. *J. Anim. Ecol.* 85(3):692-704.

Volkova, L. 1973. The effect of light intensity on the availability of food organisms to some fishes in Lake Baikal. *J. Ichthyol.* 13:591-602.

Vøllestad, L.A., J. Skurdal and T. Qvenild 1986. Habitat use, growth, and feeding of pike (*Esox lucius* L.) in SE Norway. *Arch. Hydrobiol.* 108(1):107-117.

Walrath, J.D., M.C. Quist and J.A. Firehammer. 2015. Population structure and dynamics of northern pike and smallmouth bass in Coeur d'Alene Lake, Idaho. *Northwest Sci.* 89(3):280-296.

West, S.A. 2009. *Sex Allocation*. Princeton University Press. Princeton, NJ.

Winfield, I.J., J.B. James and J.M. Fletcher. 2008. Northern pike (*Esox lucius*) in a warming lake: changes in population size and individual condition in relation to prey abundance. *Hydrobiologia* 601(1):29-40.

Winfield, I.J., J.M. Fletcher and J. Ben James. 2012. Long-term changes in the diet of pike (*Esox lucius*), the top aquatic predator in a changing Windermere. *Fresw. Biol.* 57(2):373-383.

Wright, R.M. and N. Giles. 1987. The survival, growth and diet of pike fry, *Esox lucius* L, stocked at different densities in experimental ponds. *J. Fish. Biol.* 30(5):617-629.

Wright, R.M. 1990. The population biology of pike, *Esox lucius* L., in two gravel pit lakes, with special reference to early life history. *J. Fish Biol.* 36(2):215-229.

Zaikov, A., T. Hubenova and P. Vasileva. 2006. Investigation on predator/prey body weight and length proportions in pike *Esox lucius* L. *Bulg. J. Agric. Sci.* 12:203-207.

Chapter 7

Population Genetics of Pike

Lovisa Wennerström[*1], Dorte Bekkevold[2] and Linda Laikre[3]*

Studies of genetic composition and how genetic variation is distributed in space and time provide us with information on evolutionary histories and contemporary processes of populations. Genetic data can also be used in conservation and management contexts by identifying populations appropriate for management units, estimating sustainable fishing quotas, and finding populations to use for reintroductions or supportive stocking. Several studies have addressed these aspects in pike *Esox lucius* Linnaeus, 1758, and in this chapter we summarize main achievements in pike population genetics, synthesizing information on the intraspecific biodiversity, evolutionary history and current genetic processes of this species, and exemplify how this information is of relevance for sustainable management.

Population genetic patterns are affected by a number of factors operating from ancient to contemporary time scales. Over wide geographic areas, such patterns are largely shaped by the evolutionary history of the species and by demographic history over the last few thousand years (Allendorf et al. 2013). In temperate regions specifically, repeated glaciations during the Pleistocene forced populations to retract in glacial refugia. Population bottlenecks in these refugia, and subsequent population expansion into the present day distribution, have left traces in the genomes of organisms (Hewitt 1996). Current population genetic patterns, and particularly those over restricted geographic areas, are also affected by contemporary processes. These processes include gene flow (or lack thereof) among

**Corresponding author*: [1]Stockholm University, Department of Zoology, Division of Population Genetics, SE-106 91 Stockholm, Sweden; Email: lovisa.wennerstrom@zoologi.su.se

[2]DTU Aqua, Technical University of Denmark, Section for Marine Living Resources, Vejlsøvej 39, 8600, Silkeborg, Denmark; Email: db@aqua.dtu.dk

[3]Stockholm University, Department of Zoology, Division of Population Genetics, SE-106 91 Stockholm, Sweden; Email: linda.laikre@popgen.su.se

populations and adaptations to local environments following natural selection. Further, stochastic processes such as genetic drift can result in random changes of allele frequencies, the magnitude of which is largely coupled to population size. In addition, anthropogenic activities including fishing, habitat alteration and stocking of non-native strains can alter genetic diversity and the composition of populations and species.

7.1 LOW LEVELS OF GENETIC VARIATION IN PIKE

Genetic variation, ultimately variation in the sequence of nucleotides in the DNA, can be studied using a number of different methods. Pike has been shown to have extremely low levels of variation at almost every genetic marker used, especially in North American pike populations (Miller and Senanan 2003, Rondeau et al. 2014). This means that any two pike individuals will have more DNA sequences in common (be more genetically similar) than when comparing, for example, two trout with each other. In the 1980s, population genetics in natural populations was studied using protein variants known as allozymes. However, in pike virtually no allozyme variation was found. For example, from a screening of 65 allozyme loci in North American populations, only two of them turned out to be polymorphic, i.e. showed genetic variation (Seeb et al. 1987). Slightly more variation was detected when European populations were examined, but the levels of variation were still too low for any detailed studies of population structure (Healy and Mulcahy 1980). In the 1990s, studies of mitochondrial DNA (mtDNA, see glossary in Box 7.1) and microsatellites (Box 7.1) were more successful at finding genetic variation, although these markers also showed lower variation in pike compared to other freshwater fishes (Skog et al. 2014).

Heterozygosity is a common measure used to determine how much genetic variation is harbored in individuals and populations (Skog et al. 2014). In North American pike populations the heterozygosity in microsatellite markers was below 0.30 (Senanan and Kapuscinski 2000), and in a study of French populations heterozygosity ranged between 0.18 and 0.45 (Launey et al. 2006). These estimates can be compared with average heterozygosities for freshwater and anadromous fishes at 0.46 and 0.68 respectively (DeWoody and Avise 2000). Later studies have found relatively high levels of genetic variation in pike inhabiting larger water bodies, for example, the Baltic Sea (e.g. H_e: 0.22-0.61; Bekkevold et al. 2014). The low levels of genetic variation in pike have been suggested to be caused by small effective population sizes, which is typical for top predators (Healy and Mulcahy 1980, Senanan and Kapuscinski 2000), coupled with the effects of population bottlenecks during founder events and the fluctuating availability of suitable habitats (Miller and Senanan 2003, Oullet-Cauchon et al. 2014). Both large- and small-scale genetic structure has been studied using mtDNA and microsatellites, and most of our present day knowledge on phylogeography and local population genetic structure has come from studies of these genetic markers.

Box 7.1 Glossary

Allele: Variant of a specific gene. The occurrence of separate alleles of a gene implies that there is genetic variation at this specific location or locus in the DNA.

Allozymes: Variant forms of enzymes resulting from different alleles at a particular locus.

Effective population size: The size of an ideal, panmictic population that experiences the same rate of genetic drift and loss of genetic variation as the observed population.

F_{ST}**:** A measure of genetic differentiation between populations, estimated as differences in allele frequencies.

Gene flow: Exchange of genetic material among populations through migration or human assisted stocking.

Genetic drift: Random fluctuations in allele frequencies between generations.

Haplotypes: A haplotype represents the combination of alleles that are found on a haploid DNA molecule, e.g. in mitochondria.

Heterozygosity: Commonly used as a measure of genetic variation. Indicates the proportion of individuals that carry two alleles (are heterozygote) in separate genes.

Hybridization: Crossing between individuals of different species or populations resulting in admixed genomes.

Locus: A location in the genome where a particular gene can be found (plural: loci).

Metapopulation: A system of spatially divided subpopulations that are connected to each other via gene flow.

Microsatellites: Repetitive sequences of nuclear (chromosomal) DNA; commonly used neutral genetic markers to infer population structures in population genetic studies.

Mitochondrial DNA (mtDNA): A small, circular DNA molecule found in the mitochondria of eukaryotic cells. Its sequence information is haploid in contrast to the diploid nuclear sequences of e.g. microsatellites. Commonly used maternally inherited genetic markers (e.g. cyt b or D-loop) to infer phylogenetic relationships.

Genetics recently entered the genomic era and studies using thousands of genetic markers over the entire genome (e.g. single nucleotide polymorphisms, SNPs) are nowadays becoming increasingly common. Recently, the genome of the pike was described, considerably increasing the genomic resources (Rondeau et al. 2014). Genomic studies can increase the precision and accuracy of the identification of population structure and demographic processes, as well as enable studies of the genetic mechanisms behind local adaptation – all of which are important issues in conservation genetics and management (Shafer et al. 2015). The pike genome has so far not been the focus of direct studies, but it has been used in comparative studies with salmonids, to which esocids are the most closely related group (Rondeau et al. 2014).

7.2 BROAD SCALE PHYLOGEOGRAPHY

Large-scale genetic structure of freshwater fish in the northern Hemisphere is to a substantial degree shaped by glacial history. Populations retracted into glacial refugia during repeated glaciations in the Pleistocene, and expanded and recolonized habitats in the present day distribution after the retraction of glaciers (Bernatchez and Wilson 1998). Using genetic data, especially mtDNA, it is possible to track the geographic distribution of different clades (populations originating from separate post-glacial founder events). The geographical location of the different glacial refugia can be inferred from contemporary levels of genetic variation. The maximum levels of variation are expected to be found in the area where refuge population persisted and lower levels of genetic variation are expected to occur further from the glacial refugia due to founder effects during post-glacial recolonization (e.g. Hewitt 1996).

The phylogeography of pike indicates a rapid expansion over the Holarctic region after the last ice age from only a few glacial refugia and subsequent isolation of populations. These hypotheses are based on the fact that only a small number of clades consisting of a few mtDNA haplotypes with only small differences between them have been identified (Box 7.1, Maes et al. 2003, Miller and Senanan 2003, Nicod et al. 2004, Launey et al. 2006, Skog et al. 2014, Wooller et al. 2015). Higher genetic variation in European populations compared to those in North America suggests that these refugia were located in Eurasia (Miller and Senanan 2003, Skog et al. 2014). A distant Eurasian origin would thus explain why relatively few haplotypes 'made it' to found populations on the American continent.

In the first studies of pike phylogeography, sampling designs were rather restricted geographically, and only short sequences from the mtDNA molecule were used to describe genetic structure (Maes et al. 2003, Nicod et al. 2004). Very little variation was detected and a Holarctic expansion from one single refugium was suggested (Maes et al. 2003). Maes et al. (2003) detected only three variable sites in the analyzed mtDNA sequence, suggesting that longer fragments are required to capture genetic variation (Nicod et al. 2004). Limited geographic sampling also made it impossible to pinpoint the location of glacial refugia in these early studies (Maes et al. 2003, Nicod et al. 2004). Identification of the possible location of glacial refugia in e.g. Beringia – the landmass once connecting North America and Asia – was instead based on both direct genetic evidence from contemporary pike populations, as well as on suggested locations of important glacial refugia for other species (Senanan and Kapuscinski 2000, Maes et al. 2003, Miller and Senanan 2003, Nicod et al. 2004).

In the most comprehensive study on pike phylogeography to date, Skog et al. (2014) both confirmed and refuted some earlier theories and added much needed information to the field. They used a sampling design that covered large parts of the natural range of pike and used longer fragments of mtDNA in comparison to previous studies, including two different locations on the mtDNA molecule, the cytochrome b and the D-loop, in order to increase information content. With this approach Skog et al. (2014) identified four different mitochondrial clades that all showed signs of rapid post-glacial expansion. The estimated splitting time between these clades was ~ 200,000 years before present, which predates the most recent glacial maximum (10,000-20,000 YBP). Thus, present day genetic variation cannot have arisen solely after the last glaciation. Instead, the more parsimonious explanation is that the clades inhabited different glacial refugia from which they expanded to their present day distribution.

One of the clades is extremely widespread and distributed over the entire northern Hemisphere, which indeed indicates a Holarctic colonization from one single refugium to both North America and Eurasia (Skog et al. 2014). The exact location of this refugium could not be established with certainty, but two regions that exhibit the highest genetic variation are possible candidates. The glacial refugium from which both European and North American populations expanded might have been located in eastern Asia or Beringia (Miller and Senanan 2003, Wooller et al. 2015). Another possible region, where present day populations show even higher genetic variation, is the Ural River catchment, draining into the Caspian Sea. This Ponto-Caspian region has also been suggested to have been an important glacial refugium for other freshwater species (Bernatchez and Wilson 1998).

Multiple mtDNA clades in Europe suggest expansion from additional glacial refugia (Miller and Senanan 2003, Nicod et al. 2004, Skog et al. 2014). Suggestions as to their location include the sub-alpine region and central Europe (Nicod et al. 2004), but in spite of the comprehensive sampling design the location of these potential glacial refugia could not be inferred by Skog and colleagues (2014).

7.3 POPULATION STRUCTURE

Population genetic structure on a local scale is affected by 1) colonization history (how many fish initially founded the population, which evolutionary background did they come from, and how genetically diverse were they?), 2) degree of isolation (has the population been fully isolated from immigration since founding, or does it receive immigrants from other populations contributing new genetic variants occasionally, or even regularly?), 3) population size that results from the carrying capacity of the local habitat and its variation in time (does the habitat support an adult population of tens, hundreds or thousands of fish, and does environmental variation or stochasticity sometimes lead to population reductions, crashes – or even local extinctions?), and 4) natural selection pressures (do environmental parameters act as selective drivers leading to evolutionary changes in local populations, e.g. heritable adaptation to spawning in salt water?). Each of these components affects the genetic variation of populations, and each can be evaluated using population genetic methods. Both individually and in combination, estimates of their importance inform about the genetic 'health' status of the population and allow predictions about its resilience and how it should be managed.

Apart from being a widespread top predator in lakes and ponds, the pike is also a key species in several very large ecosystems, including brackish marine systems and the huge, connected lake systems in North America (Eriksson et al. 2009, 2011). Hence, population life history and dynamics (chapter 6), migratory behavior (chapter 5) and population genetic parameters will vary greatly depending on the type of waters that the population inhabits, e.g. lotic or lentic waters. Whereas pike occur across habitats of greatly varying size, genetic studies have often focused on populations either in physically isolated (commonly smaller lakes) or in physically connected systems (commonly larger lake systems or brackish waters). Here, we therefore exemplify genetic characteristics of pike populations in two main water types: small lakes in freshwater systems and large water systems of fresh or brackish water.

7.3.1 Genetic Dynamics in Small Water Bodies

The low levels of genetic diversity observed across pike populations inhabiting more-or-less isolated lakes and streams correspond with the expectation that small populations generally lose genetic diversity quite rapidly due to random genetic drift (Wright 1931). The genetically effective population size, N_e, is a parameter that quantifies the rate of such loss and can roughly be described as the effective number of individuals that have contributed genetically to a population over time. N_e can be estimated based on genotype information, and has in several cases been used to evaluate and monitor the genetic health of populations, including in fishes (Schwartz et al. 2007). An N_e below 50 is considered a warning sign that the population is so small that it will from purely random effects lose genetic variation at a rate that may lead to negative effects associated with inbreeding, such as the lowering of population fitness (so-called inbreeding depression). Low N_e and genetic diversity also mean that the population is expected to be too limited to respond effectively to natural selective pressures, and long-term population resilience is hence expected to be at risk. Only a few studies have estimated N_e in natural pike populations, and although they apply analytical approaches that are not always directly comparable, they confirm that pike N_es are small (Jacobsen et al. 2005) and may be below 50, at least in some lakes (Lucentini et al 2009, Bekkevold et al. 2014). Miller and Kapuscinski (1997) estimated N_e to be 48 in an introduced Wisconsin population, and Aguilar et al. (2005) reported N_e at 12 for an introduced Californian population. Nonetheless, estimates are sometimes substantially higher, even for presumed small and isolated populations (Jacobsen et al. 2005, Bekkevold et al. 2014). These may in some cases be statistical artifacts caused by estimation variance or sampling effects, but may also reflect that contemporary N_e estimates track different post glacial founder events (Bekkevold et al. 2014, Skog et al. 2014). N_e estimates nevertheless support a scenario of contemporary natural populations being established in post glacial founder events, followed by climatic and habitat changes that led to increased reproductive isolation and genetic drift in local populations. Using a coalescence-based method, Jacobsen et al. (2005) estimated that contemporary N_es in pike populations from across Europe have declined to only a few percent of what characterized early post glacial populations. It is interesting to note that despite several estimates of small N_es in pike populations, the species has been able to persist in a wide variety of aquatic habitats.

The degree of genetic divergence between populations is of interest because it yields information about how demographically connected two or more populations are. Divergence is commonly estimated using the parameter F_{ST}, which can theoretically range between zero (no differentiation; samples represent a common reproductive unit) and one (populations are completely isolated and genetically fixed for different gene variants). As pike populations in smaller lakes are often small and more or less physically isolated, random genetic drift has a significant impact on divergence. This corresponds well with estimates of F_{ST} generally being large (> 0.2) when comparing lake populations (e.g. Hansen et al. 1999, Larsen et al. 2005, Jacobsen et al. 2005, Bekkevold et al. 2014). This also means that pike from individual lakes can be genetically characterized (e.g. Hansen et al. 1999,

Eschbach and Schöning 2013) and molecular approaches can be used to assess the impact of stocking with non-native broodstock and steer management decisions (e.g. Larsen et al. 2005, Launey et al. 2006, Eschbach et al. 2009). Genetic divergence, F_{ST}, can also be compared to another measure of divergence, Q_{ST}, which describes quantitative genetic divergence in order to distinguish among selection and genetic drift as responsible drivers of differentiation among populations (Leinonen et al. 2013).

Bekkevold et al. (2014) compared genetic relationships among Danish continental and insular populations and showed that information on historical founder events and gene flow is retained in the genetic signatures of contemporary populations. Although almost all analyzed populations showed marked genetic divergence, as expected under isolation, they also displayed relatively closer genetic relationships among populations within, compared to between, regional watersheds. Natural dispersal and gene flow is mainly expected to take place within river systems and watersheds. Studies like these thus underline the importance of considering the extended (meta) population, e.g. when the aim is to ascertain genetically sound management by allowing natural gene flow. In Bekkevold et al. (2014), spatial samples from lakes within river systems, separated by 22-133 km, exhibited low genetic differentiation (F_{ST} estimated between 0.004 and 0.072; average = 0.035). Nonetheless, in one case two lake populations from the same river system, separated by 69 km, exhibited strong genetic divergence at a level consistent with different regional founder events (F_{ST} = 0.210). As it is highly unlikely that either of the populations had been stocked, the interpretation is that even within rivers the populations may have originated from multiple founder events. This raises the question of which factors maintain population divergence. Is genetic isolation purely an effect of physical barriers to dispersal and gene flow? Or do pike display local adaptations and selection against dispersal? Under local adaptation, specific genotypes have higher fitness in the home environment compared to genotypes from extant populations (Kawecki and Ebert 2004). This may result in selection against dispersal or the evolution of homing behavior (Blanquart and Gandon 2014, and references therein). Local adaptation is demonstrated in an increasing number of freshwater and anadromous fishes (Fraser et al. 2011). Evidence of local adaptation has not been reported in freshwater populations of pike, but is indicated in brackish populations, see section 7.3.2.2.

7.3.2 Genetic Dynamics in Large Water Bodies

Pike occur in some large and open systems, such as the freshwater Great Lakes in North America, the brackish Caspian Sea in Asia, and the brackish Baltic Sea in northern Europe. In large open water bodies the potential for migration and gene flow among populations is much higher than in small and isolated freshwater systems. Larger waters can also sustain larger population sizes, and thus the rate of random genetic drift is expected to be smaller than in the majority of pike populations that are restricted to smaller lakes and streams. For example, in the Baltic Sea, estimates of genetic variation (heterozygosity) and effective population sizes are generally found to be larger among coastal populations than among populations in lakes and streams in northern Europe (Bekkevold et al. 2014).

Relatively low levels of genetic differentiation among populations have been estimated in both Lake Ontario in North America (Bosworth and Farrell 2006, Ouellet-Cauchon et al. 2014) and in the Baltic Sea (Laikre et al. 2005, Wennerström et al. 2013). A reduced or non-existent exchange of genes (reproduction) between individuals from different populations is a requirement for maintaining genetic differentiation, and in pike this is often the case due to migration barriers. For freshwater populations of pike, dispersal between different water systems is often not feasible, typically leading to large genetic differences among freshwater populations (Miller and Senanan 2003). In open systems, where dispersal is physically possible, genetic differentiation among populations is expected to be lower but can still be maintained by e.g. homing to natal spawning grounds.

Site fidelity or 'natal homing' (chapter 5) has been suggested to reduce genetic exchange between local pike populations (Craig 1996, Miller et al. 2001, Engstedt et al. 2014, Larsson et al. 2015). Tagging and telemetry studies have shown that pike may have small home ranges and not move over areas larger than a few square kilometers (Kobler et al. 2008). However, seasonal migration of 10-80 km has also been documented (Ovidio and Phillipart 2003, Koed et al. 2006). This variation could be dependent on the specific opportunities for movement and migration (e.g. Jacobsen et al. 2016, chapter 5). Migration is even observed among spawning sites within a single spawning season (Bosworth and Farrell 2006). Thus, although local adaptation could be coupled to site fidelity and homing behavior in pike, it is currently unknown if there is general scope for selection and which evolutionary processes lie behind observed movement behaviors. Natal homing has been shown for pike spawning in streams flowing into the Baltic Sea (Engstedt et al. 2014, Tibblin et al. 2015, 2016b, chapter 5). However, in other open systems, low levels of genetic differentiation suggest that a lack of natal homing, or at least natal homing in a less "strict" version (Oullet-Cauchon et al. 2014), indicates that homing may not always be as pervasive a trait in pike as in other anadromous fishes. It can also be hypothesized that the existence of large and continuous spawning grounds in open water systems increases gene flow and hinders the formation of genetically separated spawning populations. In addition, less genetic drift in large populations further slows down the genetic divergence among populations. These processes may together lie behind the relatively low differentiation among populations both within Lake Ontario and in the Baltic Sea, compared to pike populations in smaller freshwater systems. We here use these two examples to illustrate processes governing pike genetic structure in large water body systems.

7.3.2.1 Lake Ontario and the St. Lawrence River: Fluctuating Habitat Availability Affects Genetics

The Lake Ontario system in North America, fed by the large St. Lawrence River, is inhabited by several pike populations. The overall estimated genetic differentiation in the system is rather low with F_{ST} values ranging between 0.00 and 0.02 (Bosworth and Farrell 2006, Ouellet-Cauchon et al. 2014). However, the genetic structure varies considerably between the lake and the river. In the St. Lawrence River, differentiation among samples from different locations is almost zero, whereas geographical populations in Lake Ontario differ from each other with an average F_{ST} of about 0.02. Ouellet-Cauchon et al. (2014) applied landscape analysis together with genetic data and showed that there is a negative correlation between fluctuation in available spawning habitats and genetic differentiation among populations. Due to the varying water levels between years, some spawning habitats in the Lake Ontario-St. Lawrence River system are only available in some years, and homing to these spawning grounds would thus periodically be impossible. Pike are here forced to be more opportunistic in their use of spawning grounds and hence migrate between them. This in turn increases gene flow and lowers genetic differentiation among populations. Local spawning units in the system display a pronounced exchange of genes, and the dynamics in one area (availability of spawning habitat) directly affect the dynamics in other parts of the system (through immigration). Ouellet-Cauchon et al. (2014) proposed that pike in this area should be managed as a metapopulation (*sensu* Levins 1969).

7.3.2.2 The Baltic Sea: Spawning Strategy Influences Genetic Diversity

The Baltic Sea is one of the largest brackish water bodies in the world. It is a unique environment in which both typical freshwater and marine species can be found in close proximity to each other (Johannesson and André 2006). The Baltic Sea is very young in evolutionary terms. It has existed in its current state for only about 7,000 years. All species present there have colonized the area after the last glacial maximum, some 12,000 years ago (Voipio 1981). Many species in the Baltic Sea, both freshwater and marine, live at their physiological limit (Ojaveer and Pihu 2003). For freshwater species, spawning in streams or wetlands connected to the Baltic Sea might be advantageous, or sometimes necessary, for juvenile survival because freshwater habitats can provide a physiologically more beneficial environment than brackish water, and in some cases also reduce predation risks (Rohtla 2015). On the other hand, brackish waters generally confer high coastal production, providing superior feeding grounds to species that are physiologically capable of tolerating saltwater (Rohtla 2015).

The pike is one of the most well studied species of the Baltic Sea (Wennerström et al. 2017). Large scale genetic differentiation in pike inhabiting the Baltic Sea has been shown to be rather low. Laikre et al. (2005) detected population differences (F_{ST}) of only 0.034 among populations sampled over large parts of the Swedish coast, together with an overall genetic isolation by distance. These results were later confirmed by Wennerström et al. (2013, 2016), using samples with an even larger geographical coverage of the Baltic Sea. The magnitude of population differentiation is similar to what was found in Lake Ontario, a water body of similar size to the Baltic Sea. It can only be speculated if processes similar to those observed in Lake Ontario, including a lack of homing due to variation in the availability of spawning habitats, inhibit the formation of separate spawning populations, or if other factors are responsible for the low levels of variation in Baltic Sea pike. Samples collected outside the spawning season may consist of a mix of individuals from several spawning populations, thus obscuring signals of population differentiation (Laikre et al. 2005). The indication of individuals from multiple spawning populations mixing during feeding has indeed been suggested in the Baltic Sea (Bekkevold et al. 2015, Tibblin et al. 2015), and the genetic differentiation among sampling sites is higher in the spring, close to spawning, than during other seasons (Wennerström et al. 2016). However, levels of genetic differentiation are consistently lower among Baltic Sea populations in comparison with populations from freshwater habitats, thus suggesting that gene flow is relatively high within the Baltic Sea, possibly coupled with larger effective population sizes (Bekkevold et al. 2015).

Within-population genetic variation varies within the Baltic Sea. Genetic variation measured as allelic richness increases significantly both northwards and eastwards (Wennerström et al. 2016, Fig. 7.1). Reasons for this might include that less saline environments support larger populations, where genetic diversity is better

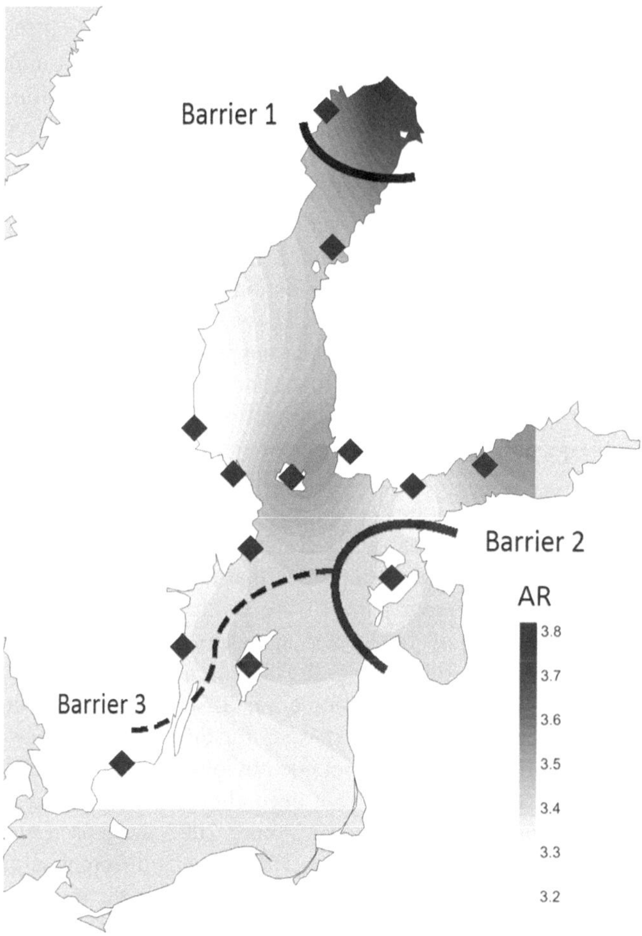

FIGURE 7.1 Genetic variation estimated as allelic richness (AR) and the three strongest barriers to gene flow in coastal populations of Baltic Sea pike, for samples collected both during and outside spawning. In the Baltic Sea, the pike shows a pattern of increased genetic variation towards the north and the east. Some peripheral samples are highly diverged from the rest of the populations (data from Wennerström et al. 2016).

retained. Present day processes are more likely to shape these patterns than post colonization patterns. This is because post-glacial colonization of the Baltic area most likely took place from the south (Skog et al. 2014), and the expected pattern of genetic diversity would be higher diversity closer to the source population if colonization was the major force shaping current patterns of genetic diversity (Widmer and Lexer 2001, Wennerström et al. 2016).

Two different reproductive behaviors exist for pike in the Baltic Sea. One type is an anadromous tactic, where feeding takes place in the brackish water and freshwater tributaries are used mainly, or only, for spawning. Homing of individual pike to the natal stream for spawning is probably rather extensive (e.g. Engstedt et al. 2010, Tibblin et al. 2015). The other type spends the entire life cycle in the coastal waters of the Baltic Sea and never enters freshwater (Westin and Limburg 2002, Jacobsen et al. 2016). The genetic differentiation pattern between these two forms appears to contrast (Wennerström et al. 2016). The genetic differentiation among spawning populations in neighboring freshwater streams are much higher than among populations sampled at spawning in coastal sea areas. F_{ST} values of up to 0.10 are observed among populations spawning in freshwater tributaries separated by ca. 50 km, as compared to F_{ST} values of about 0.03 in coastal samples (Laikre et al. 2005, Wennerström et al. 2013, 2016, Tibblin et al. 2015). In addition, stream specific differences in both juvenile and adult life history traits in geographically proximate populations (< 50 km) indicate that local adaptation to spawning habitat is present in Baltic Sea pike, even though populations live in sympatry during major parts of their life cycle (Tibblin et al. 2015, 2016a, Berggren et al. 2016). Local adaptation among freshwater spawning populations is also inferred from reciprocal translocation experiments (Berggren et al. 2016) and also suggested by significantly higher Q_{ST} values compared to F_{ST} among populations, which indicates directional selection and subsequent adaptations to different spawning habitats (Tibblin et al. 2015, 2016a).

Populations of anadromous freshwater pike separate during spawning, and subsequently mix during feeding seasons (Engstedt et al. 2014, Tibblin et al. 2015). The proportion of individuals in the Baltic Sea that are recruited from freshwater habitats seems to vary in different parts of the Baltic Sea. Rohtla et al. (2012) estimated higher proportions of freshwater spawners in Estonia than Engstedt et al. (2010) did for populations along the southern Swedish coast (82% of individuals hatched in freshwater in Estonia *vs.* 36% at the southern Swedish coast). One can only speculate as to the reasons for different frequency estimates of anadromous pike in different studies, but one obvious cause could be the varying salinities in the Baltic Sea. Pike eggs have been shown to be able to develop in salinities of up to 6-8.5 ppm (Westin and Limburg 2002, Jørgensen et al. 2010). Although there may be local adaptation for performance at different salinities, it is likely that coastal spawning is not physiologically possible in the more saline areas of the Baltic Sea. Thus, anadromy is likely the only possible life-history strategy in large parts of the Baltic Sea.

7.4 CONSERVATION GENETICS AND MANAGEMENT IMPLICATIONS

Genetic diversity can have a direct effect on species viability, resilience, and adaptive potential (Reusch et al. 2005, Barshish et al. 2013, Hellmair and Kinziger 2014). Conservation genetic efforts aim to maximize retention of the genetic diversity of populations and species to assure the potential for long term survival and evolution. A prerequisite for effective gene-level conservation is to identify and describe population genetic structure, including natural dispersal patterns and local adaptations (Allendorf et al. 2013). Human induced selection, habitat alteration or fisheries-induced selection, might have far-reaching evolutionary effects on populations (Allendorf et al. 2008), including in pike (Carlson et al. 2007, chapter 12). Similarly, human-induced gene flow following releases of fish in stocking operations can cause changes that affect population resilience (Laikre et al. 2010, chapter 9, Box 2).

The importance of protecting genetic diversity is reflected in international conservation policies such as the Convention on Biological Diversity (CBD), and it is particularly highlighted in the Aichi Target 13 of the CBD Strategic Plan for 2011-2020 (UNEP/CBD/COP/DEC/X/2; www.cbd.int/sp/targets), which focuses on the need for maintaining the genetic diversity of species of socio-economic value. Pike is a species of both high ecological and socio-economic importance (Eriksson et al. 2009, 2011, Crane et al. 2015). Thus, it is a species to which the CBD Aichi Target 13 typically applies, and for which assessment of genetic biodiversity is particularly warranted (Ljunggren et al. 2010, Karlsson et al. 2014, Wennerström et al. 2016). Globally, pike is not considered endangered, but concerns about recruitment failure and declining populations have been raised, for instance in the Baltic Sea (e.g. Andersson et al. 2000, Ljunggren et al. 2010) and in Italy (Lorenzoni et al. 2002). Conservation actions for some populations are thus warranted.

As exemplified in this chapter, current scientific knowledge indicates that spatial genetic diversity patterns in pike differ between large water bodies, which lack migration barriers and provide ample spawning habitat (e.g. Lake Ontario and the Baltic Sea; Laikre et al. 2005, Wennerström et al. 2013, 2016, Ouellet-Cauchon et al. 2014), and smaller lakes and streams, where spawning habitats are more restricted (e.g. Senanan and Kapuscinski 2000). Genetic exchange (gene flow) is indicated to be substantial in the large water bodies lacking obvious migration barriers. Such population connectivity implies that actions that affect local habitats or spawning grounds, either positively or negatively, might also affect population dynamics in other areas in the system. For Lake Ontario, Ouellet-Cauchon et al. (2014) thus suggest that pike should be managed as a metapopulation, i.e. that management should consider the system as a whole.

For pike in the Baltic Sea, a similar metapopulation approach of management might also be feasible (Wennerström et al. 2016). However, current data suggest a more complex spatial structure here. The early study of Laikre et al. (2005) suggested local management to be warranted and suggested that management units of about 150 km would be fitting. Management units of 250-400 km have later been suggested, based on both genetic and demographic data (Östman et al. 2016). The need for local management was supported by Bekkevold et al. (2014) who demonstrated that, in addition to the large scale genetic separation of populations into different drainages, which probably reflect colonization history, processes in local populations can differ substantially and management of populations should be decided case-by-case. Recent work on anadromous populations of pike in the Baltic Sea also indicate that local management is needed for freshwater spawning populations and that the management of these populations could be modeled on current practice for salmonids (Engstedt et al. 2010, Tibblin et al. 2015). The mixing of populations of both anadromous and brackish water origin outside the reproductive period means that temporal aspects need to be taken into account in management. Fishing pike outside spawning time might to some degree target mixtures of fish from multiple spawning populations, which introduces risks of overfishing of stocks with lower productivity (Hilborn 1985). A combination of metapopulation and local management is therefore warranted for Baltic pike (Wennerström et al. 2016, see also Östman et al. 2016).

Genetic diversity and divergence of Baltic Sea spawning pike appears to have remained stable over a ten year monitoring period 2001-2010 explored by Wennerström et al. (2016). These results indicate no immediate threat to the genetic integrity of Baltic Sea pike, but it should be stressed that the genes studied were relatively few and presumed selectively neutral ones; differing patterns may exist for adaptive genes that are under selection. Marine protected areas (MPAs) in the Baltic Sea cover relatively limited amount of pike coastal spawning grounds. A recent assessment showed that existing MPAs do not protect genetic diversity of pike specifically but that these areas harbour a representative subset of Baltic Sea populations (Wennerström et al. 2017).

Stocking fish from unrecorded source populations is a common practice in pike management (chapter 9), and can have significantly negative effects on naturally occurring populations (Laikre et al. 2010, Box 2). Examples from other species show that the stocking of non-native broodstocks has led to genetic homogenization among populations of e.g. alewife (*Alosa pseudoharengus* Wilson, 1811, McBride et al. 2015) and salmonids (e.g. Vasemägi et al. 2005, Pearse et al. 2011, Ozerov et al. 2016). This homogenization may negatively impact local adaptation and long term survival. For example, Lamaze et al. (2013) demonstrated that the level of introgression from stocked populations affected the expression levels of two biologically important genes in brook charr (*Salvelinus fontinalis* Mitchill, 1814), thus potentially affecting the fitness of naturally occurring populations. In addition, if a small number of individuals are used as brood stock, stocking can reduce the effective population size in wild populations, because of increased inbreeding (Ryman and Laikre 1991).

The risks associated with introductions of non-native populations imply that genetic monitoring of stocking operations is highly warranted. In practice, however, stocking activities are rarely even recorded (Laikre et al. 2006). The stocking of pike has been a common practice over the last 100 years (chapter 9), and it is likely that extensive stocking has already had an effect on the genetic composition of pike (Nicod et al. 2004, Launey et al. 2006). Stocking has had a homogenizing effect on genetic profiles of French pike populations (Launey et al. 2006), and there is evidence that stocking may even lead to hybridization between esocid species (Denys et al. 2014). Stocking pike in areas inhabited by southern species is in the latter case considered to be the greatest threat to native pike populations (Denys et al. 2014, Bianco 2014). However, there is also evidence of stocked freshwater pike contributing little or nothing genetically to brackish populations, potentially due to poor adaptation of stocked freshwater pike to life in a brackish environment (Larsen et al. 2005). Whereas stocking in that case may not have affected native gene pools, genetic analyses importantly demonstrated that there was no evidence that stocking had had any positive effects on local populations, thus, negating the prime argument for using stocking as a conservation measure (see chapter 9 for further discussion). Similar effect of little or no positive consequences of stocking has been indicated in German pike populations (Eschbach et al. 2009), as well as for other species such as pikeperch (*Sander lucioperca* Linnaeus, 1758, Eschbach et al. 2014) and trout (*Salmo trutta* Linnaeus, 1758, Hansen 2002).

Many have over the years warned against the potentially adverse effects of stocking (e.g. Ryman 1991, Hansen et al. 2001, Laikre et al. 2010, Valiquette et al. 2014, Bianco 2014, chapter 9). Unless warranted for conservation purposes, stocking should generally be avoided. If stocking is carried out, only broodstock from genetically close populations should be used (Nicod et al. 2004, Skog et al. 2014) and its effects should be critically evaluated and documented using genetic monitoring (Laikre et al. 2010) as is also recognized in management plans for the Baltic Sea marine protected area of Stora-Nassa and Svenska Högarna (Laikre et al. 2016). This includes even smaller geographic scales since the mixing of individuals from different genetic backgrounds has been shown to either affect the genetic composition of naturally occurring species and populations or be inefficient if, for example, the poor adaptation to the stocking material's new environment results in low fitness and a lack of positive effects on the recipient population (e.g. Berggren et al. 2016, Tibblin et al. 2016a). In chapter 9, this volume, further recommendations about pike stocking can be found.

7.4.1 Case study: Management Controversy of Irish Pike

Pike is assumed to have been introduced by humans to Ireland in the 16th century and is considered to be an invasive species on the island, although this has been debated (Barbe and Garrett 2016). Pike has in many ways been treated as an invasive species and actions such as the eradication of pike populations in order to protect the native brown trout have been conducted (Pedreschi 2014, Barbe and Garrett 2016). Pike populations in Ireland have remarkably low genetic variation, and this has been hypothesized as due to founder effects at introduction (e.g. Jacobsen et al. 2005). However, Pedreschi et al. (2014) reported evidence of natural colonization to Ireland. Based on Approximate Bayesian Computation population genetic methods, the pike populations in Ireland and Britain were indicated to have branched off from other European populations around 8000 years ago. In a second event, around 3500-4000 years ago, the Irish and British populations became separated. This indicates that pike colonized Ireland much earlier than previously thought, and that colonization might actually have been a natural event. Interestingly, if it could be shown that pike colonized Ireland naturally the species should be considered part of native Irish freshwater fauna and would most likely gain a high level of protection.

The theory of naturally occurring pike in Ireland was challenged by Ensing (2015), who cited evidence of high sea levels separating Ireland and Britain at the time of suggested colonization, which would have acted as a migration barrier and made colonization impossible. Ensing (2015) instead proposed that pike could have been introduced by humans several thousand years ago, as well as by a later, medieval introduction. This theory was strongly fought by Pedreschi et al. (2015) who also raised the philosophical question of when a species should be considered "natural". A species present in an ecosystem since the bronze ages might be considered native regardless of the colonization vector.

Box 7.2 Newly Described Species of Pike Supported by Genetic Data

Recently, two additional *Esox* species from southern Europe have been described: *E. cisalpinus*, Bianco and Delmastro, 2011 (also known by the junior name *E. flaviae*) in Italy, together with possible human-mediated introductions at other European locations (Lucentini et al. 2011, Bianco 2014, Denys et al. 2014), and *E. aquitanicus* in southwestern France (Denys et al. 2014). These species were identified based on differentiation in both morphology (e.g. color pattern, body shape, number of lateral scales) and in nuclear and mtDNA genetic markers. Genetically, unique populations of pike in southern Europe have been noted several times in the scientific literature (e.g. Nicod et al. 2004, Launey et al. 2006, Skog et al. 2014), but it has only recently been suggested that the differences reflect the existence of separate species. Divergence times between the different pike variants have not been estimated, but the large genetic differences suggest that they have been separated at least since the Pleistocene (Bianco 2014). The designation of new species of pike will likely have a considerable impact on management, and specifically on stocking practices. Hybridization with translocated *E. lucius* has been identified as the largest threat to both *E. cisalpinus* and *E. aquitanicus* (Bianco 2014, Denys et al. 2014). Currently, the geographic distribution of these newly described species is not completely known, and the IUCN threat status is not evaluated for any of them (www.fishbase.org; accessed October 2016).

7.5 FUTURE RESEARCH OPPORTUNITIES IN PIKE GENETICS

Recent scientific work on pike genetics has provided many new insights into the demographic history of pike and how current processes affect the genetic composition of populations. Several phylogenetic lineages have been identified, and their major distributions have been outlined (Skog et al. 2014). Smaller scale population genetic studies have shown large variation in genetic structure among populations (e.g. Bekkevold et al. 2014) and that genetic differences can be maintained by natal homing (Tibblin et al. 2015). However, in some populations, migration rates can be substantial (Wennerström et al. 2013, 2016, Oullet-Cauchon et al. 2014), and tagging studies have suggested that within populations individuals may display different migratory behaviors (Jacobsen et al. 2016, chapter 5). Access to vastly improved genomic resources for pike (Rondeau et al. 2014) now allow for elucidating the genomic architecture behind different life-history traits and how they are maintained, as is transpiring for other fish species (e.g. Barson et al. 2015). Population genetic studies of pike have so far made use of putatively neutral genetic markers. There is now ample scope for applying gene associated markers to characterize functional differences among locally adapted populations. Moreover, the screening of total genome-wide diversity has become technically and economically feasible also for non-model species, allowing for dramatically improved inference about population demographic histories and functional adaptations (Therkildsen and Palumbi 2016). Pike has even been suggested as an emerging model species for ecology and evolutionary studies (Forsman et al. 2015). Thus, future directions of genetic research on pike will likely combine more extensive sampling of populations of specific interest with the application of genomic methods to elucidate the genomic basis for divergent evolutionary trajectories. This is of particular interest for e.g. pike populations in North America, which have been less studied than European ones, and for populations adapted to more extreme environments, as in the Baltic Sea. Such integrated analyses have the potential to not only increase the knowledge of this ecologically and economically important species, but also to add to the general knowledge of the mechanisms behind local adaptation and evolution over short and longer time scales.

Acknowledgements

LL acknowledges support from the Swedish Research Council Formas and from the project BONUS BAMBI. BONUS is the joint Baltic Sea research and development programme (Art 185), funded jointly from the European Union's Seventh programme for research, technological development and demonstration and from the Swedish Research Council Formas. DB acknowledges the Danish Angling License Funds.

REFERENCES CITED

Aguilar, A., J.D. Banks, K.F. Levine and R.K. Wayne. 2005. Population genetics of northern pike (*Esox lucius*) introduced into Lake Davis, California. *Can. J. Fish. Aquat. Sci.* 62:1589-1599.

Allendorf, F.W., P.R. England, G. Luikart, P.A. Ritchie and N. Ryman. 2008. Genetic effects of harvest on wild animal populations. *Trends Ecol. Evol.* 23:327-337.

Allendorf, F.W., G. Luikart and S.N. Aitken. 2013. Conservation and the Genetics of Populations. Second Edition. Wiley-Blackwell, Chichester.

Andersson, J., J. Dahl, A. Johansson, P. Karås, J. Nilsson, O. Sandström and A. Svensson. 2000. Eliminated fish reqruitment and failing fish stocks in Kalmar County coastal waters. [Utslagen fiskrekrytering och sviktande fiskbestånd i Kalmar läns kustvatten, in Swedish]. *Swedish Board of Fisheries Report* 2000:5.

Barbe, F. and S. Garrett. 2015. The pike in Ireland: a (necessary) review. Available at http:// homepage.eircom.net/~sheelin/IPSPikelnlre.html [Accessed March 2016].

Barshish, D.J., J.T. Ladner, T.A. Oliver, F.O. Seneca, N. Taylor-Knowles and S.R. Palumbi. 2013. Genomic basis for coral resilience to climate change. *P. Natl. Acad. Sci. USA* 110:1387-1392.

Barson, N.J., T. Aykanat, K. Hindar, M. Baranski, G.H. Bolstad, P. Fiske, C. Jacq, A.J. Jensen, S.E. Johnston, S. Karlsson, M. Kent, T. Moen, E. Niemelä, T. Nome, T.F. Næsje, P. Orell, A. Romakkaniemi, H. Sægrov, K. Urdal, J. Erkinaro, S. Lien and C.R. Primmer. 2015. Sex-dependent dominance at a single locus maintains variation in age at maturity in salmon. *Nature* 528:405-408.

Bernatchez, L. and C.C. Wilson. 1998. Comparative phylogeography of Nearctic and palearctic fishes. *Mol. Ecol.* 7:431-452.

Bekkevold, D., L. Jacobsen, J. Hemmer-Hansen, S. Berg and C. Skov. 2014. From regionally predictable to locally complex population structure in a freshwater top predator: river systems are not always the unit of connectivity in Northern Pike *Esox lucius*. *Ecol. Freshw. Fish* 24:305-316.

Berggren, H., O. Nordahl, P. Tibblin, P. Larsson and A. Forsman. 2016. Testing for local adaptation to spawning habitat in sympatric subpopulations of pike by reciprocal translocations of embryos. *PLoS ONE* 11:e0154488.

Bianco, P.G. and G.B. Delmastro. 2011. *Recenti novità tassonomiche iguardanti i pesci d'acqua dolce autoctoni in Italia e descrizione di una nuova specie di luccio*. Research on wildlife conservation 2. USA: IGF publishing.

Bianco, P.G. 2014. An update on the staus of native and exotic freshwater fishes of Italy. *J Appl Ichtyol* 30:62-77.

Blanquart, F. and S. Gandon. 2014. On the evolution of migration in heterogeneous environments. *Evolution* 68:1617-1628.

Bosworth, A. and J.M. Farrell. 2006. Genetic divergence among Northern pike from spawning locations in the upper St. Lawrence River. *N. Am. J. Fish Manage.* 26:676-684.

Carlson, S. M., E. Edeline, L.A. Vollestad, T.O. Haugen, I.J. Winfield, J.M. Fletcher, J.B. James and N.C. Stenseth. 2007. Four decades of opposing natural and human-induced artificial selection acting on Windermere pike (*Esox lucius*). *Ecol. Lett.* 10:512-521.

Craig, J.F (ed.). 1996. Pike, Biology and Exploitation. Chapman & Hall. Fish and Fisheries Series 19. London.

Crane, D.P., L.M. Miller, J.S. Diana, J.M. Cassemlan. J.M. Farrell, K.L. Kapuscinski and J.K. Nohner. 2015. Muskelunge and Northern pike ecology and management: important issues and research needs. *Fisheries* 40:258-267.

Denys, G.P.J., A. Dettai, H. Persat, M. Hautecoeur and P. Keith. 2014. Morphological and molecular evidence of three species of pikes *Esox spp.* (Actinopterygii, Esocidae) in France, including the description of a new species. *CR. Biol.* 337:521-534.

DeWoody, J.A. and J.C. Avise. 2000. Microsatellite variation in marine, freshwater and anadromous fishes compared with other animals. *J. Fish Biol.* 56:461-473.

Engstedt, O., P. Stenroth, P. Larsson, L. Ljunggren and M. Elfman. 2010. Assessment of natal origin of pike (*Esox lucius*) in the Baltic Sea using Sr:Ca in otoliths. *Environ. Biol. Fish* 89:547-555.

Engstedt, O., P. Engqvist and P. Larsson. 2014. Elemental fingerprinting in otoliths reveals natal homing of anadromous Baltic Sea pike (*Esox Lucius* L.). *Ecol. Freshw. Fish* 23:313-321.

Ensing, D. 2015. Pike (*Esox lucius*) could have been an exclusive human introduction to Ireland after all: a comment on Pedreschi et al. (2014). *J. Biogeogr.* 42:604-607.

Eriksson, B.K., L. Ljunggren, A. Sandström, G. Johansson, J. Mattila, A. Rubach, S. Råberg and M. Snickars. 2009. Declines in predatory fish promote bloom-forming macroalgae. *Ecol. Appl.* 19:1975-1988.

Eriksson, B.K., K. Sieben, J. Eklöf, L. Ljunggren and J. Olsson. 2011. Effects of altered offshore food webs on coastal ecosystems emphasize the need for cross-ecosystem management. *Ambio* 40:786-797.

Eschbach, E., A. Nolte, K. Kohlmann, A. Joseph, P. Kersten, S. Schoning, T. Rapp and R. Arlingaus. 2009. Genetische Vielfalt von Zander und Hecht populationen in Deutschland: Schlussfolgerungen für die nachhaltige fischereiliche Hege durch Besatz. *Fischer & Techwirt.* 9:327-330.

Eschbach, E. and S. Schöning. 2013. Identification of high-resolution microsatellites without *a priori* knowledge of genotypes using a simple scoring approach. *Method. Ecol. Evol.* 4:1076-1082.

Eschbach, E., A.W. Nolte, K. Kohlmann, P. Kersten, J. Kail and R. Arlinghaus. 2014. Population differentiation of zander (*Sander lucioperca*) across native and newly colonized ranges suggests increasing admixture in the course of an invasion. *Evol. Appl.* 7:555-568.

Forsman, A., P. Tibblin, H. Berggren, O. Nordahl, P. Koch-Schmidt and P. Larsson. 2015. Pike *Esox lucius* as an emerging model organism for studies in ecology and evolutionary biology: a review. *J. Fish. Biol.* 87:472-479.

Fraser, D.J., L.K. Weir, L. Bernatchez, M.M. Hansen and E.B. Taylor. 2011. Extent and scale of local adaptation in salmonid fishes: review and meta-analysis. *Heredity* 106:404-420.

Hansen, M.M., J.B. Taggart and D. Meldrup. 1999. Development of new VNTR markers for pike and assessment of variability at di- and tetra nucleotide repeat microsatellite loci. *J. Fish Biol.* 55:183-188.

Hansen, M.M., D.E. Ruzzante, E.E. Nielsen and K-L.D. Mensberg. 2001. Brown trout (*Salmo trutta*) stocking impact assessment using microsatellite DNA markers. *Ecol. Appl.* 11:148-160.

Hansen, M.M. 2002. Estimating the long-term effects of stocking domesticated trout into wild brown trout (*Salmo trutta*) populations: an approach using microsatellite DNA analysis of historical and contemporary samples. *Mol. Ecol.* 11:1003-1015.

Healy, J.A. and M.F. Mulcahy. 1980. A biochemical genetic analysis of populations of the northern pike, *Esox lucius* L., from Europe and North America. *J. Fish Biol.* 17:317-324.

Hellmair, M. and A.P. Kinziger. 2014. Increased extinction potential of insular fish populations with reduced life history variation and low genetic diversity. *PLoS ONE* 9:1-10.

Hewitt, G.M. 1996. Some genetic consequences of ice ages, and their role in divergence and speciation. *Biol. J. Linn. Soc.* 58:247-276.

Hilborn, R. 1985. Apparent stock recruitment relationships in mixed stock fisheries. *Can. J. Fish. Aquat. Sci.* 42:718-723.

Jacobsen, B.H., M.M. Hansen and V. Loeschcke. 2005. Microsatellite DNA analysis of northern pike (*Esox lucius L.*) populations: insights into the genetic structure and demographic history of a genetically depauperate species. *Biol. J. Linn. Soc.* 84:91-101.

Jacobsen, L., D. Bekkevold, S. Berg, N. Jepsen, A. Koed, K. Aarestrup, H. Baktoft and C. Skov. 2016. Pike (*Esox lucius* L.) on the edge: consistent individual movement patterns in transitional waters of the Western Baltic. *Hydrobiologia* 784:143-154. doi:10.1007/s10750-016-2863-y

Johannesson, K. and C. André. 2006. Life on the margin: genetic isolation and diversity loss in a peripheral marine ecosystem, the Baltic Sea. *Mol. Ecol.* 15:2013-2019.

Jørgensen, A.T., B.W. Hansen, B. Vismann, L. Jacobsen, C. Skov, S. Berg and D. Bekkevold. 2010. High salinity tolerance in eggs and fry of a brackish *Esox Lucius* population. *Fish. Manag. Ecol.* 17:554-560.

Karlsson, M., H. Ragnarsson Stabo, E. Petersson, H. Carlstrand and S. Thörnqvist. 2014. National plan for knowledge support for sport fisheries within fish-, sea-, and water management [In Swedish]. *Aqua reports* 2014:12. Swedish University of Agricultural Sciences. Drottningholm.

Kawecki, T.J. and D. Ebert. 2004. Conceptual issues in local adaptation. *Ecol. Lett.* 7:1225-1241.

Kobler, A., T. Klefoth, C. Wolter, F. Fredrich and R. Arlinghaus. 2008. Contrasting pike (*Esox lucius* L.) movement and habitat choice between summer and winter in a small lake. *Hydrobiologia* 601:17-27.

Koed, A., K. Balleby, P. Mejlhede and K. Aarestrup. 2006. Annual movement of adult pike (*Esox lucius* L.) in a lowland river. *Ecol. Freshw. Fish* 15:191-199.

Laikre, L., M. Miller, A. Palmé, S. Palm, A.R. Kapuscinski, G. Thoresson and N. Ryman. 2005. Spatial genetic structure of northern pike (*Esox lucius*) in the Baltic Sea. *Mol. Ecol.* 14:1955-1964.

Laikre, L., A. Palmé, M. Josefsson, F. Utter and N. Ryman. 2006. Release of alien populations in Sweden. *Ambio* 35:255-261.

Laikre, L., M.K. Schwartz, R.S. Waples and N. Ryman. 2010. Compromising genetic diversity in the wild: unmonitored large-scale releases of plants and animals. *Trends Ecol. Evol.* 25:520-529.

Lamaze, F.C., D. Garant and L. Bernatchez. 2013. Stocking impacts the expression of candidate genes and physiological condition in introgressed brook charr (*Salvelinus fontinalis*) populations. *Evol. Appl.* 6:393-407.

Larsen, P.F., M.M. Hansen, E.E. Nielsen, L.F. Jensen and V. Loeschke. 2005. Stocking impact oand temporal stability of genetic composition in a brackish Northern Pike population (*Esox lucius* L.), assessed using microsatellite DNA analysis of historical and contemporary samples. *Heredity* 95:136-143.

Larsson, P., P. Tibblin, P. Koch-Schmidt, O. Engstedt, J. Nilsson, O. Nordahl and A. Forsman. 2015. Ecology, evolution, and management strategies of northern pike populations in the Baltic Sea. *Ambio* 44:S451-S461.

Launey, S., J. Morin, S. Minery and J. Laroche. 2006. Microsatellite genetic variation reveals extensive introgression between wild and introduced stocks, and a new evolutionary unit in French pike *Esox lucius* L. *J. Fish. Biol.* 68:193-216.

Leinonen, T., R.J.S. McCairns, R.B. O'Hara and J. Merilä. 2013. Q_{ST}-F_{ST} comparisons: evolutionary and ecological insights from genomic heterogeneity. *Nat. Rev. Gen.* 14:179-190.

Levins, R. 1969. Some demographic and genetic consequences of environmental heterogeneity for biological control. *B. Entomol. Soc. Am.* 15:237-240.

Ljunggren, L., A. Sandström, U. Bergström, J. Mattila, A. Lappalainen, G. Johansson, G. Sundblad, M. Casini, O. Kaljuste and B.J. Eriksson. 2010. Recruitment failure of coastal predatory fish in the Baltic Sea coincident with an offshore ecosystem regime shift. *ICES J. Mar. Sci.* 67:1587-1595.

Lorenzoni, M., M. Corboli, A.J.M. Dorr, M. Mearelli and G. Giovinazzo. 2002. The growth of pike (*Esox lucius* Linnaeus, 1798) in Lake Trsimeno (Umbria, Italy). *Fish. Res.* 59:239-246.

Lucentini, L., A. Palomba, L. Gigliarelli, G. Sgaravizzi, H. Lancioni, L. Lanfaloni, M. Natali and F. Panara. 2009. Temporal changes and effective population size of an Italian isolated and supportive-breeding managed northern pike (*Esox lucius*) population. *Fish. Res.* 96:139-147.

Lucentini, L., M.E. Puletti, C. Riccolini, L. Gigliarelli, D. Fontaneto, L. Lanfoloni, F. Biló, M. Natali and F. Panara. 2011. Molecular and phenotypic evidence of a new species of genus *Esox* (Esocidae, Esociformes, Actinopterygii): the Southern pike, *Esox flaviae*. *PLoS ONE*. 6(12):e25218. doi:10.1371/journal.pone.0025218.

McBride, M.C., D.J. Hasselman, T.V. Willis, E.P. Palcovacs and P. Bentzen. 2015. Influence of stocking history on the population genetic structure of anadromous alewife (*Alosa pseudoharengus*) in Maine rivers. *Conserv. Genet.* 16:1209-1223.

Miller, L.M., and A.R. Kapuscinski. 1997. Historical analysis of variation reveals low effective population size in a northern pike (*Esox lucius*) population. *Genetics*. 147:1249-1258.

Miller, L.M., L. Kallemeyn and W. Senanan. 2001. Spawning-site and natal-site fidelity by Northern pike in a large lake: mark-recapture and genetic evidence. *T. Am. Fish. Soc.* 130:307-316.

Miller, L.M. and W. Senanan. 2003. Review of northern pike population genetics research and its implications for management. *J. Fish. Manage.* 23:297-306.

Maes, G.E., J.K.J. vad Houdt, D. De Charleroym and F.A.M. Volckaert. 2003. Indications for a recent holarctic expansion of pike based on a preliminary study of mtDNA variaion. *J. Fish. Biol.* 63:254-259.
Nicod, J.C., Y.Z. Wang, L. Excoffier and C.R. Largiadèr. 2004. Low levels of mitochondrial DNA variation among central and southern European *Esox lucius* populations. *J. Fish. Biol.* 64:1442-1449.
Ojaveer, E. and E. Pihu. 2003. Estonian natural fish waters. pp 15-27. *In*: E. Ojaveer, E. Pihu and T. Saat (eds.). *Fishes of estonia.* Tallinn: Estonian Academy Publishers.
Östman, Ö., J. Olsson, J. Dannewitz, S. Palm and A-B. Florin. 2016. Inferring spatial structure from population genetics and spatial synchrony in demography of Baltic Sea fishes: implications for management. *Fish & Fish.* 18:324-339. doi:10.1111/faf.12182
Ouellet-Cauchon G., M. Mingelbier, F. Lecomte and L. Bernatchez. 2014. Landscape variability explains spatial pattern of population structure of northern pike (*Esox Lucius*) in a large fluvial system. *Ecol. Evol.* 4:3723-3735.
Ovidio, M. and J.C. Philipart. 2003. Long range seasonal movements of northern pike (*Esox lucius* L.) in the barbell zone of the River Ourthe (River Meuse basin, Belgium). Aquatic telemetry: advances and applications. Proceedings of the Fifth Conference on Fish Telemetry held in Europe. Ustica.
Ozerov, M.Y., R. Gross, M. Bruneauz, J-P. Vähä, O. Burimski, L. Pukk and A. Vasemägi. 2016. Genomewide introgressive hybridization patterns in wild Atlantic salmon influenced by inadvertent gene flow from hatchery releases. *Mol. Ecol.* 25:1275-1293.
Pearse, D.E., E. Martinez and J.C. Garza. 2011. Disruption of historical patterns of isolation by distance in coastal steelhead. *Conserv. Genet.* 12:691-700.
Pedreschi, D., M. Kelly-Quinn. J. Caffrey, M. O'Grady and S. Mariani. 2014. Genetic structure of pike (*Esox lucius*) reveals a complex and previously unrecognized colonization history of Ireland. *J. Biogeogr.* 41:548-560.
Pedreschi, D. and S. Mariani. 2015. Towards a balanced view of pike in Ireland: a reply to Ensing. *J. Biogeogr.* 42:607-609.
Reusch, T.B.H., A. Ehlers, A. Hammerli and B. Worm. 2005. Ecosystem recovery after climatic extremes enhanced by genotypic diversity. P. Natl. *Acad. Sci.* 102:2826-2831.
Rohtla, M., M. Vetemaa, K. Urtson and A. Soesoo. 2012. Early life migration patterns of the Baltic Sea pike *Esox lucius. J. Fish Biol.* 80:886-893.
Rohtla, M. 2015. *Otolith sclerochronological studies on migrations, spawning habitat preferences and age of freshwater fishes inhabiting the Baltic Sea.* PhD-thesis. Estonian Marine Institute and Department of Zoology, Institute of Ecology and Earth Science and Technology, University of Tartu, Estonia.
Rondeau, E.B., D.R. Minkley, J.S. Leong, A.M. Messmer, J.R. Jantzen, K.R. von Schalburg, C. Lemon, N.H. Bird and B.F. Koop. 2014. The genome and linkage map of the northern pike (*Esox lucius*): conserved synteny revealsed between the salmonid sister group and the neoteleosti. *PLoS ONE* 9:7. doi:10.1371/journal.pone.0102089.
Ryman, N. 1991. Conservation genetics considerations in fishery management. *J. Fish Biol.* 39:211-224.
Ryman, N. and L. Laikre. 1991. Effects of supportive breeding on the genetically effective population size. *Conserv. Biol.* 5:325-329.
Schwartz, M.K., G. Luikart and R.S. Waples. 2007. Genetic monitoring as a promising tool for conservation and management. *Trends Ecol. Evol.* 22:25-33.
Seeb, J.E., L.W. Seem, D.W. Oates and F.M. Utter. 1987. Genetic variation and postglacial dispersal of populations of northern pike (*Esox lucius*) in North America. *Can. J. Fish. Aquat. Sci.* 44:556-561.
Senanan, W. and Kapuscinski, A.R. 2000. Genetic relationships among populations of northern pike (*Esox lucius*). *Can. J. Fish. Aquat. Sci.* 57:391-404.

Shafer, A.B.A., J.B.W. Wolf, P.C. Alves, L. Bergström, M.W. Bruford, I. Brännström, G. Colling, L. Dalén, L. De Meester, R. Ekblom, K.D. Fawcett, S. Fior, M. Hajibabaei, J.A. Hill, A.R. Hoezel, J. Höglund, E.L. Jensen, J. Krause, T.N. Kristensen, M. Krützen, J.K. McKay, A.J. Norman, R. Ogden, E.M. Österling, N.J. Ouborg, J. Piccolo, D. Popovic, C.R. Primmer, F.A. Reed, M. Roumet, J. Salmona, T. Schenekar, M.K. Schwartz, G. Segelbacher, H. Senn, J. Thaulow, M. Valtonen, A. Veale, P. Vergeer, N. Vijay, C. Vilà, M. Weissensteiner, L. Wennerström, C.W. Wheat and P. Zieliński. 2015. Genomics and the challenging translation into conservation practice. *Trends Ecol. Evol.* 30:78-87.

Skog, A., L.A. Vøllestad, N.C. Stenseth, A. Kasumyan and K.S. Jakobsen. 2014. Circumpolar phylogeography of the northern pike (*Esox lucius*) and its relationship to the Amur pike (*E. reichertii*). *Front. Zool.* 11:67.

Therkildsen, N.O. and S.R. Palumbi. 2016. Practical low-coverage genomewide sequencing of hundreds of individually barcoded samples for population and evolutionary genomics in nonmodel species. *Mol. Ecol. Resour.* 17:194-208. doi:10.1111/1755-0998.12593.

Tibblin, P., A. Forsman, P. Koch-Schmidt, O. Nordahl, P. Johannessen, J. Nilsson and P. Larsson. 2015. Evolutionary divergence of adult body size and juvenile growth in sympatric subpopulations of a top predator in aquatic ecosystems. *Am. Nat.* 186:98-110.

Tibblin, P., H. Berggren, O. Nordahl, P. Larsson and A. Forsman. 2016a. Causes and consequences of intra-specific variation in vertebral number. *Sci. Rep.* 6:26372.

Tibblin, P., A. Forsman, T. Borger and P. Larson. 2016b. Causes and consequences of repeatability, flexibility and individual fine-tuning of migratory timing in pike. *J. Anim. Ecol.* 85:136-145.

Valiquette, E., C. Perrier, I. Thibault and L. Bernatchez. 2014. Loss of genetic integrity in wild lake trout populations following stocking: insights from an exhaustive study of 72 lakes from Québec, Canada. *Evol. Appl.* 7:625-644.

Vasemägi, A., R. Gross, T. Paaver, M-L. Koljonen and J. Nilsson. 2005. Extensive immigration from compensatory hatchery releases into wild Atlantic salmon population in the Baltic Sea: spatio-temporal analysis over 18 years. *Heredity* 95:76-83.

Voipio, A. (ed.). 1981. The Baltic Sea. Elsevier Scientific Publishing Company. Amsterdam.

Wennerström, L., L. Laikre, N. Ryman, F.M. Utter, N.I. Ab Ghani, C. André, J. DeFaveri, D. Johansson, L. Kautsky, J. Merilä, N. Mikhailova, R. Pereyra, A. Sandström, A.G.F. Teacher, R. Wenne, A. Vasemägi, M. Zbawicka, K. Johannesson and C.R. Primmer. 2013. Genetic biodiversity in the Baltic Sea: species-specific patterns challenge management. *Biodivers. Conserv.* 22:3045-3065.

Wennerström, L., J. Olsson, N. Ryman and L. Laikre. 2016. Temporally stable, weak genetic structuring in Baltic Sea northern pike (*Esox lucius*) indicates a contrasting divergence pattern relative to freshwater populations. *Can. J. Fish. Aquat. Sci.* 74:562-571. doi:10.1139/cjfas-2016-0039.

Wennerström, L., E. Jansson and L. Laikre 2017. Baltic Sea genetic biodiversity: Current knowledge relating to conservation management. *Aquatic Conserv.: Mar. Freshw. Ecosyst.* doi:10.1002/aqc.2771

Westin, L. and K.E. Limburg. 2002. Newly discovered reproductive isolation reveals sympatric populations of *Esox lucius* in the Baltic. *J. Fish Biol.* 61:1647-1652.

Widmer, A. and C. Lexer. 2001. Glacial refugia: sanctuaries for allelic richness, but not for gene diversity. *Trends Ecol. Evol.* 16:267-269.

Wooller, M.J., B. Gaglioti, T.L. Fulton, A. Lopez and B. Shapiro. 2015. Post-glacial dispersal patterns of Northern pike inferred from an 88000 year old pike (Esox cf. lucius) skull from interior Alaska. *Quat. Sci. Rev.* 120:118-125.

Wright, S. 1931. Evolution in Mendelian populations. *Genetics* 16:97-159.

Chapter 8

Trophic Interactions

Anders Persson[*1], *P. Anders Nilsson*[2] *and Christer Brönmark*[3]

8.1 INTRODUCTION

North temperate regions across the globe are characterised by a multitude of lakes, ponds and wetlands. These waterbodies are very similar with respect to origin, morphology and catchment and are subjected to similar climatic conditions. Still, some are covered with emergent macrophytes, others contain a soup of green and bluegreen algae, and yet others have clear water and submerged macrophytes. What processes are responsible for generating these differences? Questions like these became increasingly addressed when it became evident that processes both at the bottom and the top of the food web can structure natural systems (e.g. Hairston et al. 1960). It was suggested that the presence or absence of apex predators can have consequences for the structure and dynamics of entire ecosystems, and this called for a unifying theoretical framework for the role of predation for ecosystem structure and function, work that still progresses (e.g. Estes et al. 2011). This chapter deals with the complex consequences of predator-prey interactions in food webs with pike (*Esox lucius* Linnaeus, 1758) as a top predator, and, in particular, the indirect effects these interactions may have as mechanisms driving community patterns.

The pike is a specialist piscivore that can grow to substantial size and thus has capacity to consume prey of a large range of sizes (Mittelbach and Persson

**Corresponding author*: [1]Lund University, Department of Biology – Aquatic ecology, Ecology Building, 22362 Lund, Sweden; Email: anders.persson@biol.lu.se

[2]Lund University, Department of Biology – Aquatic ecology, Ecology Building, 22362 Lund, Sweden; Email: anders.nilsson@biol.lu.se

and

Karlstad University, Department of Environmental and Life Sciences – Biology, 65188 Karlstad, Sweden; Email: p.anders.nilsson@kau.se

[3]Lund University, Department of Biology – Aquatic ecology, Ecology Building, 22362 Lund, Sweden; Email: christer.bronmark@biol.lu.se

1998, chapter 2), and, hence, pike are expected to have considerable effects on prey populations and communities. However, we will show that while predator effects on the structure of prey populations often are direct and considerable, effects on community dynamics may be much more intricate, depending on the nature of the interaction, the identities of the players involved and the characteristics of their interactions. We will first provide a background to help understand what mechanisms are important for generating a trophic cascade or not. We then continue by providing empirical examples of top-down effects in the presence of pike, highlighting some of the factors and mechanisms we believe are of specific importance.

8.2 CASCADING TROPHIC INTERACTIONS

The term trophic cascade, first mentioned by Paine (1980), stems from the trophic level concept and describes the process when predators suppress their prey and thereby release the next trophic level below from predation/grazing. Hence, according to the trophic cascade hypothesis, the biomass at the bottom of the food chain depends not only on the availability of basic resources, such as nutrients and light, but also on the number of trophic levels in the food chain. The study of trophic interactions and, in particular, the phenomenon of trophic cascades received substantial attention when it was observed that considerable amounts of variation in primary producer biomass could be explained by the structure of the consumer community. This may seem obvious now, but at that time limnology had been focusing on environmental problems, such as which nutrient limits the primary productivity of aquatic systems, so the focus was more on understanding the cycling of different elements in aquatic ecosystems than on the interactions occurring among organisms further up in the food chain. Since top consumers in general contribute little directly to this cycling, it was natural to focus on water chemistry and interactions at the bottom of the food chain. This emphasis was indeed very successful and the discovery that phytoplankton biomass is positively related to phosphorus loading (e.g. Schindler 1974) had large implications for societal involvement in lake management strategies, exemplified by the huge investments in chemical treatment of wastewater in the early seventies that contributed to mitigating the increasing problems with eutrophication. However, in the late seventies and eighties, the worldview of traditional limnologists collided with that of animal ecologists when results from studies in population ecology, especially the role of predation and competition, were expanded to the community level to encompass interactions in food chains and ecosystems. Field and laboratory experiments suggested that top predators have considerable effects that are transmitted down the food web in a predictable way, causing every second trophic level to attain high biomass. The conceptual models by Hairston, Smith and Slobodkin (1960, HSS) were evaluated and theoretically developed in food web manipulations of entire ponds and lakes (e.g. Hrbácek et al. 1961, Carpenter and Kitchell 1993), convincingly demonstrating the occurrence of cascading trophic interactions in systems encompassing the complexities of natural systems. These early empirical studies received theoretical support from predator-prey models that were expanded to trophic levels (Oksanen et al. 1981) and formalised the original verbal HSS model that asked the fundamental question "Why is the world green?".

One important issue is how productivity shapes ecosystems and the amount of biomass accrued at different trophic levels. Simple food chain models based on Lotka-Volterra-like predator-prey dynamics, such as the Oksanen et al. (1981) model, typically predict that prey abundance at equilibrium is independent of productivity, and that productivity only affects top predator abundance and every second level below. Although predator-prey theories date back to the early 20th century, it was not until these theories were applied to food chains (Oksanen et al. 1981, Carpenter et al. 1985) that their predictions caused a controversy and a debate of whether bottom-up (e.g. nutrient loading) or top-down (predation) forces structure aquatic communities, a controversy likely fuelled by the explicit nature of the Oksanen et al. (1981) model and the specific predictions by Carpenter et al. (1985) of patterns in freshwater food chains with different number of trophic levels. The debate turned very much into a discussion of whether to be right for the wrong reason (mechanistically incorrect models reproducing empirical patterns) or wrong for the right reason (mechanistically correct models unable to reproduce empirical patterns). Today, the scientific community has reached the consensus view that both sides of the discussion were partly right (and wrong), but that appropriate modifications of the original simple models taking into account real life complexities are necessary to reproduce the dynamics of real ecosystems that sometimes lead to trophic cascades. The patterns found by Hansson (1992), who studied lakes along a productivity gradient in Sweden and Antarctica, illustrate this. Hansson (1992) found that the relationship between productivity and producer biomass was positive, independent of food web structure, partly contradicting Oksanen et al. (1981), but that the slope of the relationship depended on food web configuration, such that producer biomass in two-level systems (i.e., fish-free lakes in Antarctica) increased with productivity at a lower rate than three-level systems (i.e. Swedish lakes with fish). Hansson explained this pattern by the fact that slow-growing but grazing-resistant prey (phytoplankton) replaced fast-growing and grazing-susceptible prey in two-level systems. This enabled prey to respond to increases in productivity although at a slower rate compared to three-level systems. Drenner and Hambright (2002) expanded this approach to cover four-level systems with piscivores and found support for the view that the slope of the productivity-phytoplankton biomass relationship depends on the number of trophic levels, being lower for four-level systems compared to three-level systems. These findings support the present consensus view that top-down and bottom-up processes interact in shaping trophic level biomass.

8.2.1 Alternative Stable States

The Oksanen et al. (1981) model predicted that ecosystems would shift dramatically between states along the productivity gradient, driven by the inclusion of more trophic levels as productivity increases. However, already before Oksanen's paper, it was recognised that systems subjected to continuous changes could produce regime shifts and even display multiple stable states for similar conditions and settings (reviewed in May 1977). Both simulations of mathematical models (e.g. Holling 1973, May 1977, Tilman 1982) and empirical observations (e.g. Sutherland 1974, Scheffer et al. 2001) suggested that the history of events played an important role for the current state of the ecosystem, which contrasted the previous view that systems have a global attractor that the system moves towards independently of any perturbations. The term "alternative stable states" implies that a system that reaches a new state through a certain pathway (e.g. eutrophication) will be stabilised by feedback mechanisms in the new state, making the system resistant to small perturbations. Moreover, regime shifts where a system responds abruptly to a small change in an external driver, such as changes in nutrient loads, is not a sign of alternative stable states if the system can be reversed along the same pathway. Instead, a system with alternative stable states that experiences a reversal of the external driver follows a different trajectory, ending in another bifurcation point where the system shifts back to its original state, i.e., at a level of the external driver that is different from where it first changed. This is an important distinction since systems displaying alternative stable states, and where perturbation history matters, are inherently less predictable. Hence, the external driver may not affect the state of the system within certain domains, but will affect the resilience of the system and its resistance against stochastic and catastrophic perturbations, such as climatic events or fish kills (Scheffer et al. 2001).

Alternative stable states have been observed in many different types of ecosystems (Scheffer et al. 2001) and has been a common explanation for the dominance of either piscivorous or planktivorous fish in aquatic ecosystems (Walters and Kitchell 2001). Scheffer (1997) suggested that alternative stable states are especially common in shallow lakes, driven by the competition for nutrients and light between phytoplankton and submerged macrophytes. In the case of shallow lakes, the presence of submerged macrophytes covering a substantial part of the lake area provides the foundation for a number of feed-back mechanisms that maintain the system in a clear-water state and makes it resistant to moderate stochastic perturbations. The macrophytes and periphytic algae growing on the macrophytes compete with phytoplankton for soluble nutrients and light, the macrophytes make the sediment more resistant to physical resuspension of sediments that can increase the turbidity of the water, and the macrophytes add structural heterogeneity to the habitat that mainly benefits predator-sensitive invertebrates and young stages of piscivorous fish, such as Eurasian perch (*Perca fluviatilis* Linnaeus, 1758) and pike, that are able to forage relatively efficiently in structurally complex environments (Diehl 1988). If phytoplankton instead monopolise light and nutrient resources, the turbidity of the water increases, which instead will benefit planktivorous fish. Hence, it is the presence of macrophytes that makes shallow lakes resistant to change, but it is generally a drastic event such as a fish kill or extreme water level that induces the shift (Scheffer et al. 1993). The theory of alternative stable states provides the theoretical backbone for biomanipulation efforts, including stocking of pike (chapter 11).

8.3 PIKE AS A TOP PREDATOR

So how does the presence or absence of pike in natural systems comply with the trophic cascade hypothesis? Indeed, pike is commonly viewed as a voracious predator, a view that has been the focus of many studies of pike diets. These have given ample evidence that pike has a broad diet ranging from invertebrates, via fish of different sizes, to birds, amphibians and small mammals (e.g. Solman 1945, Beaudoin et al. 1999, Dessborn et al. 2011, Winfield et al. 2012, chapter 11). Further, a number of studies, using different methods, suggest that pike as a top predator has a strong effect on the density, species composition and size structure of the prey fish community. For example, analyses of long-term data on the dynamics of pike and its prey populations suggest that the pike is an important mortality source for prey fish, and that these patterns show strong responses to cannibalistic interactions (e.g. Kipling 1984, Mills and Hurley 1991). A further approach is to compare survey data from a large number of lakes with contrasting fish community structures to infer structuring effects of piscivores. An intercontinental comparison of fish communities in forest lakes of North America (Wisconsin) and northern Europe (Finland) revealed some intriguing similarities and differences in community structures and structuring processes between the regions (Tonn and Magnuson 1982, Tonn et al. 1990). Isolated lakes with severe environmental conditions (winterkills, low pH) generally lacked piscivorous predators, such as pike, and therefore could be dominated by predator-sensitive species, such as mud minnows (*Umbra limi* Kirtland, 1841, Wisconsin) or crucian carp (*Carassius carassius* Linnaeus, 1758, Finland). Isolated lakes with low productivity but with less severe environmental conditions were generally dominated by piscivorous fish, such as largemouth bass (*Micropterus salmoides* Lacépède, 1802, Wisconsin) or perch and pike (Finland), that were able to control prey populations. Finally, less isolated and productive lakes were species rich and dominated by pike (Wisconsin) or roach (*Rutilus rutilus* Linnaeus, 1758, Finland). Piscivores in Wisconsin lakes were able to drive some small, soft bodied and specialised prey fish species to extinction, leading to presence-absence communities, whereas species interactions among the more generalist Finnish species led to different species dominating different communities, without biotic interactions causing extinctions. Hence, species in Finland were able to coexist, albeit at low densities, with piscivorous predators such as pike, by attaining a size refuge (chapter 2) and occupying densely vegetated areas of the lakes (Tonn et al. 1990). This illustrates how dispersal and evolutionary history can determine the structuring role of piscivores, such as pike, in lake communities.

The importance of pike predation for prey fish was further evidenced in a study by Spens and Ball (2008), where they analysed data on fish community structure from 1029 North Swedish lakes and found that self-sustaining salmonid populations were absent from lakes with pike (cf. chapter 14). In addition, introduction of salmonids to a number of these lakes convincingly established that it was predation by pike that was the driver behind the patterns; salmonids failed to establish in pike lakes whereas in the majority of lakes where pike had been removed by rotenone before salmonid introduction, the introductions were successful. Further, a study on fish communities in coastal landlift lakes showed a similar negative co-occurrence pattern where the presence of piscivore populations (pike and/or perch) explained the absence of nine-spined (*Pungitius pungitius* Linnaeus, 1758) and three-spined sticklebacks (*Gasterosteus aculeatus* Linnaeus, 1758) and crucian carp (Englund et al. 2009). Introductions of piscivores resulted in the extinction of the sticklebacks, whereas crucian carp remained in some of the lakes. Similarly, introduction of pike led to extinction of a weakly armoured morph of three-spined stickleback also in a Canadian lake (Patankar et al. 2006), whereas a survey of ponds and small lakes in southern Sweden showed that crucian carp and tench (*Tinca tinca* Linnaeus, 1758) could coexist with piscivorous pike and perch (Brönmark et al. 1995). There were dramatic differences in density and size structure of the crucian carp and tench populations among lakes with and without piscivores, with very dense populations dominated by small individuals in ponds without piscivores, and sparse populations with a few, large individuals in piscivore ponds. These large individuals had supposedly reached a size refuge from predation (e.g. Hambright et al. 1991, Nilsson and Brönmark 2000). That predation really was the driver behind the patterns in density and size-structure was further shown in an experiment were pike was introduced into ponds with crucian carp (Brönmark and Miner 1992). Interestingly, this study also showed that crucian carp grow deeper bodied in the presence of pike, i.e., the crucian carp has an inducible morphological defence adaptation that increases its possibility to coexist, albeit at low densities, with piscivores.

Clearly, the effects of pike are particularly obvious in studies where pike are introduced into lakes (chapters 11, 14). Besides the ones cited above, numerous other studies show that pike may cause dramatic reduction or even total extirpation of the prey fish populations when introduced to lakes. In a now classic study, He and Kitchell (1990) introduced pike to a small piscivore-free lake and studied direct, lethal as well as behavioural effects of pike on the prey fish assemblage. Predation by pike reduced total prey fish density over summer with almost 50%, and soft-rayed, slender-bodied species, such as dace and minnow, were especially vulnerable to predation. Thus, introduction of pike changed overall prey densities as well as the species composition of the prey assemblage. Interestingly, the behavioural response to the threat of pike predation was greater than the direct lethal effect, manifested by the large number of various dace and minnows moving out of the lake into the outflow stream shortly after pike introduction. Other studies of pike introductions have also shown strong effects. Findlay et al. (2005) showed that experimental introductions of pike into several lakes in the Experimental Lakes Area in Canada resulted in a rapid restructuring of the piscivore naive prey fish community with a reduction of small, soft-rayed and shallow-bodied species. A revisit to the same lakes 19-25 years after the introduction revealed that prey fish populations had gone completely extinct and the lakes were by then pike-only lakes (Nicholson et al. 2015). Strong effects of pike introduction have also been found in other systems, e.g. Arizona reservoirs (Flinders and Bonar 2008). Venturelli and Tonn (2005) took a somewhat different approach and introduced pike to lakes in the boreal plains in Canada that were completely fishless due to winterkills. Comparisons of macroinvertebrate data previous to pike introductions as well as with data from unmanipulated reference lakes revealed that pike preyed heavily on conspicuous macroinvertebrates, such as odonates, coleopterans and leeches, that were top predators before pike introduction. Introductions of pike and the concomitant negative effect on prey populations is also of much concern in areas where the pike is invasive and where it may have strong negative effects on the native fish community (chapter 14). A useful method to estimate consumption rate of piscivores and their impact on prey populations is bioenergetics modelling (Hanson et al. 1997). Several studies have used this approach and shown that pike can have a strong effect on e.g. stocked rainbow trout (*Oncorhynchus mykiss* Walbaum, 1792) populations (up to 100% mortality; Flinders and Bonar 2008) and yellow perch (*Perca flavescens* Mitchill, 1815, Paukert et al. 2003).

It is clear that pike may have strong effects on prey populations, but could these effects also cascade to lower trophic levels? A whole-lake experiment where large pike were selectively removed to decrease cannibalism on smaller pike resulted in an increase in the total density of pike, especially small and medium-sized individuals, and this in turn resulted in a reduced population density of the major prey fish, roach (Sharma and Borgstrøm 2008). The diet of the few remaining roach shifted to include a higher proportion of large cladoceran zooplankton, which was suggested to be due to a lower predation pressure and hence a higher availability of cladoceran zooplankton. It was further argued that this increase in zooplankton biomass and, hence, grazing pressure should cause increasing water transparency due to lower algal biomass. Increasing water transparency was the goal of a whole-lake experiment, where Prejs et al. (1994) introduced pike fry in three consecutive years to a lake with algal blooms. This resulted in a strong reduction, near failure, in the recruitment of the main zooplanktivorous fish in the lake; roach, white bream (*Blicca bjoerkna* Linnaues, 1758), and sunbleak (*Leucaspius delineatus* Heckel, 1843). In spite of this strong reduction in the populations of the major zooplanktivorous fish, there was no effect on the zooplankton community and thus no cascading effects on algal biomass (Prejs et al. 1997), supposedly due to the remaining populations of large-sized roach and white bream, outside of the gape-size window of the stocked pike, resulting in a continued strong predation pressure on zooplankton (chapter 11). In other examples, strong negative effects on algal biomass, following the predictions of the trophic cascade hypothesis, have been found after pike introductions (Findlay et al. 2005). However, although there was a shift to larger bodied cladoceran zooplankton, as predicted, there was still no effect on total zooplankton biomass and no relationship between zooplankton and phytoplankton. Instead, it was suggested that the changes in algal biomass was driven by a reduction in phosphorus excretion by yellow perch, which were reduced when pike were introduced. Further, Byström et al. (2007) followed the invasion of pike into an alpine lake in northern Sweden, where it altered fish community composition by replacing the native Arctic char (*Salvelinus alpinus* Linnaeus, 1758) as top predator and reduced the density of sticklebacks, the main fish prey. The altered fish community composition had consequences further down the food web, causing increased zooplankton and macroinvertebrate abundances. Evidently, the effect of Arctic char on sticklebacks in the absence of pike was much lower than the interaction strength of the invading pike on sticklebacks. When comparing morphological and behavioural traits, pike is indeed better adapted to a piscivorous lifestyle than Arctic char, which is better described as an ontogenetic omnivore that needs several years to attain a piscivorous diet (Mittelbach and Persson 1998), if at all. Frequent introductions of pike into alpine systems, often dominated by salmonids, have led to extinctions of the latter, and pike is therefore viewed as a pest in parts of the world (chapter 14). Further, future warming is predicted to make it possible for pike to invade alpine systems presently too cold for pike to persist. This has raised concerns of the possible consequences of global warming in alpine systems typically being species poor and situated in remote, pristine areas.

In addition to effects of pike introductions we may see effects on lower trophic levels when pike go extinct from a system, e.g. due to winterkill acting selectively on pike. Tonn and Magnuson (1982) showed that predator-sensitive minnows are only common in lakes with frequent winterkills that eliminate pike and other piscivores. Brönmark and Weisner (1992) suggested that if winterkills were particularly severe on pike, effects should cascade to lower trophic levels including a shift from a dominance of macrophytes to a dominance of phytoplankton. The main mechanism for the transition towards phytoplankton dominance in pike-free systems was argued to be an increase in benthivorous fish abundance that would result in a decrease in the abundance of invertebrates, such as snails grazing on epiphytes. Epiphytes growing on macrophytes would thereafter increase and shade macrophytes, which should ultimately reduce macrophyte abundance and depth distribution. However, testing these predictions in a survey of food-chain patterns in natural pond systems showed that the trophic effects of pike were not necessarily strong enough, even though pike is viewed as a specialist piscivore (Brönmark and Weisner 1996). It was suggested that the trophic effect of pike was uncoupled at the level of the intermediate consumer, i.e. the prey of pike. In the studied ponds, tench and crucian carp dominated the prey community, two species that may reach a size refuge from pike predation, as opposed to sticklebacks and minnows that remain small their entire life. Hence, in systems with pike predation, tench and crucian carp densities were much lower than in pike-free systems, but the survivors experienced a faster growth rate and individual size was much larger, thereby partly compensating for the higher loss rate of small individuals. This illustrates that the identity of the species in the predator-prey game is of crucial importance for the outcome and the consequences of the interaction. Moreover, theories on trophic interactions rely on the assumption that a trophic level will respond as a population. It is easy to see how this assumption is violated in many natural systems, since species composition of a trophic level depends on the risk of mortality from top predators. Hence, the strong interactions predicted from predator-prey models are often not observed in nature because predator-sensitive species are replaced by species with more efficient defence strategies, and because individuals change strategy in response to risk level, blurring the effects of piscivores such as pike on lower trophic levels. Next, we consider how interaction strength affects food web dynamics.

8.4 INTERACTION STRENGTH AND FOOD-CHAIN DYNAMICS

If top predators like pike will have effects down the food web or not is dependent on the interaction strength between the predators and their prey. Predators may affect lower trophic levels in multiple ways, such as consuming prey, altering prey behaviour, or changing prey life history. All these effects may cascade to lower trophic levels or indirectly affect competitors in ways that alter the structure and function of aquatic food webs. Ecological theory specifies the strength of interactions between predator and prey populations in the functional response that describes the per capita predator effect on the size of the prey population. As outlined in chapter 2, consumption rate of an individual predator increases with prey density in a non-linear way determined by predator search rate and handling time of prey. But there are also numerous models that account for e.g. habitat dependence, predator interference, evolutionary consequences on prey susceptibility to predation, switching of predator diets, etc., i.e. when the per capita consumption rate is dependent on the traits (e.g. sizes, behavioural response) and densities of the players involved as well as the context (e.g. habitat, season) in which the interaction takes place. Such factors may either stabilise or destabilise predator-prey dynamics and make interaction strength dependent on both predator and prey densities. Hence, understanding the nature of the interaction between predators and prey is necessary to be able to understand the consequences such interactions would have on the structure and dynamics of entire communities. Below, we will outline some examples leading to both weaker and stronger interactions, ultimately affecting the strength of the trophic cascade.

8.4.1 Pike Density-Dependent Cannibalism, Kleptoparasitism and Interference

The shape of the functional response should affect how predators affect prey, and hereby also the potential for and strength of cascading trophic interactions (e.g. Holling 1965, Carpenter and Kitchell 1993, Fryxell and Lundberg 1998, chapter 2). Theoretically, densities of predators and prey should drive interaction strengths and consumer-resource dynamics, albeit such patterns have been difficult to observe in nature. This may be due to the effects of a range of different behaviours among both predators and prey. For example, prey may respond to predation risk by expressing different anti-predatory behaviours, with consequences for predation rates and thereby the interaction strength between predators and prey (see section 8.4.5). Moreover, behavioural interactions among predators, such as cannibalism or interference, may also affect interaction strength, partly due to direct mortality effects from cannibalism, partly due to indirect effects on functional responses as predators are forced to devote attention to avoid becoming victims when cannibal densities are high. Predator and prey behaviours can hereby affect the functional response.

Although cannibalism should be avoided for obvious reasons, it is still relatively frequent in pike populations (Grimm 1981). Intraspecific predation inevitably has consequences for pike mortality and hereby densities. Successful cannibalism reduces pike densities, and should hereby decrease predation rates on prey, both by cannibals feeding on conspecifics instead of prey, as well as from victims to cannibalism ceasing foraging. Further, the interference, kleptoparasitic and cannibalistic behaviours should have consequences for predator-prey interactions and the intraspecific effects of these behaviours should be density dependent (Giles et al. 1986, Nilsson and Brönmark 1999, Nilsson et al. 2006, chapter 2). Theory on density-dependent intraspecific effects and interactions among predators should hence apply also to pike functional responses. Individual pike may reduce the risk of kleptoparasitism and cannibalism by reducing foraging activity (chapter 2) and the decrease in intake rates should be both prey- and predator-density dependent, affecting the shape of the functional response by decreasing per capita intake rates, particularly at low prey densities, according to interference theory (Fryxell and Lundberg 1998, Giraldeau and Caraco 2000, Fig. 2.1). A reduced foraging activity should directly decrease predation rates on prey, and thereby potentially reduce top-down trophic effects from pike predation. Moreover, changes in the shape of the functional response due to interference has implications for consumer-resource dynamics (Fryxell and Lundberg 1998, Nilsson et al. 2007). Briefly, interference is predicted to have stabilizing effects, if the predator zero-growth isocline crosses the prey isocline where the prey isocline has a negative slope (see e.g. Fryxell and Lundberg 1998 for details, Fig 8.1). This means that avoidance behaviours in individual pike to reduce the risk of kleptoparasitism and cannibalism will result in interference phenomena that reduce the overall impact of pike predation on prey and, further, it stabilizes this reduced effect and decreases the potential for top-down trophic cascades from pike apex predation (e.g. Nilsson 2001). On the flip side, foraging facilitation, that can occur in e.g. social foraging groups of perch (Nilsson et al. 2006), can produce situations where predator and prey isoclines cross where the prey isocline has a positive slope, with unstable or limit-cycle dynamics as a response (Fig. 8.1).

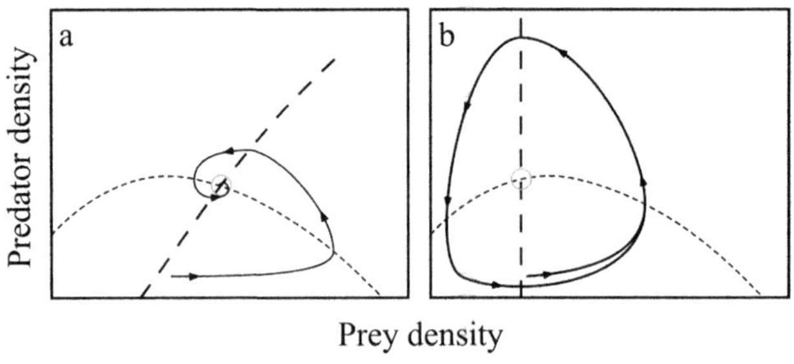

FIGURE 8.1 Conceptual population trajectories (solid) for predators and prey, with (a, stable) and without (b, unstable) predator interference. Zero isoclines for predators (long dash) and prey (short dash), and non-trivial equilibria (grey circles) are shown. Stability arises when the prey zero isocline has a negative slope at the intersection with the predator isocline, while unstable dynamics arise for slopes > 0. Inspired by Fryxell and Lundberg (1998) and Nilsson et al. (2007).

8.4.2 Food Chains, Intra-Guild Predation, and Size-Structured Interactions

Food chains can be viewed and analysed from different perspectives. Early food-chain theory considered each trophic level as a homogenous entity, such that, from a top-down perspective, all individuals at a specific trophic level have the same potential effect on individuals at the trophic level below (Fig. 8.2a). Such model food chains can be used to predict and evaluate effects of food-chain length or changes in densities at different trophic levels, such as e.g. how increases in top predator density could cause cascading trophic effects down the food chain and alter primary producer densities (see section 8.2). Food chains are here regarded as strict chains, with no trophic levels deviating from the chain. However, nature is often more complex than this. For instance, piscivores (top predators) may be both predators and competitors to their prey (consumers) if they at least at some point during their ontogeny feed on the same food source as their prey (Fig. 8.2b). This places the top predators at the same trophic level as their prey, creating a so called Intra-Guild Predation (IGP) system. For apex fish predator species, IGP can occur if e.g. the top predator is a generalist and not a strict piscivore. It can also arise if the top-predator goes through several ontogenetic niches to reach the piscivorous stage and therefore have to compete with their future prey during their early life stages before becoming piscivorous and ending up at the top trophic level. For example, juvenile piscivore species may compete with their future zooplanktivore prey species for the zooplankton resource before they have reached a size when they can start feeding on fish. IGP has received considerable attention both empirically and theoretically (e.g. Polis et al. 1989, Holt and Polis 1997, de Roos and Persson 2013). The predation by adult piscivores in IGP systems can benefit their own, not yet piscivorous, young by releasing them from competition from their future prey, a mechanism sometimes referred to as cultivation. However, if the abundance of piscivores does not suffice to control the prey, increased competition between prey and non-piscivorous stages of the predator serves as a feedback stabilising the prey-dominated stage, a mechanism often termed depensation. This may lead to alternative stable equilibria dominated by either the piscivore or its prey.

Competitive cultivation-depensation systems (Walters and Kitchell 2001) seem common among piscivores and may lead to juvenile bottlenecks (Fig. 8.2c) and/or stunted populations of individuals not reaching the piscivorous niche, as is sometimes the case for perch. It has, moreover, been suggested that it is difficult for predator populations that undergo ontogenetic diet shifts to control prey populations if intermediate predator stages have to compete for resources with their future prey (Mittelbach et al. 1988, Persson et al. 1998, van Leeuwen et al. 2013). This seems particularly applicable for generalist predators such as perch. Pike avoid competition with future prey by spawning in cold waters in early spring well before their prey initiate spawning, a strategy to spatially and temporally avoid competition with future prey. Also, when pike reach 10-12 cm body length they become predominately piscivorous and start feeding on newly hatched prey fish (Raat 1988, Craig 1996, Skov and Nilsson 2007).

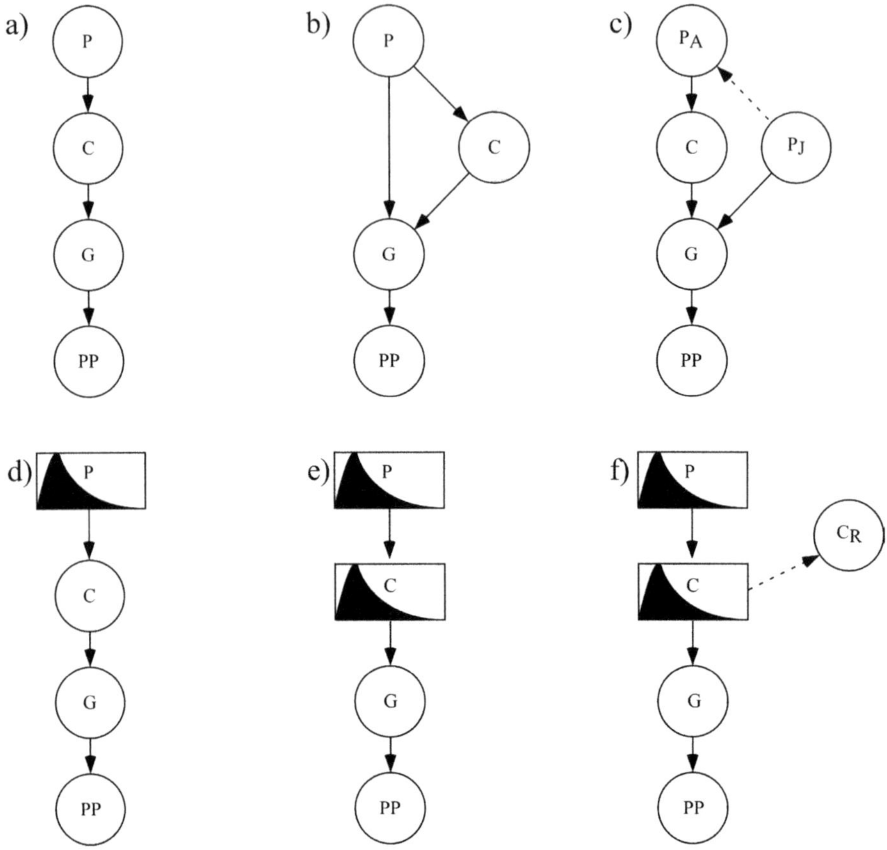

FIGURE 8.2 Illustrations of different food web configurations. a) Food chain with four trophic levels including a top predator (P), consumer (C), grazer (G) and a primary producer (PP) (Oksanen et al. 1981). b) An intraguild predation system where consumers are both prey and competitors with the top predator (Holt and Polis 1997). c) An intraguild predation system where the top predator experiences a bottleneck between juvenile (P_J) and adult predators (P_A) caused by competition between juvenile predators and their future prey (Byström et al. 1998). d) A size-structured predator population (de Roos and Persson 2013), and e) a system with size-structured predator and consumer populations (Persson et al. 1999). f) When both predator and prey populations are size structured, large consumers can reach a size refuge from predation (C_R) (Brönmark and Weisner 1996, Hambright et al. 1991).

A special case of intra-guild predation is when the prey is predator on early life stages of the predator, referred to as predatory cultivation-depensation (Walters and Kitchell 2001, Gårdmark et al. 2014). In the coastal areas of the Baltic Sea, the abundance of piscivorous fish has decreased considerably, something attributed to increases in three-spined stickleback and sprat (*Sprattus sprattus* Linnaeus, 1758). Sticklebacks are potential prey to pike, but sticklebacks may also be predators on the earliest life stages of pike. A reduction in the offshore cod (*Gadus morhua* Linnaeus, 1758) population has released sticklebacks from predation in offshore areas of the Baltic Sea, with consequences also in coastal areas where the spawning and nursery grounds of sticklebacks overlap with that of pike and perch. Experimental studies excluding piscivores as well as correlative field studies in the Baltic Sea furthermore suggest a negative association between abundance of piscivorous fish and ephemeral, filamentous algae (Eriksson et al. 2009, Östman et al. 2016), suggesting cascading trophic interactions between piscivores and algae. In all, the negative effects of the increasing stickleback population on pike populations cascade to lower trophic levels including filamentous algae that threaten to outcompete perennial macroalgae and seagrass, ultimately modifying the coastal habitat.

The populations of sticklebacks and coastal pike only overlap spatially during the spawning season, which causes pike to benefit relatively little from predation on an increasing population of sticklebacks. Instead, stickleback predation on early stages of pike creates a positive feedback by decreasing pike survival. This reasoning is supported by field data showing the absence of negative effects of sticklebacks on pike populations when stickleback and pike spawning areas are spatially separated (Byström et al. 2015).

The above reasoning around IGP is based on ontogenetic changes in population structure, and in fish this is of course associated with indeterminate growth through ontogeny. The early predator-prey models assumed that all individuals of predators or prey had similar traits, while for size-structured populations, such as fish populations (Fig. 8.2d, e), this assumption is seriously violated. Fish change size several orders of magnitude during their life and most traits involved in vital rates, such as foraging, consumption, digestion, reproduction and risk of predation, are strongly correlated with size. The development of models based on the size structures of populations has advanced the field considerably. Such models account for continuous ontogenetic changes in traits related to energy gain and loss, but the models demand numerous parameters, and consequently require considerable analytical power that only recently has become available. Size-structured interactions have revealed numerous intricate consequences at population (Ebenman and Persson 1988, chapter 6) and community levels (de Roos and Persson 2013), with likely consequences also for trophic interactions.

Size-structured piscivore populations exert a size-structured predation effect on size-structured prey populations (Ebenman and Persson 1988, Brönmark et al. 1995, Persson et al. 1999). For example, gape-size limited piscivores are restricted in their prey choice by the relative size between predator and prey, and prey vulnerability to predation generally decreases with increasing prey size, with implications for trophic processes (Hambright et al. 1991, Persson et al. 1999, Nilsson and Brönmark 2000). Moreover, gape-size limitation in piscivores may decouple the trophic cascade if enough prey individuals reach a size where they are outside the prey-size window of the piscivores (Fig. 8.2f), as these large prey individuals are no longer vulnerable to predation and can continue foraging on lower trophic levels (Fig. 8.2f).

Pike populations are size structured and they feed on size-structured prey. Pike gape-size limits restrict prey choice, especially in small individuals, while only large pike individuals are free to forage on most prey sizes (Nilsson and Brönmark 2000). Similarly, risk of cannibalism is greater for smaller pike individuals, and avoidance of cannibals should hence be more pronounced in smaller sized pike (Bry et al. 1992, Nilsson 2006). Consequently, direct and indirect effects of intraspecific interactions also depend on the size structure of the pike population, and predictions of top-down trophic cascades from pike predation should take this into consideration.

If we assume that pike did not show any intraspecific interactions, pike would act as a size-structured predator population predating on size-structured prey, with prey-size limits determined by individual pike gape size and prey size (Fig. 8.3, Nilsson and Brönmark 2000). If so, only a small portion of the prey population would enjoy a size refuge from predation, and most prey sizes would be vulnerable to pike predation. This could in theory create strong top-down trophic cascades, by reducing prey fish population densities of all sizes, thereby releasing zooplankton from predation, allowing zooplankton to graze efficiently on phytoplankton, with consequences for system composition (Fig. 8.3). However, as discussed above, pike show strong intraspecific interactions; they are cannibals and kleptoparasites. Avoiding mortality from intraspecific predation or avoiding having your food stolen by other pike should be beneficial for individuals. As pike avoid intraspecific interactions by reducing foraging propensity (chapter 2 and above), individual foraging rates decrease under risk of interaction. The potential for trophic effects is hence reduced compared to if pike individuals did not have to avoid intraspecific interactions (Fig. 8.3). Moreover, the risk of intraspecific interactions among pike individuals increases during manipulation and handling

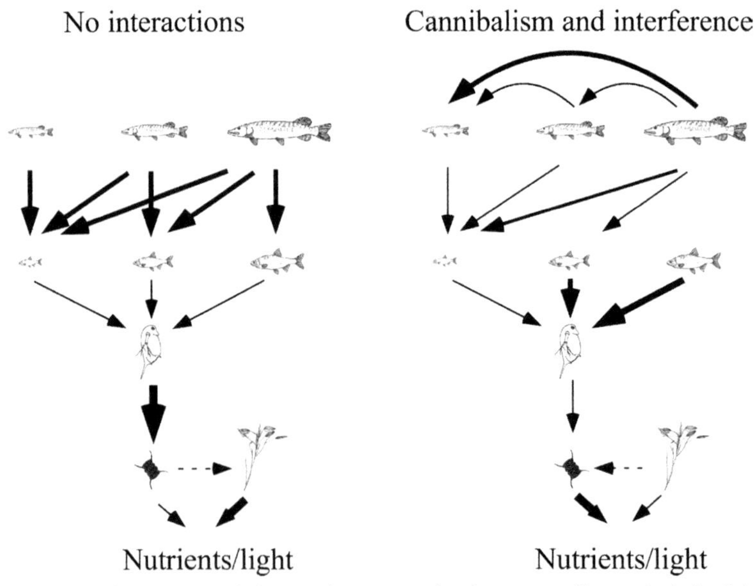

FIGURE 8.3 Size-structured interaction strengths between pike and their fish prey, illustrated without pike intraspecific interactions (left) and with size-structured cannibalism and interference among pike (right). Arrow thickness indicates interaction strength. Dashed arrows indicate the possible transitions between phytoplankton- and macrophyte-dominated communities. Inspired by Nilsson (2000).

of prey and pike consequently prefer smaller prey in the presence of conspecifics (Nilsson et al. 2000, chapter 2). A shift in prey-size selection towards smaller prey would release more prey from being vulnerable/susceptible to pike predation, resulting in expanded, behaviourally mediated size refuges from predation (Fig. 8.3). In spite of the reduced activity and altered prey-size preference, pike do fall victim to cannibals, reducing pike densities. Decreased foraging activity, prey size window, and pike population density inevitably reduce the potential for pike predation to create top-down trophic cascades (Fig. 8.3). Ultimately, pike intraspecific behaviours can release zooplanktivores from predation, maintaining low zooplankton densities, with high probability for a phytoplankton-dominated stable state in lakes (Fig 8.3, chapter 11).

8.4.3 Cross-System Subsidies

Lakes are an integral part of the landscape and although we often view them as closed systems they are of course open to transport of resources across ecosystem boundaries, and this may have strong effects on the structure and dynamics of their food webs. A tremendous amount of work has been performed on how import/export of nutrients and detritus affect freshwater ecosystems, whereas the importance of movement of organisms across habitat borders is less studied, especially in a cascading trophic interactions perspective. However, already Polis et al. (1997) argued that, to understand the dynamics of local food webs, we need to consider the importance of spatial resource subsidies, i.e., the transport of nutrients, detritus and organisms across habitat boundaries, from a donor habitat to a recipient habitat. For example, import of terrestrial prey organisms into recipient, freshwater system might lead to an increase in the population density of aquatic predators which then results in an increase in the strength of the trophic cascade as the subsidized predator population can further decrease herbivores and indirectly increase plant biomass. Research on trophic interactions in stream ecosystems provide examples on the importance of subsidies from the terrestrial habitat, but also the reverse, i.e., that insects that emerge as adults after spending their larval period in the water are eaten by terrestrial predators and thus provide an important subsidy to terrestrial food webs (e.g. Nakano and Murakami 2001, Baxter et al. 2005). Further, it has been shown that changes in the density of apex predators with concomitant effects on cascading trophic interactions may transcend ecosystem barriers and affect consumers in adjacent habitats. For example, introduction of piscivorus cutthroat trout (*Oncorhynchus clarkii* Richardson, 1836) to experimental ponds with three-spined sticklebacks led to a trophic cascade that resulted in increased biomass and size of the adult insects that emerged from the pond and this, in turn, resulted in an increased foraging activity of insectivorous bats (Rudman et al. 2016). However, aquatic organisms may not only have a strong subsidising effect on terrestrial food webs by providing terrestrial predators with a food source, but may also have negative effects if the emerging organisms are important predators (e.g. dragonflies) on terrestrial prey. In a comparison of ponds with and without fish, Knight et al. (2005) showed that differences in cascading trophic interactions in the ponds spilled over and affected the terrestrial system: fish fed on dragonfly larvae and thereby reduced adult dragonfly densities which in turn reduced the predation rate of dragonflies on bees and other insects. Thus, it is obvious that food web structure in freshwater systems, especially the presence/absence of predatory fish, may have strong repercussions on the prevalence and direction of cross-system subsidies. However, further manipulative studies of trophic cascades in freshwater habitats also involving piscivores, such as pike, are warranted if we are to increase a general understanding of the importance of subsidies across the terrestrial-aquatic border.

8.4.4 Multiple Predators

Studying interactions among multiple species is a difficult task that involves methodological trade-offs in temporal and spatial resolution. In assessing the interaction strength between multiple predators and prey, stomach analysis is of limited value. In fact, what occurs in the diet of a predator may be what is left after the effect of predation has eliminated more valuable prey. You may view it as the ghost of predation past. Hence, interactions are instead often studied using an experimental approach. Several small-scale experiments have shown that multiple predators may have complex effects on prey communities, ranging from risk reduction (Vance-Chalcraft and Soluk 2005) to neutral, additive or even multiplicative effects (Sih et al. 1998, Eklöv and Van Kooten 2001, Carey and Wahl 2010). Predator facilitation is most likely to occur when predators have different foraging modes, suggesting that complementarity in foraging strategies overrides predator interference (Soluk 1993, Sih et al. 1998). For example, Eklöv and VanKooten (2001), in a study with piscivorous perch and pike, showed that prey fish (roach) responded to the threat from perch in the open water by hiding in the vegetation, a habitat where the pike were lurking. Consequently, pike benefitted from perch presence by prey being more readily available. Carey and Wahl (2010) performed a similar experiment with the cruising predator largemouth bass and the ambush predator muskellunge (*Esox masquinongy* Mitchill, 1824) feeding on bluegill (*Lepomis macrochirus* Rafinesque, 1810) prey, but here it was the cruising largemouth bass that experienced enhanced growth when both predators were together. Both studies suggest that two predators with complementary foraging strategies and habitat use have stronger effects on prey populations than when only a single predator species is present. The discrepancy between the above two studies regarding which predator species that benefitted most was explained by differences in prey anti-predator behaviour, where roach were more strongly associated with the vegetation in presence of predators than were bluegills.

But could experimental studies in mesocosms, run over short time periods, explain population and community dynamics? Persson et al. (2007) studied a whole-lake system with similar fish community structure as in the experiment by Eklöv and Van Kooten (2001) above. Roach where introduced to lakes with either perch only or with both perch and pike, and predator and prey populations where then followed for several years. The results showed that roach were only able to invade systems with perch if pike were also present. In the absence of pike, high predation pressure from piscivorous perch efficiently removed roach invaders and new recruits and this ultimately made roach colonisation impossible. In lakes with pike, on the other hand, selective predation on perch by pike created an invasion window for roach. The pike effect on perch was twofold. First, pike reduced the numbers of perch, and, second, pike presence altered perch behaviour creating habitat pockets were roach could survive their first years. Hence, the presence of pike was a prerequisite for coexistence, a conclusion quite different from the results of the small-scale experimental study by Eklöv and VanKooten (2001). Thus, while conclusions from small-scale experiments may be useful to discern mechanisms over short timescales, the outcome may be reversed when viewed over longer time scales that allow numerical responses in the predator populations. To assess such long-term dynamics, modelling tools in combination with monitoring of entire ecosystems become especially powerful.

It may seem counterintuitive that the less specialised perch would be able to drive roach towards extinction whereas the highly specialised piscivore pike would not. In this respect it is important to distinguish between processes operating at different levels, i.e., at the individual versus the population level. The pike is a specialist piscivore while the perch is a generalist forager and more opportunistic. Hence, on an individual level, pike would theoretically have a greater impact on prey fish than would perch. However, perch is more of a team player whereas pike is individualistic. Hence, perch (at least of similar sizes) cooperate and could hunt down larger and faster prey when together with conspecifics, whereas pike are kleptoparasitic and cannibalistic, i.e., have negative effects on conspecifics (Nilsson et al. 2006). Thus, there may be a positive feedback of perch density on the hunting success of individual perch, whereas the opposite is true for the cannibalistic and kleptoparasitic pike. Perch may also be cannibalistic, but their relatively smaller gape sizes make cannibalism a more restricted event in perch compared to in pike populations.

8.4.5 Behavioural Cascades

Textbook examples of predator-prey interactions, based on Lotka-Volterra model dynamics, describe a very close relation between predator and prey dynamics often generating cycles and unstable dynamics when predators overexploit prey. However, the dynamics predicted from such models are rarely observed in nature because in natural systems individuals respond to changes in the trade-off between growth and mortality risk by altering behaviour, life history or even morphology. Hence, such responses in prey serve to compensate for effects of increased predation risk and thereby stabilises predator-prey dynamics. Prey commonly respond to increased predation risk by seeking shelter in refuge habitats or by reducing activity. This increases prey survival and thereby has a negative effect on the predator's numerical response and ultimately prevents overexploitation.

The presence of a top predator may still have far-reaching consequences at multiple trophic levels even when the actual mortality caused by the top predator is small. This is because of costs associated with altered behaviour to become more vigilant, decrease activity, or increase the use of refuge habitats. This reduces prey foraging rates which in turn has an indirect positive effect on resource populations, i.e., we have a behavioural trophic cascade. Evidence across taxa and in both terrestrial and aquatic environments show that such trait-mediated indirect interactions may be just as, or even more, important for community dynamics as the direct consumptive effect of a predator (Werner and Peacor 2003). For example, in the absence of predators, planktivorous fish typically use open water habitats and have a negative effect on zooplankton densities. However, Turner and Mittelbach (1990) showed that the addition of piscivorous fish resulted in a habitat change in the planktivores moving into the safety of the littoral zone. The population density of planktivores did not change considerably as a result of the piscivore addition, but the habitat shift resulted in a reduced predation pressure on zooplankton in the open water mass and a concomitant trophic cascade. Further, He and Kitchell (1990) showed that the addition of pike to a small lake resulted in a strong trophic cascade, but here the pike effect on the planktivore population was mainly a non-consumptive, behavioural effect; planktivores simply emigrated out of the lake into a stream. Since prey fish susceptibility to predation correlates with light level, this may cause diel migratory patterns similar to diel vertical migration from refuge habitats in daytime to open water habitats during night-time (e.g. Lindegren et al. 2012). Romare and Hansson (2003) demonstrated such a behavioural cascade in a system with an open and a refuge habitat and pike as top predator. Hence, this demonstrates that risk of pike predation can cause behaviourally induced trophic cascades and spatial altering of where consumptive effects of consumers are directed. Romare and Hansson (2003) did not only demonstrate the occurrence of a behavioural cascade, but they also showed that migratory patterns, and hence interactions among trophic levels, changed over the season as prey size increased and their metabolic demand and susceptibility to predation was altered. This brings us to more recent advancements in seasonal dynamics of piscivore-prey fish interactions.

8.5 SEASONAL DYNAMICS

Freshwater fish are poikilotherms and physiological processes are temperature dependent. This means that growth rate and metabolic costs depend on temperature and consequently drives the seasonal variation in the trade-off between costs of predation risk and benefits of foraging returns. In summer, when food availability and thus growth potential is high, prey organisms are willing to take the cost of a high predation risk, whereas in winter the benefits are reduced (low food availability) while predators still are feeding. Prey fish in temperate environments therefore use different means of escaping predation when temperature decreases during autumn and growth prospects diminish. Such escape behaviours include shoaling in darker environments in deep lakes, or migration to connected streams in shallow lakes. Hence, it should be recognised that environmental factors such as temperature are of importance for the interactions strength between pike and their prey. In cold environments, the relative benefits of foraging vs. metabolic demand may be low compared to warmer environments because of reduced growth prospects for prey fish as the temperature and resource production drops. A particularly well-studied example of this is the predator-prey interactions in Lake Krankesjön, southern Sweden, a shallow lake with pike as the main predator and roach as their main prey. Every autumn, parts of the roach population migrate to connected streams and wetlands, and this migration is driven by a trade-off between avoiding predation risk and attaining growth opportunity (Brönmark et al. 2008). As long as growth prospects are high, i.e., during summer, roach prefer the lake, but when zooplankton densities diminish in autumn, roach leave the lake and migrate to stream and wetland habitats where predation risk from pike is minimal. In spring, roach typically migrate back to the lake when temperature and thereby zooplankton biomass and growth prospects increase again.

All roach individuals, however, do not respond in the same way to the risk of predation from pike. Only part of the population migrates and the remaining part stays resident in the lake. The proportion and timing of migration in roach have been shown to relate to the fitness cost of predation. According to the ecology of fear concept (Brown et al. 1999), the cost of predation may be divided in three parts: predation risk, fitness cost, and the marginal fitness value. Hence, individuals that were experimentally exposed to a higher perceived risk of predation from pike where more prone to migrate (Hulthén et al. 2015). Moreover, roach that had a bold personality, and were thus expected to be at higher risk from predation by pike in the lake, migrated to a higher extent than more shy individuals (Chapman et al. 2011). As pike is a gape-limited predator it may also be expected that large prey that have reached a size refuge from predation should migrate less, a prediction that was confirmed in a study of size-dependent migration patterns in common bream (*Abramis brama* Linnaeus, 1758, Skov et al. 2011). Furthermore, roach individuals that were given supplementary food in an experiment and, hence, were in good condition in autumn, were more prone to migrate than individuals in poorer condition, possibly because the marginal value of staying in the lake was lower for individuals in good condition (Brodersen et al. 2008a). Consequently, given the well-known effects of cascading trophic interactions on lake ecosystem dynamics, feeding conditions for fish during one year may affect ecosystem dynamics the following year through changes in individual behavioural decisions, which in turn may affect population-level migration patterns.

This large-scale migration, where a considerable proportion of the population of the major zooplanktivorous fish is absent from the lake during a large part of the year, may have effects on the strength of cascading trophic interactions in the lake, which in turn may have implications for the stability of alternative stable states in shallow lakes. In addition to the proportion of the roach population that migrates, individual characteristics of roach (e.g. body condition, Brodersen et al. 2008a) may also affect the timing of migration. The proportion and timing of migration of the dominant zooplanktivore in the lake in turn affect the spring dynamics of the pelagic food chain, specifically the spring peak abundance and size structure of zooplankton (Hansson et al. 2007, Brodersen et al. 2008b). The timing of zooplankton and intensity of zooplankton grazing pressure during spring affects the probability of grazing-induced limitation of phytoplankton, and thereby the likelihood of a spring clear-water phase (Scheffer 1997, Hansson et al. 2007), and this is crucial for many shallow eutrophic lake ecosystems as it provides a window of opportunity for establishment of submerged macrophytes (e.g. Van Donk and Otte 1996). Thus, factors affecting the timing and fraction of resident/migratory zooplanktivorous fish, such as the risk of predation from pike, may provide feedback loops affecting the resilience of alternative stable states (macrophyte dominated versus phytoplankton dominated) in shallow eutropic lakes (Brönmark et al. 2010).

Predators such as pike may also take advantage of a heterogeneous temperature environment within the system (Pierce et al. 2013). For example, the vertical temperature profile of deep lakes in summer may cause predators to perform daily migration between shallow, warm environments where they forage and deeper, colder environments where digestion is slow, in order to "make the food last longer". Conversely, if food is abundant in relatively cold environments, predators may choose to digest the food in warmer habitats, such as shallow, sheltered bays, to maximise growth rate. Temperature may not only affect the growth-predation risk trade-off, but also the interaction in itself through temperature-dependent escape and attack performances of prey and predators, respectively. Thus, under such situations it is imperative to view the relative performance of predators and prey rather than the absolute one (Öhlund et al. 2014). For example, the performance of pike relative to the cold-water species Arctic char is likely to favour pike in warmer and char in colder waters, whereas the opposite pattern is likely the case for the interaction between pike and the warm-water species roach.

8.6 ECO-EVOLUTIONARY DYNAMICS

A basic assumption of studies of cascading trophic interactions is that evolution has no effect on the strength of these interactions, simply because evolutionary and ecological processes work at completely different temporal scales. However, in recent years, a number of studies have shown that evolution of ecologically important traits may be surprisingly rapid and that such contemporary evolution might result in eco-evolutionary feed back processes (Hairston et al. 2005, Schoener 2011, de Meester and Pantek 2014). A much cited study on eco-evolutionary dynamics in freshwater food webs involves lakes in eastern North America where populations of the major zooplanktivorous predator, the alewife *Alosa pseudoharengus* Wilson, 1811 differ in their ecology based on if they are able to migrate to the sea or not (e.g. Palkovacs and Post 2008). In their ancestral migratory pattern young-of-the-year fish spend their first summer in the lake before migrating to sea where they live as adult alewife and only return to the lake to spawn. In dammed rivers, however, alewifes become landlocked and this results in intense, year-round predation pressure by alewives, that eliminate large-bodied zooplankton and creates a zooplankton community dominated by small-bodied zooplankton. This, in turn, creates a strong selection pressure for adaptations related to foraging on small zooplankton and has resulted in the evolution of smaller mouth size and gill raker spacing in landlocked alewife populations. Further, besides such top-down eco-evolutionary effects, recent research has also shown bottom-up effects, i.e., ecological and evolutionary effects on the main top predator in the system, the chain pickerel *Esox niger* Lesueur, 1818 (Brodersen et al. 2015). Pickerel in landlocked lakes have a more fusiform body shape, a body morphology optimal for foraging in pelagic habitats. Hence, the alewife system clearly shows how environmental changes, i.e., dam building, can drive an evolutionary change in intermediate consumers in the food chain and may alter the selective landscape for other organisms, causing eco-evolutionary interactions that propagate up and down the food chain and influence the strength of cascading trophic interactions. The understanding of such processes is especially important in face of strong anthropogenic disturbances on the environment.

8.7 FUTURE PERSPECTIVES

A number of factors suggest that the future of pike is bright and that pike will continue to be a keystone predator in many systems. First, increased temperatures, driven by on-going climatic changes, are likely to allow pike to invade new environments in more alpine areas that currently lack pike (Byström et al. 2007). In Sweden alone, Arctic char is predicted to loose 73% of its current range by 2100 and a major cause is pike invasions to small lakes (Hein et al. 2012). Increased temperature may also decrease the risk of winterkill as ice cover will be shorter, which may increase pike distribution also in low altitude areas. Second, a current trend in stream management is to improve connectivity within rivers by removing migration obstacles. Such activities are likely to further increase pike ability to invade new systems, because a major constraint on pike dispersal is stream slope (Spens et al. 2007). Third, there is also a trend to reconstruct wetlands, especially in agriculture areas where wetlands previously have been sacrificed to create arable land. Sometimes such wetlands are even constructed with the specific purpose of being pike factories (chapter 10). Possible threats to pike in the future include extreme eutrophication and impoverished light regime due to brownification that may reduce the availability of suitable habitats, as well as fishing practices that directly or indirectly target pike (chapters 12, 13). As outlined in this chapter, the extensive research, especially in recent years, on pike ecology and the role of pike in communities has now made managers well equipped to deal with pressing questions concerning e.g. conservation and ecosystem functioning. This research has also generated possible tools of how to drive ecosystems in directions that provide the services that are desired by society and mitigate the undesired consequences pike presence sometimes have. We therefore foresee continued and expanded research interest in pike and the effects of pike on trophic interactions in the future.

REFERENCES CITED

Baxter, C.V., K.D. Fausch and C.W. Saunders. 2005. Tangled webs: reciprocal flows of invertebrate prey link streams and riparian zones. *Freshw. Biol.* 50:201-220.

Beaudoin, C.P., W.M. Tonn, E.E. Prepas and L.I. Wassenaar. 1999. Individual specialization and trophic adaptability of northern pike (*Esox lucius*): an isotope and dietary analysis. *Oecologia* 120:386-396.
Brodersen, J., P.A. Nilsson, C. Skov, L-A. Hansson and C. Brönmark. 2008a. Condition-dependent individual decision-making determines cyprinid partial migration. *Ecology* 89:1195-1200.
Brodersen, J., E. Ådahl, C. Brönmark and L-A. Hansson. 2008b. Ecosystem effects of partial fish migration in lakes. *Oikos* 117:40-44.
Brodersen, J., J.G. Howeth and D.M. Post. 2015. Emergence of a novel prey life history promotes contemporary sympatric diversification in a top predator. *Nat. Commun.* 6:8115. doi:10.1038/ncomms9115.
Brown, J.S., J.W. Laundre and M. Gurung. 1999. The ecology of fear: Optimal foraging, game theory, and trophic interactions. *J. Mammal.* 80(2):385-399.
Bry, C., E. Basset, X. Rognon and F. Bonamy. 1992. Analysis of sibling cannibalism among pike, *Esox lucius*, juveniles reared under semi-natural conditions. *Env. Biol. Fish.* 35:75-84.
Brönmark, C. and J.G. Miner. 1992. Predator-induced phenotypical change in body morphology in crucian carp. *Science* 258:1348-1350.
Brönmark, C. and S.E.B. Weisner. 1992. Indirect effects of fish community structure on submerged vegetation in shallow, eutrophic lakes: an alternative mechanism. *Hydrobiologia* 243/244:293-301.
Brönmark, C., C. Paszkowski, W.M. Tonn and A. Hargeby. 1995. Predation as a determinant of size structure in populations of crucian carp (*Carassius carassius*) and tench (*Tinca tinca*). *Ecol. Freshw. Fish* 4:85-92.
Brönmark, C. and S.E.B. Weisner. 1996. Decoupling of cascading trophic interactions in a freshwater, benthic food chain. *Oecologia* 108:534-541.
Brönmark, C., C. Skov, J. Brodersen, P.A. Nilsson and L-A. Hansson. 2008. Seasonal migration determined by a trade-off between predator avoidance and growth. *PLoS ONE* 3:e1957.
Brönmark, C., J. Brodersen, B.B. Chapman, A. Nicolle, P.A. Nilsson, C. Skov and L.-A. Hansson. 2010. Regime shifts in shallow lakes: the importance of seasonal fish migration. *Hydrobiologia* 646(1):91-100.
Byström, P., L. Persson and E. Wahlström. 1998. Competing predators and prey: juvenile bottlenecks in whole-lake experiments. *Ecology* 79:2153-2167.
Byström, P., J. Karlsson, P. Nilsson, T. Van Kooten, J. Ask and F. Olofsson. 2007. Substitution of top predators: effects of pike invasion in a subarctic lake. *Freshw. Biol.* 52(7):1271-1280.
Byström, P., U. Bergström, A. Hjälten, S. Stål, D. Jonsson and J. Olsson. 2015. Declining coastal piscivore populations in ther Baltic Sea: where and when do sticklebacks matter? *Ambio* 44:S462-S471.
Carey, M.P. and D.H. Wahl. 2010. Interactions of multiple predators with different foraging modes in an aquatic food web. *Oecologia* 162(2):443-452.
Carpenter, S.R., J.F. Kitchell and J.R. Hodgson. 1985. Cascading trophic interactions and lake productivity. *Bioscience* 35:634-639.
Carpenter, S.R. and J.F. Kitchell. 1993. The Trophic Cascade in Lakes. Cambridge, UK: Cambridge University Press.
Chapman, B., K. Hulthén, D. Blomqvist, L-A. Hansson, J-Å. Nilsson, J. Brodersen, P.A. Nilsson, C. Skov and C. Brönmark. 2011. To boldly go: Individual differences in boldness influence migratory tendency. *Eco. Let.* 14:871-876.
Craig, J.F. (ed.). 1996 . Pike: Biology and Exploitation. London: Chapman & Hall.
de Meester, L. and J. Pantek. 2014. Eco-evolutionary dynamics in freshwater systems. *J. Limnol.* 73:193-200.

de Roos, A.M. and L. Persson. 2013. Population and Community Ecology of Ontogenetic Development. Princeton, NJ, Princeton Univ. Press.

Dessborn, L., J. Elmberg and G. Englund. 2011. Pike predation affects breeding success and habitat selection of ducks. *Freshw. Biol.* 56:579-589.

Diehl, S. 1988. Foraging efficiency of three freshwater fishes: effects of structural complexity and light. *Oikos* 53:207-214.

Drenner, R.W. and K.D. Hambright. 2002. Piscivores, trophic cascades, and lake management. *Sci. World J.* 2:284-307.

Ebenman, B. and L. Persson. 1988. Size-structured Populations. Berlin: Springer-Verlag.

Eklöv, P. and T. VanKooten. 2001. Facilitation among piscivorous predators: effects of prey habitat use. *Ecology* 82:2486-2494.

Englund, G., F. Johansson, P. Olofsson, J. Salonsaari and J. Öhman. 2009. Predation leads to assembly rules in fragmented fish communities. *Ecol. Lett.* 12:663-671.

Eriksson, B.K., L. Ljunggren, A. Sandström, G. Johansson, J. Mattila, A. Rubach, S. Råberg and M. Snickars. 2009. Declines in predatory fish promote bloom-forming macroalgae. *Ecol. Appl.* 19:1975-1988.

Estes, J.A., J. Terborgh, J.S. Brashares, M.E. Power, J. Berger, W.J. Bond, S.R. Carpenter, T.E. Essington, R.D. Holt, J.B.C. Jackson, R.J. Marquis, L. Oksanen, T. Oksanen, R.T. Paine, E.K. Pikitch, W.J. Ripple, S.A. Sandin, M. Scheffer, T.W. Schoener, J.B. Shurin, A.R.E. Sinclair, M.E. Soulé, R. Virtanen and D.A. Wardle. 2011. Trophic downgrading of planet earth. *Science* 333:301-306.

Findlay, D.L., M.J. Vanni, M. Paterson, K.H. Mills, S.E.M. Kasian, W. Findlay and A.G. Salki. 2005. Dynamics of a boreal lake ecosystem during a long-term manipulation of top predators. *Ecosystems* 8:603-618.

Flinders, J.M. and S.A. Bonar. 2008. Growth, condition, diet, and consumption rates of northern pike in three Arizona reservoirs. *Lake Reserv. Manag.* 24(2):99-111.

Fryxell, J.M. and P. Lundberg. 1998. Individual Behavior and Community Dynamics. London: Chapman & Hall.

Giles, N., M.R. Wright and M.E. Nord. 1986. Cannibalism in pike fry, *Esox lucius* L.: some experiments with fry densities. *J. Fish. Biol.* 29:107-113.

Giraldeau, L.-A. and T. Caraco. 2000. *Social foraging theory*. New Jersey: Princeton University Press.

Grimm, M.P. 1981. Intraspecific predation as a principal factor controlling the biomass of northern pike (*Esox lucius* L.). *Fish. Manage.* 12:77-79.

Gårdmark, A., M. Casini, M. Huss, A. van Leeuwen, J. Hjelm, L. Persson and A. de Roos. 2014. Regime shifts in exploited marine food webs: detecting mechanisms underlying alternative stable states using size-structured community dynamics theory. *Phil. Trans. R. Soc.* 370:20130262.

Hambright, K.D., R.W. Drenner, S.R. McComas and N.G. Jr. Hairston. 1991. Gape-limited piscivores, planktivore size refuges, and the trophic cascade hypothesis. *Arch. Hydrobiol.* 121:389-404.

Hairston, N.G., F.E. Smith and D. Slobodkin. 1960. Community structure, population control, and competition. *Am. Nat.* 94:421-425.

Hairston, N.G., S.P. Ellner, M.A. Geber, T. Yoshida and J.A. Fox. 2005. Rapid evolution and the convergence of ecological and evolutionary time. *Ecol. Lett.* 8(10):1114-1127.

Hanson, P.C., T.B. Johnson, D.E. Schindler and J.F. Kitchell. 1997. Fish bioenergetics 3.0 Software program and manual. University of Wisconsin-Madison Center for Limnology and University of Wisconsin Sea Grant Institute.

Hansson, L-A. 1992. The role of food web composition in shaping algal communities. *Ecology* 73:241-247.

Hansson, L-A., A. Nicolle, J. Brodersen, P. Romare, C. Skov, P.A. Nilsson and C. Brönmark. 2007. Consequences of fish predation, migration, and juvenile ontogeny on zooplankton spring dynamics. *Limnol. Oceanogr.* 52:696-706.

He, X. and J.F. Kitchell. 1990. Direct and indirect effects of predation on a fish community: a whole lake-experiment. *Trans. Am. Fish. Soc.* 119:825-835.

Hein, C.L., G. Öhlund and G. Englund. 2012. Future distribution of Arctic char Salvelinus alpinus in Sweden under climate change: effects of temperature, lake size and species interactions. *Ambio* 41:303-312.

Holling, C.S. 1965. The functional response of predators to prey density and its role in mimicry and population regulation. *Mem. Entomol. Soc. Can.* 45:1-60.

Holling, C.S. 1973. Resilience and stability of ecological systems. *Ann. Rev. Ecol. Syst.* 4:1-23.

Holt, R.D. and G.A. Polis. 1997. A theoretical framework for intraguild predation. *Am. Nat.* 149:745-764.

Hrbácek, J., M. Dvorakova, V. Korinek and L. Procházková. 1961. Demonstration of the effect of the fish stock on the species composition of zooplankton and the intensity of metabolism of the whole plankton association. *Verh. Int. Ver. Theor. Ang. Limn.* 14:192-195.

Hulthén, K., B.B. Chapman, P.A. Nilsson, J. Vinterstare, L-A. Hansson, C. Skov, J. Brodersen, H Baktoft and C. Brönmark. 2015. Escaping peril: perceived predation risk affects migratory propensity. *Biol. Lett.* 11:20150466. DOI: 10.1098/rsbl.2015.0466

Kipling, C. 1984. A study of perch (*Perca fluviatilis* L.) and pike (*Esox lucius* L.) in Windermere from 1941 to 1982. *ICES J. Mar. Sci.* 41:259-267.

Knight, T.M., M.W. McCoy, J.M. Chase, K. McCoy and R.D. Holt. 2005. Trophic cascades across ecosystems. *Nature*, 437:880-883.

Lindegren, M., P. Vigliano and P.A. Nilsson. 2012. Alien Invasions and the Game of Hide and Seek in Patagonia. *PLoS ONE* 7(10).

May, R. 1977. Thresholds and breaking points in ecosystems with a multiplicity of stable states. *Nature* 6:471-477.

Mills, C.A. and M.A. Hurley. 1991. Long-term studies on the Windermere populations of perch (*Perca fluviatilis*), pike (*Esox lucius*) and Arctic charr (*Salvelinus alpinus*). *Freshw. Biol.* 23:119-136.

Mittelbach, G.G., C.W. Osenberg and M.A. Leibold. 1988. Trophic relations and ontogenetic niche shifts in aquatic ecosystems. pp 219-235. *In*: B. Ebenman and L. Persson (eds.). *Size-structured populations*. Berlin: Springer-Verlag.

Mittelbach, G.G. and L. Persson. 1998. The ontogeny of piscivory and its ecological consequences. *Can. J. Fish. Aquat. Sci.* 55:1454-1465.

Nakano, S. and M. Murakami. 2001. Reciprocal subsidies: dynamic interdependence between terrestrial and aquatic food webs. *Proc. Nat. Acad. Sci.* 98:166-170.

Nicholson, M.E., M.E. Renne and K.H. Mills. 2015. Apparent extirpation of prey fish communities following the introduction of Northern pike (*Esox lucius*). *Can. Field-Nat.* 129:165-173.

Nilsson, P.A. and C. Brönmark. 1999. Foraging among cannibals and kleptoparasites: effects of prey size on pike behavior. *Behav. Ecol.* 10:557-566.

Nilsson, P.A. 2000. Pikeivory: behavioural mechanisms in northern pike piscivory. PhD Thesis, Department of Ecology, Animal Ecology, Lund University, Sweden.

Nilsson, P.A. and C. Brönmark. 2000. Prey vulnerability to a gape-size limited predator: behavioural and morphological impacts on northern pike piscivory. *Oikos* 88:539-546.

Nilsson, P.A., K. Nilsson and P. Nyström. 2000. Does risk of intraspecific interactions induce shifts in prey-size preference in aquatic predators? *Behav. Ecol. Sociobiol.* 48:268-275.

Nilsson, P.A. 2001. Predator behaviour and prey density: evaluating density-dependent intraspecific interactions on predator functional responses. *J. Anim. Ecol.* 70:14-19.

Nilsson, P.A. 2006. Avoid your neighbours: size-determined spatial distribution patterns among northern pike individuals. *Oikos* 113:251-258.

Nilsson, P.A., H. Turesson and C. Brönmark. 2006. Friends and foes in foraging: intraspecific interactions act on foraging-cycle stages. *Behaviour* 143:733-745.

Nilsson, P.A., P. Lundberg, C. Brönmark, A. Persson and H. Turesson. 2007. Behavioral interference and facilitation in the foraging cycle determine the functional response. *Behav. Ecol.* 18:354-357.

Öhlund, G., P. Hedström, S. Norman, C.L. Hein and G. Englund. 2014. Temperature dependence of predation depends on the relative performance of predators and prey. *Proc. R. Soc. Lond. B.* 282:201442254. DOI: 10.1098/rspb.2014.2254

Oksanen, L., S.D. Fretwell, J. Arruda and P. Niemela. 1981. Exploitation ecosystems in gradients of primary productivity. *Am. Nat.* 118:240-261.

Östman, Ö., J. Eklöf, B. Klemens Eriksson, J. Olsson, P-O. Moksnes and U. Bergström. 2016. Top-down control as important as enrichment for eutrophication effects in North Atlantic coastal ecosystems. *J. Appl. Ecol.* 53:1138-1147.

Paine, R.T. 1980. Food webs: Linkage, interactions strength and community infrastructure. *J. Anim. Ecol.* 49:666-685.

Palkovacs, E.P. and D.M. Post. 2008. Eco-evolutionary interactions between predators and prey: can predator induced changes to prey communities feed back to shape predator foraging traits? *Evol. Ecol. Res.* 10:699-720.

Patankar, R., F.A. von Hippel and M.A. Bell. 2006. Extinction of a weakly armoured threespine stickleback (*Gasterosteus aculeatus*) population in Prator Lake, Alaska. *Ecol. Freshw. Fish* 15:482-487.

Paukert, C. P., W. Stancill, T.J. Debates and D.W. Willis. 2003. Predatory effects of Northern pike and largemouth bass: bioenergetics modelling and ten years of fish community sampling. *J. Freshw. Ecol.* 18:13-24.

Persson, L., K. Leonardsson, A.M. de Roos, M. Gyllenberg and B. Christensen. 1998. Ontogenetic scaling of forgaing rates and the dynamics of a size-structured consumer-resource model. *Theor. Popul. Biol.* 54:270-293.

Persson, L., P. Byström, E. Wahlström, J. Andersson and J. Hjelm. 1999. Interactions among size-structured populations in a whole-lake experiment: size- and scale-dependent processes. *Oikos* 87:139-156.

Persson, L., A.M. De Roos and P. Byström. 2007. State-dependent invasion windows for prey in size-structured predator-prey systems: whole lake experiments. *J. Anim. Ecol.* 76(1):94-104.

Pierce, R.B., A.J. Carlson, B.M. Carlson, D. Hudson and D.F. Staples. 2013. Depths and thermal habitat used by large versus small northern pike in three Minnesota lakes. *Trans. Am. Fish. Soc.* 142:1629-1639.

Polis, G.A., C.A. Myers and R.D. Holt. 1989. The ecology and evolution of intraguild predation: potential competitors that eat each other. *Ann. Rev. Ecol. Sys.* 20:297-330.

Polis, G.A., W.B. Anderson and R.D. Holt. 1997. Toward an integration of landscape and food web ecology: the dynamics of spatially subsidized food webs. *Ann. Rev. Ecol. Sys.* 28:289-316.

Prejs, A., A. Martyniak, S. Boron, P. Hliwa and P. Koperski. 1994. Food web manipulation in a small, eutrophic Lake Wirbel, Poland: effect of stocking with juvenile pike on planktivorous fishes. *Hydrobiologia* 275/276:65-70.

Prejs, A., J. Pijanowska, P. Koperski, A. Martyniak, S. Boron and P. Hliwa. 1997. Food-web manipulation in a small, eutrophic Lake Wirbel: long-term changes in fish biomass and basic measures of water quality. A case study. *Hydrobiologia* 342/343:383-386.

Raat, A.J.P. 1988. Synopsis of the biological data on the northern pike, *Esox lucius* Linnaeus, 1758. *FAO Fisheries Synopsis*: No. 30, Rev. 2, 178 p.
Romare, P. and L-A. Hansson. 2003. A behavioural cascade: top-predator induced behavioural shifts in planktivorous fish and zooplankton. *Limnol. Oceanogr.* 48:1956-1964.
Rudman, S.R., J. Heavyside, D.J. Rennison and D. Schluter. 2016. Piscivore addition causes a trophic cascade within and across ecosystem boundaries. *Oikos* 125:1782-1789.
Scheffer, M., S.H. Hosper, M-L. Meijer, B. Moss and E. Jeppesen. 1993. Alternative equilibria in shallow lakes. *Trends Ecol. Evol.* 8:275-279.
Scheffer, M. 1997. The Ecology of Shallow Lakes. Chapman & Hall, London.
Scheffer, M., S. Carpenter, J.A. Foley, C. Folke and B. Walker. 2001. Catastrophic shifts in ecosystems. *Nature* 413:591-596.
Schindler, D.W. 1974. Whole-lake eutrophication experiments with phosphorus, nitrogen and carbon. *Verh. Int. Ver. Theor. Ang. Limn.* 19:3221-3231.
Schoener, T.W. 2011. The newest synthesis: Understandning the interplay of evolutionary and ecological dynamics. *Science* 331:426429.
Sharma, C.M. and R. Borgstrøm. 2008. Shift in density, habitat use, and diet of perch and roach: An effect of changed predation pressure after manipulation of pike. *Fisher. Res.* 91:98-106.
Sih, A., G. Englund and D. Wooster. 1998. Emergent impacts of multiple predators on prey. *Trends Ecol. Evol.* 13:350-355.
Skov, C. and P.A. Nilsson. 2007. Evaluating stocking of YOY pike *Esox lucius* L. as a tool in the restoration of shallow lakes. *Freshw. Biol.* 52:1834-1845.
Skov, C., H. Baktoft, J. Brodersen, C. Brönmark, B.B. Chapman, L-A. Hansson and P.A. Nilsson. 2011. Sizing up your enemy: individual predation vulnerability predicts migratory probability. *Proc. R. Soc. Lond. B* 278:1414-1418.
Solman, V.E.F. 1945. The ecological relations of pike, *Esox lucius* L., and waterfowl. *Ecology* 26:157-170.
Soluk, D.A. 1993. Multiple predator effects: predicting combined functional response of a stream fish and invertebrate predators. *Ecology* 74:219-225.
Spens, J., G. Englund and H. Lundqvist. 2007. Network connectivity and dispersal barriers: using geographical information system (GIS) tools to predict landscape distribution of a key predator (*Esox lucius*) among lakes. *J. Appl. Ecol.* 44:1127-1137.
Spens, J. and J.P. Ball. 2008. Salmonid or nonsalmonid lakes: predicting the fate of northern boreal fish communities with hierarchical filters relating to a keystone piscivore. *Can. J. Fish. Aquat. Sci.* 65:1945-1955.
Sutherland, J.P. 1974. Multiple stable points in natural communities. *Am. Nat.* 108:859-873.
Tonn, W.M. and J.J. Magnuson. 1982. Patterns in the species composition and richness of fish assemplages in northern Wisconsin lakes. *Ecology* 63:1149-1166.
Tonn, W.M., J.J. Magnuson, M. Rask and J. Toivonen. 1990. Intercontinental comparison of small-lake assemblages: te balance between local and regional processes. *Am. Nat.* 136:345-375.
Tilman, D. 1982. Resource Competition and Community Structure. Princeton University Press. Princeton, NJ. 296 p.
Turner, A.M. and G.G. Mittelbach. 1990. Predator avoidance and community structure: interactions among piscivores, planktivores, and plankton. *Ecology* 71:2241-2254.
Vance-Chalcraft H.D., D.A. Soluk and N. Ozburn. 2004. Is prey predation risk influenced more by increasing predator density or predator species richness in stream enclosures? *Oecologia* 139:117-122.
Vance-Chalcraft H.D. and D.A. Soluk. 2005. Multiple predator effects result in risk reduction for prey across multiple prey densities. *Oecologia* 144:472-480.

van Donk, E. and A. Otte. 1996. Effects of grazing by fish and waterfowl on the biomass and species composition of submerged macrophytes. *Hydrobiologia* 340:285-290.

van Leeuwen, A., M. Huss, A. Gårdmark, M. Casini, F. Vitale, J. Hjelm, L. Persson and A. de Roos. 2013. Predators with multiple ontogenetic niche shifts have limited potential for growth and top-down control of their prey. *Am Nat.* 182:53-66.

Venturelli, P.A. and W.M. Tonn. 2005. Invertivory by northern pike (*Esox lucius*) structures communities of littoral macroinvertebrates in small boreal lakes. *J. North Am. Benthol. Soc.* 24(4):904-918.

Walters, C. and J.F. Kitchell. 2001. Cultivation/depensation effects on juvenile survival and recruitment: implications for thew theory of fishing. *Can. J. Fish. Aquat. Sci.* 58:39-50.

Werner, E.E. and S.D. Peacor. 2003. A review of trait-mediated indirect interactions in ecological communities. *Ecology* 84(5):1083-1100.

Winfield, I.J., J.M. Fletcher and J.B. James. 2012. Long-term changes in the diet of pike (*Esox lucius*), the top aquatic predator in a changing Windermere. *Freshw. Biol.* 57:373-383.

Section III

Management and Fisheries

Chapter 9

Stocking for Pike Population Enhancement

Nicolas Guillerault[*1], *Daniel Hühn*[2], *Julien Cucherousset*[3], *Robert Arlinghaus*[4] *and Christian Skov*[5]

There are three basic strategies in the fisheries managers' toolbox that can be used to protect or enhance fish populations (Arlinghaus et al. 2016). The first is enacting fisheries regulations, which manage fishing mortality (i.e., demand, chapter 12) and the other two are habitat management/restoration (chapter 10) and stocking, which attempt to enhance recruitment (i.e., supply). Stocking is the release of cultured or wild-captured fish into a different water body. It is widely used in inland and coastal fisheries for multiple reasons spanning from enhancement of recreational or commercial fisheries to conservation purposes (Cowx 1994, Arlinghaus et al. 2002, Lorenzen et al. 2012). This chapter focuses on stocking of northern pike *Esox lucius* Linnaeus, 1758 (hereafter pike) as a way to enhance populations, i.e., increase pike abundance, biomass or catches. Stocking of pike for lake quality restoration purposes (e.g. to improve water clarity, Mehner et al. 2004, Brönmark and Hansson 2005) is reviewed in chapter 11 of this book. Here, we first provide an overview of pike stocking, e.g. reasons why pike are stocked, typology and some historical and quantitative aspects of pike stocking. We then review studies that have assessed the effectiveness of pike stocking for population enhancement and discuss important factors affecting the outcomes of stocking as well as the risks that may be associated with stocking pike. We end the chapter with recommendations for the future use of pike stocking as a fisheries management tool.

[1]see Authors' addresses at the end of this chapter.

9.1 DEFINING PIKE STOCKING

9.1.1 Typology and Motivations for Stocking

Depending on the objective and the status of the natural population, stocking-based population management can generally be classified into five basic types following Lorenzen et al. (2012) and Arlinghaus et al. (2015), which also apply to pike stocking.

1. Stocking for culture-based fisheries (equal to maintenance stocking in the terminology of Cowx [1994]) depends on the release of cultured fish to produce (and maintain) a fishery under conditions where there is a lack of natural recruitment and the fishery would cease to exist without stocking. The release of fish in culture-based fisheries follows the exclusive goal of improving or establishing fisheries. Stocking to establish culture-based fisheries is common for the pike congenerics muskellunge *Esox masquinongy* Mitchill, 1824 and 'tiger muskellunge' *E. lucius* x *E. masquinongy* (sterile hybrids) in the USA (e.g. Rust et al. 2002, Wingate and Younk 2007) yet remains rare for pike in both the USA and Europe. Muskellunge is a species of very high demand in the USA, but reproductive failures are very common (Johnson 1981, Kerr 2011). One of the main causes of poor recruitment in muskellunge is the high egg mortality related to anoxic conditions at the substrate-water interface where muskellunge preferentially spawn (Zorn et al. 1998). Pike, however, stick eggs above the substrate (preferably on submerged vegetation, chapter 3) where there are higher oxygen concentrations (Crane et al. 2015). In addition, reproductive traits of pike, e.g. staggered spawning and plasticity in spawning ground selection (Crane et al. 2015, Raat 1988), allow pike to reproduce in most waterbodies where they occur.
2. Stocking for stock enhancement is the continued release of hatchery-origin or wild-captured fish into naturally reproducing fish stocks. It implies that released fish have to cope with intra-specific competition from wild conspecifics. Stock enhancements are conducted to compensate for exploitation or habitat-induced recruitment shortages and are aimed at fostering fisheries (i.e., catch and/or yield; Lorenzen et al. 2012). Stock enhancement continues to be very common in pike management (e.g. Arlinghaus et al. 2015).
3. Restocking or stock rebuilding aims at supporting severely depleted populations (e.g. after fish kills) by the release of high densities of wild or hatchery-originated fish over a short period of time. The goal of stock rebuilding is conservation, but often with the intent that the stock can be used by fisheries after reestablishment. Stock rebuilding occurs regularly for pike populations, e.g. after winterkills (e.g. Kerr and Lasenby 2001).
4. Stocking for supplementation is the continued release of fish to reduce the risk of extinction in very small and declining fish stocks. It usually has a conservation focus rather than a fisheries focus. Releases occur in low to moderate densities to avoid further depression of the declining stock. Supplementation programs, where wild captured fishes are artificially raised and released, is widespread

in Atlantic salmon *Salmo salar* Linnaeus, 1758 and other salmonids but is very rare in pike, probably because pike successfully reproduce in many water bodies.

5. Introduction and translocation stocking is the release of fish into a water body from which the species is absent (outside or within its native range, respectively) with the aim of (re)-establishing a self-sustaining population and fishery after which stocking ceases. In Denmark, translocation of pike has, for example, been frequently conducted in newly established lakes (e.g. gravel pits) to create a fish community comparable to small natural water bodies (Skov et al. 2006). Similarly, in France, translocation of pike in gravel pits often occurs to establish and maintain fisheries (Zhao et al. 2016). Although introductions of pike have been common in the past (e.g. Pedreschi et al. 2014), nowadays they are increasingly rare due to conservation concerns. However, some anglers continue to illegally introduce pike into pike-less lakes to establish a fishery (Johnson et al. 2009; chapter 14).

Stocking for fisheries enhancement is often based on the deeply rooted belief among stakeholders that stocking elevates stock size and catches (Cowx 1994, Connelly et al. 2000, van Poorten et al. 2011, von Lindern and Mosler 2014). Stocking may also be chosen as a strategy based on the belief that it stabilizes population sizes with limited ecological risk and, hence, is a measure of conservation (Arlinghaus et al. 2015). The ultimate decision when implementing a stocking program depends on many social, ecological and economic components of the social-ecological system (e.g. the type of property rights, the degree of social norms, fishing pressure and the ecological status of the water body; Lorenzen 2014, Arlinghaus et al. 2015, 2016). Stocking is often preferred over alternative management approaches by dissatisfied, consumptive and avid angler types (Arlinghaus and Mehner 2005, van Poorten et al. 2011). However, in some countries, anglers prefer to catch wild pike (e.g. France, Armand et al. 2002), which may reduce the anglers' demand for stocking. In other countries, anglers lack distinct preferences for wild over hatchery-originated pike in their catches (e.g. Germany, Arlinghaus et al. 2014). Coupled with the difficulty of discriminating stocked and wild pike after capture, this indifference may foster a reliance on stocking, notably among more locally managed fisheries in Central Europe, where decisions about stocking are exclusively made by angling clubs with little external input from authorities (van Poorten et al. 2011, Arlinghaus et al. 2015). Stocking is also popular among clubs and agencies for political reasons and has become a ritualized habit that is often conducted due to the difficulty of engaging in alternatives (e.g. fisheries regulation or habitat enhancement) for social or economic reasons (Lorenzen et al. 2010, Arlinghaus et al. 2015).

9.1.2 Historical Aspects and Pike Stocking Practices

In its native range, stocking of pike mostly relates to the stock enhancement type (#2 mentioned above), i.e., stocking of fish into wild, naturally recruiting stocks to increase yield above, or population abundance beyond, the level that can be sustained by (often anthropogenically impaired) natural recruitment (Raat 1988, Cowx 1994, Welcomme and Bartley 1998, Lorenzen et al. 2012). Accordingly, there is a long history of pike stocking for enhancement in many industrialized countries. Evidence of pike stocking within its native range exists from the late 19th – early 20th century. For instance, Jones (1963) reported the existence of pike stocking in lakes located in Nebraska (USA) in 1889, and Charpy (1948) mentioned stocking occurring in the Untersee of Lake Constance (Germany) as early as 1902. One of the first pike hatcheries was created in 1892 in Denmark (Rasmussen and Geertz-Hansen 1998) where pike stocking for fisheries enhancement reached a peak in the 1930's and remained important until the beginning of this century (Jacobsen et al. 2004). In France, the concerns of pike anglers about a global decline in wild populations led to the creation of a national grant for pike culture in the early 20th century (Chimits 1947). It is only since the mid-20th century that pike stocking has been more quantitatively documented (e.g. Krohn 1969, Snow 1974, Raat 1988, Halverson et al. 2008, Pierce 2012, Arlinghaus et al. 2015), and information about pike culture for stocking has been extensively published (e.g. Heuschmann 1940, Beyerle and Williams 1973, Bry et al. 1995, Nilsson et al. 2008).

Despite increasing evidence for the lack of positive stocking outcomes (see further below), pike stocking continues to be a popular management practice worldwide, especially in Europe and North America (FAO 2005). Most pike stocking has been performed with age-0 fish, usually free-swimming fry or fingerlings (1-30 cm in Total Length [TL]). For instance, in Michigan (USA) from 1979 to 2005, most pike (69%) were stocked as fry whereas stocking of adults remained marginal (0.06%, Diana and Smith 2008). Early life-stages have been used in pike stocking for three main reasons. First, releasing young fish is believed to be more cost-effective (see Johnston et al. 2015 for counter examples) because it is relatively expensive to raise predatory fish species such as pike to a larger juvenile or adult stage, mainly due to large losses resulting from cannibalism (Huet 1948, Bry and Souchon 1982; chapters 11 and 13). Second, early life stages are often assumed to be less domesticated and, therefore, better adapted to survive under natural conditions than later life stages that have been exposed to culture conditions for longer periods (Lorenzen et al. 2012, Hühn et al. 2014). However, good empirical evidence for this statement is so far lacking (but see Szczepkowoski et al. 2012). Third, early life stages are preferably used for pike stocking because the loss of spawning grounds is often outlined as one of the main sources of decline of wild populations (Casselman and Lewis 1996, Farrell et al. 2006, Arlinghaus et al. 2015, chapter 10).

It is impossible to provide precise estimates of the number of waterbodies around the globe where pike have been or are currently stocked as data is scarce. We here provide some examples that can help showcase some figures. In Poland, about 85% of the lakes (average number of lakes studied: 2,453) were stocked with pike between 2001 and 2007 making pike the most frequently stocked species (Mickiewicz and Wołos 2012). In Scandinavia, where stocking primarily targets salmonids, pike is the seventh most stocked species. Pike are stocked in about 1,500 lakes in Norway (0.4% of Norwegian lakes), about 625 lakes in Sweden (1.1% of Swedish lakes) and about 200 lakes in Finland (0.7% of Finnish lakes)

(Tammi et al. 2003). In Germany, the total volume of pike stocked by organized anglers was about 124 t or 4.7 million pike in 2010 (Pagel, unpublished data). Pike contributed about 3% (according to both biomass and abundance) of the fish stocked in lotic water bodies, and about 4% (biomass) and 10% (abundance) of the fish stocked in lentic water bodies (Arlinghaus et al. 2015). In the USA, Fish and Game Agencies and the U.S. Fish and Wildlife Services stocked 9,994.6 t of pike in 2004 representing about 0.6% of the total biomass of fish stocked (Halverson 2008). Despite pike not being the most stocked species across the northern hemisphere, the amount of pike released and thus the financial cost of pike stocking is still substantial. In this context, concerns questioning the efficiency and usefulness of pike stocking have been repeatedly raised in many countries (e.g. Schreckenbach 1996, Pierce 2012, Mickiewicz 2013, Hühn et al. 2014).

9.2 OUTCOMES OF PIKE STOCKING FOR STOCK ENHANCEMENT

9.2.1 Population Dynamics and the Role of Fishing Pressure

Thorough understanding of pike stocking demands a careful understanding of population dynamics and processes, in particular density dependence. In many fish, such as pike, ontogenetic changes in size and morphology relate to shifts in key stage-specific population regulatory mechanisms (Lorenzen 2005). The key regulatory processes are size- and density-dependent mortality as well as density-dependent growth (chapter 6). Mortality is size-dependent throughout the life cycle of fish and is generally inversely proportional to length (McGurk 1986, Lorenzen 1996b, 2000), particularly in esocids (Lorenzen 2000, Haugen et al. 2007). It is generally assumed that eggs and larvae suffer high density-independent changes of vital rates (e.g. hatching, growth and survival rates; Myers and Cadigan 1993a, Leggett and DeBlois 1994). For instance, Eckmann et al. (1988) found that both temperature and wind were the main factors affecting the variance in whitefish *Coregonus lavaretus* Linnaeus, 1758 larvae abundance in Lake Constance (Germany, Switzerland, Austria) and, as mentioned earlier in this chapter, low oxygen concentration at the water-substrate interface is a source of high mortality in muskellunge eggs (Zorn et al. 1998, Crane et al. 2015). Juveniles mainly suffer density-dependent mortality directly from predation or starvation (Elliott 1994, Hazlerigg et al. 2012) or, indirectly, from the interplay between size-dependent mortality and density-dependent growth (Shepherd and Cushing 1980, Post et al. 1999). In larger juveniles and, in particular, in adults, the density dependence is often manifested in regulation of growth and condition rather than in mortality (Walters and Post 1993, Post et al. 1999, Lorenzen and Enberg 2002, Lorenzen 2005, 2008). As fish grow, there is also a transition from intra-cohort to inter-cohort density dependence (Walters and Post 1993, Lorenzen 1996a). These changes are certainly gradual but, in practice, distinct phases of intra-cohort density-dependent mortality and inter-cohort density-dependent growth are often assumed. Recruitment can be defined as the transition from one phase to the next (e.g. Lorenzen 2005). Below, in this section, recruitment refers to this meaning since it deals with population dynamic principles. Yet, in fisheries and throughout this chapter (apart from this section), recruitment is defined as the size when the fish enters the fishery.

In fish, the high variability in egg production and mortality of eggs and larvae is generally believed to account for a large part of the overall variability in recruitment (Beyer 1989, Rothschild 2000). However, density-dependent mortality of juveniles tends to mitigate the variability created at early life stages (Myers and Cadigan 1993b, Elliott 1994), and, therefore, also strongly affect outcomes of stocking fry or juveniles (Walters and Juanes 1993, see Hühn et al. [2014] for an example in pike). Using models, Minns et al. (1996) showed that the recruitment rate of juvenile pike is usually most constraining for adult stock size suggesting that measures increasing juvenile habitat should be more sustainable to increase the pike stock size relative to stocking juveniles. Density dependence at the recruited stage affects the current biomass of the recruited stock and the production (and possibly the quality) of eggs (Lorenzen 1996a, Lorenzen and Enberg 2002, Lorenzen 2005, see Edeline et al. [2007] for examples in pike). In highly variable fish populations, strong year classes are often followed by weak recruitment and *vice-versa* in response to multiple dimensions of density dependence (Marshall and Frank 1999; chapter 6). However, pike populations seem to show more stable dynamics overall compared to other piscivores, most likely due to strong inter-cohort population regulation through cannibalism (Persson et al. 2004, 2006, van Kooten et al. 2010; chapters 6 and 8). The effectiveness of stocking for enhancement should thus increase as more advanced life stages are stocked (Rogers et al. 2010), but even in the adult stages, compensatory processes are bound to act and ultimately limit the extent to which abundance and biomass can be enhanced (Secor and Houde 1998, Lorenzen 2005).

Fishing-induced mortality affects fish population size, size structure, and dynamics through density dependence and demographic changes (e.g. Darimont et al. 2009, Arlinghaus et al. 2010, Sutter et al. 2012). Therefore, local fishing pressure and the resulting harvest intensity has the potential to strongly affect the outcomes of stocking (e.g. Lorenzen 2005, 2008, Camp et al. 2017, Garlock et al. 2017). This might be especially true in pike which are very vulnerable to harvest (e.g. 12-50% [McCarraher 1957], 3-39% [Snow 1974] and 63.9% [Beyerle 1980] annual harvest rate; chapter 12). High local harvest pressure affects the structure and dynamics of the stock-enhanced population which may both promote or hinder the outcomes of stocking (Arlinghaus et al. 2016). Harvesting can reduce the mortality of stocked fish by removing predators or cannibalistic conspecifics and, thus, positively affect stocking outcomes of the target species (Botsford and Hobbs 1984). Harvesting can also increase the growth of fish that are stocked by reducing inter- and intra-specific competition for food. As a result, successful stocking can increase the net productivity of fisheries (often expressed as yield) beyond the level achievable by harvesting the natural component of the population alone (Lorenzen 2014). However, fishing can also reduce the efficiency of stocking by quickly removing stocked fish (e.g. Baer et al. 2007) then reducing the duration of additive effects, or affecting future natural recruitment negatively by removing large spawners (Botsford and Hobbs 1984, Hixon et al. 2014). In some cases, the quick exploitation of stocked fish can also be advantageous. For instance, in the case of unintended overstocking (i.e., when the number of released fish is far beyond the carrying capacity of the system), fast removal of stocked fish through harvest can help the population return to a healthy equilibrium. When stocking mature fish, strong harvest of stocked fish also has the potential to reduce risks of interbreeding between stocked and wild fish (Lorenzen et al. 2012) if stocked fish are more vulnerable to harvest than their wild conspecifics. Interestingly, Beyerle (1980) reported harvest rates of stocked pike (61.9-70.5%) twice that of wild pike. To sort out the positive or negative potentials of locally high fishing pressure, appropriate release and harvest experiments, considering the socio-economical context of the fisheries (e.g. commercial *vs.* recreational fisheries), are needed (Botsford and Hobbs 1984, Lorenzen 1995, Lorenzen 2005, Johnston et al. 2015), but these have rarely been conducted in pike fisheries (see Hühn et al. [2015a] for an exception).

9.2.2 Methods to Estimate the Enhancing Effects of Stocking

Pike stocking was, and locally continues to be, a common measure used to manage pike populations. Therefore, it is of crucial importance for fisheries management to assess the outcomes of pike stocking in terms of its contribution to the stock, catches or yield (Lorenzen 2005, Rogers et al. 2010). Ideally, the evaluation of enhancements should be done by comparing measures of abundance or relative abundance (e.g. Catch Per Unit of Effort [CPUE] or effort-standardized catches before and after stocking in stocked and control water bodies/sites ([Before-After-Control-Impact approach, BACI], Smith 2002). However, for many reasons (e.g. technical and financial) such optimal study design has rarely been implemented (Hilborn 1999, see Hühn et al. [2015b] for an exception related to pike stocking). An alternative way to estimate the efficiency of stock enhancement is the analysis of (long) time series where stocking periods alternate with non-stocking periods resulting in before-after-designs without temporal controls. For instance, Jansen et al. (2013) gathered 62 years of angling CPUE data using angler logbooks to assess the effect of stocking pike fry and commercial fisheries on the pike population in Lake Esrom (Denmark). However, this approach is often limited by the number of volunteers willing to engage in the surveys, and by the difficulties in maintaining rigor in data collection over a long-term period. For this reason, most studies on pike stocking outcomes are unreplicated single-ecosystem studies (e.g. Dorow 2005, Klein 2011).

Mark-recapture studies are a further alternative to study stocking outcomes; they usually report the recapture ratio (number of stocked fish caught divided by number stocked) or the share of stocked fish as a proportion of the total (stocked and wild) fish catches as measures of stocking success (e.g. Kerr and Lasenby 2001, Grønkjær et al. 2004). Based on five pike stocking experiments in North American water bodies, Kerr and Lasenby (2001) observed that anglers could expect to harvest 3.2% to 65.3% of stocked pike. Snow (1974) found that 6.6% of the pike (26.5-58.0 cm) stocked were caught by anglers in Murphy Flowage (USA). Beyerle (1980) reported that stocked fingerlings represented 15.8% to 36.3% of angler's catches in

Long Lake (USA) and Snow (1964) reported that pike stocked in 1963 represented 34.2% of the total catches over five years (annual catch rate decreased from 64.9% to 0.5% after stocking) in Murphy Flowage (USA). Krohn (1969) reported that the proportion of stocked pike in total pike population (estimated by spring and fall netting) ranged from 33.3% to 94.0% in Murphy Pleasant Lake (USA), 12.6% to 94.1% in Silver Lake (USA), and 0% to 75% in Golden lake (USA). These measures provide some insights into stocking success and fitness-related traits of stocked fish. For instance, catch rate provides a minimum estimate of survival and, if size data are collected, information on growth. However, such mark-recapture studies do not normally provide information about the additive effects arising from stocking (Hilborn 1999) or if recruitment is impaired (Hühn et al. 2014), although this information is central to evaluate both the ultimate effectiveness of stocking as a fisheries enhancement measure and its impact on the wild population.

Overall, there are surprisingly few empirical studies quantifying how pike population abundance and biomass are affected by stocking, even though enhancement is often the main purpose of pike stocking. Lack of non-stocked control sites and pre-stocking information, insufficient replication or contrasts, and inappropriate experimental design often limit the strength of inference that can be drawn from most published evaluations of stock enhancements using stocking (Hilborn 1999, Walters and Martell 2004).

9.2.3 Outcomes of Stocking in a Size-Specific Context

Lorenzen (2000, 2005) proposed that most stocking experiments can be analysed and interpreted on the basis of allometric mortality–size relationships and that simple models can be constructed to study the outcomes of stocking before actually engaging in the practice (see Rogers et al. [2010], Johnston et al. [2015], and Garlock et al. [2017] for applications). Because the size of stocked fish crucially determines the additive effects to be expected from stocking (Lorenzen 2005, Rogers et al. 2010), we will below review results of pike population enhancement experiments in relation to the size (or life stage) at stocking. Several terms describe the different phases of pike ontogeny (e.g. eleutheroembryo, protopterygiolarvae, perygiolarvae; Raat 1988). In practice and in the stocking literature, these terms are often simplified to the detriment of clarity (e.g. it is often unclear what pike fry or pike fingerling means in terms of size). For our review, we adopted terminology used by the Food and Agricultural Organization (FAO) in their glossary for aquaculture (http://www.fao.org/faoterm/en/). Accordingly, fry describe fish from the beginning of exogenous feeding (i.e., from the end of the hanging phase, ~ 1 cm) to the advanced juvenile form where the phenotype resembles adults (marked by complete differentiation of organs and scalation, ~ 7 cm). Fingerlings describe age-0 fish older than fry to age-1, regardless of size (~ 7-30 cm). The term 'adult' is used in the below review for sexually mature individuals. In pike, age and size at maturation varies with growth rate (and hence geographical latitude and fishing mortality due to relaxed density-dependence in exploited stocks; Diana 1983). Hence, one-year-old pike can be sexually mature (e.g. in some lakes in the Netherlands, 19 cm for males and 30 cm for females; Raat 1988). Therefore, for simplicity, we considered pike to be adult at a TL > 30 cm, although such individuals may have been referred to as fingerlings in some papers.

9.2.3.1 Outcomes of Stocking Fry

Vuorinen et al. (1998) reported that stocking of pike fry (~ 1 cm TL) in Lakes Rahtijärvi and Siilinjärvi (Finland), which hosted a natural stock of pike, did not increase juvenile pike density. Jansen et al. (2013) found that pike fry stocking in the self-sustained population of Lake Esrom (Denmark) did not increase angler catches between 1949 and 1969. Klein (2011) analysed the pike yield between 1990 and 2009 of three fry-stocked and three un-stocked large German pre-alpine lakes and found that mean pike yield gained by commercial fisheries varied across lakes but

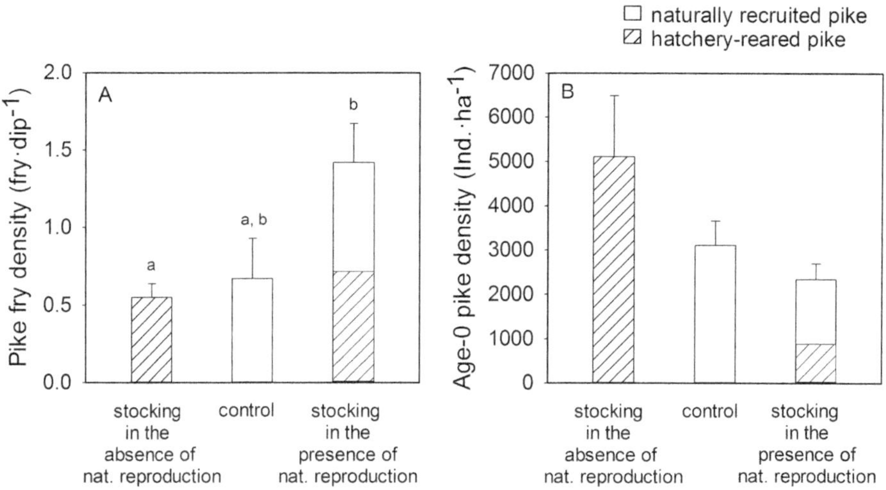

FIGURE 9.1 The additive effects of stocking on the density of pike fry over natural reproduction (control) three weeks post-stocking measured by electrofishing (A) and the age-0 pike density in the same ponds, four months post stocking, assessed by draining (B). In one treatment group, a naturally reproducing pike population was established and additionally stocked with hatchery-reared pike fry (stocking in the presence of natural reproduction). This treatment was contrasted to a second treatment group (control) characterized by a naturally reproducing pike population without stocking, and a third group of stocking pike fry in populations that failed to reproduce naturally (stocking in the absence of natural recruitment) as an example for restocking of pike (or culture based stocking). All treatment groups were replicated four times ($N = 12$ ponds total). The figure shows the absence of additive effects in populations with natural reproduction (stocking in the presence of natural reproduction *vs.* control) and that stocking of pike fry in systems with negligible or lacking natural reproduction can generate a year class strength comparable to that produced by natural reproduction (stocking in the absence of natural reproduction *vs.* control). Figures modified from Hühn et al. (2014)..

a and b indicate significant differences between treatment groups. Error bars represent standard errors.

was unaffected by stocking at a biomass density of 0.1-1.2 kg/ha. Grimm (1982) studied shallow lakes and found that three years of fry (4-6 cm fork length, FL) stocking did not increase the abundance of pike up to 41 cm (FL) relative to the abundance of larger pike reported from four years pre-stocking in the same shallow lakes Jan Verhoefgracht, Fortgracht, Kleine Wielen and Parkeerterreinsloot (Netherlands). Hühn et al. (2014) reported from experimental ponds that stocking fry into naturally reproducing pike stocks failed to generate additive effects, i.e., there was a failure to elevate year class strength over unstocked controls after few weeks post-stocking (Fig. 9.1). In Lake Halle (Denmark), Skov et al. (2011) reported that stocked pike fry were replaced by wild pike over the duration of the season, indicating larger mortality of stocked fry relative to wild recruits. In fact, the proportion of stocked pike in the population dropped from about 65% soon after stocking to about 15% four months after stocking. Hühn et al. (2014) also found that the stocked pike fry established in the juvenile pike populations without producing additive effects at the year-class level, thereby reporting a partial replacement of 31.4% of natural recruits by stocked fry pike in late summer (Fig. 9.1B). A replacement effect can also be hypothesized in the studies by Grimm (1982, 1983) as the author did not find any additive effect of stocking even though stocked pike amounted to among 3 to 67% of the total cohort.

In conclusion, it appears that stocking fry in waters with established pike populations has very short-term effects on pike population density and rarely succeeded in increasing pike stocks, a finding also mentioned by others (Margenau et al. 2008, Larsen et al. 2005; chapter 11). It can, thus, be concluded that stocking of pike fry has limited potential for stock enhancement.

9.2.3.2 Outcomes of Stocking Fingerlings

Relative to pike fry, the stocking literature about pike fingerlings is scarce. Grimm (1983) mentioned that the contribution of stocked fingerlings (18-23 cm FL) to the population of Lake Parkeerterreinloot (Netherlands) was absent. In small lakes (< 12 ha) in Germany with naturally reproducing pike populations, Hühn et al. (2015b) reported that stocking large fingerlings (average TL: 208 mm ± 29 Standard Deviation [SD]) increased the age-1 cohort in spring following stocking (at age-1), but stock size decreased one year post-stocking to levels comparable to the pre-stocking situation and to unstocked controls (Fig. 9.2A). The lack of additive effects of enhancement by fingerlings was further confirmed one and a half year post-release in the age-2 cohort (Fig. 9.2B). The lack of additive effects was independent of the habitat quality of lakes in terms of availability of structure (Hühn et al. 2015b). However, similar to the pond study (using fry) mentioned above (Hühn et al. 2014), some of the stocked fingerlings established in the populations without increasing the pike stock, again documenting a partial replacement of wild recruits by stocked pike. Stocked fingerlings have been reported in other studies to prevail in receiving populations to some extent. For example, Beyerle and Williams (1973) observed that marsh-reared stocked pike fingerlings (6.4-8.9 cm and 10.2-45.2 cm) remained, on average, longer than the wild pike of the same year-classes and represented 62% and 76% (ranging from 50% to 77% and 66.7 to

81.7%, respectively) of pike collected over three autumn electrofishing samplings in Long Lake (USA).

In conclusion, stocking fingerlings into naturally recruiting populations seems to result in a short-term increase of the stocked cohort, but additive effects at the population level are unlikely. In addition, the stocking may imply replacement effects where stocked pike (and their genes; chapter 7) are established in the population. Therefore, similar to the case of pike fry, the stocking of pike fingerlings seems to have limited potential for fisheries enhancement.

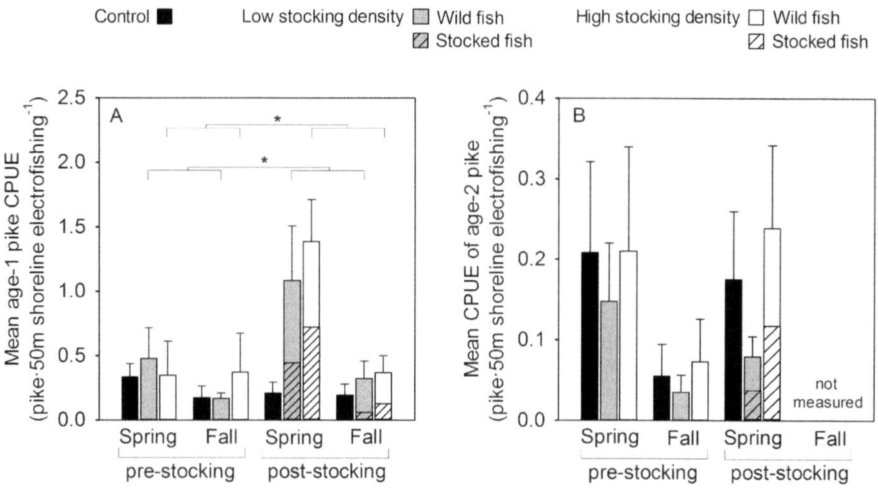

FIGURE 9.2 Mean catch-per-unit-effort (CPUE, number of pike caught per 50 m shoreline electrofishing) of age-1 pike (A) and age-2 pike (B) in the year pre- and post-stocking (following a before-after-control-impact design). The bars show the mean CPUE of three treatments and also the composition of the cohort in terms of wild and stocked individuals. Small artificial lakes (< 12 ha) with naturally reproducing pike populations were stocked with a low stocking density (35 age-0 individuals/ha) or high stocking density (70 age-0 inividuals/ha) of pike fingerlings, alongside an unstocked control group (N = 6 lakes per treatment). Age-0 pike abundance was assessed twice a year, in spring and fall, before and after stocking at the age-1 level (A), and again at the age-2 level (B). The experiments were conducted in 18 small gravel pits in Germany.

*indicates significant differences of treatment groups relative to controls. Error bars represent standard errors. Figures modified from Hühn et al. (2015a).

9.2.3.3 *Outcomes of Stocking Adults*

Similar to the case with fingerlings, there are only a few studies on the outcomes of stocking adult pike for stock enhancement. In Murphy flowage (Wisconsin, USA), Snow (1974) reported that stocking pike adults (25-55 cm TL) into a self-sustaining pike population doubled pike density, resulting in increases in angler catches (from 3.9 adults/100 angling hours before stocking to 7.2 adults/100 angling hours after stocking) in the first season after stocking. However, in the absence of continued stocking, the pike stock declined two years after stocking to reach values recorded before stocking (or even below). The authors also reported detrimental effects of stocking on total harvest and abundance of large pike (> 66 cm), presumably due to density-dependent effects on growth and mortality. In particular, the strongly elevated density increased mortality (annual natural mortality rate of 22% before stocking and 81% after stocking), presumably caused by a parasitic infection believed to have been introduced at the time of stocking, as well as emigration of stocked pike. In another study, Carlander (1958) observed that 20% of the stocked adult pike (~ 25-41 cm) were captured within the first six weeks of the next fishing season, and Snow (1974) also observed that 78% of the stocked adult pike (26.5-58 cm) were reported in the catches by anglers in the first year after stocking. These values are substantially higher than the recapture rates previously reported for pike fry stocking (e.g. Dorow 2005) suggesting that stocking adults could strongly affect angler catches in the short term after release. Snow (1974) observed that two years after stocking (at a population size comparable to before stocking), stocked adult pike still represented 21% in abundance and 29% in biomass of the total pike stock suggesting partial replacement of wild pike. In the river Lot, Guillerault et al. (2012) found that the majority of anglers' captures of stocked pike (28-82 cm TL) occurred within a year after stocking, but some fish stayed up to three years in the river after stocking. Stocking of adult pike thus has the potential to increase catches in the short term and maintain or possibly increase the abundance of pike spawners in the long term. However, in a single-lake experiment with mature pike (36-70 cm TL), Hühn et al. (2015b) observed that the probability of spawning successfully in the first spawning season post- stocking was lower for stocked pike compared to resident controls. Moreover, Hühn et al. (2015b) reported that the per capita offspring production in the wild was lower for pike stocked from foreign sources relative to the offspring production of resident pike. The reduced reproductive fitness of non-local and stocked pike relative to resident pike (which was approximated 57% lower fitness, Hühn et al. 2015b) is in line with studies on salmonids (e.g. Araki et al. 2007) suggesting that stocking non-local adults may ultimately harm offspring production in the pike population.

In conclusion, stocking of adult pike is likely to be successful for restocking and short-term enhancement of catches and may possibly also elevate adult stock sizes. However, strongly boosting stock sizes beyond natural carrying capacity may lower growth rates and could affect individual fish condition by reducing egg sizes and reproductive output. Further studies are needed to test these predictions in the wild.

9.2.4 Synthesis of Pike Stocking for Stock Enhancement

Stocking pike into self-recruiting populations often fails to increase the pike stock, especially when fry or fingerlings are stocked, but even then replacement effects are likely. The likelihood of generating additive effects of pike stock enhancement is higher when large adult pike are released. Because larger pike have strongly increased survival compared to smaller conspecifics (Haugen et al. 2007), larger pike prevail longer in the ecosystem after stocking compared with smaller conspecifics

(Fig. 9.3). However, even for larger pike, density-dependent processes will reduce the stock-enhanced pike population in the long term (e.g. altering fish growth-rate and fish condition; Lorenzen 2005). Still, a successful stock enhancement with stocking of adults could result in elevated densities that prevail longer than is the case with the release of smaller conspecifics, but this is most likely to eventually fall back to a population density defined by the ecosystem's carrying capacity (e.g. due to elevated hunger and higher catchability of fish and, generally, compensatory mechanisms).

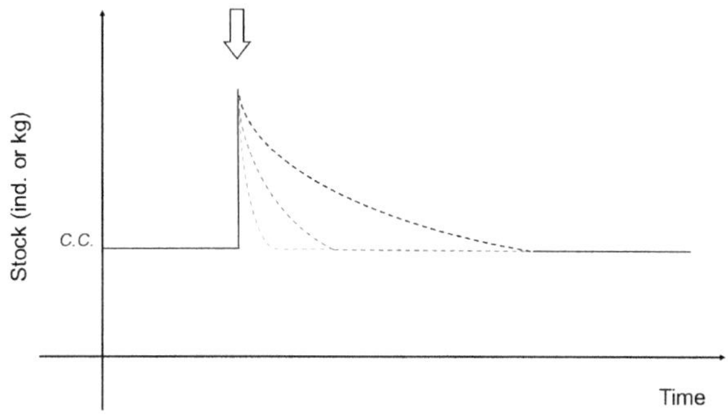

FIGURE 9.3 Effect of size at stocking on the pike stock (solid line) in naturally recruiting populations (C.C. = carrying capacity). The arrow represents the stocking event. The darker the dashed lines the larger the fish stocked (e.g. light grey: fry, dark grey: fingerlings, black: adults). The duration of additive effects increases with increasing fish size at stocking.

9.3 FACTORS AFFECTING PIKE STOCKING OUTCOMES

The above review of pike stocking outcomes has revealed the overarching importance of size at stocking for the additive benefits from stocking. The reviewed studies have also revealed other ecological issues that can affect stocking outcome. The following section synthesizes the available pike literature and presents factors/variables that affect the success (and failure) of pike stocking. Many of these variables relate to limiting post-stocking mortality which is of crucial importance for improving fisheries and population management in stock enhancements (Lorenzen et al. 2012). Four keys issues are presented in a hierarchical order as proposed by Arlinghaus et al. (2015):

1. degree of natural recruitment,
2. degree of ecological and genetic adaptation (e.g. degree of domestication and genetic background) of stocked fishes,
3. size at stocking and stocking density, and
4. handling, transport and acclimatization.

9.3.1 Natural Recruitment

Empirical and modelling studies (Lorenzen 2005, Rogers et al. 2010, Hühn et al. 2014, Johnston et al. 2015, Garlock et al. 2017) support the proposal that stocking pike is most likely to generate additive effects when the fish are stocked in ecosystems where natural recruitment is very low or absent. Souchon (1980) demonstrated that, at moderate stocking rates (up to 5 individuals per m^2), the biomass and density of pike was a direct function of the initial fry stocking density in ponds that otherwise lacked pike. Relatedly, Hühn et al. (2014; Fig. 9.1) showed that stocking pike fry in ponds which lack natural reproduction resulted in age-0 pike densities comparable to levels emerging from natural recruitment in the absence of stock enhancement. In natural lakes, Sutela et al. (2004) reported successful stocking of pike fry (~ 10-15 mm TL) in terms of increase of fry density if natural recruits were extremely scarce or absent. Similarly, the stocking of adult pike in ponds or lakes otherwise lacking a pike population repeatedly resulted in the production of juveniles and the establishment of a pike population (chapter 14). For instance, Bry and Souchon (1982) reported that even few individuals (one female and two males, 170-750 g) produced ample juveniles in ponds although the production of young pike was less consistent than when relying on stocking of pike fry.

As reviewed in previous sections, in the cases of pike fry and fingerling stocking, the additive effects of stocking are much more uncertain and, in most cases, non-existent when wild conspecifics are already present in the ecosystem (Wysujack et al. 2001, Lorenzen 2005, Schreckenbach 2006, Hühn et al. 2014, 2015a). For instance, Beyerle (1971) reported that post-stocking survival of fingerlings in two lakes (5 ha and 2.3 ha) in Michigan (USA) decreased from 44-60% during the first stocking in the absence of natural recruitment to only 0.8-9.2% when the stocking was repeated three years later after establishment of a pike stock. Similarly, in Danish lakes where pike fry were stocked for biomanipulation purposes, survival was less than 10% after 40 days (Skov and Nilsson 2007, chapter 11), and survival was highest in lakes where no wild pike were present (Skov et al. 2006). The reason for the lack of long-term stock enhancement effects of stocking juvenile pike in reproducing stocks probably relates to the strong compensatory mortality of juveniles (Lorenzen 2005, Hühn et al. 2014). The resulting recruitment bottlenecks are strongly size-dependent in cannibalistic pike (Grimm and Klinge 1996, Haugen et al. 2007) which is why stocking of juveniles in self-reproducing stocks is usually bound to fail and lead to very high mortality of the stocked pike (Skov et al. 2011). For instance, Hühn et al. (2014) demonstrated stocked pike fry to generally have lower fitness (measured as growth rate and survival) when forced into competition with natural recruits of the same genetic origin and size.

A range of ecological factors affect natural pike recruitment and, hence, the prospect for successful pike stocking, most notably the presence of structured habitat used for spawning and for shelter and food availability (Raat 1988, Casselman and Lewis 1996, Grimm and Klinge 1996, chapter 10). Whereas Grimm (1981b, 1983) suggested a limited effect of prey fish density on pike density in natural waters, trophic resource availability (i.e., abundance of prey) has the potential to strongly affect intraspecific competition and cannibalism, notably when the timing of stocking does not match with the natural production of prey (Skov et al. 2003, Skov and Nilsson 2007). Maloney and Schupp (1977) reported that pike stocking may fail when prey fish density is low (in this case, perch *Perca flavescens* Mitchill, 1814) especially in the presence of other predatory species (in this case, walleye *Sander vitreus* Mitchill, 1818). Flickinger and Clark (1978) also reported that survival of stocked pike fry (50 mm) was dependent on forage fish availability. In addition, in Cave Run Lake (USA), Axon (1981) found that fluctuations in a stocked muskellunge population were primarily due to changes in the abundance of the primary prey species (in this case, gizzard shad *Dorosoma cepedianum* Lesueur, 1818). Skov et al. (2003) showed in pond experiments that cannibalism among fry (20-31 mm TL) was markedly more frequent in the absence of alternative prey, at least when the variation in stocking length was relatively high. The type of food available also influences pike survival. For instance, Beyerle (1978) showed that pike stocked in ponds with bluegill (*Lepomis macrochirus* Rafinesque, 1819) had, on average, a lower survival rate than pike stocked in ponds with fathead minnows and golden shiners (*Pimephales promelas* Rafinesque, 1820 and *Notemigonus crysoleucas* Mitchill, 1814, respectively). Apart from these examples, studies that highlight the importance of food type and availability for pike stocking outcomes remain scarce. Timing of stocking is then an important variable by affecting the availability of refuge, food, and the relative size of stocked and wild pike. Grønkjær et al. (2004) reported that pike stocked early in the season showed significantly higher survival than their conspecifics stocked three weeks later and suggested that the late stocked fry were outcompeted or predated upon by larger wild conspecifics. The situation may be different when adults that are no longer (or much less) under cannibalistic control (Haugen et al. 2007) are stocked. However, the additive effect of stocking adults in naturally-reproducing stocks is likely to increase catches but will not necessarily increase recruitment as the new recruits are, again, forced through density- and size-dependent bottlenecks (Lorenzen 2005).

The presence of structured habitats, essential in each life-stage of pike (chapters 2, 3 and 10), is of main importance for the survival of young pike. Shore vegetation, especially macrophyte cover, provides shelter against cannibalism (Grimm 1981a, 1983) and predation (including predation by piscivorous birds; Cucherousset et al. 2007) and serves as habitat for prey (Casselman and Lewis 1996). In lentic systems, the littoral area generally limits recruitment and, thus, stock size (Grimm 1983, 1989, Pierce and Tomcko 2005). Grimm (1994b) reported that an optimal recruitment requires 15-30% of the lake area covered by submerged or emergent vegetation (either temporarily or permanently). Grimm and Klinge (1996) reported, from studies in shallow vegetated ponds and lakes, that any increase in the population (from stocking) above the carrying capacity of the system will quickly be removed by cannibalism or displacement, rendering the likelihood of a positive stocking outcome, in terms of elevation of recruitment, unlikely if sufficient recruits are available. While vegetation cover and vegetation structure affects cannibalism among pike, allowing segregation or overlap between habitats of different size classes (Grimm and Backx 1990, chapters 3 and 5), the degree of cannibalism broadly depends on the biomass of larger conspecifics (Grimm 1981b, 1983, 1994a, Persson et al. 2006, Sharma and Borgstrøm 2008). Interestingly, Pierce (2010) could not find that conservation of large fish decreased the abundance of smaller conspecifics, possibly because he studied large systems where juveniles and adults may no longer share the same habitats, to the same extent, as in smaller systems (e.g. shore pike vs. lake pike, Grimm 1994a). Hühn et al. (2015a) reported that in lakes that offer poor habitat, stocking of juveniles did not elevate stock size of age-2 fish due, most likely, to cannibalism. In fact, the natural pike population sizes varied with the availability of structured pike habitat, but the lack of additive effects of juvenile pike stocking was independent of habitat structure; in "poor" or "good" pike lakes, the final pike stock size was independent of stocking intensity (Hühn et al. 2015a).

In conclusion, an increased pike density can only be achieved by stocking if the natural stock is low (e.g. due to high fishing pressure or weak natural reproduction) and, at the same time, a sufficient forage base and a sufficient quantity of appropriate and unsaturated habitats is available (Grimm 1981a, Skov et al. 2011). Prolonged additive effects on year-class strength are highly unlikely when relying on stocked fry or juveniles in naturally reproducing populations (Hühn et al. 2014a, 2015a) unless there is weak recruitment and, consequently, few natural young pike. If one relies only upon stocking adults, adult stock size might be elevated (see size section below for details) but not necessarily result in increased future recruitment.

9.3.2 Eco-Genetic Adaptation

Ecological and genetic adaptation to the conditions of the receiving environment is fundamental for successful stocking by affecting survival and recruitment success post-stocking. Rearing in aquaculture facilities, such as hatcheries, induces a domestication syndrome in the cultured organism that involves both phenotypic developmental responses and genetic selection (Lorenzen et al. 2012). Aquaculture often leads to the development of domesticated strains, notably in the case of captive brood stock or when certain traits (e.g. growth, colouration patterns) are selected. However, domestication effects and consequent loss of fitness in the wild can, and usually does, occur even when intentional selection is avoided (Lorenzen et al. 2012). While natural evolutionary forces favour local adaptation in wild populations (Eschbach et al. 2015, Bekkevold et al. 2016), natural (and artificial) selection in aquaculture systems, as well as developmental adaptation to rearing environments, tends to increase fitness in the aquaculture system but leads to maladaptation of stocked fishes to natural conditions and, therefore, reduced fitness in the wild (Price 2002, Lorenzen 2006, Lorenzen et al. 2012).

Independent of genetic selection and adaptation, rearing conditions and time in artificial environments (hatching jars, tanks and raceways, and presumably, to a lesser degree, in ponds) strongly influence the survival of stocked fish by affecting their ability to cope with their new environment. Hühn et al. (2014) showed that pike that were artificially spawned from the same wild-living brood stock as those naturally spawning in ponds had lower growth and elevated mortality when stocked with wild fry (Fig. 9.4). This occurred despite that the stocked individuals were reared in captivity, even for a very short period (until the free-swimming phase). No differences in fitness components were seen when the fry were stocked without naturally-spawned fry (Fig. 9.4) indicating that the forced competition invokes the fitness depression to emerge. The reduced fitness of stocked fish is

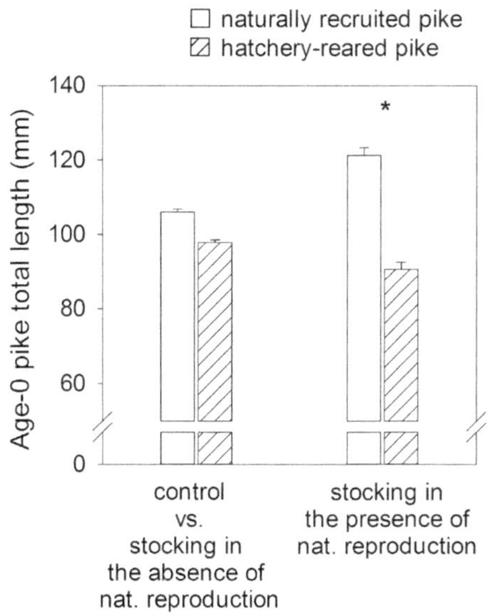

FIGURE 9.4 Mean total length of naturally-recruited and hatchery-reared age-0 pike from a replicated pond experiment. In the two bars on the left, growth of naturally-recruited pike without additional adding of hatchery-reared pike fry (control) is compared with fish from ponds with hatchery-reared fry only (stocking in the absence of natural recruitment). To the right, fish from ponds containing both naturally-recruited and hatchery-reared age-0 pike of the same genetic background in a competitive situation (stocking in the presence of natural recruitment) is compared. Fish are sampled in July.

*indicates significant difference. Error bars represent standard errors. Figure modified from Hühn et al. (2014). See caption of Fig. 9.1. for further details on the study design.

driven by multiple, morphological, physiological and behavioural factors, which can be partially mitigated by husbandry approaches such as rearing in semi-natural conditions, enrichment of the aquaculture environment, or life skills training (Lorenzen et al. 2012, Näslund and Johnsson 2014). Moreover, in hatcheries, non-natural selection forces lead to very high survival rates and successful hatching of individuals that have naturally low fitness. Finally, circumventing sexual selection can lead to fitness depression and genetic effects that might reduce survival in the wild as shown for salmonids (Thériault et al. 2011). However, this statement needs a proper empirical assessment in pike. Most importantly, behaviours associated with food selection, food acquisition and social interactions are unlikely to properly develop during the early ontogeny in hatchery conditions as observed in several salmonids (e.g. Brockmark et al. 2007, Brockmark and Johnsson 2010). Szczepkowski et al. (2012) showed in an experiment in ponds that growth of hatchery-reared pike, from the same genetic strain, was inversely related with duration of time held in the tank. This suggests that longer durations of time in captivity reduces performance of stocked pike in the wild as proposed in salmonids (e.g. Baer 2008). Franklin and Smith (1963) also found that stocked pike produced in marshes containing a large diversity of prey displayed a higher survival than pike reared under standard hatchery conditions. This suggests that pike reared in ponds prior to stocking may have higher survival after stocking than those reared in artificial aquaculture facilities. In support of this statement, Gillen et al. (1981) reported that pellet-reared congeneric muskellunge were slow to shift their diet toward forage fish during ontogeny. Although rearing pike used for stocking on pellets is not common (as most rearing happens in ponds), the example shows how husbandry methods that do not promote the development of natural behavioural and other ecological traits can contribute to low fitness of stocked pike compared with wild conspecifics. Decreased foraging efficiency of stocked pike in the wild might then lead to suboptimal growth and subsequent increased mortality of stocked pike when forced into competition with similarly-sized conspecifics (Skov 2002, Skov et al. 2011).

Besides (non-genetic) domestication effects, genetic maladaptation of translocated fish to conditions in the receiving environment is also likely to affect stocking outcomes, but it is often difficult to differentiate purely genetic from ecological maladaptation, hence the term 'eco-genetic' adaptation in this section. For example, from 1993 to 2002, nearly 400,000 translocated freshwater pike fry were stocked to compensate for population declines in brackish waters of Denmark. Using a molecular-based approach, Larsen et al. (2005) found that only 0.3% of the genetic material originated from stocking of pike fry (population admixture analysis), documenting poor survival of stocked fish in their new environment. The low fitness of stocked fish was presumably caused by maladaptation of the freshwater stocked pike to the receiving ecosystem (brackish water), but it is unclear whether the results were caused by poor tolerance to high salinity (genetically-based physiological adaptation) or other factors such as the presence of strong natural recruitment.

Hühn et al. (2015b) reported that stocked large mature pike that were translocated from two lakes located only a few kilometers from the stocked water body showed strongly reduced reproductive success after stocking compared with their local wild conspecifics. The lower fitness might prevent (or lower) genetic replacement effects (as in the case of Larsen et al. 2005), but Hühn et al. (2015b) also found that the foreign stocks produced viable hybrids despite having a lower per capita reproductive fitness. If these hybrids survive and reproduce themselves, the reproductive performance of the entire pike population could be altered by replacement of wild pike by less well-adapted foreign fish (e.g. if pike are stocked larger than their wild conspecifics) or hybrids that can have lower reproductive performance (see Chilcote et al. [2011] and Araki et al. [2007] for an example, in salmonids). Both Eschbach et al. (2015) and Bekkevold et al. (2016) show that local pike stock often shows divergent genetic signatures from those expected from the catchment suggesting local loss of native biodiversity regularly happens due to stocking. Hybridization between strains has been observed for other species which often shapes colonization dynamics and evolutionary trajectories (Araki et al. 2007, Fraser et al. 2011). However, hybridization can also be advantageous outside the natural range or in altered environments by induced evolutionary novelty (Abbott et al. 2013, Eschbach et al. 2014). Despite this ongoing discussion, it is safe to assume that the fitness of stocked fishes is usually lower for non-local strains. Therefore, a clear recommendation is to rely on offspring from the same ecosystem or the same catchment to avoid genetic mixing and maximize the likelihood that stocked fishes have high fitness (Arlinghaus et al. 2015).

9.3.3 Stocking Size and Stocking Density

As reviewed above, the size at stocking strongly affects stocking outcome and the probability of generating additive effects particularly by affecting mortality post-stocking. Stocked individuals may face wild conspecifics that are often competitively superior (e.g. Skov et al. 2011), either because of higher competitive ability (resource-holding potential), or by motivation for defense of refuges (value asymmetry, i.e., prior residency advantage; e.g. Huntingford and de Leaniz 1997, Rhodes and Quinn 1998). Post-stocking survival of pike should increase with individual body size at stocking (e.g. Cucherousset et al. 2007), as in many other fish species (Lorenzen 2000), since in pike, cannibalism is strongly size-dependent (Grimm 1981b, 1994, Haugen et al. 2007) and even small size differences matter (e.g. Grimm and Backx 1990, Skov et al. 2011). Grønkjær et al. (2004) demonstrated that stocked pike fry may well survive competition with wild recruits, in particular, when the stocked individuals are larger than their wild conspecifics. Skov et al. (2011) demonstrated that the movement of stocked pike fry after release was inversely related to the size at stocking suggesting that larger pike fry out-competed smaller pike by restricting them to suboptimal habitat, social stress, and risk of cannibalism. Snow (1974) also reported that intense stocking resulted in emigration of stocked pike because of competition with wild pike. Skov et al. (2003) found that mortality of stocked pike increased with the level of body size heterogeneity within the cohorts as larger pike forage on smaller individuals (i.e., cannibalism). Overall, fish that are stocked larger and are developmentally advanced, relative to the naturally produced pike of the same cohort, have better chances of survival (Beyerle and Williams 1973, Grimm 1982, Raat 1988). Interestingly, earlier spawning and emergence of larvae is often outlined as a reason of pike dominance on sympatric congeneric muskellunge (Inskip 1986). Note, however, that an increased survival of stocked fishes must not necessarily elevate the probability of generating additive (stock-enhancing) effects. By contrast, elevated survival of stocked fishes, particularly when releasing juveniles, might simply mean that the degree of replacement of wild fish by stocked fish increases without necessarily elevating year-class strength (Fig. 9.1 and 9.2). Moreover, poor ecological adaptation can reverse the size-advantage on survival, as, for example, documented when comparing the survival of glass eel with the survival of (larger) stocked farm eels (*Anguilla anguilla* Linnaeus, 1758; Simon and Dörner 2014). The fish size at stocking seems to be of less importance when natural recruitment is lacking. Under such conditions, the stocking of pike fry can generate year class strengths according to prevailing ecological conditions.

Stocking density directly affects the level of intraspecific competition for food and habitat, the risk of cannibalism, and post-stocking mortality (Bry and Gillet 1980, Wright and Giles 1987, Edeline et al. 2010). Pike stocking densities reported in the literature are highly variable (e.g. for fry: 0.11 individuals/m^2 [van

Donk et al. 1989] or 6.3 individuals/m^2 [Skov et al. 2003] and, for fingerlings: 0.001 individuals/m^2 [Beyerle and Williams 1973] or 0.22 individuals/m^2 [Szczepkowski et al. 2012]) depending on the life-stage of stocked pike and habitat availability. For fry, a stocking density of 5-6 individuals/m^2 is commonly used (Bry and Souchon 1982, Wright and Giles 1987, Skov et al. 2011, Hühn et al. 2014). The highest stocking densities are often reported from stocking as a lake restoration measure (see chapter 11 for further discussions). For fingerlings, values reported in the literature often suggest to release about 30 individuals/ha (Baer et al. 2007). However, all populations of fishes, in particular pike, are density-regulated (Lorenzen 2005, Haugen et al. 2007) such that an increased density of stocked juveniles usually means elevated density-dependent mortality and strong self-regulation towards carrying capacity. The situation can be different when recruited (large) fish are stocked (Johnston et al. 2015) where the population size is mainly regulated by density-dependent growth and less by density-dependent mortality (i.e., cannibalism; Lorenzen 2005). Under these circumstances, additive effects of stocking and, at least, increased catches in the short-term could also be conceivable (Lorenzen 2005) in pike (Fig. 9.3).

9.3.4 Handling, Transport and Acclimatization

Finally, the stocking procedure, itself, can strongly affect the success of stocking (Wahl 1999). Stress caused by handling and transportation (e.g. duration of transport, density in holding tanks, temperature differences among transport tanks and ecosystem, unsuitable environmental conditions in the tank) can generate direct and indirect mortality (e.g. Gomes et al. 2003, Braun et al. 2010). For example, Hühn et al. (2015b) demonstrated that stocking stress associated with capture, transport, netting and release into a novel environment resulted in a stress-induced mortality of at least 10% even when adult pike (36-70 cm TL) were stocked. The mortality happened within the first winter following release independent of the origin of fish (Fig. 9.5). However, other studies have revealed that esocids are rather resilient to a range of stressors during transport and that acclimatization procedures (e.g. previous exposure to natural food) do not substantially improve post-release survival in the congeneric muskellunge. Mather and Wahl (1988) found elevated mortality (3.3% ± 5.8 SD) of pike (21 cm mean TL) when stocked at temperatures above 25°C. At the same time, the authors reported little mortality even after a rapid temperature increase of 10°C between transport tank and the temperature of the lake. Yet, when the temperature differences was about 15°C, almost a complete mortality (98%) occurred after stocking suggesting that severe thermal stress is more important for the post-stocking survival of 0+ pike than either handling or transport-induced stress (Mather et al. 1986, Mather and Wahl 1988).

To limit indirect mortality associated with handling and transportation, acclimation of fish to the water temperature and water chemistry, prior to release, has been suggested in a range of species (see Wahl [1999], Brennan et al. [2006], Baer and Brinker [2008] for examples in other fish species), but no research is available in pike. Research in the congeneric muskellunge and tiger muskellunge has also failed to find significant survival benefits when fishes were previously exposed to predators (in this case, largemouth bass *Micropterus salmoides* Lacepède, 1802; Wahl et al. 2012). Furthermore, survival post-release was the same when fishes were released during the day or during the night (assuming less predation mortality during the night-time; Stein et al. 1981). However, it was found that feeding experience increased survival post-release compared to naïve (only pellet-fed) esocids (muskellunge or tiger muskellunge; Szendrey and Wahl 1995), and that adding of artificial vegetation to rearing tanks improved behaviour consistent with increased survival in the wild (Einfalt et al. 2013). Hence, acclimatization procedures prior to stocking and reduced stress during transport promises to increase the likelihood of positive stocking outcomes. To minimize the immediate mortality post-release, it can be suggested to distribute pike along the bank in shallow (< 2-3 m, or very shallow (15-30 cm) for fry) highly vegetated areas to provide shelter, when predators are less active (e.g. during the evening to avoid avian predation) (e.g. Kerr and Lasenby 2001), despite the lack of evidence in support of these recommendations.

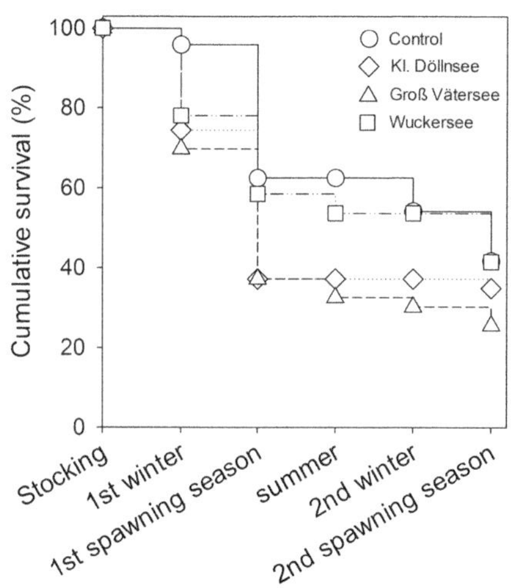

FIGURE 9.5 Survival of control fish (native pike) compared to a native group of pike (Kl. Döllnsee) that were handled and transported to simulate stocking stress and two groups of pike originating from two foreign nearby lakes (Groß Vätersee and Wuckersee) after stocking in fall. Control fish were caught, tagged and released immediately in the native water body. All fish (adult pike) were tagged with acoustic transmitters to estimate survival. Comparisons were made over different observation periods post-stocking. Figure modified from Hühn et al. (2015b).

9.4 ECOLOGICAL AND CONSERVATION RISKS ASSOCIATED WITH PIKE STOCKING

Although stocking of pike could offer multiple benefits for fisheries and conservation, the practice can also have negative impacts on the receiving ecosystems as well as the users. For instance, high expectations by anglers about stocking-induced increases in stock sizes can increase fishing pressure and reduce angler satisfaction through overexploitation of the fish stock or overcrowding of stocked sites (Krohn 1969, Post et al. 2002, Fayram et al. 2006, Lorenzen et al. 2012). Stocking might also modify ecosystem functioning, and this is particularly true in the case of pike which is a keystone predator in many ecosystems (chapters 8, 11 and 14). Stocked pike compete with other predatory fish for prey, and successful increases in stock sizes (for example due to the release of recruited fishes), may reduce the abundance and growth of wild conspecifics and other predatory fish through predation (Snow et al. 1974, Maloney and Schupp 1977). Increased competition with stocked pike can also increase the vulnerability of other species to angling as demonstrated for walleye (Wesloh and Olson 1962). In addition, stocked fish may carry pathogens. For example, Snow (1974) reported that stocking of pike led to high mortality of wild pike from parasitic infections (*Myxobolus* sp.). The infections affected both stocked and wild pike, but larger wild pike were disproportionally affected as these fish did not emigrate after stocking.

Another key problem relates to genetic pollution by stocked individuals as surviving individuals can contribute to the wild population by reproducing with wild conspecifics, thereby, introgressing genes into local stocks. Introgression of foreign genes into local gene pools can lead to the loss of the local stock due to genetic swamping by numerically abundant stocked fishes spawning with numerically less abundant wild conspecifics (Laikre et al. 2010). In many European countries, stocking is commonly performed by local stakeholders in a self-organized manner where the source population used for stocking is largely uncontrolled and determined by supply-demand factors in commercial hatcheries (Arlinghaus et al. 2015). This is despite the fact that recent investigations have shown the existence of different genetic pike strains, at least at catchment levels in Europe (Pedreschi et al. 2014, Eschbach et al. 2015, Bekkevold et al. 2016; chapter 7). For instance, Launey et al. (2006) conducted a study in 11 rivers in France and demonstrated the occurrence of genetic introgression from farmed pike (originating from Poland and Czech Republic, which are among the main pike-exporting countries in Europe) in the natural populations. In addition, pike stocking may alter the genetic integrity of other species of pike. Denys et al. (2014) showed evidence of hybridization between *E. lucius* and *Esox aquitenacus* [Denys et al. 2014] in France, and Gandolfi et al. (2017) showed evidence of hybridization between *E. lucius* and *Esox cisalpinus* (Bianco and Delmastro 2011) in Italy. By contrast, in Finland, most pike fry produced come from eggs of fish caught on the spawning grounds in the wild (FAO 2005). In Denmark, age-0 pike used in stocking programmes are the progeny of wild parents (25 males and 25 females crossed) captured in lakes within the geographical vicinity (< 200 km) of the target ecosystem to conserve local adaptation of fish and conserve local genetic composition (Skov et al. 2011).

9.5 CONCLUSIONS

Our review suggests that pike stocking produces very different outcomes depending on the state of the ecosystem and the genetic source and size of the pike that are stocked. One of the key factors for stocking success from a fisheries perspective is that pike have to be stocked at sizes large enough to bypass strong density-dependence mortality particularly in the juvenile life-stage. Pike stocking for enhancement has shown very poor long-term efficiency, especially when relying on juveniles as stocking material. Juveniles and small adults of pike (< 50 cm) are associated with structured vegetation where strong competition for food and shelter, predation, and cannibalism drive pike abundance. Hence, in naturally recruiting stocks, stocking pike smaller than 50 cm in stock enhancement projects will rarely lead to long-term increases in the stock and may, if at all, only boost catches in the short term. Releasing juveniles in recruiting stocks can, thus, no longer be recommended without careful appraisal of risks and potential benefits. By contrast, stocking in waters naturally lacking pike (i.e., introductory stocking, re-stocking or culture-based stocking) is often very successful in establishing a fishery no matter which size classes are stocked (see also chapter 14). However, immediate success in establishing a population or year class does not imply that the newly established pike will provide good quality fisheries over a long-term period unless the reasons for the poor recruitment are addressed through habitat improvement. The situation is different with stocking large (recruited) pike if the goal is to maintain high catches (e.g. Snow et al. 1974). Stocking of adults seems to be the only tool that can elevate catches in self-recruiting stocks (Johnston et al. 2015). Even then, no positive effect on future recruitment is to be expected because of compensatory mechanisms, lower fitness of stocked fishes and their offspring, and because the new recruits will, again, be forced through the same habitat bottlenecks that were likely responsible for motivating the stocking exercise. Therefore, stocking of adult pike seems to be a sensible option when a significant local fishing pressure is present to take advantage of the strongly elevated stock size by elevated catches or for restocking programs after fish kills to "buy" time for a new natural recruitment.

To conclude, it appears that in ecosystems where pike naturally reproduce, stocking may be superfluous because of (*i*) the absence of additive effects in terms of increasing stock size and subsequently fisheries yield, especially when stocking young-of-the-year fish, and (*ii*) the partial replacement of wild pike by stocked individuals, implying the risk of genetic introgression and loss of local gene pools (Hühn et al. 2014, 2015b). In general, we believe, based on the data reviewed in this chapter, that in ecosystems were natural pike populations are depressed, stocking pike (i.e., larger adult pike) is a way to treat symptoms and, if stocking is meant to support fisheries, maintain catches. However, if managers are looking for the "full cure" they should direct their attention to fisheries regulations (chapter 12) and/or habitat improvement (chapter 10) as alternatives to pike stocking (Arlinghaus et al. 2016). This conclusion is, of course, not new and has previously been expressed by Grimm (1983) more than 30 years ago. He evaluated several stocking attempts of both fingerlings and adult pike in Dutch lakes and concluded that "habitat-engineering rather than stocking may be the answer to the problem of maximizing pike populations" (chapter 10). Based on our review, this seems, indeed, to be the most promising way forward.

In systems were pike stocking is chosen as a management tool we suggest paying attention to the following suggestions.

- Pike should be stocked as large as needed to outgrow recruitment bottlenecks.
- Stocked pike may best originate from complex natural/semi-natural environments (i.e., lakes and ponds rather than tanks) because such fishes are likely more adapted to natural settings.
- Stocked pike may best originate from the same gene pool as wild individuals to maintain the genetic integrity of wild populations and avoid the disruption of local adaptation.
- It is of high importance to account for the health and sanitary status of fish used for stocking. Transportation and handling should be minimized and proper acclimation prior to release considered.

Acknowledgements

The authors thank the German Ministry for Education and Research for supporting the Besatzfisch project, which was instrumental for conducting several of the empirical studies that formed the basis of this review (grant no. 01UU0907 to RA, www.besatz-fisch.de). Christian Skov thanks the Danish Rod and Net Fishing Funds for support.

REFERENCES CITED

Abbott, R., D. Albach, S. Ansell, J.W. Arntzen, S.J.E. Baird, N. Bierne, J. Boughman, A. Brelsford, C.A. Buerkle, R. Buggs, R.K. Butlin, U. Dieckmann, F. Eroukhmanoff, A. Grill, S.H. Cahan, J.S. Hermansen, G. Hewitt, A.G. Hudson, C. Jiggins, J. Jones, B. Keller, T. Marczewski, J. Mallet, P. Martinez-Rodrigez, M. Möst, S. Mullen, R. Nichols, A.W. Nolte, C. Parisod, K. Pfennig, A.M. Rice, M.G. Ritchie, B. Seifert, C.M. Smadja, R. Stelkens, J.M. Szymura, R. Väinölä, J.B.W. Wolf and D. Zinner. 2013. Hybridisation and speciation. *J. Evol. Biol.* 28:229-246.

Araki, H., B. Cooper and M.S. Blouin. 2007. Genetic effects of captive breeding cause a rapid, cumulative fitness decline in the wild. *Science* 318:100-103.

Arlinghaus, R., T. Mehner and I.G. Cowx. 2002. Reconciling traditional inland fisheries management and sustainability in industrialized countries, with emphasis on Europe. *Fish Fish.* 3:261-316.

Arlinghaus, R. and T. Mehner. 2005. Determinants of management preferences of recreational anglers in Germany: habitat management versus fish stocking. *Limnologica* 35:2-17.

Arlinghaus R., S. Matsumara S. and U. Dieckmann. 2010. Conservation and fishery benefits of saving large pike (*Esox lucius* L.) by harvest regulations in recreational fishing. *Biol. Conserv.* 146:1444-1459.

Arlinghaus, R, B. Beardmore, C. Riepe, J. Meyerhoff and T. Pagel. 2014. Species-specific preferences of German recreational anglers for freshwater fishing experiences, with emphasis on the intrinsic utilities of fish stocking and wild fishes. *J. Fish Biol.* 85:1843-1867.

Arlinghaus, R., E.M. Cyrus, E. Eschbach, M. Fujitani, D. Hühn, F. Johnston, T. Pagel and C. Riepe. 2015. *Hand in Hand für eine nachhaltige Angelfischerei: Ergebnisse und Empfehlungen aus fünf Jahren praxisorientierter Forschung zu Fischbesatz und seinen Alternativen.* Berichte des IGB 28. Berlin, Germany: Berichte des IGB.

Arlinghaus, R., K. Lorenzen, B.M. Johnson, S.J. Cooke and I.G. Cowx. 2016. Management of freshwater fisheries. pp 557-579. *In*: J.F. Craig (ed.). *Freshwater fisheries ecology*. Chichester, U.K.: John Wiley & Sons, Ltd.

Armand, C., F, Bonnieux and T. Changeux. 2002. Evaluation économique des plans de gestion piscicole. *Bull. Fr. Peche Piscic.* 365/366:565-578.

Axon, J.R. 1981. Development of a muskellunge fishery at Cave Run Lake, Kentucky, 1874-1979. *N. Am. J. Fish. Manage.* 1:134-143.

Baer, J., K. Blasel and M. Dieckmann. 2007. Benefits of repeated stocking with adult, hatchery-reared brown trout, *Salmo trutta*, to recreational fisheries? *Fish. Manag. Ecol.* 14:51-59.

Baer, J. 2008. *Untersuchungen zur Optimierung des Besatz- und Bestandsmanagements von Bachforellen* (*Salmo trutta* L.). Doctoral dissertation. Berlin. Germany: Humboldt-Universität zu Berlin.

Baer, J. and A. Brinker. 2008. Pre-stocking acclimatisation of brown trout *Salmo trutta*: effects on growth and capture in a fast-flowing river. *Fish. Manag. Ecol.* 15:119-126.

Bekkevold, D., L. Jacobsen, J.H. Hansen, S. Berg and C. Skov. 2016. From regionally predictable to locally complex population structure in a freshwater top predator: river systems are not always the unit of connectivity in northern pike *Esox lucius*. *Ecol. Freshw. Fish* 24:305-316.

Beyer, J.E. 1989. Recruitment stability and survival: simple size-specific theory with examples from the early life dynamics of marine fish. *Dana* 7:45-147.

Beyerle, G.B. 1971. A study of two northern pike-bluegill populations. *Trans. Am. Fish. Soc.* 100:69-73.

Beyerle, G.B. and E.J. Williams. 1973. Contribution of northern pike fingerlings raised in managed marsh to the pike population of an adjacent lake. *Prog. Fish Cult.* 35:99-103.

Beyerle, G.B. 1978. Survival, growth and vulnerability of northern pike and walleye stocked as fingerlings in small lakes with bluegills or minnows. *Am. Fish. Soc. Special Publication* 11:135-139.

Beyerle, G.B. 1980. *Contribution to the angler's creel of marsh-reared northern pike stocked as fingerlings in Long Lake, Barry County*. Fisheries Research Report 1876. Ann Arbor, Michigan, USA: Michigan Department of Natural Resources.

Bianco, P.G. and G.B. Delmastro. 2011. *Recenti novità tassonomiche iguardanti i pesci d'acqua dolce autoctoni in Italia e descrizione di una nuova specie di luccio*. Research on wildlife conservation 2. USA: IGF publishing.

Botsford, L.W. and R.C. Hobbs. 1984. Optimal fishery policy with artificial enhancement through stocking: California's white sturgeon as an example. *Ecol. Mod.* 23:293-312.

Braun, N., R.L. De Lima, B. Baldisserotto, A.L. Dafre and A.P.O. Nuñer. 2010. Growth, biochemical and physiological responses of *Salminus brasiliensis* with different stocking densities and handling. *Aquaculture* 301:22-30.

Brennan N.P., M.C. Darcy and K.M. Leber. 2006. Predator-free enclosures improve post-release survival of stocked common snook. *J. Exp. Mar. Biol. Ecol.* 335:302-311.

Brockmark, S. and J.I. Johnsson. 2010. Reduced hatchery rearing density increases social dominance, postrelease growth, and survival in brown trout (*Salmo trutta*). *Can. J. Fish. Aquat. Sci.* 67:288-295.

Brockmark, S., L. Neregård, T. Bohlin, B.T. Björnsson and J.I. Johnsson. 2007. Effects of rearing density and structural complexity on the pre- and postrelease performance of Atlantic salmon. *Trans. Am. Fish. Soc.* 136:1453-1462.

Brönmark, C. and L. A. Hansson. 2005. The Biology of Lakes and Ponds, 2nd Ed.. New York, New York, USA: Oxford University Press

Bry, C. and C. Gillet. 1980. Reduction of cannibalism in pike (*Esox lucius*) by isolation of full-sib families. *Reprod. Nutr. Dev.* 20:173-182.

Bry, C. and Y. Souchon. 1982. Production of young northern pike families in small ponds: natural spawning versus fry stocking. *Trans. Am. Fish. Soc.* 111:476-480.

Bry, C., F. Bonamy, J. Manelphe and B. Duranthon. 1995. Early life characteristics of pike, *Esox lucius*, in rearing ponds: temporal survival pattern and ontogenetic diet shifts. *J. Fish Biol.* 46:99-113.

Camp, E.V., S.L. Larkin, R.N.M. Ahrens and K. Lorenzen. 2017. Trade-offs between socioeconomic and conservation management objectives in stock enhancement of marine recreational fisheries. *Fish. Res.* 186:446-459.

Carlander, K.D. 1958. Disturbance of the predator-prey balance as a management technique. *Trans. Am. Fish. Soc.* 87:34-38.

Casselman, J.M. and C.A. Lewis. 1996. Habitat requirements of northern pike (*Esox lucius*). *Can. J. Fish. Aquat. Sci.* 53:161-174.

Charpy, R. 1948. Note au sujet de divers essais de repeuplements en brochets réalisés en Allemagne avant 1942. *Bull. Fr. Peche Piscic.* 148:113-120.

Chilcote, M.W, K.W. Goodson and M. Falcy. 2011. Reduced recruitment performance in natural populations of anadromous salmonids associated with hatchery-reared fish. *Can. J. Fish. Aquat. Sci.* 68:511-522.

Chimits, P. 1947. Note sur le repeuplement artificiel du brochet. *B. F. Peche Piscic.* 146:16-24.

Connelly, N.A., T.L. Brown and B.A. Knuth. 2000. Do anglers and fishery professionals think alike? *Fisheries.* 25:21-25.

Cowx, I.G. 1994. Stocking strategies. *Fish. Manag. Ecol.* 1:15-30.

Crane, D.P., L.M. Miller, J.S. Diana, J.M. Casselman, J.M. Farrell, K.L. Kapuscinski and J.K. Nohner. 2015. Muskellunge and northern pike ecology and management: important issues and research needs. *Fisheries* 40:258-267.

Cucherousset, J., J.-M. Paillisson, and J.-M. Roussel. 2007. Using PIT technology to study the fate of hatchery-reared YOY northern pike released into shallow vegetated areas. *Fish. Res.* 85:159-164.

Darimont, C.T., S.M. Carlson, M.T. Kinnison, P.C. Paquet, T.E. Reimchen and C.C. Wilmers. 2009. Human predators outpace other agents of trait change in the wild. *Proc. Natl. Acad. Sci. USA* 106:952-954.

Denys, G.P.J., A. Dettai, H. Persat, M. Hautecœur and P. Keith. 2014. Morphological and molecular evidence of three species of pikes *Esox* spp. (Actinopterygii, Esocidae) in France, including the description of a new species. *C. R. Biol.* 337:521-534.

Diana, N. 1983. Growth, maturation, and production of northern pike in three Michigan lakes. *Trans. Am. Fish. Soc.* 112:38-46.

Diana, N. and K. Smith. 2008. Combining ecology, human demands, and philosophy into the management of northern pike in Michigan. *Hydrobiologia* 601:125-135.

Dorow, M. 2005. Evaluierung der Hechtbesatzmaßnahmen im Peenestrom. pp 44-50. *In*: Mitteilungen der Landesforschungsanstalt für Landwirtschaft und Fischerei Mecklenburg-Vorpommern (eds.) *Beiträge zur Fischerei aus den Bereichen Binnenfischerei, Küstenfischerei und Aquakultur* - Heft 28. Germany: Gülzow-Prüzen.

Eckmann R., U. Gaedke, H.J. Wetzlar. 1988. Effects of climatic and density-dependent factors on year-class strength of Coregonus lavaretus in Lake Constance. *Can. J. Aquat. Sci.* 45: 1088-1093.

Edeline E, S.M. Carlson, L.C. Stige, I.J. Winfield, J.M. Fletcher, B.J. James, T.O. Haugen, A.L. Vøllestad and N.C. Stenseth. 2007. Trait changes in a harvested population are driven by a dynamic tug-of-war between natural and harvest selection. *Proc. Natl. Acad. Sci. USA* 104:15799-15804.

Edeline, E., T.O. Haugen, F.A. Weltzien, D. Claessen, I.J. Winfield, N.C. Stenseth and L.A. Vøllestad. 2010. Body downsizing caused by non-consumptive social stress severely depresses population growth rate. *Proc. R. Soc. B* 277:843-851.

Einfalt, L.M., D.B. Wojcieszak and D.H. Wahl.2013. Behavior, growth and habitat selection of hatchery esocids reared with artificial vegetation. *Trans. Am. Fish. Soc.* 142:345-352.

Elliott, J.M. 1994. *Quantitative Ecology and The Brown Trout*. Oxford University Press. Oxford.

Eschbach, E., A.W. Nolte, K. Kohlmann, P. Kersten, J. Kail and R. Arlinghaus. 2014. Population differentiation of zander (*Sander lucioperca*) across native and newly colonized ranges suggests increasing admixture in the course of an invasion. *Evol. Appl.* 7:555-568.

Eschbach, E., A.W. Nolte, K. Kohlmann, J. Kail, J. Alós, P. Kersten, S. Schöning and R. Arlinghaus. 2015. Genetische Vielfalt von Zander- und Hechtpopulationen in Deutschland. pp 32-37. *In*: R. Arlinghaus, E.-M. Cyrus, E. Eschbach, M. Fujitani, D. Hühn, F. Johnston, T. Pagel and C. Riepe (eds.). *Hand in Hand für eine nachhaltige Angelfischerei: Ergebnisse und Empfehlungen aus fünf Jahren praxisorientierter Forschung zu Fischbesatz und seinen Alternativen*. Berlin, Germany: Berichte des IGB.

FAO - Food and Agriculture Organization of the United Nations. 2005. *World inventory of fisheries. Stocking techniques for increased production*. World inventory of fisheries - Issues Fact Sheets. Rome, Italy: FAO Fisheries and aquaculture department. [online].

Farrell, J.M., J.V. Mead and B.A. Murry. 2006. Protracted spawning of St. Lawrence River northern pike (*Esox lucius*): simulated effects on survival, growth, and production. *Ecol. Freshw. Fish* 15:169-179.

Fayram, A.H., M.J. Hansen and T.J. Ehlinger. 2006. Influence of walleye stocking on angler effort in Wisconsin. *Hum. Dim. Wildlife* 11:129-141.

Flickinger, S.A. and J.H. Clark. 1978. Management evaluation of stocked northern pike in Colorado's small irrigation reservoirs. *Am. Fish. Soc. Special Publication* 11:284-291.

Franklin, D.R. and L.L. Smith. 1963. Early life history of the northern pike, *Esox lucius* L., with special reference to the factors influencing the numerical strength of year classes. *Trans. Am. Fish. Soc.* 92:91-110.

Fraser, D.J., L.K. Weir, L. Bernatchez, M.M. Hansen and E.B. Taylor. 2011. Extent and scale of local adaptation in salmonid fishes: review and meta-analysis. *Heredity* 106:404-420.

Gandolfi, A., C. Ferrari, B. Crestanello, M. Girardi, L. Lucentini and A. Meraner. 2017. Population genetics of pike, genus *Esox* (Actinopterygii, Esocidae), in Northern Italy: evidence for mosaic distribution of native, exotic and introgressed populations. *Hydrobiologia* X:1-20.

Garlock, T.M., E.V. Camp and K. Lorenzen. 2017. Using fisheries modeling to assess candidate species for marine fisheries enhancement. *Fish. Res.* 186:460-467

Gillen, A.L., R.A. Stein and R.F. Carline. 1981. Predation by pellet-reared tiger muskellunge on minnows and bluegills in experimental systems. *Trans. Am. Fish. Soc.* 110:197-209.

Gomes, L.C., C.A.R.M. Araujo-Lima, R. Roubach, A.R. Chippari-Gomes, L.N. Lopes and E.C. Urbinati. 2003. Effect of fish density during transportation on stress and mortality of juvenile tambaqui *Colossoma macropomum*. *J. World Aquacult. Soc.* 34:76-84.

Grimm, M.P. 1981a. The composition of northern pike (*Esox lucius* L.) population in four shallow waters in the Netherlands, with special reference to factors influencing 0+ pike biomass. *Fish. Manage.* 12:61-76.

Grimm, M.P. 1981b. Intraspecific predation as a principal factor controlling the biomass of northern pike (*Esox lucius* L.). *Fish. Manage.* 12:77-79.

Grimm, M.P. 1982. The evaluation of the stocking of pike fingerlings. *Hydrobiol. Bull.* 16:285-286.

Grimm, M.P. 1983. Regulation of biomasses of small (< 41 cm) northern pike (*Esox lucius*) with special reference to the contribution of individuals stocked as fingerlings (4-6 cm). *Fish. Manage.* 14:115-134.
Grimm, M.P. 1989. Northern pike (*Esox lucius* L.) and aquatic vegetation, tools in the management of fisheries and water quality in shallow waters. *Hydrobiol. Bull.* 23:59-65.
Grimm, M.P. and J. Backx. 1990. The restoration of shallow eutrophic lakes and the role of northern pike, aquatic vegetation and nutrient concentration. *Hydrobiologia* 200/201:557-566.
Grimm, M.P. 1994a. The influence of aquatic vegetation and population biomass on recruitment of 0+ and 1+ northern pike (*Esox lucius* L.). pp 226-234. *In*: I.G. Cowx (ed.). *Rehabilitation of freshwater fisheries*. Oxford, U.K.: Fishing News Bools.
Grimm, M.P. 1994b. The characteristics of the optimum habitat of northern pike (*Esox lucius* L.). pp 235-243. *In*: I.G. Cowx (ed.). *Rehabilitation of freshwater fisheries*. Oxford, U.K.: Fishing News Bools.
Grimm, M.P. and M. Klinge. 1996. Pike and some aspects of its dependence on vegetation. pp 125-156. *In*: J.F. Craig (ed.). *Pike: Biology and exploitation*. London: Chapman & Hall.
Grønkjær, P., C. Skov and S. Berg. 2004. Otolith-based analysis of survival and size-selective mortality of stocked 0+year pike related to time of stocking. *J. Fish Biol.* 64:1625-1637.
Guillerault, N., A. Martineau L'Hotis, F. Azémar, A. Compin and F. Santoul. 2012. *Etude des poissons carnassiers du Lot. Rapport final/version courte*. Cahors. France: FDPPMA 46.
Halverson, M.A. 2008. Stocking trends: a quantitative review of governmental fish stocking in the United States, 1931 to 2004. *Fisheries* 33:69-75.
Haugen, T.O., I.J. Winfield, L.A. Vøllestad, J.M. Fletcher, J.B. James and N.C. Stenseth. 2007. Density dependence and density independence in the demography and dispersal of pike over four decades. *Ecol. Monogr.* 77:483-502.
Hazlerigg, C.R.E., K., Lorenzen, P., Thorbek, J.R., Wheeler and C.R., Tyler. 2012. Density-dependent processes in the life history of fishes: evidence from laboratory populations of zebrafish *Danio rerio*. *PLoS ONE* 7(5) e37550.
Heuschmann, O. 1940. Die Hechtzucht. pp 23-199. *In*: R. Demoll and H.N. Maier (eds.). *Handbuch der binnenfischerei mitteleuropas*. Verlagsbuchhandlung Schweizbart'sche.
Hilborn, R. 1999. Confessions of a reformed hatchery basher. *Fisheries.* 24:31-33.
Hixon, M.A., D.W. Johnson and S.M. Sogard. 2014. BOFFFFs: on the importance of conserving old-growth age structure in fishery populations. *ICES J. Mar. Sci.* 71:2171-2185.
Huet, M. 1948. Esociculture: la production de brochetons. *Bull. Fr. Peche Piscic.* 148:121-124.
Hühn, D., K. Lübke, C. Skov and R. Arlinghaus. 2014. Natural recruitment, density-dependent juvenile survival, and the potential for additive effects of stock enhancement: an experimental evaluation of stocking northern pike (*Esox lucius*) fry. *Can. J. Fish. Aquat. Sci.* 71:1508-1519.
Hühn, D., K. Lübke, C. Skov, T. Pagel and R. Arlinghaus. 2015a. Ist Besatz mit Hechtbrut bzw. –Jungfischen in natürlich reproduzierenden Beständen fischereilich gesehen erfolgreich? pp 75-80. *In*: R. Arlinghaus, E.-M. Cyrus, E. Eschbach, M. Fujitani, D. Hühn, F. Johnston, T. Pagel and C. Riepe (eds.). *Hand in Hand für eine nachhaltige Angelfischerei: Ergebnisse und Empfehlungen aus fünf Jahren praxisorientierter Forschung zu Fischbesatz und seinen Alternativen.* Berlin, Germany: Berichte des IGB.
Hühn, D., E. Eschbach, R. Hagemann, T. Mehner, D. Bekkevold and R. Arlinghaus. 2015b. Ist Besatz mit Laichhechten in natürlich reproduzierenden Beständen fischereilich gesehen erfolgreich? pp 80-84. *In*: R. Arlinghaus, E.-M. Cyrus, E. Eschbach, M. Fujitani, D. Hühn, F. Johnston, T. Pagel and C. Riepe (eds.). *Hand in Hand für eine nachhaltige Angelfischerei: Ergebnisse und Empfehlungen aus fünf Jahren praxisorientierter Forschung zu Fischbesatz und seinen Alternativen.* Berlin, Germany: Berichte des IGB.

Huntingford, F.A. and C.G. De Leaniz. 1997. Social dominance, prior residence and acquisition of profitable feeding sites in juvenile Atlantic salmon. *J. Fish Biol.* 54:469-472.

Inskip, P.D. 1986. Negative associations between abundances of muskellunge and northern pike: evidence and possible explanations. *Am. Fish. Soc. Special Publication* 15:135-150.

Jacobsen, L., S. Berg and C. Skov. 2004. Management of lake fish populations and lake fisheries in Denmark: history and current status. *Fish. Manag. Ecol.* 11:219-224.

Jansen, T., R. Arlinghaus, T.D. Als and C. Skov. 2013. Voluntary angler logbooks reveal long-term changes in a lentic pike, *Esox lucius*, population. *Fish. Manag. Ecol.* 20:125-136.

Johnson, L.D. 1981. *Comparison of Muskellunge (Esox masquinongy) populations in a stocked lake and unstocked lake in Wisconsin, with notes on the occurrence of northern pike (Esox lucius).* Research Report 110. Madison, Wisconsin, USA: Wisconsin Department of Natural Resources.

Johnson, B.M., R. Arlinghaus and P.J. Martinez. 2009. Are we doing all we can to stem the tide of illegal fish stocking? *Fisheries* 34:389-394.

Johnston, F., B. Beardmore, C. Riepe, T. Pagel, D. Hühn and R. Arlinghaus. 2015. Kosten-Nutzen praxisüblicher Besatzmaßnahmen am Beispiel von Hecht und Karpfen. pp 95-111. *In*: R. Arlinghaus, E.-M. Cyrus, E. Eschbach, M. Fujitani, D. Hühn, F. Johnston, T. Pagel and C. Riepe (eds.). *Hand in Hand für eine nachhaltige Angelfischerei: Ergebnisse und Empfehlungen aus fünf Jahren praxisorientierter Forschung zu Fischbesatz und seinen Alternativen.* Berlin, Germany: Berichte des IGB.

Jones, D.J. 1963. *A history of Nebraska's fishery resources.* Paper 31. Lincoln, Nebaska, USA: Nebraska Game and Parks Commission Publications.

Kerr, S.J. and T.A. Lasenby. 2001. *Esocids stocking: an annotated bibliography and literature review.* Peterborough, Ontario, Canada: Ontario Ministry of Natural Resources.

Kerr, S.J. 2011. *Distribution and management of muskellunge in North America: an overview.* Peterborough, Ontario, USA: Ontario Ministry of Natural Resources.

Klein, M. 2011. Zur Sinnhaftigkeit von Hechtbesatz. *Fischer & Teichwirt* 62:252-253.

Krohn, D.C. 1969. *Summary of northern pike stocking investigations in Wisconsin.* Research Report 44. Madison, Wisconsin, USA: Wisconsin Department of Natural Resources.

Laikre, L., M.K. Schwartz, R. Waples, N. Ryman and the GeM Working Group. 2010. Compromising genetic diversity in the wild: unmonitored large-scale release of plants and animals. *Trends Ecol. Evol.* 26:520-529.

Larsen, P.F., M.M. Hansen, E.E. Nielsen, L.F. Jensen and V. Loeschcke. 2005. Stocking impact and temporal stability of genetic composition in a brackish northern pike population (*Esox lucius* L.), assessed using microsatellite DNA analysis of historical and contemporary samples. *Heredity* 95:136-143.

Launey, S., J. Morin, S. Minery and J. Laroche. 2006. Microsatellite genetic variation reveals extensive introgression between wild and introduced stocks, and a new evolutionary unit in French pike *Esox lucius* L. *J. Fish Biol.* 68:193-216.

Leggett, W.C. and E. DeBlois. 1994. Recruitment in marine fish: is it regulated by starvation and predation in egg and larval stages. *Neth. J. Sea Res.* 32:119-134.

Lorenzen, K. 1995. Population dynamics and management of culture-based fisheries. *Fish. Manag. Ecol.* 2:61-73.

Lorenzen, K. 1996a. A simple von Bertalanffy model for density- dependent growth in extensive aquaculture, with an application to common carp (*Cyprinus carpio*). *Aquaculture* 142:191-205.

Lorenzen, K. 1996b. The relationship between body weight and natural mortality in fish: a comparison of natural ecosystems and aquaculture. *J. Fish Biol.* 49:627-647.

Lorenzen, K. 2000. Allometry of natural mortality as a basis for assessing optimal release size in fish-stocking programmes. *Can. J. Fish. Aquat. Sci.* 57:2374-2381.

Lorenzen, K. and K. Enberg. 2002. Density-dependent growth as a key mechanism in the regulation of fish populations: evidence from among-population comparisons. *Proc. R. Soc. B* 269:49-54.

Lorenzen, K. 2005. Population dynamics and potential of fisheries stock enhancement: practical theory for assessment and policy analysis. *Philos. Trans. Roy. Soc. B* 360:171-189.

Lorenzen, K. 2006. Population management in fisheries enhancement: gaining key information from release experiments through use of a size-dependent mortality model. *Fish. Res.* 80:19-27.

Lorenzen, K. 2008. Fish population regulation beyond 'stock and recruitment': the role of density-dependent growth in the recruited stock. *Bull. Mar. Sci.* 83:181-196.

Lorenzen, K., K.M. Leber and H.L. Blankenship, 2010. Responsible approach to marine stock enhancement: an update. *Rev. Fish. Sci.* 18:189-210.

Lorenzen, K., M.C.M Beveridge and M. Mangel. 2012. Cultured fish: integrative biology and management of domestication and interactions with wild fish. *Biol. Rev.* 87:639-660.

Lorenzen, K. 2014. Understanding and managing enhancements: why fisheries scientists should care. *J. Fish Biol.* 85:1807-1829.

Maloney, J. and D.H. Schupp. 1977. *Use of winter rescue northern pike in maintenance stocking.* Fisheries Investigational Report 345. St. Paul, Minnesota, USA: Minnesota Department of Natural Resources

Margenau, T.L., S.P. AveLallemant, D. Giehtbrock and S.T. Schram. 2008. Ecology and management of northern pike in Wisconsin. *Hydrobiologia* 601:111-123.

Marshall, C.T. and K.T. Frank. 1999. Implications of density dependent juvenile growth for compensatory recruitment regulation of haddock. *Can. J. Fish. Aquat. Sci.* 56:356-363.

Mather, M.E., R.A. Stein and R.F. Carline. 1986. Experimental assessment of mortality and hyperglycemia in tiger muskellunge due to stocking stressors. *Trans. Am. Fish. Soc.* 115:762-770.

Mather, M.E. and D.H. Wahl. 1988. Comparative mortality of three esocids due to stocking stressors. *Can. J. Fish. Aquat. Sci.* 46:214-217.

McCarraher, D.B. 1957. The natural propagation of northern pike in small drainable ponds. *Prog. Fish Cult.* 19:185-187.

McGurk, M.D. 1986. Natural mortality of marine pelagic fish eggs and larvae: the role of spatial patchiness. *Mar. Ecol. Progr. Ser.* 37:227-242.

Mehner, T., R. Arlinghaus, S. Berg, H. Dörner, L. Jacobsen, P. Kasprzak, R. Koschel, T. Schulze, C. Skov, C. Wolter and K. Wysujack. 2004. How to link biomanipulation and sustainable fisheries management: a step-by-step guideline for lakes of the European temperate one. *Fish. Manag. Ecol.* 11:261-275.

Mickiewicz, M. and A. Wołos. 2012. Economic ranking of the importance of fish species to lake fisheries stocking management in Poland. *Arch. Pol. Fish.* 20:11-18.

Mickiewicz, M. 2013. Economic effectiveness of stocking lakes in Poland. *Arch. Pol. Fish.* 21:323-329.

Minns, C.K., R.G. Randall, J.E. Moore and V.W. Cairns. 1996. A model simulating the impact of habitat supply limits on northern pike (*Esox lucius*) in Hamilton Harbour, Lake Ontario. *Can. J. Fish. Aquat. Sci.* 53:20-34.

Myers, R.A. and N.G. Cadigan. 1993a. Is juvenile natural mortality in marine demersal fish variable? *Can. J. Fish. Aquat. Sci.* 50:1591-1598.

Myers, R.A. and N.G. Cadigan, 1993b. Density-dependent juvenile mortality in marine demersal fish. *Can. J. Fish. Aquat. Sci.* 50:1576-1590.

Näslund, J. And J.I. Johnsson. 2014. Environmental enrichment for fish in captive environments: effects of physical structures and substrates. *Fish Fish.* 17:1-30.

Nilsson, P.A., C. Skov and J.M. Farrell. 2008. Current and future directions for pike ecology and management: a summary and synthesis. *Hydrobiologia* 601:137-141.

Pedreschi, D., M. Kelly-Quinn, J. Caffrey, M. O'Grady and S. Mariani. 2014. Genetic structure of pike (*Esox lucius*) reveals a complex and previously unrecognized colonization history of Ireland. *J. Biogeogr.* 41:548-560.

Persson, L., A.M. de Roos and A. Bertolo. 2004. Predicting shifts in dynamics of cannibalistic field populations using individual-based models. *Proc. R. Soc. B* 71:2489-2493.

Persson, L., A. Bertolo and A.M. de Roos. 2006. Temporal stability in size distributions and growth rates of three *Esox lucius* L. populations. A result of cannibalism? *J. Fish Biol.* 69:461-472.

Pierce, R.B. and C.M. Tomcko. 2005. Density and biomass of native northern pike populations in relation to basin-scale characteristics in north-central Minnesota lakes. *Trans. Am. Fish. Soc.* 134:231-241.

Pierce, R.B. 2010. Long-term evaluations of length limit regulations for northern pike in Minnesota. *N. Am. J. Fish. Manage.* 30:412-432.

Pierce, R.B. 2012. Northern Pike: Ecology, Conservation, and Management History. Minneapolis, Minnesota, USA: University of Minnesota Press.

Post, J.R., E.A. Parkinson and N.T. Johnston. 1999. Density dependent processes in structured fish populations: interaction strengths in whole-lake experiments. *Ecol. Monogr.* 69: 155-175.

Post, J.R., M. Sullivan, S. Cox, N.P. Lester, C.J. Walters, E.A. Parkinson, A.J. Paul, L. Jackson and B.J. Shuter. 2002. Canada's recreational fisheries: the invisible collapse? *Fisheries* 27:6-15.

Price, E.O. 2002. *Animal domestication and behavior*. Wallingford, UK: CABI Publishing.

Raat, A.J.P. 1988. Synopsis of biological data on the northern pike *Esox lucius* Linnaeus, 1758. FAO Fisheries Synopsis 30.

Rasmussen, G. and P. Geertz-Hansen. 1998. Stocking of fish in Denmark. *In*: I.G. Cowx (ed.). *Stocking and introduction of fish*. London, UK: Fishing News Books. pp. 14-21.

Rhodes, J.S. and T.P. Quinn. 1998. Factors affecting the outcome of territorial contest between hatchery and naturally reared coho salmon parr in the laboratory. *J. Fish Biol.* 53:1220-1230.

Rogers, M., M.S. Allen, P. Brown, T. Hunt, W. Fulton and B.A. Ingram. 2010. A simulation model to explore the relative value of stock enhancement versus harvest regulations for fishery sustainability. *Ecol. Model.* 221:919-926.

Rothschild, B.J. 2000. Fish stocks and recruitment: the past thirty years. *ICES J. Mar. Sci.* 57, 191-201.

Rust, A.J., J.S. Diana, T.L. Margenau and C.J. Edwards. 2002. Lake characteristics influencing spawning success of muskellunge in northern Wisconsin lakes. *N. Am. J. Fish. Manage.* 22:834-841.

Schreckenbach, K. 2006. Förderung von Hechten und Zandern. *VDSF Schriftenreihe Fischerei und Gewässerschutz* 2:21-28.

Secor, D.H. and E.D. Houde. 1998. Use of larval stocking in restoration of Chesapeake Bay striped bass. *ICES J. Mar. Sci.* 55:228-239.

Sharma, C.M. and R. Borgstrøm. 2008. Increased population density of pike *Esox lucius*: a result of selective harvest of large individuals. *Ecol. Freshw. Fish* 17:590-596.

Shepherd, J.G. and D.H. Cushing. 1980. A mechanisms for density-dependent survival of larval fish as the basis of a stock–recruitment relationship. *ICES J. Mar. Sci.* 39:160-167.

Simon, J. and H. Dörner. 2014. Survival and growth of European eels stocked as glass- and farm-sourced eels in five lakes in the first years after stocking. *Ecol. Freshw. Fish* 23:40-48.

Smith, E.P. 2002. BACI design. pp 141-148. *In*: A.H. El-Shaarawi and W.W. Piegorsch (eds.). *Encyclopedia of environmetrics Volume 1*. Chichester, U.K.: Wiley.

Skov, C. 2002. *Stocking 0 + pike (Esox lucius L.) as a tool in the biomanipulation of shallow eutrophic lakes*. Ph.D. thesis. Copenhagen, Denmark: University of Copenhagen.

Skov, C., O. Lousdal, P.H. Johansen and S. Berg. 2003. Piscivory of 0+ pike (*Esox lucius* L.) in a small eutrophic lake and its implication for biomanipulation. *Hydrobiologia* 506:481-487.

Skov, C., L. Jacobsen, S. Berg, J. Olsen and D. Bekkevold. 2006. *Udsætning af geddeyngel i danske søer: Effektvurdering og Perspektivering*. DFU-rapport 161-06. Silkeborg, Denmark: Danmarks Fiskeriundersøgelser.

Skov, C. and P.A. Nilsson. 2007. Evaluating stocking of YOY pike *Esox lucius* as a tool in the restoration of shallow lakes. *Freshw. Biol.* 52:1834-1845.

Skov, C., A. Koed, L. Baastrup-Spohr and R. Arlinghaus. 2011. Dispersal, growth, and diet of stocked and wild northern pike fry in a shallow natural lake, with implications for the management of stocking programs. *N. Am. J. Fish. Manage.* 31:1177-1186.

Snow, H.E. 1974. *Effects of stocking northern pike in Murphy Flowage*. Technical Bulletin 79. Madison, Wisconsin, USA: Wisconsin Department of Natural Resources.

Souchon, Y. 1980. Effet de la densité initiale de peuplement sur la survie et la croissance du brochet (*Esox lucius* L.) élevé jusqu'au stade de brocheton (45 jours). pp 309-316. *In*: R. Billard (ed.). *La pisciculture en étang*. Paris, France: Institut National de la Recherche Agronomique publications.

Stein, R.A., R.F. Carline, and R.S. Hayward 1981. Largemouth bass predation and physiological stress sources of mortality for stocked tiger muskellunge. *Trans. Am. Fish. Soc.* 110:604-612.

Sutela, T., P. Korhonen and K. Nyberg. 2004. Stocking success of newly hatched pike evaluated by radioactive strontium (^{85}Sr) marking. *J. Fish Biol.* 64:653-664.

Sutter, D.A.H., C.D. Suski, D.P. Philipp, T. Klefoth, D.H. Wahl, P. Kersten, S.J. Cooke and R. Arlinghaus. 2012. Recreational fishing selectively captures individuals with the highest fitness potential. *Proc. Natl. Acad. Sci.* 109:20960-20965.

Szczepkowski, M, Z. Zakęś, A. Kapusta, B. Szczepkowska, M. Hopko, S. Jarmołowicz, A. Kowalska, M. Kozłowski, K. Partyka, I. Piotrowska and K. Wunderlich. 2012. Growth and survival in earthen ponds of different sizes of juvenile pike reared in recirculating aquaculture systems. *Arch. Pol. Fish.* 20:267-274.

Szendrey, T.A. and D.H. Wahl. 1995. Effect of feeding experience on growth, vulnerability to predation, and survival of esocids. *N. Am. J. Fish. Manage.* 15:610-620.

Tammi, J., M. Appelberg, U. Beier, T. Hesthagen, A. Lappalainen and M. Rask. 2003. Fish status survey of Nordic lakes: effects of acidification, eutrophication and stocking activity on present fish species composition. *Ambio.* 32:98-105.

Thériault, V., G.R. Moyer, L.S. Jackson, M.S. Blouin and M.A. Banks. 2011. Reduced reproductive success of hatchery coho salmon in the wild: insights into most likely mechanisms. *Mol. Ecol.* 20:1860-1869.

van Donk, E., R.D. Gulati and M.P. Grimm. 1989. Food web manipulation in Lake Zwemlust: positive and negative effects during the first two years. *Hydrobiol. Bull.* 23:19-34.

van Poorten, B., R. Arlinghaus, K. Daedlow and S. Haertel-Borer. 2011. Social-ecological interactions, management panaceas, and the future of wild fish. *Proc. Natl. Acad. Sci. USA.* 108:12554-12559.

van Kooten, T., J. Andersson, P. Byström, L. Persson and A.M. de Roos. 2010. Size at hatching determines population dynamics and response to harvesting in cannibalistic fish. *Can. J. Fish. Aquat. Sci.* 67:401-416.

von Lindern, E. and H.J. Mosler. 2014. Insights into fisheries management practices: using the theory of planned behavior to explain fish stocking among a sample of Swiss anglers. *PLoS ONE* 9(12) e115360.

Vuorinen, P.J., K. Nyberg K. and H. Lehtonen. 1998. Radioactive strontium (^{85}Sr) in marking newly hatched pike and success of stocking. *J. Fish Biol.* 52:268-280.

Wahl, D.H. 1999. An ecological context for evaluating the factors influencing muskellunge stocking success. *N. Am. J. Fish. Manage.* 19:238-248.

Wahl, D.H., L.M. Einfalt and D.B. Wojcieszak. 2012. Effect of experience with predators on the behavior and survival of muskellunge and tiger muskellunge. *Trans. Am. Fish. Soc.* 141:139-146.

Walters, C.J. and F. Juanes. 1993. Recruitment limitation as a consequence of natural selection for use of restricted feeding habitats and predation risk taking by juvenile fishes. *Can. J. Fish. Aquat. Sci.* 50:2058-2070.

Walters, C.J. and J.R. Post. 1993. Density-dependent growth and competitive asymmetries in size-structured fish populations: a theoretical model and recommendations for field experiments. *Trans. Am. Fish. Soc.* 122:34-45.

Walters, C.J. and S.J. Martell. 2004. *Fisheries ecology and management.* Princeton, New Jersey, USA: Princeton University Press.

Welcomme, R.L. and D.M. Bartley. 1998. Current approach to the enhancement of fisheries. *Fish. Manag. Ecol.* 5:351-382.

Wesloh, M.L. and D.E. Olson. 1962. *The growth and harvest of stocked yearling northern pike* (*Esox lucius*) *in a Minnesota lake*. Fisheries Investigational Report 242. St. Paul, Minnesota, USA: Minnesota Department of Natural Resources

Wingate, P.J. and J.A. Younk. 2007. A program for successful muskellunge management: a Minnesota success story. *Environ. Biol. Fish.* 79:163-169.

Wright, R.M. and N. Giles. 1987. The survival, growth and diet of pike fry (*Esox lucius*) stocked at different densities in experimental ponds. *J. Fish Biol.* 30:617-629.

Wysujack, K., U. Laude, K. Anwand and T. Mehner. 2001. Stocking, population development and food composition of pike *Esox lucius* in the biomanipulated Feldberger Haussee (Germany) – implications for fisheries management. *Limnologica* 31:45-51.

Zhao T., G. Grenouillet, T. Pool, L. Tudesque and J. Cucherousset. 2016. Environmental determinants of fish community structure in gravel pit lakes. *Ecol. Freshw. Fish* 25:412-421.

Zorn, S.A., T.L. Margenau, J.S. Diana and C.J. Edwards. 1998. The influence of spawning habitat on natural reproduction of Muskellunge in Wisconsin. *Trans. Am. Fish. Soc.* 127:995-1005.

AUTHOR ADDRESSES

**Corresponding author*: [1]EcoLab, Université de Toulouse, CNRS, INPT, UPS, 118 route de Narbonne, 31062 Toulouse, France; Email: nicolas.guillerault@univ-tlse3.fr

and

Station d'Ecologie Expérimental du CNRS à Moulis, Lab. USR 2936, 09100 Moulis, France

[2]Institute of Inland Fisheries Potsdam-Sacrow, Im Königswald 2, 14469 Potsdam, Germany; Email:daniel.huehn@ifb-potsdam.de

and

Leibniz-Institute of Freshwater Ecology and Inland Fisheries, Müggelseedamm 310, 12587 Berlin, Germany

[3]CNRS, Université Toulouse III Paul Sabatier, ENFA, UMR5174 EDB (Laboratoire Évolution & Diversité Biologique), 118 route de Narbonne, 31062 Toulouse, France; Email: julien.cucherousset@univ-tlse3.fr

[4]Leibniz-Institute of Freshwater Ecology and Inland Fisheries, Müggelseedamm 310, 12587 Berlin, Germany; Email: arlinghaus@igb-berlin.de

and

Humboldt-Universität zu Berlin, Philippstrasse 13, Haus 7, 10115 Berlin, Germany

[5]National Institute of Aquatic Resources, Technical University of Denmark, Vejlsøvej 39, 8600, Silkeborg, Denmark; Email: ck@aqua.dtu.dk

Chapter 10

Habitat Restoration – A Sustainable Key to Management

Olof Engstedt[*1], *Jonas Nilsson*[2] *and Per Larsson*[3]

10.1 INTRODUCTION

A vegetated spawning habitat with contiguous vegetated nursery areas has been shown to be crucial for optimal pike (*Esox lucius* Linnaeus, 1758) reproduction (Bry 1996, Casselman and Lewis 1996). The considerable importance of vegetation in the reproduction and early life of the pike has long been recognized, and the presence of live or decaying vegetation is crucial for a successful spawning, as pike eggs and yolk-sac larvae are adapted to adhere to vegetation (Raat 1988, see also chapter 3). Numerous plant taxa have been observed in pike spawning grounds, ranging from flooded terrestrial vegetation to completely submerged aquatic vegetation (Bry 1996, Casselman and Lewis 1996, Grimm and Klinge 1996). Pike prefer to spawn in shallow vegetated areas, which heat up rapidly in spring (Bry 1996). These kinds of optimal spawning areas could be found in temporary wetlands created by a high spring water run-off. Some pike populations migrate up streams to spawn in such areas. This phenomenon is known both from river-lake systems (Miller et al. 2001) and river-brackish sea systems (Müller 1986, Nilsson 2006, Larsson et al. 2015, chapter 5).

Wetlands, i.e., potential pike spawning habitats, have decreased on a global scale over the last two centuries (Casselman and Lewis 1996, Paludan et al. 2002).

**Corresponding author*: [1]Linnaeus University, Department of Biology and Environmental Science, Barlastgatan 11, SE-39182 Kalmar, Sweden; Email: olof.engstedt@lnu.se
and
Swedish Anglers Association, Svartviksslingan 28, SE-16739 Bromma, Sweden;
Email: olof.engstedt@sportfiskarna.se

[2]Linnaeus University, Department of Biology and Environmental Science, Barlastgatan 11, SE-39182 Kalmar, Sweden; Email: jonas.nilsson@lnu.se

[3]Linnaeus University, Department of Biology and Environmental Science, Barlastgatan 11, SE-39182 Kalmar, Sweden; Email: per.larsson@lnu.se

This decrease is mainly caused by agricultural activities, but is to some extent also due to urban development (Casselman and Lewis 1996, Paludan et al. 2002, Sandstrom et al. 2005). In some areas in Sweden, up to 90% of the wetlands have been drained (Fig. 10.1). Watercourses, drainage and flooding naturally work together, establishing relative equilibrium in watershed character and distribution. Human interference, by draining wetlands and agricultural land as well as straightening and deepening streams and rivers, has caused faster transport of water downstream and thereby increased problems with transport of nutrients and sediments. Climate change may also affect the hydrological regimes in freshwater streams (Graham 2004). Models point to a higher winter run-off as a result of increased winter precipitation and lower spring and summer discharge due to the reduced winter snow storage and an increase of evapotranspiration (Neumann 2010, Wake 2012). As a consequence of both extensive draining and the climate

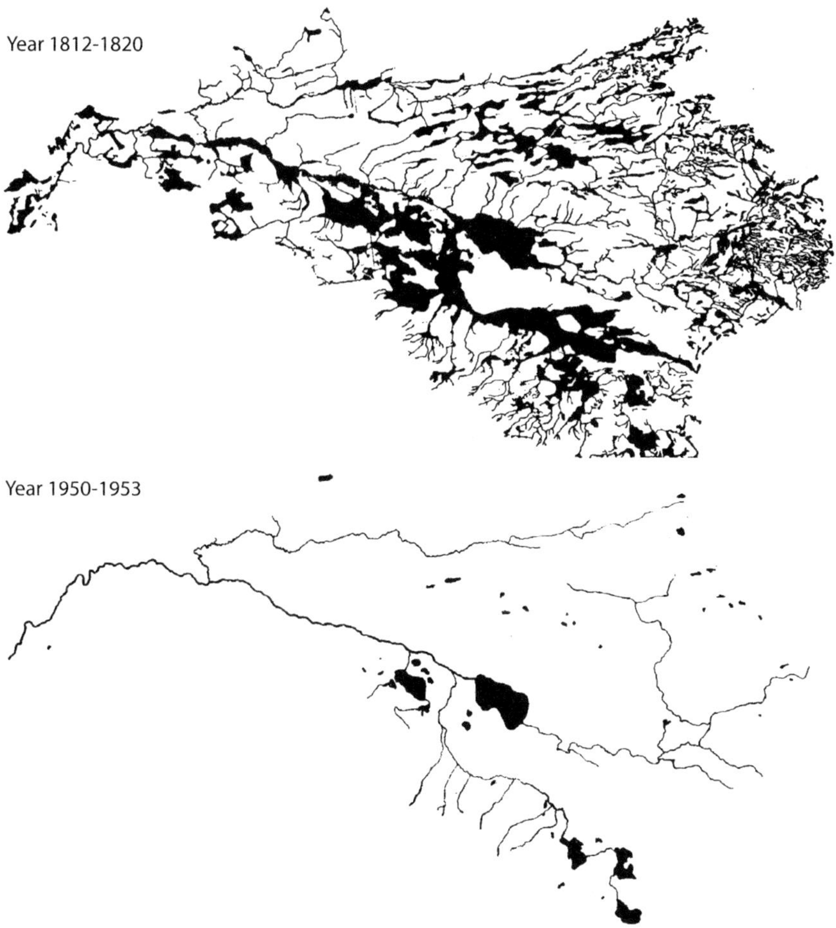

FIGURE 10.1 Example from River Kävlingeån, southern Sweden, showing the loss of wetlands in the last hundred years (Wolf 1956). Most river systems today in populated areas have a similar negative change in hydrology.

effects, potential pike spawning areas may be filled up with water and flooded earlier than before. On the other hand, wetlands may also be drained more quickly, as the timing of peak flows would occur earlier in the season. This general decrease in retention time and subsequent loss of wetlands may have a severe negative impact on pike populations that are dependent on these habitats for reproduction (Threinen 1969, Casselman and Lewis 1996, Nilsson et al. 2014). A most obvious negative effect is when former spawning and nursery areas are no longer flooded and natural succession leads to overgrowth by terrestrial vegetation. Less obvious is when spawning areas are filled with water but dry out before the fertilized pike eggs hatch or before the larvae have emigrated (Raat 1988). The drastic historical loss of wetlands has created a serious need for habitat restoration measures of former pike reproduction habitats to strengthen weakened pike populations.

10.2 HABITAT RESTORATION

There has been a long tradition of improving pike populations through habitat restoration in North America (e.g. Fago 1977, Raat 1988, Casselman and Lewis 1996, Farrell et al. 2016), and restoration efforts have also been carried out in Sweden and Denmark during the last two decades (Skov et al. 2006, Nilsson et al. 2014). Casselman and Lewis (1996) reviewed habitat restoration programmes conducted in the Great Lakes Basin, North America. The authors investigated both spawning and nursery requirements of fry, and created a classification system that ranks important environmental variables for pike habitat. Vegetation was considered to be the most important habitat variable for spawning, with moderately dense grasses and sedges at the highest rank, and cattails and floating aquatic plants at the lowest rank. Water level was the second most important spawning habitat variable, and a water depth of 10-70 cm was considered optimal. Water level should fluctuate, with increasing levels prior to spawning, be stable until larva switch to exogenous feeding and until they start to migrate from spawning areas, and then gradually decrease during the emigration period. A highly productive spawning habitat should also be protected from dominant winds and receive sunlight from the south or west, providing rapid warming in early spring. Waterways connecting the spawning habitat to the river/lake are necessary both for the migrating spawners and for a successful emigration of juveniles. For nursery habitats, the proximity to the spawning habitat was ranked as being of the highest relative importance, together with a dense cover (40-90%) of submerged and emergent vegetation. Restoration projects to improve pike spawning habitat differ considerably in, for example, dredging channels, excavating riverbanks, planting of emergent plants, installing water control structures and excluding other fish species (Casselman and Lewis 1996). Quite noticeable is that only a few of these restoration projects were classified as successful, implying that it is of great importance to carry out restorations where the physical and biological requirements for a successful spawning and nursery are met (Nilsson et al. 2014, Larsson et al. 2015). This means identifying the bottlenecks for conservation for each specific place of interest as well as trying to identify the actual reason behind the low pike population densities. If spawning areas are lacking, then focus should be put on this. If nursery areas are in poor shape or missing, perhaps restoring vegetation is the better possible measure.

10.2.1 Spawning Habitat Restoration

Efforts to restore pike populations through habitat manipulation have usually focused on spawning habitat – for example, controlling the water level to create managed pike spawning marshes (Franklin and Smith 1963, Forney 1968, Fago 1977). Such managed marshes were a way to increase pike stocks in lakes in North America. A water-controlling structure dammed up natural marshes or ponds that were filled with water, in some cases naturally but in many cases by pumping water into the marsh using electric pumps. Adult fish were caught in the lake prior to spawning (or taken from some other lake) and released in the managed spawning area. Water levels were held high to enhance the development of juvenile pike in sizes often ranging 30-100 mm (Williams and Jacob 1971, Beyerle and Williams 1973, Fago 1977). The number of fingerling pike released from the managed spawning marshes differed widely, from 1 fingerling to about 12,000 per year and hectare (Fago 1977). Although these managed spawning marshes produced high numbers of pike in some instances, stocking them with adult fish that were close to spawning involved a large amount of human intervention in the spawning. In ideal habitat restoration measures, human intervention should be kept to a minimum. The efforts to create and maintain managed rearing marshes have decreased since the 1980s. In 1985 in Minnesota, USA, 143 controlled spawning areas were operated, but by 1995 they had decreased to 28 (Pierce 2012).

In Casselman and Lewis (1996), several restoration projects from the Great Lakes Basin are described, focusing on both spawning and nursery habitats. Recent restoration measures have however not been of the 'managed wetland' type. Habitat improvement measures involved creating channels in cattail marshes and controlling water levels, as well as excluding carp to allow re-establishment of vegetation. Habitat creation measures involved planting emergent vegetation, excavating upland channels and re-establishing wetland by creating lagoons close to the riverbank. Specific examples include the creation of two 1 ha lagoons by excavating the riverbank, and these lagoons were used extensively by spawning pike the year after restoration. Another example comprised dredging a 480 m long 1.2-2.5 m deep and 10-15 m wide channel complex. These environments created both spawning and nursery areas, and juvenile pike were found in the channel complex the summer following the dredging (Casselman and Lewis 1996). However, several of the applied measures are still not fully evaluated, or seem to not increase juvenile pike production to any greater extent (Casselman and Lewis 1996), and we return to such shortcomings in the discussion later in this chapter (section 10.3).

A large multi-partnered project in the St Lawrence River in the Thousand Islands area, North America has tried to develop, implement and evaluate conservation strategies for enhancing native fish species with emphasis on spawning habitat for pike, but also other fish species. An altered hydrological regime in the St Lawrence River has increased the growth of *Typha* species (Farrell et al. 2010), which is unfavourable for pike recruitment in several ways. First, the thick *Typha* stands are not optimal pike spawning vegetation, and, second, they impede access to more suitable sedge meadows (Farrell 2001). Recently, restoration projects have focused on excavating channels through the dense *Typha* stands, creating migration pathways for pike to more suitable spawning habitat (Crane et al. 2015, Farrell et al. 2016, Fig. 10.2). Overall, the channels were of benefit for the fish community, and pike used the channels for migration but also for spawning. Using historic aerial photos, the channels could be placed where former migration pathways existed prior to the invasion of the thick *Typha* stands, improving the connectivity to spawning grounds and between other bays (Farrell et al. 2016).

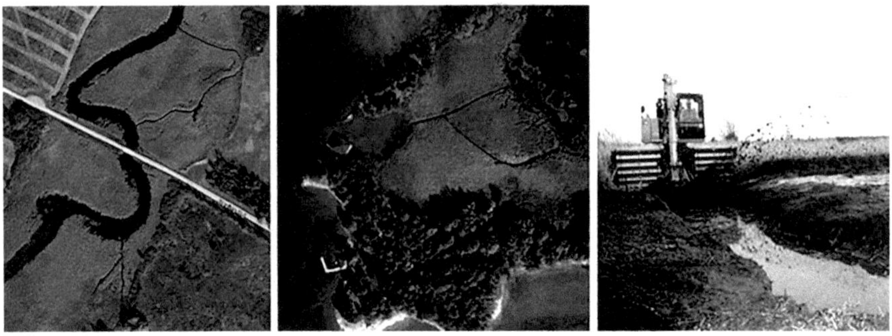

FIGURE 10.2 Former routes for spawning migrating pike are recreated by excavating channels in vegetation, in this case thick monocultures of *Typha* spp. Aquatic excavator (right) is owned and operated by the US Fish and Wildlife Service New York Field Office in Cortland New York, USA.

Under the same project, so-called spawning pools were created in two areas. These are small water pockets, connected with channels to the river, which restore open water areas in overgrown areas (Fig. 10.3). The pools are deeper and wider than the channels, and have a longer lifespan than channels. After three years' monitoring, they required little or no maintenance to function. It is, however, thus far unknown at what temporal interval spawning pools may require management to ensure continued provision of favourable habitat for pike reproduction (Farrell et al. 2016).

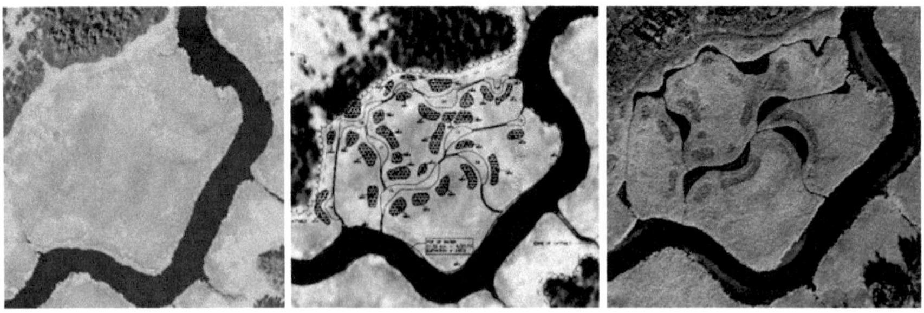

FIGURE 10.3 Spawning pool complex at French Creek, St Lawrence River, NY, USA: before the project (left), planning (centre), and after implementation (right). Implementation was managed by Ducks Unlimited Inc. and design provided by Sarah Fleming

Creating the spawning pools is more complicated and expensive than excavating channels, but the overall benefit to fish and other aquatic organisms will likely justify their continued construction and use. Results on production of age-0 pike were evaluated in various restoration measures, where spawning pools gave the highest production per hectare (about 650 age-0 pike ha^{-1}), compared to excavating channels (about 50 age-0 pike ha^{-1}) and managed marshes (about 3 age-0 pike ha^{-1}) during the three years of monitoring (Farrell et al. 2016).

In the Danish capital Copenhagen, measures to increase pike densities in lakes in order to improve water quality have been of interest to managers and researchers (Skov et al. 2006). Especially three artificial lakes in central Copenhagen have very few natural shallow water areas, since they are heavily modified depending on their use through history (e.g. water reservoir and fortification). Specifically, there is no real littoral zone with shallow, sloping bottom. Instead, a man-made concrete wall constitutes the perimeter of the lake (Fig. 10.4). One of the restoration measures was to establish shallow water areas in order to enhance juvenile pike survival and to stimulate spawning. This was done by introducing natural vegetation on three 250 m^2 areas with a depth varying between 0.1-0.5 m (Fig. 10.4). Another method was to create artificial spawning habitats by placing spruce trees in the shallow near-shore areas. Investigations revealed that the spruce trees were used as nursery areas (see also Skov and Berg 1999) and the introduced vegetation was used as spawning and nursing habitats, as both eggs and juvenile pike were found, albeit in low numbers, probably reflecting the relative low number of spawners in the lake. However, the longevity of the introduced vegetated areas was limited, i.e., only a few years, probably due to intense grazing by herbivorous birds (Skov, personal communication). Hence, such measures clearly need maintenance in order to prevail.

FIGURE 10.4 Introduced vegetated areas in Northern basin of Lake Sortedam, Copenhagen, Denmark during construction in March/April 2005 (left) and a year after establishment (July 2006). (Photos by Søren Berg and Christian Skov)

Swedish researchers initiated a research project in the mid 2000s to describe and scientifically evaluate the restoration of pike spawning areas in south-east Sweden (Nilsson et al. 2014). The most successful restoration measure, with the highest pike production, was when grass meadows were flooded with high spring water, creating a temporary wetland. Such temporarily flooded areas have previously been described in Threinen (1966), and are referred to as *pike factories*.

10.2.1.1 Pike Factories – Natural Wetland Restoration

This following section about pike factories and other recent wetland restorations is from the Baltic Sea. Before presenting these examples, a short introduction about the pike in the Baltic Sea is provided in order to better understand that local measures can have a large impact in terms of management.

Sympatric populations of pike in the Baltic Sea have at least two different reproductive strategies. They either spawn in the brackish sea or migrate to spawn in freshwater streams with adjacent wetlands and small lakes (Müller 1986, Westin and Limburg 2002, Engstedt et al. 2010, Rohtla et al. 2012). The populations share a common coastal habitat during the majority of their life cycle (Tibblin et al. 2015). Along the Swedish Baltic Sea coast there are about equal shares of non-anadromous and anadromous pike, as revealed by strontium quotients in otoliths from pike caught in the sea (Engstedt et al. 2010). This may well differ between coastal areas in the Baltic, but the results show that both strategies are important in the coastal stock (Rohtla et al. 2012). The spawning migration of pike to freshwater areas is extensive along the Baltic Sea coast (Müller and Berg 1982, Engstedt et al. 2010, Rohtla et al. 2012, Tibblin et al. 2016). The main reason for pike being anadromous is probably the same as for anadromous salmonids, i.e., enhanced opportunity for predation avoidance and foraging for larvae and juveniles in the freshwater system, and thus an increased probability of success when leaving for the coastal brackish areas (Gross et al. 1988). The migration to specific streams and wetlands has resulted in natal homing in the Baltic pike (Tibblin et al. 2016, chapter 5). Being iteroparous, individual pike return to the same spawning areas every year. Such homing behaviour leads to spawning-area-specific gene pools, and, consequently, to unique populations (Engstedt et al. 2013, Larsson et al. 2015, Tibblin et al. 2015). The spawning migration may include several hundred mature pike within a period of a few weeks, and commonly starts with smaller males arriving in the freshwater area, followed by females and larger males (Tibblin et al. 2016). The spawning of migratory pike takes place in March to May, with the earliest spawning activities in the south (Müller 1986, Larsson et al. 2015). Spawning time can, however, differ on a smaller spatial scale, i.e., between nearby coastal streams, and may well be an adaptation to various environmental variables, such as hydrological regimes and temperature (Larsson et al. 2015, chapter 3 and 5)

Pike populations have generally decreased in the Baltic Sea during recent decades (Nilsson et al. 2004, Ljunggren et al. 2010). Studies have suggested that the decline is caused by poor recruitment, which in turn could be related to intense egg predation by three-spined sticklebacks (*Gasterosteus aculeatus* Linnaeus, 1758), eutrophication and negative physical changes in spawning and nursery habitats (Nilsson et al. 2004, Nilsson 2006, Lehtonen et al. 2009, Ljunggren et al. 2010). The identified recruitment problem gave rise to a need for measures to improve the pike populations in the area, and a joint project was initiated by the Swedish Environmental Protection Agency and Swedish universities in 2006.

In 2007-2008, wetlands were restored or established within 200-700 m distance from the sea in association with three coastal freshwater streams in south-east Sweden. All three streams are known for spawning migration of anadromous pike in numbers ranging from several hundred to over a thousand adult pike (Engstedt et al. 2013). Each stream is characterized by a mean annual water discharge of less than 0.5 $m^3 s^{-1}$. The streams are 2-3 m wide and nutrient-enriched by agricultural activities. The restoration measures were different depending on stream-specific physical properties.

In one of the restorations, the stream was widened to approximately 75 m by digging a shallow pond covering 1.5 ha and removing 200 m of the northern bank (Fig. 10.5). The new wetland had an average water depth of 0.5 m, with gently sloping littoral areas, but almost completely lacked vegetation. In the second restoration, an existing wetland, mainly covered with dense stands of common reed (*Phragmites australis* Cavanilles 1840), was restored (Fig. 10.5). The reeds were removed together with the root mat and approximately 0.2 m of sediment; the sediment was deposited in the wetland in the form of small islands. After restoration, the total area available for fish increased from 1.5 to 3 ha at an average depth of about 1 m. Besides the reed, submerged plants such as *Elodea canadensis* Michaux 1803 and *Myriophyllum spicatum* Linnaeus 1758 dominated the wetland. The year after restoration, the total cover of submerged macrophytes decreased by more than 90% and no new vegetation established during the study of the former reed area. As it turned out, this initial lack of vegetation in both of these two restorations was not beneficial for the pike reproduction.

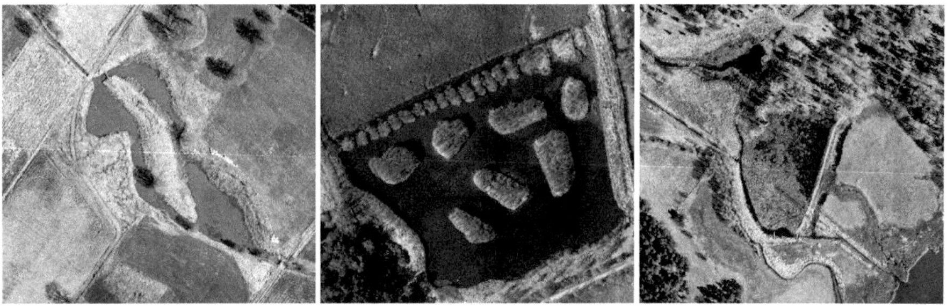

FIGURE 10.5 The three restoration measures in the research project in southeast Sweden. On the left the widened stream, in the middle the wetland with removed vegetation, and to the right the flooded meadow – the pike factory.

In the third restoration, the *pike factory*, approximately 3.2 ha of formerly flooded grasslands, was restored by constructing an adjustable weir, including a fish passage, near the outlet to the sea. Before the measure was implemented, about 0.5 ha of the area, dominated by *Phragmites australis*, was naturally flooded on an annual basis, although these flooding events generally only occurred over shorter periods. Digging was restricted to a new streambed that meandered through the restored wetland. The grasslands, consisting mainly of grass and sedge hummocks, were inundated with stream water (Fig. 10.5). The depth was 0.2-0.5 m in the flooded areas, and 0.5-1.5 m in the stream. To maintain dominant terrestrial vegetation, the wetland was flooded only from March to June each year. Soon after restoration, submerged vegetation started to grow in parts of the new stream.

Juvenile pike production was measured a year before each of the three restoration measures (Fig. 10.5) and consecutive years after in all restored wetlands. The third restoration example, the pike factory, was the only habitat restoration project that had a significantly positive effect on pike production. The total annual emigration increased from about 3,000 juveniles before restoration to > 100,000 in the years after restoration (Nilsson et al. 2014). Five years after the restoration, emigration was > 300,000 (Nilsson 2013, Larsson et al. 2015). The fact that increased production followed directly after restoration indicated that production was not limited by the number of spawning pike, but was released from limitation by the newly created spawning and nursery habitats in the pike factory (Casselman and Lewis 1996). Before restoration, the two other restored wetlands had similar numbers of emigrating pike (Nilsson et al. 2014), and after restoration, the total numbers of emigrating juveniles decreased slightly. Both of these restoration projects increased the wetland area, though the loss of vegetation did not initially favour reproduction (e.g. Casselman and Lewis 1996). It is possible that pike production may increase in both these wetlands after some years if vegetation is established.

The distribution of yolk-sac pike larvae in different habitats in the pike factory was surveyed after restoration. Yolk-sac larvae were found in the shallowest areas (0.1-0.4 m deep) in all vegetated habitats (i.e., flooded hummocks of sedges and grasses, emerged *Phragmites australis* and submerged vegetation), but most abundant and frequently found around the flooded vegetation. No yolk-sac larvae were found in the non-vegetated areas, indicating the importance of vegetation during the spawning period (see also chapter 3). When the larvae turned from endogenous to exogenous feeding they were equally abundant among the flooded terrestrial vegetation and *Phragmites* belts (Nilsson et al. 2014). Flooded and emerged vegetation offer a complex and rich habitat for development of zooplankton suitable as food for juvenile pike (Wright and Giles 1987, Bry 1996). Surveys in the pike factory revealed that the highest densities of copepods and cladocerans were found in the flooded vegetation and *Phragmites* belts (Nilsson et al. 2014). The stomach contents of juvenile pike showed that zooplankton was the major food source for pike up to 35 mm long. For larger pike juveniles, insects and fish were the dominant food source (Nilsson et al. 2014). Emigration of pike larvae starts immediately after they become free swimming and 80-95% of the juveniles emigrate at a size of less than 6 cm and by 1 month of age. This early emigration probably represents an adaption to naturally decreasing water levels in the watercourses and wetlands, but may also be a way to avoid cannibalism (Nilsson et al. 2014). Cannibalism has been shown for pike as small as 21 mm, but more likely this behaviour occurs in pike over 35 mm in length (Fago 1977).

The main goal with constructing pike factories is to strengthen the adult population. In 2013, five years after restoration, the adult population migrating up for spawning in the pike factory was counted by catching in fyke nets and population size was calculated with mark/recapture. Before restoration the population that came for spawning consisted of about 1400 adult pike. Five years after the restoration the spawning population was estimated at 3100 adult individuals (Nilsson 2013, Larsson et al. 2015).

Inspired by the results from the pike factory research by Nilsson et al. (2014), the Swedish Anglers' Association started to construct wetlands for pike reproduction along the Swedish Baltic coast. After an extensive survey of possible restoration areas, it was realized that the need for wetland restoration was urgent and extensive. The surveys were methodical, visiting streams and ditches from north to south in each county searching for wetlands and possible spawning areas. A majority of the streams visited had been deepened and straightened, with a resulting loss of pike spawning habitats. Former wetlands could be identified in the field and through map investigations, and suitable restoration objects could hereby be chosen. The Swedish Anglers' Association started establishing pike factories in 2010, and by 2016 more than 20 pike factories had been constructed. The restoration work is an ongoing process, with 10 fulltime employees working only with wetlands for pike reproduction, funded from both Swedish authorities and the European Union. The cost for restoring wetlands on the east coast of Sweden was estimated to at least 360 million Swedish kronor (SEK) – around 30 million euros (EUR) (Ljunggren et al. 2012). The cost of constructing a pike factory as described in this chapter depends on the size of the wetland, the length of the embankment, weir construction and soil characteristics. The wetlands constructed by the Swedish Anglers' Association have a mean size of about 3 ha and a mean total cost of 30,000 euros.

10.2.1.2 How to Construct a Pike Factory

A pike factory is a sheltered and shallow vegetated area commonly composed of terrestrial vegetation that is temporarily flooded during spring with water from a connected stream or ditch (Threinen 1966, Nilsson et al. 2014, Figs 10.6 and 10.7). These flooded areas create optimal conditions for pike spawning and for the pike's early life stages (Forney 1968, Bry 1996, Casselman and Lewis 1996). Depending on the size of the watercourse there are two slightly different ways to construct a pike factory. In small watercourses, like agricultural ditches and brooks with a limited spring water flow, the pike factory can be created by letting the entire watercourse flow through it. In streams and rivers, i.e., larger watercourses, the pike factory is better placed in proximity to the watercourse with only a part of the water flowing through. Pike factories are often constructed in former natural wetlands, but it is also possible to use other areas as long as the criteria for an optimal pike spawning and nursery habitat can be fulfilled (Casselman and Lewis 1996, Nilsson et al. 2014, Larsson et al. 2015). The following section is based on our experiences from Nilsson et al. (2014) and more than 20 other recent constructed pike factories in Sweden.

To optimize juvenile pike production, water depth should be 10-70 cm, averaging 20-50 cm. The water level should gradually increase prior to spawning and then be stable or slowly decrease until the fry emigrate to the lake or sea. These requirements could be met by building embankments along the wetland as needed, and by placing an adjustable weir at the outlet structure. The slope profile should generally be flat with the top of the embankment not exceeding more than 0.5-1 m, resulting in a 'natural' object in the landscape once vegetation is re-established. It is also important that the embankment is waterproof to maintain the function of the pike factory and simultaneously prevent adjacent farmlands from flooding. Vegetation should be re-established on the embankment as soon as possible to prevent erosion and thereby the spreading of sediment in the wetland.

FIGURE 10.6 Pictures from Swedish pike factories. Top left: Flooded terrestrial vegetation (Photo: Jonas Nilsson). Centre: Weir and fishway (Photo: Lars Vallin). Top right: Spawning pike (Photo: Olof Engstedt).

The weir or embankment must be constructed with fish passages allowing pike and other migrating fish to safely swim over or around it, and then continue to move freely up and down the watercourse (Figs 10.6 and 10.7). Fish passages can be constructed in many ways and we recommend following national guidelines when available. It is important that the fish passage is made lower than both the weir and the embankment, giving a way for excess run-off water to leave. The variable-height weir could be built in concrete, metal, rock, wood or a combination of these materials. At pike factories, the water level is controlled by this outlet structure, which can be a box with removable wooden stop logs, allowing water levels to be adjusted easily (Figs 10.6 and 10.7). Weirs must be designed to pass the maximum probable flow, and it is important that the top of the weir is placed a little lower than the wetland embankment to function as a spillway in periods of high water run-off. The construction with embankment, weir and fish passage has also been shown to efficiently prevent smaller egg predators, like the three-spined stickleback (Nilsson 2006), from entering the spawning area. The weir could also be made of soil and mud, re-creating a former wetland with no water-level-controlling function. This means that the water in the wetland stays for a longer time. This could be a functioning solution in cases where the wetland is remote and cannot be maintained with a controllable weir, or where the use of the land does not demand a dry up in summertime.

FIGURE 10.7 A schematic design of a pike factory. Left: filled up during spawning. Right: dried up during summertime (Illustration: Erik Ohlsson)

Digging in the pike factory should be restricted to prevent loss of spawning habitat and spreading of sediment in the spawning areas, since pike eggs are highly sensitive to siltation. In watercourses that carry high amounts of particles it may be necessary to trap these particles before they accumulate on spawning vegetation or are deposited on pike eggs. This could be done by gently digging a deeper zone in the inlet area where sediment can accumulate. In some pike factories, it could be necessary to dig a new streambed that meanders through the restored wetland. The new stream bed should be about 0.5-1.0 m deeper than the surrounding flooded area, allowing establishment of submerged aquatic vegetation as well as stands of emergent vegetation like *Phragmites* sp., which could serve as nursery areas for larger juveniles. This deeper area also provides adult pike with a place to hide from predators before spawning or in periods when spawning activity is low.

To maintain the function of a pike factory, and trigger the emigration of remaining fry, one or more of the stop logs must be removed in late spring. To maximize the number of emigrating fry, this should be done directly after fry become free swimming, or if the purpose is to produce fewer but larger juveniles, stop logs can be removed later in the season. More stop logs are successively removed as the water level in the wetland decreases and this is done continuously over a few weeks until the entire flooded area is drained. The former flooded area will then dry out during summer and autumn, and preferably be grazed by cattle, preventing it from being overgrown by other unwanted terrestrial vegetation. The stop logs are then put back in the weir during winter so the water level will rise and fill the wetland before the following spawning period. The drainage of the wetland is important to avoid establishment of reed (*Phragmites* sp.) or cattail (*Typha* sp.) that are often seen in both constructed and natural wetlands. In the case of a natural weir with no water controlling function, juvenile pike will be trapped in the wetland until coming periods of precipitation. This is natural and will lead to cannibalism and thereby lower numbers of emigrating juveniles compared to if the weir is controllable, allowing timed release of juveniles.

10.2.1.3 Other Ways to Enhance Pike Production

In some cases where a pike factory is impossible to construct, other measures could be put in place to increase juvenile production. In many watercourses and wetlands constructed for nutrient retention or irrigation, passage is often physically obstructed for pike (see also chapter 5). Sometimes the only necessary measure to increase pike production in such cases is to construct a passage for the fish. Migration routes can also be impassable due to thick stands of reed, so cutting this vegetation or gently digging an access route to spawning areas could also increase pike production. The cutting of the reed should be done under the water level leaving the remaining reed straws inundated in order to achieve a lasting effect (Russell and Kraaij 2008). If shallow vegetated areas are established when wetlands are constructed for nutrient retention, such wetlands could also be used to produce juvenile pike (Nilsson et al. 2014). Improving shallow areas and thereby potential pike spawning habitats could also be done in already existing wetlands, for example by re-establishing vegetation and altering embankment slopes. Modifying traditional agricultural ditches or brooks into two-stage ditches (Powell et al. 2007) could favour pike reproduction. A two-stage ditch consists of a deeper section in the middle of the ditch, with shallow vegetated sections suitable for pike spawning along both sides. Constructing a two-stage ditch requires more excavation than traditional ditch maintenance, but the ditch maintains original drainage and will benefit both agriculture and ecological functions (Powell et al. 2007). Creating shallow shoreline inundations along watercourses could be another useful measure if two-stage ditches are impossible to construct.

10.2.1.4 Controlling Water Levels in Lake and River Systems

It is quite common that lakes have some kind of regulation of the water level in their outflow. This could for instance be related to hydropower or for other human consumption. It has been shown in many cases that fish populations are negatively affected by water levels going up and down irregularly (Gaboury and Patalas 1984, Sutela et al. 2002). For instance, if the water level in a lake drops one metre during the pike spawning season, the majority of the year class of pike recruits could be wasted. If water in a regulated lake could be maintained at a stable and high level during the pike spawning period (often around March-April), the production of pike juveniles could be maximized. In a large Finnish study (Keto et al. 2008), 49 out of 105 regulated lakes had such low water levels that suitable sedge spawning habitats were completely dry during pike spawning time. Several of the lakes in the study even had a lower level of water when spring flood arrived compared to summer water levels. In contrast, all unregulated lakes in the study had inundated sedge zones during spawning time. Several Finnish lakes are regulated, primarily to meet the needs of hydropower production and flood protection. Water is used during wintertime to produce electricity, resulting in the lowest annual water levels when pike are spawning. Extensive research projects have been launched to investigate and eliminate the negative effects of water level regulation (Marttunen et al. 2001). In the St Lawrence River in North America, negative long-term water level management of the river denies pike access to spawning areas, forcing them to spawn in deeper waters (as deep as 2-3 m), substantially reducing hatching success (Farrell 2001). A study analysing historical vegetation in the tributary of the St Lawrence river showed diverse and flooded habitat in 1948, before the regulation of the river. Post-regulation photos reveal that various species of *Typha* were dominant instead, which is unfavourable for pike spawning (Farrell et al. 2010). Maintaining lakes and rivers unregulated, or minimizing the negative effects of water control in regulated lakes and rivers, could therefore be viewed as habitat restoration beneficial for pike production in the system.

10.2.2 Nursery, Juvenile and Adult Pike Habitat Restoration

Habitat requirements vary between the different life stages of pike. Juvenile pike grow fast and increase their movement and activity, and their need for habitat therefore increases (Casselman 1996). This increase in activity exposes juvenile pike to potential predation (including cannibalism), which is why vegetation cover is necessary (Grimm and Klinge 1996, see chapters 2 and 3). Juvenile pike typically prefer submerged vegetation mixed with emergent and floating vegetation. There is a strong correlation between increasing depth and increasing total length of pike during their first year of life (Casselman 1996). This correlation was corroborated in the established shallow water areas in the inner lakes of Copenhagen, where electrofishing revealed larger juveniles caught at greater depth (Skov et al. 2006). When pike grow, their preference for vegetation also changes from emergent, submergent and floating for young pike, to more submerged vegetation for adult pike (Casselman 1996). Further information regarding the pikes vegetation needs during different life stages is described in chapter 3 of this book.

Habitat restoration for pike has in general focused on spawning habitats. However, the understanding of pike life-stage-specific vegetation dependencies should allow life-stage-specific restoration incentives. Modifying and optimizing spawning habitat is feasible since the areas are relatively small. Manually modifying vegetation and habitat for 0+ up to adult pike is more difficult and expensive, and perhaps only possible if the system is very small. The efforts to achieve positive changes in vegetation and habitat for pike should be holistic with a system-level approach, preferably integrated with incentives for e.g. nutrient retention. This may for instance be achieved by creating wetlands in the inflow of lakes, or by creating protective zones close to or in streams to reduce the effects of nutrient runoff from agriculture. Grimm (1989) recommended a reduction in external nutrient loading to shallow European lakes in order to enhance conditions for native aquatic vegetation suitable for 0+ pike. It was moreover stressed that submerged vegetation should not be cut during June and July, to avoid loss of habitat ideal for age-0 pike. Emergent vegetation should also be conserved as it provides protection and cover for age-0 and age-1 pike in late autumn, winter and early spring when submerged vegetation is limited (Grimm 1989).

Restoration of macrophytes suitable for pike nursery areas can also be approached by improving water quality in eutrophicated lakes through various types of biomanipulation. Manipulation measures include removing cyprinids, with the idea of increasing zooplankton densities and grazing on phytoplankton, aiming for increased water transparency and thereby securing growth opportunity for submerged vegetation (Mehner et al. 2004, chapter 11). Similarly, stocking lakes with pike could reduce cyprinid densities via increased predation, and create a top-down trophic cascade that alters system composition (see chapters 8 and 11).

There is a generally high level of human exploitation of shallow water areas possibly suitable as pike spawning and nursery areas in lakes, rivers and brackish waters such as the Baltic Sea (Lotze et al. 2006, Sundblad and Bergström 2014). A recent study showed that shoreline constructions (e.g. marina development, construction of jetties or dredging) are concentrated to spawning and nursery areas of pike. The degradation rate of these habitats was calculated at 0.5% of available habitat per year and about 1% in areas close to larger cities (Sundblad and Bergström 2014). Avoiding such habitat destruction is essential to secure successful spawning and recruitment of pike.

10.3 DISCUSSION AND FUTURE PERSPECTIVES

Constructing pike factories is an efficient way of strengthening pike populations if spawning habitat is limited. It is a way to restore former spawning areas and could be applied in all water systems where pike are found. The key feature distinguishing this measure from others is an increased retention of water combined with a flooding of terrestrial vegetation (Nilsson et al. 2014). Constructing pike factories to improve pike populations has similarities with the work done in North American rearing marshes in terms of vegetation type, stable water level, and high growth of fry (Franklin and Smith 1963, Forney 1968, Fago 1977), and both types of restored environments are drained during the summer. The pike factory, however, is a more natural way to produce fry compared to rearing marshes, where adult pike are usually stocked for spawning (Williams and Jacob 1971, Fago 1977). In the Swedish example, the mature pike, induced by natural cues such as day length, water discharge, or temperature, migrate into the spawning areas from coastal zones. After spawning and embryo development, the fry may emigrate from the pike factory at their own inclination governed by ecological and environmental variables. In this way, natural selection is maintained for e.g. survival, growth, and timing of migration to the sea (Nilsson et al. 2014). Fago (1977) summarized data on pike production from managed rearing marshes in six states of the United States during the 1960s and 1970s. With data from 50 rearing marshes with a total of 87 production years, the average annual number of juveniles produced was 1042 juveniles ha^{-1} (highest 11823 and lowest 1 individual ha^{-1}). The average size of the produced pike at the time of draining the marshes was 63 mm, with a size range 18-155 mm (Fago 1977). Pike production in natural marshes in the United States is lower than in the managed spawning and rearing marshes (Fago 1977). The reason for better production in the rearing marshes is most likely due to a combination of maintaining high water levels and eliminating predators and competitors (Beyerle and Williams 1973). The estimated average production of pike juveniles in the pike factory described in Nilsson et al. (2014) varied between 22,000 and 30,000 ha^{-1} in the two years following restoration. Five years after restoration, production exceeded 75,000 ha^{-1} – way above the highest production for rearing marshes reported by Fago (1977) or compiled in Raat (1988). However, the average total length of fry at emigration given in Nilsson (2014) was much shorter. The smaller size of emigrating juveniles from the pike factory was probably due to the continuous outflow of pike juveniles, whereas in the closed marshes juveniles were retained and allowed to emigrate later and therefore at larger sizes. The lower production and larger average size in rearing marshes was probably due to cannibalism, which is a major mortality factor affecting total juvenile production (Fago 1977, Skov and Koed 2004, chapter 6). The use of managed rearing marshes demands intense human intervention with water level maintenance and stocking with adult spawning fish. Such efforts could work where other measures are not applicable, while restoring to more natural situations demanding little or preferably no human intervention and incorporating natural flow regimes should be prioritized (Poff et al. 1997).

Other restoration measures, such as excavating channels in dense *Typha* stands and/or creating connected spawning pools, seem to also improve spawning and nursery habitats and pike success (Farrell et al. 2016). One concern with such measures is whether the altered habitat is self-sustaining or requires regular maintenance. If maintenance proves necessary, it is central to develop management recommendations for both construction and maintenance for best practice.

To enhance pike recruitment, nursery areas are more important than spawning areas, and should always be considerably larger. Nursery areas for 0+ and 1+ pike are, however, more difficult to manage than spawning areas (Casselman and Lewis 1996). Efficient recruitment depends on survival of pike larvae into older size classes. Casselman and Lewis (1996) describe a comprehensive study of 22-year classes of pike in the Bay of Quinte, Lake Ontario. Interestingly, year class strengths were not correlated with spring water elevation, which has been shown in other work (Johnson 1957). The variation in water level was relatively extreme in the 22-year period, whereas year class strengths were more related to summer temperature associated with young pike and nursery habitat. Minns et al. (1996) showed similar results in a modelling study of habitat limits of pike, indicating that fry, juvenile and adult habitats are more limiting than spawning habitat. Sundblad et al. (2014) came to a similar conclusion in a study of perch (*Perca fluviatilis* Linnaeus, 1758) and pikeperch (*Sander lucioperca* Linnaeus, 1758). The availability of suitable habitats for various life stages, nursery areas in particular, may limit the production of fish (Sundblad et al. 2014). Casselman and Lewis (1996) further discuss that restoration of spawning habitat is important, albeit not the major limiting factor for pike in the Great Lakes. Restoring macrophyte habitat and cover is more important, although no effective and easy restoration method seems to exist to achieve this. Macrophyte cover provides protection in nursery areas for juvenile pike but also for various species of prey. The optimal vegetation density for pike nursery habitats ranges from 40-90% (Casselman and Lewis 1996). There is, however, no doubt that the extensive loss of marshes and wetlands in the Great Lakes area has almost depleted pike recruitment in certain areas. Restoration of wetlands is necessary in many areas in order to regain healthy pike populations (Casselman and Lewis 1996). In the example with pike factories from Sweden, Nilsson et al. (2014) showed that the returning spawning population went from 1400 to 3100 adults five years after restoration. This indicates that the spawning habitat restoration had a substantial effect on that pike population. The pike population on the coast of the brackish Swedish Baltic Sea had decreased dramatically during the 20 years prior to construction of those pike factories (Ljunggren et al. 2010). As nursery areas for 0+ and 1+ pike probably did and still exist, focusing on spawning habitats in freshwater wetlands close to the sea should thus enhance the opportunity for pike recovery. As much of such restored spawning habitat also provides nursery habitat for 0+ pike, recreating wetlands for pike is highly recommended.

Homing has been demonstrated for many pike populations in river, sea and lake habitats (Miller et al. 2001, Engstedt et al. 2013, Tibblin et al. 2016, chapter 5). Such homing behaviour enhances the success of spawning-habitat restoration. Restoration projects should be initiated in wetlands where pike formerly used to spawn, even if only a few per cent of the original population remain in the system. The homing behaviour ensures that the recruits return to the spawning site, creating the potential to build up a functional population in only a few years. Furthermore, the natal homing of anadromous pike that has been shown in the Baltic Sea, and probably in other systems as well, creates unique subpopulations (Larsson et al. 2015). This is of great importance and must be taken in consideration when it comes to management of pike populations.

Wetlands optimized for nutrient removal from rivers are often placed near the river mouths to maximize retention effects (Paludan et al. 2002). The wetlands retain phosphorous, adsorbed to particles and precipitated, and also reduce nitrogen by microbial processes transforming nitrate to nitrogen gas. The construction of wetlands in rivers entering the Baltic Sea has the potential to reduce eutrophication of the Baltic Sea, and the need to intensify wetland construction has been identified (Jansson and Dahlberg 1999, Paludan et al. 2002). A wetland for nutrient removal could in most cases quite easily be combined with pike factory characteristics. Both nitrogen reduction and pike recruitment are higher in shallow areas that heat up quickly, enabling fast hatching of pike larvae and enhancing the nitrification process. Nutrients caught in the wetland also provide the base for high food production for juvenile pike (Casselman and Lewis 1996). Pike factories may also favour other migrating fish species such as roach (*Rutilus rutilus* Linnaeus, 1758) and ide (*Leuciscus idus* Linnaeus, 1758) and may improve biodiversity and ecosystem functions (de Groot et al. 2002).

According to Casselman and Lewis (1996), successful habitat restoration must include clear objectives, measurable criteria and a well-planned assessment programme. For most restoration measures, however, proper evaluation is missing or not fully planned. Restoration measures should therefore include evaluation of whether all objectives are met. Casselman and Lewis (1996) further state that most restoration projects seem to favour both spawning and nursery habitats, sometimes probably unintentionally. This creates possible difficulties for mechanistic assessment of the restoration effort, depending on the original bottleneck for production of pike (Casselman and Lewis 1996). Moreover, follow-up surveys of young-of-the-year pike is difficult, since they are often caught in very low or extremely varying numbers (Casselman 1996). Such difficulties for restoration assessment should preferably be overcome in future work.

Protection of existing and functioning spawning and nursery areas should, if possible, precede restoration attempts. Such protection includes creating natural reserve areas, or prohibiting fishing during spawning. Conservation and protection work involves various organizations, and securing the long-term financing, competence and local connection in the collaboration between e.g. conservation organizations, administrative boards, municipalities and land owners is crucial for establishing self-sustaining conservation measures such as wetlands for pike reproduction.

Future research on habitat restoration should focus on identifying bottlenecks for pike recruitment. Reviewed knowledge in this chapter suggests that spawning habitat is commonly the limiting factor in degraded systems. In systems with functioning spawning habitat, nursery habitats for 0+ and 1+ pike seem to be the limiting factor. Further identification of limiting factors should be pursued in comparisons between healthy and depressed pike populations.

REFERENCES CITED

Beyerle, G.B. and J.E. Williams. 1973. Contribution of northern pike fingerlings raised in a managed marsh to pike population of an adjacent lake. *Pro. Fish Cul.* 35(2):99-103.

Bry, C. 1996. Role of vegetation in the life cycle of pike. pp 45-67. *In*: J.F. Craig (ed.). *Pike: Biology and exploitation.* London: Chapman & Hall.

Casselman, J.M. 1996. Age, growth and environmental requirements of pike. pp 69-102. *In*: J.F. Craig (ed.). *Pike: Biology and exploitation.* London: Chapman & Hall.

Casselman, J.M. and C.A. Lewis. 1996. Habitat requirements of northern pike (*Esox lucius*). *Can. J. Fish. Aquat. Sci.* 53(Suppl. 1):161-174.

Crane, D.P., L.M. Miller, J.S. Diana, J.M. Casselman, J.M. Farrell, K.L. Kapuscinski and J.K. Nohner. 2015. Muskellunge and Northern Pike Ecology and Management: Important Issues and Research Needs. *Fisheries* 40(6):258-267.

de Groot, R.S., M.A. Wilson and R.M.J. Boumans. 2002. A typology for the classification, description and valuation of ecosystem functions, goods and services. *Ecol. Econ.* 41(3):393-408.

Engstedt, O., P. Stenroth, P. Larsson, L. Ljunggren and M. Elfman. 2010. Assessment of natal origin of pike (*Esox lucius*) in the Baltic Sea using Sr:Ca in otoliths. *Environ. Biol. Fishes* 89(3-4):547-555.

Engstedt, O., R. Engkvist and P. Larsson. 2013. Elemental fingerprinting in otoliths reveals natal homing of anadromous Baltic Sea pike (*Esox lucius* L.). *Ecol. Freshw. Fish.* 23(3):313-321.

Fago, D.M. 1977. Northern pike production in managed spawning and rearing marshes. *In*: Technical bulletin No.96: Wisconsin Department of Natural Resources. Medison, WI.

Farrell, J.M. 2001. Reproductive Success of Sympatric Northern Pike and Muskellunge in an Upper St. Lawrence River Bay. *Trans. Am. Fish. Soc.* 130(5):796-808.

Farrell, J.M., B.A. Murry, D.J. Leopold, A. Halpern, M.B. Rippke, K.S. Godwin and S.D. Hafner. 2010. Water-level regulation and coastal wetland vegetation in the upper St. Lawrence River: inferences from historical aerial imagery, seed banks, and Typha dynamics. *Hydrobiologia* 647(1):127-144.

Farrell, J.M., J.P. Leblanc, B.L. Brown and G.A. Avruskin. 2016. The St. Lawrence River Fish Habitat Conservation Strategy: Evaluation of Habitat Enhancements and Development of Novel Restoration Approaches. *In*: Fish Enhancement and Mitigation Research Fund Project. Final Report: US Fish and Wildlife Service, New York Field Office, Cortland NY: 317 p.

Forney, J. 1968. Production of young northern pike in a regulated marsh. *New York Fish Game J.* 15(2):143-154.

Franklin, D. and L. Smith. 1963. Early life history of the northern pike (*Esox lucius* L.) with special reference to the factors influencing the numerical strength of year classes. *Trans. Am. Fish. Soc.* 92:91-110.

Gaboury, M.N. and J.W. Patalas. 1984. Influence of Water Level Drawdown on the Fish Populations of Cross Lake, Manitoba. *Can. J. Fish. Aquat. Sci.* 41(1):118-125.

Graham, L.P. 2004. Climate change effects on river flow to the Baltic Sea. *Ambio* 33(4-5):235-41.

Grimm, M.P. 1989. Northern pike (*Esox lucius* L.) and aquatic vegetation, tools in the management of fisheries and water quality in shallow waters. *Hydrobiol. Bull.* 23(1): 59-65.

Grimm, M.P. and M. Klinge. 1996. Pike and some aspects of its dependence on vegetation. pp 125-156. *In*: J.F. Craig (ed.). *Pike: Biology and exploitation*. London: Chapman & Hall.

Gross, M.R., R.M. Coleman and R.M. McDowall. 1988. Aquatic productivity and the evolution of diadromous fish migration. *Science* 239(4845):1291-1293.

Jansson, B.O. and K. Dahlberg. 1999. The environmental status of the Baltic Sea in the 1940s, today, and in the future. *Ambio* 28(4):312-319.

Johnson, F.H. 1957. Northern Pike Year-Class Strength and Spring Water Levels. *Trans. Am. Fish. Soc.* 86(1):285-293.

Keto, A., A. Tarvainen, M. Marttunen and S. Hellsten. 2008. Use of the water-level fluctuation analysis tool (Regcel) in hydrological status assessment of Finnish lakes. *Hydrobiologia* 613(1):133.

Larsson, P., P. Tibblin, P. Koch-Schmidt, O. Engstedt, J. Nilsson, O. Nordahl and A. Forsman. 2015. Ecology, evolution, and management strategies of northern pike populations in the Baltic Sea. *Ambio* 44(3):451-461.

Lehtonen, H., E. Leskinen, R. Selen and M. Reinikainen. 2009. Potential reasons for the changes in the abundance of pike, *Esox lucius*, in the western Gulf of Finland, 1939-2007. *Fish. Manag. Ecol.* 16(6):484-491.

Ljunggren, L., A. Sandstrom, U. Bergstrom, J. Mattila, A. Lappalainen, G. Johansson, G. Sundblad, M. Casini, O. Kaljuste and B.K. Eriksson. 2010. Recruitment failure of coastal predatory fish in the Baltic Sea coincident with an offshore ecosystem regime shift. *ICES J. Mar. Sci.* 67(8):1587-1595.

Ljunggren, N., O. Engstedt, N. Hellenberg, S. Söderman and J. Norlin. 2012. Swedish Anglers Association. Conservation need for pike and perch along the Swedish Baltic coast. *Local report* 2012:5 71 pp [In Swedish].

Lotze, H.K., H.S. Lenihan, B.J. Bourque, R.H. Bradbury, R.G. Cooke, M.C. Kay, S.M. Kidwell, M.X. Kirby, C.H. Peterson and J.B.C. Jackson. 2006. Depletion, Degradation, and Recovery Potential of Estuaries and Coastal Seas. *Science* 312(5781):1806-1809.

Marttunen, M., S. Hellsten and A. Keto. 2001. Sustainable development of lake regulation in Finnish lakes. *Vatten* 57:29-37.

Mehner, T., R. Arlinghaus, S. Berg, H. Dörner, L. Jacobsen, P. Kasprzak, R. Koschel, T. Schulze, C. Skov, C. Wolter and K. Wysujack. 2004. How to link biomanipulation and sustainable fisheries management: a step-by-step guideline for lakes of the European temperate zone. *Fish. Manag. Ecol.* 11(3-4):261-275.

Miller, L.M., L. Kallemeyn and W. Senanan. 2001. Spawning-site and natal-site fidelity by northern pike in a large lake: Mark-recapture and genetic evidence. *Trans. Am. Fish. Soc.* 130(2):307-316.

Minns, C.K., R.G. Randall, J.E. Moore and V.W. Cairns. 1996. A model simulating the impact of habitat supply limits on northern pike, *Essox lucius*, in Hamilton Harbour, Lake Ontario. *Can. J. Fish. Aquat. Sci.* 53(S1):20-34.

Müller, K. and E. Berg. 1982. Spring migration of some anadromous fresh-water fish species in the northern Bothnian Sea. *Hydrobiologia* 96(2):161-168.

Müller, K. 1986. Seasonal anadromous migration of the pike (*Esox-Lucius* L) in coastal areas of the northern Bothnian Sea. *Arch. Hydrobiol.* 107(3):315-330.

Neumann, T. 2010. Climate-change effects on the Baltic Sea ecosystem: A model study. *J. Mar. Syst.* 81(3):213-224.

Nilsson, J., J. Andersson, P. Karas and O. Sandstrom. 2004. Recruitment failure and decreasing catches of perch (*Perca fluviatilis* L.) and pike (*Esox Lucius* L.) in the coastal waters of southeast Sweden. *Boreal Environ. Res.* 9:295-306.

Nilsson, J. 2006. Predation of northern pike (*Esox lucius* L.) eggs: A possible cause of regionally poor recruitment in the Baltic Sea. *Hydrobiologia* 553:161-169.

Nilsson, J. 2013. Followup of the pike factory at Kronobäck in Mönsterås Municipality spring 2013. *Linnaeus University Report* (2013:10): 7 p [in Swedish].

Nilsson, J., O. Engstedt and P. Larsson. 2014. Wetlands for northern pike (*Esox lucius* L.) recruitment in the Baltic Sea. *Hydrobiologia* 721:145-154.

Paludan, C., F.E. Alexeyev, H. Drews, S. Fleischer, A. Fuglsang, T. Kindt, P. Kowalski, M. Moos, A. Radlowki, G. Stromfors, V. Westberg and K. Wolter. 2002. Wetland management to reduce Baltic sea eutrophication. *Water Sci. Technol.* 45(9):87-94.

Pierce, R.B. 2012. *Northern Pike: Ecology, Conservation, and Management History.* University of Minnesota Press. 208 pp.

Poff, N.L., J.D. Allan, M.B. Bain, J.R. Karr, K.L. Prestegaard, B.D. Richter, R.E. Sparks and J.C. Stromberg. 1997. The Natural Flow Regime. *Bioscience* 47(11):769-784.

Powell, G.E., A.D. Ward, D.E. Mecklenburg and A.D. Jayakaran. 2007. Two-stage channel systems: Part 1, a practical approach for sizing agricultural ditches. *J. Soil Water Conserv.* 62(4):277-286.

Raat, A. 1988. *Synopsis of Biological Data on the Northern pike Esox lucius Linnaeus, 1758. FAO, ROME*: 178 pp.

Rohtla, M., M. Vetemaa, K. Urtson and A. Soesoo. 2012. Early life migration patterns of Baltic Sea pike *Esox lucius*. *J. Fish Biol.* 80(4):886-893.

Russell, I.A. and T. Kraaij. 2008. Effects of cutting Phragmites australis along an inundation gradient, with implications for managing reed encroachment in a South African estuarine lake system. *Wetl. Ecol. Manag.* 16(5):383-393.

Sandström, A., B.K. Eriksson, P. Karas, M. Isaeus and H. Schreiber. 2005. Boating and navigation activities influence the recruitment of fish in a Baltic Sea archipelago area. *Ambio* 34(2):125-130.

Skov, C. and S. Berg. 1999. Utilization of natural and artificial habitats by YOY pike in a biomanipulated lake. *Hydrobiologia* 408(0):115-122.

Skov, C. and A. Koed. 2004. Habitat use of 0+ year pike in experimental ponds in relation to cannibalism, zooplankton, water transparency and habitat complexity. *J. Fish Biol.* 64(2):448-459.

Skov, C., S. Berg and J.S. Olsen. 2006. *Stocking of pike fry and the establishment of a pike stock in Copenhagen Inner Lakes 2002-2006*. Danish Fisheries Research, Div. for freshwater fishing, Silkeborg. Report 39 pp. [In Danish]

Sundblad, G. and U. Bergström. 2014. Shoreline development and degradation of coastal fish reproduction habitats. *Ambio* 43(8):1020-1028.

Sundblad, G., U. Bergström, A. Sandström and P. Eklöv. 2014. Nursery habitat availability limits adult stock sizes of predatory coastal fish. *ICES J. Mar. Sci.* 71(3):672-680.

Sutela, T., A. Mutenia and E. Salonen. 2002. Relationship between annual variation in reservoir conditions and year-class strength of peled (*Coregonus peled*) and whitefish (*C. lavaretus*). *Hydrobiologia* 485(1):213-221.

Threinen, C.W. 1966. *The northern pike, life history, ecology, and management*: Wisconsin Conservation Department 235: 16 pp.

Threinen, C.W. 1969. *An evaluation of the effect and extent of habitat loss on northern pike populations and means of prevention of losses*: Wisconsin Conservation Department 28: 25 pp.

Tibblin, P., A. Forsman, P. Koch-Schmidt, O. Nordahl, P. Johannessen, J. Nilsson and P. Larsson. 2015. Evolutionary Divergence of Adult Body Size and Juvenile Growth in Sympatric Subpopulations of a Top Predator in Aquatic Ecosystems. *Am. Nat.* 186(1): 98-110.

Tibblin, P., A. Forsman, T. Borger and P. Larsson. 2016. Causes and consequences of repeatability, flexibility and individual fine-tuning of migratory timing in pike. *J. Anim. Ecol.* 85(1):136-145.

Wake, B. 2012. Modelling: Climate and Baltic Sea nutrients. *Nature Clim. Change* 2(6): 394-394.

Westin, L. and K.E. Limburg. 2002. Newly discovered reproductive isolation reveals sympatric populations of *Esox lucius* in the Baltic. *J. Fish Biol.* 61(6):1647-1652.

Williams, J.E. and B.L. Jacob. 1971. *Management of spawning marshes for northern pike*. Michigan Department of Natural Resources. Research and Development report (242): 20 p.

Wolf, P. 1956. Utdikad civilisation [Drained Civilization]. *Svenska Lax och Laxöringfören-ingen* [In Swedish].

Wright, R.M. and N. Giles. 1987. The survival growth and diet of pike fry *Esox lucius* L. stocked at different densities in experimental ponds. *J. Fish Biol.* 30(5):617-630.

Chapter 11

Pike Stocking for Lake Restoration

Christian Skov[1]

11.1 INTRODUCTION

Stocking measures are often used to enhance or maintain pike (*Esox lucius* Linnaeus, 1758) populations (chapter 9). Stocking of pike, and piscivores in general, can also be used as a biomanipulation tool, e.g. to restore turbid, eutrophic lakes (sometimes also inclined to toxic cyanobacterial blooms) into lakes with clearer water and well-developed macrophyte cover (Søndergaard et al. 2007, Gulati et al. 2008, Bernes et al. 2015). The theoretical backbone to pike stocking for biomanipulation is the alternative stable state theory (e.g. Scheffer 2009, chapter 8). More specifically, stocking of pike for biomanipulation of freshwater lake systems has the overarching aim to initiate a top-down regulation that ultimately should "push" the lake from a turbid water state, characterised by high phytoplankton production, low density of submerged macrophytes, few large cladocerans, high densities of planktivorous and benthivorous fish, into a clear-water state, characterised by low phytoplankton density, more large cladocerans, low densities of planktivorous and benthivorous fish and a high share of piscivorous fish (Brönmark and Hansson, 2005, Jeppesen et al. 2000, Scheffer 2009, chapter 8). The rationale is that stocking of pike increases the abundance of piscivores, intensifying the predation pressure on zooplanktivorous fish, which should relax predation on zooplankton, allowing them to in turn increase grazing on phytoplankton (Fig. 11.1, chapter 8). The reduction of phytoplankton biomass should increase water clarity and enhance the development of a submerged macrophyte community in shallow areas of the lake. The establishment of macrophytes would help stabilize the lake in the clear-water state, which ultimately may improve the lake's environmental status, biodiversity and often also increase its recreational value (Brönmark and Hansson, 2005).

[1]DTU Aqua, Technical University of Denmark, Section for Inland Fisheries and Ecology, Vejlsøvej 39, 8600, Silkeborg, Denmark; Email: ck@aqua.dtu.dk

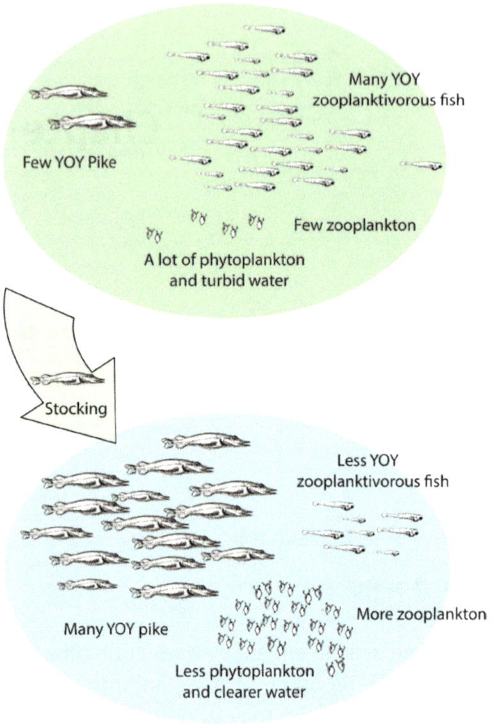

FIGURE 11.1 Schematic illustration of the intended ecosystem changes that stocking of young-of-the-year (YOY) pike should infer. Stocking pike should result in an increased density of piscivores that feed on the YOY zooplanktivorous fish. Lower zooplanktivore densities should release zooplankton from predation pressure and result in more zooplankton, grazing on and reducing phytoplankton densities, ultimately leading to clearer water. See text for further explanation. Drawing by Lene Jacobsen

If the stocked pike survive long enough to increase the density of older, larger pike, stocking may also result in increased predation pressure on larger benthivorous fish, thereby reducing possible macrophyte uprooting and resuspension of fine particles. In theory, pike stocking for lake restoration purposes can hereby have far-reaching positive effects on lake ecosystem composition. However, as will become clear in this chapter, pike stockings have rarely had these impacts.

Pike have been stocked for lake restoration purposes throughout the northern hemisphere, albeit most often in Europe and especially in Holland and Denmark (Bernes et al. 2015), most commonly as pike fry or small juveniles up to 5 cm body length (e.g. Prejs et al. 1994, Berg et al. 1997, Skov and Nilsson 2007). However, also larger juveniles, subadults and adults have been used as stocking material (i.e., Navarro and Johnson 1992, Benndorf 1995, Wysujack et al. 2001, Lappalainen et al. 2013). In Europe, stocking of pike was especially frequent in the 1980s and early 1990s, where stocking experiments from Poland (Prejs et al. 1994) and Denmark (Berg et al. 1997) showed very promising results on lake ecosystem responses. Moreover, several pike stocking experiments for biomanipulation have been reported from North America since the 1990s (e.g. Navarro and Johnson 1992, Kidd et al. 1999, Lathrop et al. 2002). There are however very few reports of pike stocking for lake restoration in recent years, which may reflect a reduced interest among managers or scientists to scientifically report such studies, or it may reflect an actual decrease in the frequency of stocking attempts for lake restoration purposes. The latter is the case in Denmark where no lake restoration stocking has taken place since 2006, when an evaluation of more than 30 case studies showed disappointing results, i.e., very few examples of successful outcomes of stocking of young-of-the-year (hereafter YOY) pike (Skov et al. 2006).

11.2 EFFICIENCY OF PIKE STOCKING AS A LAKE RESTORATION MEASURE

A successful pike stocking for lake restoration should result in clear changes in the trophic structure of the lake, i.e., reduced density of zooplanktivores, higher abundance of especially larger sized zooplankton species, less phytoplankton, and ultimately a higher water visibility and increased abundance of submerged macrophytes (e.g. Brönmark and Hansson 2005, Fig 11.1). The efficiency and success of pike stockings for lake restoration should hence be (and have been) evaluated in relation to one or several of these ecosystem components. Precise evaluation of pike stocking as a measure for lake restoration can however be difficult, as evaluation is often complicated by additional, simultaneous management measures that may create confounding effects. Pike stocking is often performed in combination with stocking of other piscivorous fish, such as pikeperch *Sander lucioperca* Linnaeus, 1758 (e.g. Benndorf et al. 1988), perch *Perca fluviatilis* Linnaeus, 1758 (e.g. Benndorf et al. 1984), walleye *Sander vitreus* Mitchill, 1818 and others (e.g. Drenner and Hambright 1999, Seda et al. 2000, Lathrop et al. 2002, Scharf 2007), or in combination with manual removal (i.e., fishing with trawls, gillnets etc.) of planktivorous fish (e.g. Drenner and Hambright 1999, Søndergaard et al. 2007, Bernes et al. 2015). In addition, variation in stocking characteristics (e.g. size of stocked pike, seasonal timing of stocking) and differing biotic and abiotic conditions among lakes where pike stockings have been conducted (Drenner and Hambright 1999) add further complexity. Truly replicated studies are therefore overall missing from the scientific literature (but see Hühn et al. 2014). The approach in this chapter is therefore to highlight studies where confounding effects from other measures, such as concurrent fishing/mass removal of planktivorous fish and stocking of other species of piscivores, are as limited as possible. Therefore, this chapter focuses on the limited amount of studies where pike was the only piscivore stocked.

The size of pike at stocking can influence their size-structured impact on targeted zooplanktivorous prey (chapters 2, 8). For example, YOY pike normally prey on smaller fish such as YOY cyprinids (e.g. Berg et al. 1997), whereas the stocking of larger pike would result in predation of larger prey (e.g. Elser et al. 1998). The longevity of the impact of pike stocking most likely also increases with increasing body sizes of the stocked pike (chapter 9). When stocking adult pike (i.e. > 30 cm), as well as stocking larger YOY pike in autumn, the main purpose is normally to build up a permanent higher pike biomass in the lake (e.g. Wusyjack et al. 2001). Stocking smaller YOY pike in spring rather creates a temporary, strong population of small pike over summer, which will eventually crash when the vegetation declines during fall and exposes stocked pike to fish, avian and cannibal piscivores (Berg et al. 1997). Hence, in the spring following stocking, the density of 1+ pike will often reflect the carrying capacity of the lake, as defined by habitat availability in the lake at the end of the first growth season (Grimm 1981, chapter 6). Due to such size-specific implications, evaluations of stocking of small YOY pike, i.e., larvae, fry and fingerlings, as well as stocking of older pike (1+ and older, > 30 cm), are presented separately below.

11.2.1 YOY Pike

Stocking of YOY pike (often < 5 cm) has been used in lake restoration projects in both North America (e.g. Lathrop et al. 2002) and Europe (e.g. Prejs et al. 1994, Berg et al. 1997). As the majority of published case studies stem from Europe (e.g. Bernes et al. 2015), the focus below is on European studies.

The rationale behind YOY stockings is that pike, even as fry, are ferocious piscivores (e.g. Søndergaard et al. 2007). Stocking YOY pike in very high densities should therefore increase the predation pressure on zooplanktivorous juveniles or fry, in Europe often roach (*Rutilus rutilus* Linnaeus, 1758) and bream (*Abramis brama* Linnaeus, 1758), which are generally very abundant and important predators on zooplankton in shallow eutrophic lakes (e.g. Hansson et al. 1998, Brönmark and Hansson 2005, Fig. 11.1). Consequently, at the end of the season, the year class strength of these YOY planktivores should be low, and by repeating the YOY pike stockings in subsequent years, an overall reduction in density of planktivores could be expected. YOY pike stocking has often been used in combination with removal of adult zooplanktivorous and benthivorous fish, for example, through intensive fishing by trawling or other methods (Hansson et al. 1998, Drenner and Hambright 1999, Mehner et al. 2004, Søndergaard et al. 2007, Bernes et al. 2015). In the case of combined zooplanktivore removal and pike stocking, the aim is for stocked YOY pike to reduce the recruitment from remaining zooplanktivorous fish, and hereby secure maintained reduction in planktivore densities (Hansson et al. 1998). Interestingly, it is normally expected that only few of the stocked YOY pike survive their first winter (Berg et al. 1997). The intention of the stocking is to create a density of YOY pike that remains during summer, i.e., until they have significantly reduced the year class of YOY planktivores. Thereafter, it is expected that inter- and intraspecific predation and competition will reduce pike density to carrying capacity, for example defined by the availability of refuge habitats for young pike (see also chapter 6).

When evaluated as a stand-alone measure, there are only few examples where stocking of YOY pike has had an impact on density and distribution of zooplanktivores (e.g. Prejs et al. 1994) and/or other species (Colby et al. 1987 and references therein). In line with this, Berg et al. (1997) stocked YOY pike each spring in the eutrophic Danish Lake Lyng (9.9 ha, max. depth 7.6 m, mean depth 2.4 m) from 1990-1993 in densities between 515 and 3616 pike ha^{-1}. As a result, the density of roach, rudd (*Scardinius erythrophthalmus* Linnaeus, 1758) and ruffe (*Gymnocephalus cernua* Linnaeus, 1758) fry decreased significantly by 64 to 97%, and there was a linear negative relationship between stocking density of pike in May/June and the abundance of juvenile zooplanktivores in the littoral zone in August. However, examples of YOY pike stockings that showed cascading effects to lower trophic levels or system effects, such as water clarity, are very few. One rare example is the abovementioned Danish Study in Lake Lyng where Secchi depths increased and chlorophyll a declined along with declines in YOY roach, caused by predation from stocked pike (Søndergaard et al. 1997). There are on the other hand numerous studies that report little or no impact of stocked YOY pike on small zooplanktivorous fish (Van Donk et al. 1989, Raat 1990, Goldyn et al. 1997, Meijer and Hosper 1997, Wysujack et al. 2001, Skov et al. 2002) or lower trophic levels (e.g. Prejs et al. 1997, Seda et al. 2000). Between 1995 and 2004, 400,000 to 800,000 YOY pike (2-4 cm) were stocked annually in Danish lakes as a tool for lake restoration (Jacobsen et al. 2004). An evaluation of a substantial part of these stockings revealed that clear positive effects could be seen in only one out of 34 lakes (> 10 ha) (Skov et al. 2006). In line with this, Søndergaard et al. (2007), in their review of more than 70 lake restoration projects from Denmark and the Netherlands, point out that stocking of YOY pike as a tool in lake restoration has been a disappointment with very few positive results. Some reasons for this will be presented later in this chapter.

11.2.2 Stocking Larger Pike

When stocking larger adult pike (> 30 cm), the intention is often to build up a higher piscivore biomass to increase predation pressure on a variety of prey size classes, including adults (e.g. Drenner and Hambright 2002). Stockings of larger pike have been shown to impact lower tropic levels in both European as well as North American lakes (e.g. Elser et al. 1998, Kidd et al. 1999, Lappalainen et al. 2013). Anderson and Schuup (1986) reported that pike stocking at irregular intervals (1969-1979) in a 340 ha lake in Minnesota was followed by a short-lasting increase in pike density. The variation in pike density over years was inversely related to the density of yellow perch (*Perca flavescens* Mitchill, 1814), that decreased in abundance with pike increases, and *vice versa*. Another example from Skov et al. (2002) showed that, although no initial effect of stocking of YOY pike was seen on the YOY planktivorous fish, seven years of YOY pike stocking eventually resulted in an increased density of larger pike (> 50 cm), that appeared to have a substantial impact on adult zooplanktiovorous/benthivorous fish in the shallow Lake Udbyover (Denmark). The authors argue that a distinct reduction of large crucian carp (*Carassius carassius* Linnaeus, 1758) resulted in a reduction in the amount of resuspended material in the lake, and hence improved Secchi depth. In contrast, Kidd et al. (1999) saw no effect on lower trophic levels or system responses despite a decline in cyprinid density after stocking of adult pike.

Navarro and Johnson (1992) stocked pike in two wetlands in 1985 and 1986 (West Marsh, 81 ha, and Darr Marsh, 18 ha) along western Lake Erie, Ohio, with the aim to determine the feasibility of pike as a biological control of carp (*Cyprinus carpio* Linnaeus, 1758). The wetlands were characterized by shallow water (0.5-1.0 m) and high turbidity caused by carp activity and wind/wave action. Mainly adult pike (> 36 cm) were stocked into these previously pike free wetlands. The stocked pike experienced high mortality, but stomach analyses of samples of the surviving pike revealed that they preferred soft-rayed prey fish and carp in particular. No changes in turbidity were reported following the stocking, but Navarro and Johnson (1992) conclude that the feeding preference for carp could imply a potential for pike top-down control of invasive common carp, and that studies should be conducted to explore this further. Unfortunately, no such studies have since been reported.

Very few studies demonstrate clear effects on water quality by stocking of adult pike. The best examples are from North America. For example, Elser et al. (2000) stocked adult pike (average weight 0.67 kg) into a 5 ha lake at a high density (26 $kg.ha^{-1}$), wherafter minnows, the main group of zooplanktivorous fish, more or less disappeared from the lake. Consequently, zooplankton biomass and especially large-bodied daphnids increased markedly and phytoplankton declined. Another example is from a 9 ha lake in Ontario, where Findlay et al. (1994), following a stocking of pike (400-500 g), observed a change in the trophic structure of the lake, i.e. decreases in yellow perch and an increase in large daphnids. Further, 16 years after the stocking was initiated in the lake, long-term results were evaluated (Findlay et al. 2005) and the authors found that phytoplankton biomass decreased after pike were introduced and increased during a period when pike were partially removed from the lake. However, they saw no effect on total zooplankton biomass, and no relationship between zooplankton and phytoplankton. Instead, it was suggested that the changes in algal biomass was driven by a reduction in phosphorus excretion by yellow perch, that decreased in density when pike were introduced (Findlay et al., 2005). Interestingly, no piscivores were present in either of the above study lakes before pike were stocked, and it seems to be a general pattern that the major post-stocking trophic changes are observed in waters where pike were not present before stocking. Lappalainen et al. (2013) corroborate this pattern by demonstrating an impact of stocked pike on the density of crucian carp in a European lake not containing pike prior to stocking, and in the study by Anderson and Schupp (1986), mentioned above, prior density of native pike was very low, presumably due to limited availability of spawning grounds for pike.

11.2.3 Pike Introductions and Trophic Cascades

Pike can be a major nuisance and invasive in regions where it is introduced (chapter 14), why it is not surprising that stocking of pike into previously pike free waters, i.e., pike introductions, can have clear and long-lasting cascading effects on lower trophic levels (e.g. Nicholson et al. 2015). An example of trophic effects following a pike invasion is presented by DeBates et al. (2003), who observed a marked change in the fish community structure, i.e., a significant reduction of the abundances of bluegill sunfish (*Lepomis macrochirus* Rafinesque, 1819), largemouth bass (*Micropterus salmoides* Lacepède, 1802), and yellow perch, in a shallow 25 ha lake in Nebraska, within four years following an invasion of pike. Based on the above and further examples given in chapter 8, it appears that the strongest trophic cascade effects from pike predation should be expected in lakes where pike are introduced, i.e. were not present beforehand. This is probably because the presence of a native pike population likely prevents stocked pike from boosting the pike population, due to, for instance, density-dependent mechanisms (chapters 2, 6, 9) and/or intraguild predation mechanisms (chapter 8). Hence, pike stocking in lakes containing pike at equilibrium densities are unlikely to result in permanent, or even transitional, increases in pike population densities, and therefore unlikely to significantly affect trophic composition, at least in the long term. This will be discussed further in the sections below with special focus on stocking of YOY pike.

11.3 REASONS BEHIND INEFFICIENCY OF YOY PIKE STOCKINGS

Reported findings in the literature indicate that stocking attempts are particularly unsuccessful when small YOY pike are stocked. This chapter section therefore goes into a more detailed discussion of reasons behind why YOY pike stockings for lake restoration often fail to have the desired biomanipulation effects (e.g. Søndergaard et al. 2007, Gulati et al. 2008). One obvious reason may relate to low stocking density, as highlighted by Goldyn et al. (1997) and Seda et al. (2000). Skov and Nilsson (2007) used a model approach to demonstrate that the timing of the stocking in relation to the hatching time of the YOY planktivorous fish likely plays a central role for success as it influences pike survival and predation potential. In fact, poor pike survival and/or a lack of planktivorous YOY fish in the diet of the pike are likely the most important reasons behind failures of stocking YOY pike for restoration purposes (Skov 2002, Søndergaard et al. 2007, Gulati et al. 2008). Hence, variability in survival rates due to handling, predation by conspecifics and other species, habitat complexity, behavior, and environmental conditions as well as variability in consumption rates are explored below.

11.3.1 Poor Pike Survival

In chapter 9 of this volume, several reasons are presented for why pike stockings in general may have poor efficiency. Naturally, most if not all of those reasons are also applicable to pike stocking for lake restoration. However, since the densities of YOY pike stocked for lake restoration purposes are much higher than in other types of stockings (e.g. Berg et al. 1997), it is relevant to discuss poor survival related to biomanipulation stocking separately. Consequently, the majority of the literature cited below stems from lake restoration stocking experiments.

Indeed, several studies report very low survival of YOY pike stocked for biomanipulation. Van Donk et al. (1989) evaluated two years of 0+ pike (4-5 cm) stocking and found survival rates (May-October) of just 19% and 6%, respectively. In a cross analysis of eight lakes, Skov and Nilsson (2007) found an average survival of less than 10% only 30 days after stocking in May/June (Fig. 11.2).

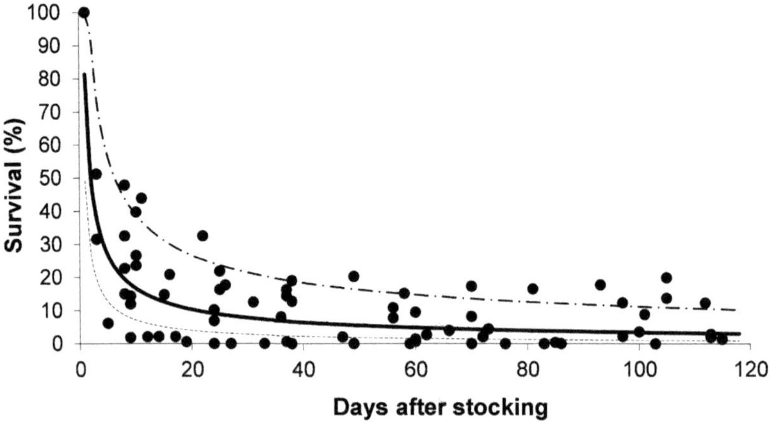

FIGURE 11.2 Average probability of survival of YOY pike stocked in eight Danish lakes based on electrofishing surveys in the littoral zones of the lakes (solid line) up to four months after stocking (dashed lines represent 95% confidence intervals). Redrawn from Skov and Nilsson (2007).

11.3.1.1 Handling and Stocking Timing

Low survival can relate to aspects of the stocking procedures such as poor handling during transportation or to poor timing of the stocking event. Grønkjær et al. (2004) discuss that YOY pike stocked too early in the season are likely to starve, and pike that are stocked too late in the season are at risk of being eaten by wild and larger YOY pike. In addition to cannibalism from larger wild YOY pike there may also be cannibalism among the stocked pike (e.g. Walker 1994, Skov et al. 2003a). In relation to this, the heterogeneity in stocking size between the stocked individuals is crucial, as the greater the initial size difference between the stocked individuals, the more cannibalism should be expected (Bry and Gillet 1980, Skov et al. 2003a).

11.3.1.2 Inter- and Intracohort Cannibalism

Cannibalism by older pike can play a major role for post stocking mortality of YOY pike (review by Raat 1988, Grimm and Klinge 1996). Likewise, Grimm (1981, 1983) argues that the biomass of small pike < 41 cm is controlled by larger pike between 41 cm and 54 cm, and Grimm (1994) demonstrates a negative correlation between densities of small pike (15-27 cm) and densities of larger pike (28-43 cm). Hence, in order to obtain a large population of smaller pike, which is intended in many lake restoration projects, Grimm and Backx (1990) suggest removal of pike larger than > 41 cm, especially when the dominant vegetation consists of emergent plants which offers less protection to smaller pike than do submerged macrophytes.

Other studies have also argued that older pike should be removed in order to facilitate optimal results of pike stockings for biomanipulation (Prejs et al. 1994, Wysujack et al. 2001), whereas others argue that a high density of older and larger pike may result in higher impact on the planktivore and benthivore populations (Skov et al. 2002, Lathrop et al. 2002) as large piscivores with large gape sizes (e.g. pike) are important regulators of larger prey (Benndorf 1990). When it comes to determining stocking size, the best approach is hence likely to be system specific as well as depending on the target species. If the aim is to increase predation on YOY cyprinids, then a high density of YOY pike is required. If the target prey species are larger planktivorous fish, then stocking larger adult pike would also be advantageous, although this would come with the cost of an increased likelihood of intercohort cannibalism.

Skov et al. (2006) evaluated pike stockings in 34 Danish lakes, and demonstrated that the highest survival was found in lakes where pike were not present before the first stocking event. Likewise, a recent German replicated pond study showed that ponds where pike fry were stocked on top of natural recruits did not result in elevated year-class strength compared to non-stocked controls (Hühn et al. 2014). This implies that YOY pike stockings may have the strongest potential to boost population densities, and consequently for impacting lower trophic levels, in lakes without an existing pike population prior to stocking, i.e., where cannibalism from a native population is low. It is hence not surprising that one of the most cited and convincing studies of a successful YOY pike stocking with cascading effects to lower trophic levels took place in Lake Lyng, Denmark, where no pike were present before the stocking in 1990 (Berg et al. 1997, Søndergaard et al. 1997).

11.3.1.3 Lack of Vegetation Cover

The abundance and species distribution of vegetation seems tightly linked to cannibalism as vegetation may give cover against conspecifics. Hence, YOY pike can be found in vegetated areas, and especially YOY pike prefer habitats with complex structures (chapter 3) and cannibalism is likely to increase when the habitats of different pike size classes overlap. Therefore, in order to minimize the risk of cannibalism and predation, stocked YOY pike need access to dense stands of vegetation (Grimm and Backx 1990), and lack of suitable habitats for the stocked pike has been argued as one of the main reason for low post-release survival in Dutch and German studies (Van Donk et al. 1989, Meijer et al. 1994, Walker 1994, Meijer and Hosper 1997, Wysujack et al. 2001).

The strong dependency of YOY pike on vegetated areas may imply that stockings of YOY pike are more likely to be successful in water bodies with plenty of shallow water and a high relative cover of emergent vegetation or submerged vegetation. However, since the lakes in need of biomanipulation and pike stocking are eutrophic and have high turbidity, submerged vegetation will rarely be available, at least not until the effects of the biomanipulation lead to reduced turbidity. The lack of natural habitats may to some extent be compensated by introducing artificial habitats. For example, Skov and Berg (1999) placed spruce trees in the littoral zone of a 16 ha shallow lake and showed that the stocked YOY pike used these as habitats. However, creating extensive coverage of artificial habitats may turn out to be a logistically substantial task, perhaps especially in larger lakes.

11.3.1.4 Predation

Predation from fish and birds can also affect survival of stocked pike. Hunt and Carbine (1951) examined 345 yellow perch (60-120 mm) and found that 21% had one or more pike in their stomachs, illustrating the potential role of percid predators. Similarly, in an analysis of eight Danish lakes, Skov (unpublished results) found a negative relationship between the density of larger, and therefore potentially piscivorous Eurasian perch, and the density of stocked pike in the littoral zone (Fig. 11.3).

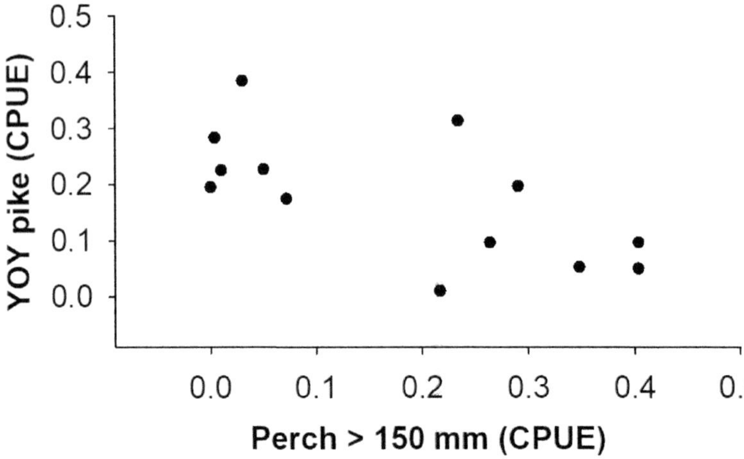

FIGURE 11.3 The relationship between CPUE of piscivorous perch (> 150 mm) and stocked YOY pike in the littoral zone of eight Danish lakes where YOY pike were stocked for lake restoration at a size of 2-4 cm in mid May-early June. Sampling took place between 1999 and 2004, and some of the lakes are represented by two sampling years. Sampling was done by point abundance electrofishing in the littoral zone (see Skov and Berg 1999) and each point is based on the average of 2-3 sampling events during the weeks after stocking (May and June). Spearman's rank correlation: $r = -0.57$, $p = 0.03$ (Skov, unpublished results).

Avian predators such as great cormorant (*Phalacrocorax carbo* Linnaeus, 1758), grey heron (*Ardea cinerea* Linnaeus), and great-crested grebe (*Podiceps cristatus* Linnaeus, 1758) have also been reported to predate on pike (Raat 1988), and could, in theory, affect survival of stocked pike. Depending on stocking size, also various kinds of large macroinvertebrates may influence survival of newly stocked pike. Monten (1948) and Le Louarn and Cloarec (1997) found that larvae of red-eyed damselfly (*Erythromma najas* Hansemann, 1823), broad-bodied chaser (*Libellula depressa* Linnnaeus, 1758), emperor dragonfly (*Anax imperator* Leach, 1815), *Dysticus marginalis* Linnaeus, 1758 (a water beetle), adult *Notonecta glauca* Linneaeus, 1758 (a backswimmer) and *Ilyocoris cimicoides* Linnaeus, 1758 (a saucer bug) to predate on 3-30 day old pike.

11.3.1.5 Size Matters

The size of stocked pike range from small YOY fish (e.g. Van Donk et al. 1989, Prejs et al. 1994, Walker 1994, Berg et al. 1997) to larger adult fish (e.g., Benndorf, et al. 1988, Goldyn et al. 1997, Elser et al. 1998, Wysujack et al. 2001). In relation to this, it is plausible to expect that survival of stocked individuals will increase with size as the number of potential predators decrease with size, as has been shown for muskellunge (McKeown et al. 1999). Similarly, there is a likely relationship between stocking size and survival for pike (chapter 9). However, choosing to stock larger pike to optimize survival is a trade off with production cost as the price of commercially reared YOY pike increase with size (chapter 9).

11.3.1.6 Concluding Remarks on Survival

To conclude from this section on survival it appears that despite the use of very high stocking densities it may be difficult, if not impossible, to increase the density of a natural pike population for prolonged periods of time. If YOY pike are stocked for biomanipulation with the only aim to reduce the YOY planktivores, it can be argued that prolonged survival is not necessarily a prerequisite, i.e. in theory pike only have to survive until the end of the first summer. However, if YOY pike are stocked during spring, studies suggest that massive mortality can occur long before the summer ends, reducing the probability that the stocked pike remain in the lake long enough to significantly reduce the year class of YOY planktivores (Fig. 11.2). Moreover, the argument also implies that the stocked YOY pike prey mainly upon YOY planktivores, but as will become evident in section 11.3.2 below, this may far from always be the case.

11.3.2 Low Consumption of YOY Zooplanktivores

The size at which fish become the dominant prey for YOY pike differs greatly between studies: e.g. 30-40 mm (Frost 1954), 65 mm (McCarraher 1957), 70 mm (Hunt and Carbine 1951), 80 mm (Holland and Huston 1984), > 85 mm (Franklin and Smith 1963), > 120 mm (Grimm and Klinge 1996). These differences could be linked to prey abundance, with pike predating on the more abundant prey (e.g. Franklin and Smith 1963, Holland and Huston 1984). In lakes where restoration is relevant, the density of suitable fish prey should be very high, and consequently also the frequency of piscivory, even among the smallest pike. However, there are strong exceptions to this pattern. Skov and Nilsson (2007) analysed 1692 YOY pike stomachs sampled from eight Danish lakes stocked with pike for restoration. All eight lakes were highly eutrophicated and held strong populations of YOY planktivorous fish. It was therefore expected that the stomachs of the sampled YOY pike would reflect this by a high occurrence of fish prey. However, the probability of piscivory was surprisingly low, especially for smaller pike (Fig. 11.4). In fact, less than half of 10 cm pike were piscivorous. Combined with the high mortality of the smallest YOY pike shortly after stocking (Fig. 11.2, Skov and Nilsson 2007), a remarkably low predation pressure on zooplanktivore recruits was established in the eight lakes.

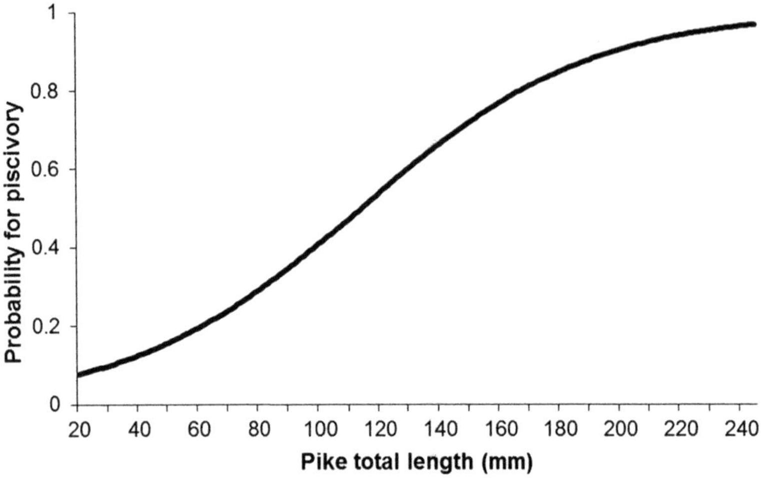

FIGURE 11.4 Size-dependent probability of piscivory (presence of fish in the diet) for pike through early juvenile ontogeny. Data are based on stomach analyses of 1621 pike sampled from eight lakes where pike stocking for biomanipulation was conducted (Redrawn from Skov and Nilsson 2007).

A straightforward explanation for this phenomenon could relate to a high abundance of non-fish prey in the lakes. Diana (1996) argues that feeding opportunity is the major determinant of pike predation, and pike can be assumed to feed on organisms in direct proportion to their availability (Holland and Huston 1984). Therefore, if the abundance of non-fish prey exceeds the abundance of fish prey, low fish consumption may occur. Another explanation could relate to system-specific differences in elements of the foraging cycle (see chapter 2), and specifically variation in the first three stages of the foraging cycle (i.e., the success of search, encounter and/or attack influenced by behavioral mechanisms and/or environmental conditions as exemplified below).

11.3.2.1 Antipredator Behaviors

Low fish consumption may result from infrequent prey encounters due to antipredator behaviors of prey fish (chapter 2). For example, Skov et al. (2003b) found that as piscivory increased in YOY pike, the dominant prey fish, YOY roach, decreased their use of the littoral area (i.e., the preferred habitat of the YOY pike). Likewise, He and Kitchell (1990) showed that prey fish avoided areas with pike following a pike introduction.

Low prey fish consumption could also be a result of antipredator behavior in pike themselves. Nilsson and Brönmark (1999) showed that foraging on larger prey increases pike risk of falling victim to cannibals and also increases the chance of escape for the prey. Prey preference among pike is hence likely determined by the cost associated with post-attack manipulation events (e.g. risk of predation or kleptoparasitic interactions). This may be one reason why pike prefer smaller prey if available (Nilsson and Brönmark 2000, chapter 2). If pike choose to feed on small, easily captured and handled non-fish prey, consumption of fish prey would decrease. In general, pike behaviors that reduce the risk of kleptoparasitism and cannibalism can reduce pike predation rates and the overall impact of pike predation on prey, and hereby decrease the potential for top-down trophic cascades from pike apex predation (chapter 8). Pike and prey interactions could further be influenced by the type of vegetation in lakes. Skov (2002) gives an example of how the vegetation type (i.e., its structural complexity) may influence predator-prey encounters. Four years of electrofishing for YOY pike and its potential fish prey in different vegetation types in shallow biomanipulated Lake Udbyover illustrated that the likelihood of encountering both pike and prey at the same electrofishing spot was significantly higher in structurally more complex vegetation (i.e., common bogbean *Menyanthes trifoliata* Linneaus, 1758) compared to less structured vegetation (i.e. common reed *Phragmites australis* Cavanilles). This could imply that the likelihood of pike encountering prey is higher if more structurally complex vegetation is available.

11.3.2.2 Stocking Timing

Pike are gape-size limited predators that sometimes prefer smaller prey than expected (chapter 2). Clearly, the larger the stocked pike are in relation to their prey, the more prey they can consume. Skov and Nilsson (2007) used a simple model approach to illustrate this and showed that the later in the season the YOY pike are stocked following the hatching of cyprinids, the fewer YOY cyprinid prey are likely to be predated. A mismatch of just a few days can severely reduce the consumption potential of the stocked pike. So in an optimal situation pike stocking should take place just a few days after the cyprinids hatch when the cyprinids are less mobile and relatively small in size compared to the stocked pike. However, Skov and Nilsson (2007) discuss that several factors make this theory very impractical, i.e., different cyprinid species hatch at different periods during summer and the same cyprinid species may actually hatch several times or during extended periods. In addition, the time of cyprinid hatching will probably differ between years depending on the temperature regime during spring, and preparing a well-timed stocking would therefore require considerable amount of work by the lake manager, i.e., a daily monitoring of water temperature as well as previous knowledge (e.g. historical protocols) of cyprinid hatching (Skov and Nilsson 2007).

11.3.2.3 Environmental Conditions

YOY pike foraging (e.g. encounter and attack) may be impaired in waters with low visibility (Bry 1996, Lammens 1999, chapters 2 and 3), and Raat (1990) questions the value of stocking YOY pike in biomanipulation projects where water visibility is limited. As the lakes where lake restoration is warranted most often are eutrophic, or even hypereutrophic, and therefore have reduced water visibility, the suggestion by Raat (1990) then questions the overall usefulness of using YOY pike as a lake restoration measure. On the other hand, studies by Skov (2002) suggest that YOY pike feeding is not impaired in low visibility water. Skov et al (2002) found equal foraging success of YOY pike (9-13 cm) in low and high water transparency, and Skov et al. (2003a) found faster growth of YOY pike (3-5 cm) in low transparency waters, which suggests that poor water visibility does not impair YOY pike foraging. See further discussion of the role of water visibility on the foraging mechanisms in pike in chapters 2, 3 and 6 of this volume.

11.3.2.4 Maladaptation in Stocked Pike

A final explanation for the low prey fish consumption relates to maladaptation in stocked pike. Skov et al. (2011) showed that stocked pike had feeding and growth patterns inferior to those of wild, native pike, and argued that this could be related to a genetically based local maladaptation among the stocked fish. For example, the stocked YOY pike were initially larger than the wild YOY pike, but at the end of the growth season the stocked pike were smaller than the wild ones, and stomach analyses revealed that the stocked pike ingested less diverse prey items and had higher fractions of empty stomachs throughout the study period. Such possible maladaptation could severely reduce the effects of pike stockings, and should be considered in restoration attempts.

11.4 CONCLUSION AND FUTURE PERSPECTIVES

In order for lake restoration with piscivore stocking to be successful, the ratio of piscivores to planktivores should increase following the stocking (Drenner and Hambright 2002). Pike is a top predator and its presence can have direct as well as indirect impacts on lower trophic levels (chapter 8). A boost in pike density through stocking measures can therefore be predicted to increase this impact, consequently pushing the lake from a turbid-water state to a clear-water state. Theory predicts that the cannibalistic nature of pike as well as the strong dependence on vegetation as shelter from predation makes it difficult to increase the density of especially smaller size classes of pike for prolonged periods of time. This chapter supports this by illustrating that the most successful stockings for lake restoration (both adults and YOY) so far have been reported from waters where pike were not present prior to stocking, i.e., pike introductions. Therefore, in waters where pike are already present, impacts of pike stocking are likely to be minor and very transitional. The latter, i.e., the period of any additive impact, likely depends on the size of the stocked pike. The effect of stocking small YOY pike likely lasts for a short period (weeks) whereas older pike can have effects for longer periods (perhaps even years) (chapter 9). Hence, stocking adult pike in high densities, and subsequently protecting them from fisheries mortality, should in theory result in an increase in piscivore density with a relatively long duration. It however remains to be thoroughly demonstrated that such elevation of adult pike can result in significantly increased predation pressure on zooplanktivorous and benthivorous fish, and it could be argued that density-dependent mechanisms such as interference between adult pike can prevent the predation rate to increase linearly with increased density (chapters 2, 8). In addition, stocked fish may be adapted to prey compositions other than in the focal lakes. Further studies could explore these potential restrictions.

European studies suggest that stocking of YOY pike as a "stand-alone measure" to reduce YOY cyprinid cohorts during summer has very rarely proven efficient due to a combination of high mortality among the stocked individuals and limited piscivory of the few survivors. It could however be worth exploring if stocking juveniles larger than previously used (i.e. > 6-8 cm) in high densities could have additive effects on pike density for a longer period, for example, up to several weeks, and thereby increase predation pressure on YOY zooplanktivores. Hence, if the aim of the biomanipulation is restricted to reduce YOY year classes, stocking larger YOY pike during spring at the time where the major part of the YOY cyprinid hatch could improve the chances of success (see also Skov and Nilsson 2007). However, finding the right stocking timing may be tricky as hatching time may show high inter- and intraspecific as well as system-specific temporal variability. This is why optimal stocking will require a thorough understanding of cyprinid recruitment mechanisms on a lake-specific basis. Finally, it remains to be proven that repeated reductions of the YOY planktivores, e.g. through pike predation from stocked individuals, can result in long term changes in trophic structure and eventually a permanent tipping of the lake into a clear water state (chapter 8). In relation to this, stocking pike in combination with fish removal is an appealing approach as a tool to reduce potential compensatory population growth of cyprinid fish. However, more studies are encouraged to fully demonstrate that such stockings have a truly additive effect, e.g. by significantly reducing the density of planktivore recruits and/or by preventing the development of strong year classes.

Acknowledgments
I thank Anders Nilsson, Lene Jacobsen, Kristine Dunker for input to this chapter and the Danish Rod and Net Fishing Funds for support.

REFERENCES CITED

Anderson, D.W. and D.H. Schupp. 1986. Fish community responses to northern pike stocking in Horseshoe Lake, Minnesota. *Minnesota Department of Natural Resources, Division of Fish and Wildlife Investigational Report Number* 387, St. Paul.

Benndorf, J., H. Kneschke, K. Kossatz, and E. Penz. 1984. Manipulation of the pelagic food web by stocking with predacious fishes. *Int. Revue ges. Hydrobiol.* 69:407-428.

Benndorf, J., H. Schultz, A. Benndorf, R. Unger, E. Penz, H. Kneschke, K. Kossatz and R. Dumke. 1988. Food-web manipulation by enhancement of piscivorous fish stocks: Long-term effects in the hypertrophic Bautzen Reservoir. *Limnologica* 19:97-110.

Benndorf, J. 1990. Conditions for effective biomanipulation: Conclusions derived from whole-lake experiments in Europe. *Hydrobiologia.* 200/201:187-203.

Benndorf, J. 1995. Possibilities and limits for controlling eutrophication by biomanipulation. *Int. Rev. ges. Hydrobio.* 80:519-534.

Berg, S., E. Jeppesen and M. Søndergaard. 1997. Pike (*Esox lucius* L.) stocking as a biomanipulation tool 1. Effects on the fish population in Lake Lyng, Denmark. *Hydrobiologia* 342/343:311-318.

Bernes, C., S.R. Carpenter, A. Gårdmark, P. Larsson, L. Persson, C. Skov, J.D.M. Speed and E.V. Donk. 2015. What is the influence of a reduction of planktivorous and benthivorous fish on water quality in temperate eutrophic lakes? A systematic review. *Environ. Evid.* 4:1-28.

Brönmark, C. and L-A. Hansson. 2005. *Biology of lakes and ponds*. 2nd Ed. Oxford. Oxford University press. pp 285.

Bry, C. and C. Gillet. 1980. Reduction of cannibalism in pike (*Esox lucius*) fry by isolation of full-sib families. *Reprod. Nutr. and Développ.* 20:173-182.

Bry, C. 1996. Role of vegetation in the life cycle of pike. pp 45-67. *In*: J.F. Craig (ed.). *Pike: Biology and exploitation*. London: Chapman & Hall.

Colby, P. J., P.A. Ryan, D.H. Schupp, and S.L. Serns. 1987. Interactions in north-temperate lake fish communities. *Can. J. Fish. Aquat. Sci.* 44(2): 104-128.

DeBates, T.J., C.P. Paukert and D.W. Willis. 2003. Fish Community Responses to the Establishment of a Piscivore, Northern Pike (*Esox lucius*), in a Nebraska Sandhill Lake, *J. Freshw. Ecol.* 18(3):353-359.

Diana, J.S. 1996. Energetics. pp 103-124. *In*: J.F. Craig (ed.). *Pike: Biology and exploitation*. London: Chapman & Hall.

Drenner, R.W. and K.D. Hambright. 1999. Biomanipulation of fish assemblages as a lake restoration technique. *Arch. Hydrobiol.* 146:129-65.

Drenner, R.W. and K.D. Hambright. 2002. Piscivores, trophic cascades, and lake management. *Sci. World J.* 2:284-307.

Elser, J.J., T.H. Chrzanowski, R.W. Sterner and K.H. Mills. 1998. Stoichiometric Constraints on Food-Web Dynamics: A Whole-Lake Experiment on the Canadian Shield. *Ecosystems* 1:293-307.

Elser, J.J., R.W. Sterner, A.E. Galford, T.H. Chrzanowski, D.L. Findlay, K.H. Mills, M.J. Paterson, M.P. Stainton, and D.W. Schindler. 2000. Pelagic C:N:P stoichiometry in a eutrophied lake: responses to a whole-lake food-web manipulation. *Ecosystems* 3:293-307.

Franklin, D.R. and L.L. Smith. 1963. Early life history of the northern pike (*Esox lucius* L.). *Trans. Am. Fish. Soc.* 92:91-110.

Findlay, D.L., S.E.M. Kasian, L.L. Hendzel, G.W. Regehr, E.U. Schindler, and J.A. Shearer. 1994. Biomanipulation of lake 221 in the Experimental Lakes Area (ElA): effects on phytoplankton and nutrients. *Can. J. Fish. Aquat. Sci.* 51:2794-2807.

Findlay, D.L., M.J. Vanni, M. Paterson, K.H. Mills, S.E.M. Kasian, W. Findlay, and A.G. Salki. 2005. Dynamics of a boreal lake ecosystem during a long-term manipulation of top predators. *Ecosystems* 8:603-618.

Frost, W.E. 1954. The Food of Pike *Esox lucius* L. in Windermere. *J. Anim. Ecol.* 23:339-360.

Goldyn, R., A. Kozak and W. Romanowicz. 1997. Food-web manipulation in the Maltanski Reservoir. *Hydrobiologia* 342/343:327-333.

Grimm, M.P. 1981. The composition of northern pike (*Esox lucius* L.) population in four shallow waters in the Netherlands, with special reference to factors influencing 0+ pike biomass. *Fish. Manage.* 12:61-76.

Grimm, M.P. 1983. Regulation of biomasses of small (< 41 cm) northern pike (*Esox lucius*) with special reference to the contribution of individuals stocked as fingerlings (4-6 cm). *Fish. Manage.* 14:115-134.

Grimm, M.P. and J.J.G.M. Backx. 1990. The restoration of shallow eutrophic lakes, and the role of northern pike, aquatic vegetation and nutrient concentration. *Hydrobiologia* 200/201:557-566.

Grimm, M.P. 1994. The influence of aquatic vegetation and population biomass on recruitment of 0+ and 1+ northern pike (*Esox lucius* L.). pp 226-234. *In*: I.G. Cowx (ed.). *Rehabilitation of freshwater fisheries*. Oxford: Fishing News Books, Blackwell Science.

Grimm, M.P. and M. Klinge. 1996. Pike and some aspects of its dependence on vegetation. pp 125-156. *In*: J.F. Craig (ed.). *Pike: Biology and exploitation*. London: Chapman & Hall.

Grønkjær, P., C. Skov and S. Berg. 2004. Otolith-based analysis of survival and size-selective mortality of stocked 0+year pike related to time of stocking. *J. Fish Biol.* 64:1625-1637.

Gulati R.D., L.M.D. Pires and E. Van Donk. 2008. Lake restoration studies: failures, bottlenecks and prospects of new ecotechnological measures. *Limnologica*. 38:233-247.

Hansson, L.-A., J. Annadotter, E. Bergman, S.F. Hamrin, E. Jeppesen, T. Kairesalo, E. Luokkanen, P.A. Nilsson, M. Søndergaard and J. Strand. 1998. Biomanipulation as an application of food-chain theory: constraints, synthesis, and recommendations for temperate lakes. *Ecosystems* 1:558-574.

He, X. and J.F. Kitchell. 1990. Direct and indirect effects of predation on a fish community: A whole-lake experiment. *Trans. Am. Fish. Soc.*119:825-835.

Holland, L.E. and M.L. Huston. 1984. Relationship of Young-of-the-Year northern pike to aquatic vegetation types in backwaters of the upper Mississippi River. *N. Am. J. Fish. Manage*. 4:514-522.

Hunt, R.L. and W.F. Carbine. 1951. Food of young pike, *Esox lucius* L., and associated fishes in Peterson's ditches, Houghton Lake, Michigan. *Trans. Am. Fish. Soc* 80:67-83.

Hühn, D., K. Lübke, C. Skov and R. Arlinghaus. 2014. Natural recruitment, density-dependent juvenile survival, and the potential for additive effects of stock enhancement: an experimental evaluation of stocking northern pike (*Esox lucius*) fry. *Can. J. Fish Aquat. Sci*. 71: 1508-1519.

Jacobsen, L., S. Berg and C. Skov. 2004. Management of lake fish populations and lake fisheries in Denmark: history and current status. *Fish. Manag. Ecol.* 11: 219-224.

Jeppesen, E., J.P. Jensen, M. Søndergaard, T. Lauridsen and F. Landkildehus. 2000. Trophic structure, species richness and biodiversity in Danish lakes: changes along a phosphorus gradient. *Freshw. Biol.* 45:201-218.

Kidd, K.A., M.J. Paterson, R.H. Hesslein, D.C.G. Muir and R.E. Hecky. 1999. Effects of northern pike (*Esox lucius*) additions on pollutant accumulation and food web structure, as determined by δ13C and δ15N, in a eutrophic and an oligotrophic lake. *Can. J. Fish. Aquat. Sci.* 56:2193-2202.

Lappalainen, J., M. Vinni and T. Malinen. 2013. Consumption of crucian carp (*Carassius carassius* L., 1758) by restocked pike (*Esox lucius* L., 1758) in a lake with frequent winter hypoxia. *J. Appl. Ichthyol.* 29(6):1286-1291.

Lammens, E.H. 1999. The central role of fish in lake restoration and management. *Hydrobiologia* 395/396:191-198.

Lathrop, R.C., B.M. Johnson, T.B. Johnson, M.T. Vogelsang, S.R. Carpenter, T.R. Hrabik, J.F. Kitchell, J.J. Magnuson, L.G. Rudstam and R.S. Stewart. 2002. Stocking piscivores to improve fishing and water clarity: a synthesis of the Lake Mendota biomanipulation project. *Freshw. Biol.* 47:2410-2424.

Le Louarn, H. and A. Cloarec. 1997. Insect predation on pike fry. *J. Fish Biol.* 50:366-370.

McCarraher, D.B. 1957. The natural propagation of Northern pike in small ponds. *Prog. Fish Cult.* 19:185-187.

McKeown, P.E., J.L. Forney and S.R. Mooradian. 1999. Effects of stocking size and rearing method on muskellunge survival in Chautauqua Lake, New York. *N. Am. J. Fish. Manage*. 19:249-257.

Mehner, T., R. Arlinghaus, S. Berg, H. Dörner, L. Jacobsen, P. Kasprzak, R. Koschel, T. Schulze, C. Skov, C. Wolter and K. Wysujack. 2004. How to link biomanipulation and sustainable fisheries management: a step-by-step guideline for lakes of the European temperate zone. *Fish. Manag. Ecol.* 11: 261-275.

Meijer, M-L., E. Jeppesen, E. Van Donk, B. Moss, M. Scheffer, E. Lammens, E. Van Nes, J.A. Van Berkum, G.J.de. Jong, B.A. Faafeng and J.P. Jensen. 1994. Long-term responses to fish-stock reduction in small shallow lakes: Interpretation of five-year results of four biomanipulation cases in The Netherlands and Denmark. *Hydrobiologia* 275/276:457-466.

Meijer, M-L. and H. Hosper. 1997. Effects of biomanipulation in the large and shallow Lake Wolderwijd, The Netherlands. *Hydrobiologia* 342-343:335-349.

Monten, E. 1948. Research on the biology and related problems of the northern pike. *Skrifter Utgivna av Södra Sveriges Fiskeriförening* 1:3-38. (in Swedish).

Navarro, J.E. and D.L. Johnson. 1992. Ecology of stocked northern pike in two Lake Erie controlled wetlands. *Wetlands* 12(3):171-177.

Nicholson, M.E., M.D. Rennie and K.H. Mills. 2015. Apparent extirpation of prey fish communities following the introduction of Northern Pike (Esox lucius). *Can. Field Nat.* 129(2):165-173.

Nilsson, P.A. and C. Brönmark. 1999. Foraging among cannibals and kleptoparasites: effects of prey size on pike behavior. *Behav. Ecol.* 10:557-566.

Nilsson, P.A. and C. Brönmark. 2000. The role of gastric evacuation rate in handling time of equal-mass rations of different prey sizes in northern pike. *J. Fish Biol.* 57:516-524.

Prejs, A., A. Martyniak. S. Boron, P. Hliwa and P. Koperski. 1994. Food web manipulation in a small, eutrophic Lake Wirbel, Poland: Effects of stocking with juvenile pike on planktivorous fish. *Hydrobiologia* 275/276:65-70.

Prejs, A., J. Pijanowska, P. Koperski, A. Martyniak, S. Boron and P. Hliwa. 1997. Food-web manipulation in a small, eutrophic Lake Wirbel, Poland: long-term changes in fish biomass and basic measures of water quality. A case study. *Hydrobiologia* 342/343:383-386.

Raat, A.J.P. 1988. Synopsis of biological data on the northern pike *Esox lucius* Linnaeus, 1758. *FAO Fisheries Synopsis 30*. Rome, 178 pp.

Raat, A.J.P. 1990. Production, consumption and prey availability of northern pike (*Esox lucius*), pikeperch (*Stizostedion lucioperca*) and European catfish (*Silurus glanis*): A bioenergetics approach. *Hydrobiologia* 200-201:497-509.

Seda, J., J. Hejzlar and J. Kubecka. 2000. Trophic structure of nine Czech reservoirs regularly stocked with piscivorous fish. *Hydrobiologia* 429:141-149.

Scharf, W. 2007. Biomanipulation as a useful water quality management tool in deep stratifying reservoirs. *Hydrobiologia* 583:21-42.

Scheffer, M. 2009. *Critical Transitions in Nature and Society*. Princeton, (NJ): Princeton University Press. 400 pp.

Skov, C. and S. Berg. 1999. Utilization of natural and artificial habitats by 0+ pike in a biomanipulated lake. *Hydrobiologia* 408:115-122.

Skov, C. 2002. *Stocking 0+ pike (Esox lucius L.) as a tool in the biomanipulation of shallow eutrophic lakes*. Doctoral dissertation. University of Copenhagen, Copenhagen. 203 pp.

Skov, C., M.R. Perrow, S. Berg and H. Skovgaard. 2002. Changes in the fish community and water quality during seven years of stocking piscivorous fish in a shallow lake. *Freshw. Biol.* 47:2388-2400.

Skov, C., L. Jacobsen and S. Berg. 2003a. Post-stocking survival of 0+ pike in ponds as a function of water transparency, habitat complexity, prey availability and size heterogeneity. *J. Fish Biol.* 62:311-322.

Skov, C., O. Lousdal, P.H. Johansen and S. Berg. 2003b. Piscivory of 0+ pike (*Esox lucius* L.) in a small eutrophic lake and its implication for biomanipulation. *Hydrobiologia* 506:481-487.

Skov, C., L. Jacobsen, S. Berg, J. Olsen and D. Bekkevold. 2006. Udsætning af geddeyngel i danske søer: Effektvurdering og Perspektivering (Stocking of YOY pike in Danish lakes: Effects and perspectives) *DFU-rapport* 161-06. 96 pp (In Danish)

Skov, C., and P.A. Nilsson. 2007. Evaluating stocking of YOY pike *Esox Lucius* as a tool in the restoration of shallow lakes. *Freshw. Biol.* 52:1834-1845.

Skov, C., A. Koed, L. Baastrup-Spohr and R. Arlinghaus. 2011. Dispersal, growth and diet of stocked and wild northern pike fry in a shallow natural lake, with implications for management of stocking programs. *N. Am. J. Fish. Manage.* 31(6):1177-1186.

Søndergaard, M., E. Jeppesen and S. Berg. 1997. Pike (*Esox lucius* L.) stocking as a biomanipulation tool 2. Effects on lower trophic levels in Lake Lyng, Denmark. *Hydrobiologia* 342:319-325.

Søndergaard, M., E. Jeppesen, T.L. Lauridsen, C. Skov, E.H. Van Nes, R. Roijackers, E. Lammens and R Portielje. 2007. Lake restoration: successes, failures and longterm effects. *J. Appl. Ecol.* 44:1095-105.

Van Donk, E., R.D. Gulati and M.P. Grimm. 1989. Food web manipulation in Lake Zwemlust: Positive and negative effects during the first two years. *Hydrobiol. Bull.* 23:19-34.

Walker, P.A. 1994. Development of pike and perch populations after biomanipulation of fish stocks. pp 376-390. *In*: I.G. Cowx (ed.). *Rehabilitation of freshwater fisheries*. Oxford: Fishing News Books, Blackwell Science.

Wysujack, K., U. Laude, K. Anwand and T. Mehner. 2001. Stocking, population development and food composition of pike *Esox lucius* in the biomanipulated Feldberger Haussee (Germany) - Implications for fisheries management. *Limnologica* 31:45-51.

Chapter 12

Recreational Piking – Sustainably Managing Pike in Recreational Fisheries

Robert Arlinghaus[*1], Josep Alós[2], Ben Beardmore[3], Ángela M. Díaz[4], Daniel Hühn[5], Fiona Johnston[6], Thomas Klefoth[7], Anna Kuparinen[8], Shuichi Matsumura[9], Thilo Pagel[10], Tonio Pieterek[11] and Carsten Riepe[12]*

12.1 IMPORTANCE OF PIKE FOR RECREATIONAL FISHERIES

Recreational fishers preferentially target top predators (Donaldson et al. 2011). In this context, northern pike (*Esox lucius* Linnaeus, 1758, hereafter pike for simplicity) constitutes a preferred target species of many freshwater and coastal recreational fishers across much of its natural circumpolar distribution in the northern hemisphere (e.g. Paukert et al. 2001, Wolter et al. 2003, Arlinghaus et al. 2008a, Crane et al. 2015). In the German states Berlin and Brandenburg, for example, the majority of anglers primarily target pike (Ensinger et al. 2016). Pike are not only regularly targeted in Germany, but are also consumed by many anglers (Fig. 12.1), although voluntary catch-and-release seems to be on the rise among more specialized angler segments. Pike are also regularly eaten in other countries in central and Eastern Europe (e.g. Poland) as well as Russia (Mickiewicz and Wołos 2012). The species is also commonly targeted by anglers in Scandinavia, the UK, the Netherlands, other countries in central Europe, and many states in the USA (e.g. Wisconsin, Minnesota) (Paukert et al. 2001, Diana and Smith 2008, Pierce 2012, Stålhammar et al. 2014). In some of these countries (e.g. UK, the Netherlands, some states of the USA) pike are often voluntarily released after capture (Pierce et al. 1995, Paukert et al. 2001, Margenau et al. 2003, Stålhammar et al. 2014). Similarly, in the USA

[1]see Authors' addresses at the end of this chapter.

the congeneric muskellunge (*Esox masquinongy* Mitchill, 1824) is a highly desired target of anglers and often managed for its trophy qualities based on voluntary catch-and-release angling (Crane et al. 2015). As top predators, both pike and muskellunge occur in low densities and are toothy, large-bodied fishes with a distinct body shape and ecology. These traits are unique among freshwater fishes, likely explaining the attractiveness of esocids to recreational fishers (Johnston et al. 2013).

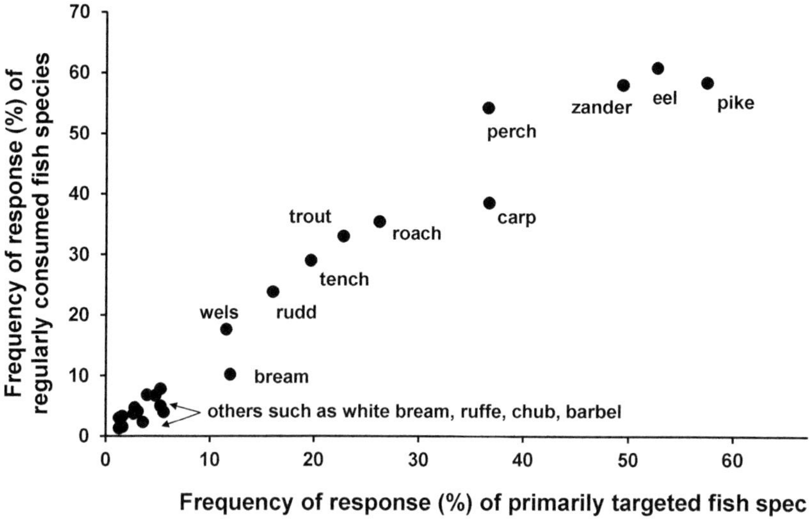

FIGURE 12.1 Relative frequency of anglers of the German city state of Berlin indicating that pike is the preferred angling species and also regularly consumed in Germany (modified from Wolter et al. 2003).

12.2 GEAR TYPES AND HUMAN DIMENSIONS

Recreational fishers use a range of gear types to capture pike. By far the most common gear is rod-and-reel-type fishing with artificial lures or natural bait. Depending on local custom, gill-nets (e.g. in Finland) or spear fishing under ice (e.g. darkhouse spearing in Minnesota) are also used by recreational fishers targeting pike. The behaviour and preferences of recreational fishers targeting pike with different gear types vary. For example, work in Minnesota has revealed that anglers and darkhouse spearers differ strongly in their preferences and attitudes to fishing (Schroeder and Fulton 2014). Darkhouse spearers are more harvest oriented, while pike anglers often engage in voluntary catch-and-release (Pierce et al. 1995, Pierce and Cook 2000, Margenau et al. 2003), and prefer to target large-sized pike (Schroeder and Fulton 2014). Although both darkhouse spearers and anglers prefer lakes with liberal harvest regulations, darkhouse spearers are reported to have a particularly strong aversion to lakes managed by protected slot-length limits (Schroeder and Fulton 2014). This regulatory preference is most likely related to the higher consumptive orientation of darkhouse spearers and may be fostered by the perceived difficulty of precisely targeting pike of a given length under ice with a spear. However, attitudes towards regulations also vary within pike anglers as a function of harvest orientation. Schroeder and Fulton (2013) found that as the intention to keep fish among pike anglers increased, so did their aversion to stricter harvest regulations. By contrast, pike anglers with a strong inclination to catch large fish increasingly accepted highly restrictive harvest regulations, including total catch-and-release, in contrast to the more consumptive darkhouse spearers (Schroeder and Fulton 2013, 2014). It is very likely that these patterns also hold for other populations of pike anglers along the gradient from consumptive to trophy orientation. When comparing pike anglers in Minnesota with anglers targeting other top-predator fish species, Schroeder and Fulton (2013) found that pike anglers had a particularly strong orientation for catching large, trophy fish. A study in Canada also supported that consumptive anglers mainly targeted walleye, and that pike were increasingly retained when the primary target species walleye was absent from the catch (Hunt et al. 2002). One can conclude that pike anglers are often trophy oriented, but that consumption of fish also commonly occurs among some pike angler segments.

There is substantial heterogeneity in the preferences and other human dimensions within pike anglers. Based on the angler specialization framework, Johnston et al. (2010) differentiated general pike anglers who fish for pike as well as other species, consumptive pike anglers who like to target pike for dinner, and highly specialized pike anglers who often practice voluntary catch-and-release after photographing their trophy catch. The three angler types broadly align along a specialization continuum from the general, non-avid angler who regularly consumes its catch, to a gear specialized trophy pike angler that dislikes the idea to take fish home for consumption and voluntary release fish to conserve rare trophy fishes in the stock (Bryan 1977). In reality, there are no distinct angler types and the progression from one angler segment to another is gradual. However, differently committed and specialized anglers are known to differ strongly in preferences related to setting and attitudes towards management and expectations for catch attributes (Bryan 1977, Beardmore et al. 2011, Johnston et al. 2015).

Data about the preferences for pike angling experiences in angling club waters were used to examine whether the above-mentioned angler classification could be reproduced for pike anglers in Germany. To that end, we reanalysed stated preference data collected from German club anglers in Lower Saxony (Arlinghaus et al. 2014). We used the raw data and specifically searched for differences in preferences among pike angler types for attributes of the pike fishing experience. The latent class analyses reproduced three angler types given the sample size (note that a larger sample size would probably have resulted in a greater number of "distinct" angler segments). The results in terms of the among-angler segment variation in preferences for attributes of the pike angling experience are visualized in Figs. 12.2 and 12.3. Interpretation of preferences (bars and lines) allows characterizing the angler types. Accordingly, the largest segment (63% of all anglers) can be characterized as a moderately specialized pike angler, the intermediately-sized pike angler segment (24%) as the least committed pike angler group, and the third and smallest group (13%) as a highly specialized trophy pike anglers. The reason for this characterization follows (Figs. 12.2 and 12.3).

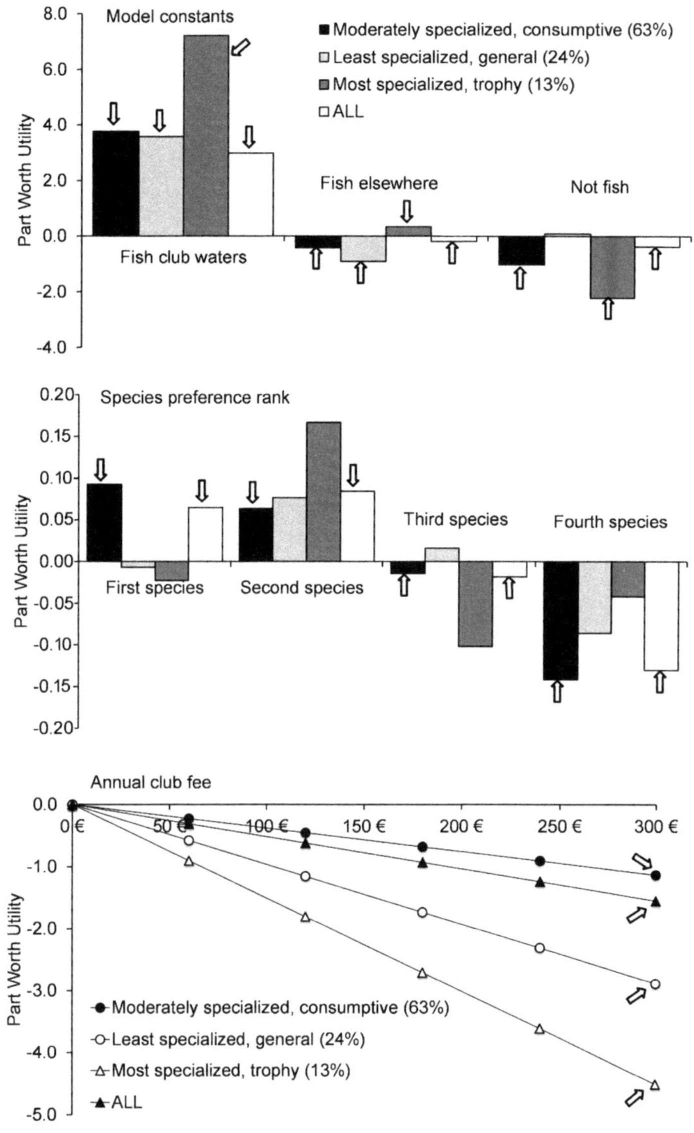

FIGURE 12.2 Results of a choice model for pike anglers in Lower Saxony, Germany. The data presented in Arlinghaus et al. (2014) were reanalysed specifically for pike anglers fitting a standard choice model to the entire sample ("all") and subsequently estimating coefficients for three "classes" (representing three angler types) of anglers using a statistical latent class approach. The terminology given to the three angler types (classes) already integrates interpretation of the regression coefficients (which indicate preferences). Values in parenthesis in the legend indicate the frequency of each angler type in the population. The importance (part-worth-utility) of fishing (as opposed to not fishing), target species, and club fee to anglers are shown. Significant coefficients (where the slope is significantly different from zero) are indicated by an arrow. The uncertainty around the coefficient is not shown to aid in visual interpretation. See text for further explanation.

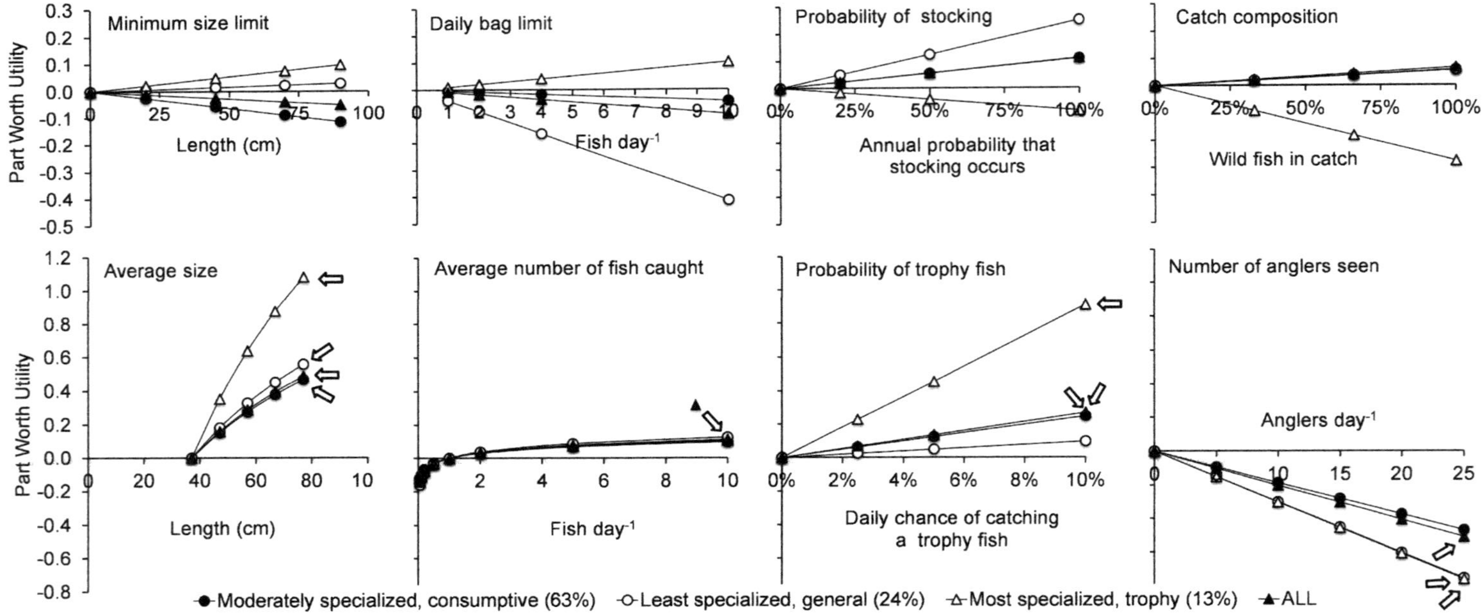

FIGURE 12.3 Results of a choice model for pike anglers in Lower Saxony, Germany. The choice model of Arlinghaus et al. (2014) was reanalysed specifically for pike anglers fitting a general choice model to the entire sample ("all") and then estimating coefficients for three classes (i.e., types) of anglers using a latent class approach. The terminology given to the three angler types (classes) already integrates interpretation of the regression coefficients (which indicate preferences). Values in parenthesis in the legend indicate the frequency of each angler type in the population. Anglers' preferences (as measured by part-worth-utilities) for various managerial, catch-oriented and social attributes are shown. Significant coefficients (slope significantly different from zero) are shown with arrows. The uncertainty around the estimates is not shown to aid in visual interpretation. See text for further explanation.

Trophy anglers were found to benefit strongly from going fishing rather than doing something else in their leisure time (Fig. 12.2). This indicates their great commitment to fishing as a leisure activity. Accordingly, German trophy pike anglers attached a large and significant disutility to the option to not fish and instead preferred to fish in angling club waters (Fig. 12.2). Counterintuitively, perhaps, we found that trophy anglers had the largest aversion to high club fees among all three angler types (Fig. 12.2). Usually, highly committed anglers accept large costs, simply because angling benefits them so much. The reason why we saw trophy anglers disliking high club fees (as a measure of fishing costs) likely relates to the club context of the choice experiment. Note that the specialized trophy anglers were the only group with a positive preference for waters outside angling clubs. Collectively, this suggests that trophy pike anglers do not have a strong attachment to a specific angling club: if angling club fees become too high trophy anglers will fish elsewhere rather than not fish at all (Fig. 12.2). As the name suggests, trophy pike anglers benefited strongly from the probability of catching large, trophy fishes (Fig. 12.3).

The least specialized pike anglers also preferred to fish in club waters, and this angler group disliked fishing elsewhere (Fig. 12.2). However, these anglers did not have a significant disutility for not fishing, indicating their lower attachment to fishing overall as a leisure activity compared to the trophy anglers (Fig. 12.2). The least specialized anglers were also the only group that did not gain utility from a high probability of catching large, trophy pike, which is known to be an indicator of a low degree of specialization in anglers (Bryan 1977). The lowly committed anglers also strongly disliked high club fees, in this case likely indicative of a low psychological attachment to fishing in general (Fig. 12.2).

The largest pike angler segment can be described as generic consumptive anglers of intermediate specialization degree. This angler segment revealed preferences in between the other two angler groups. However, the much less negative aversion to high club fees (Fig. 12.2) suggests this angler type is psychologically more strongly attached to fishing that the least specialized pike anglers. This was also indicated by their preference for fishing as opposed to the option not to fish (Fig. 12.2). Although moderately specialized pike anglers preferred catching larger rather than smaller pike, the preference for pike size was much less pronounced compared to trophy pike anglers (Fig. 12.3). Moderately specialized pike anglers were also more tolerant to crowding compared to the other two anglers types (Fig. 12.3).

In terms of similarities, all angler types preferred catching on average large fish and trophy pike, although as mentioned above the preference for exceptionally large fishes varied among angler types (Fig. 12.3). Significant coefficients (i.e., significant slopes for preferences being different from zero = indifference) were found for the average pike size for all angler types and for the probability of catching a trophy fish for the two more specialized pike angler segments (Fig. 12.3). Also, all anglers showed diminishing marginal utility returns with respect to increasing pike catch rates (e.g. large increases in utility occur at low catch rates but the utility gain is greatly reduced at high catch rates; Fig. 12.3). In this context, it is noteworthy that the catch rate of pike was an insignificant attribute for all three pike angler types, suggesting that other characteristics of the catch (e.g. the size of the fish) produced more consistent utility to anglers than catch rates. All angler types disliked crowding, although this attribute was not significant for the medium specialized pike angler (Class 1) (Fig. 12.3).

Collectively, our empirical data supported the more heuristic categorization of pike angler heterogeneity developed by Johnston et al. (2010). One of the key variable that differentiates the pike anglers is the preference for pike size, which in turn will affect the behaviours expressed by anglers in response to changes in local conditions. Johnston et al. (2013) developed a generic bioeconomic model calibrated to the life-history of pike where three angler types (in their terminology generalist, consumptive and trophy) jointly exploited the pike stock. Interestingly, it was found that at equilibrium the relative frequency of trophy anglers was higher than the frequency of other angler types, whereas other angler types were more attracted to other life history prototypes (i.e., species), such as perch (*Perca fluviatilis* Linnaeus, 1758) or brown trout (*Salmo trutta* Linnaeus, 1758) (Fig. 12.4). Apparently, the pike life history offers certain qualities (e.g. rapid growth, early

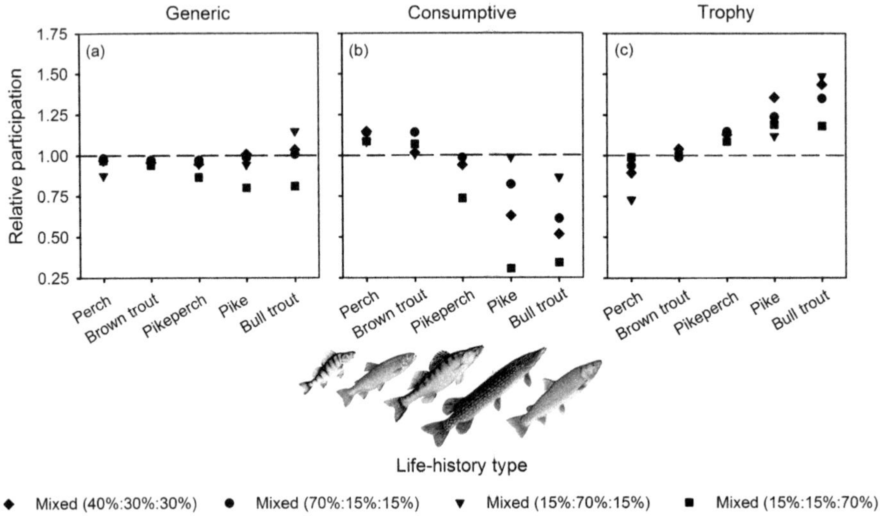

FIGURE 12.4 Relative participation of three angler types under socially optimal input and output regulations across a range of fish species (called life-history types) as an output of the bio-economic model of Johnston et al. (2013). Four mixed angler populations (indicated by differently shaped symbols) targeting one of the five fish life-history types are shown. The relative frequency of the three anglers types (in that order, generic (a), consumptive (b) and trophy anglers(c)) are shown in the legend at the start of the simulation. The relative participation plotted at equilibrium indicates how prevalent a given angler type was at the end relative to its frequency in the mixed angler population at the beginning. Alternatively framed, the relative participation means the ratio between the proportion of the fishing effort attributed to an angler type, and the corresponding proportion of that angler type in the mixed angler population at the onset of the simulation (shown in the legend). Therefore, values > 1 suggests that the relative frequency of a given angler type at equilibrium is greater than expected by the prevalence of that same angler type in the beginning of the simulation, which in turn represents an attraction effect to a given fish species. See Johnston et al. (2013) for model details.

maturation, large terminal size) that are particularly attractive to anglers with trophy-catch orientation. The theoretical model thus supports empirical observations that pike is a particularly attractive species to anglers with a high desire for catching trophy fish (Arlinghaus et al. 2014, Beardmore et al. 2015). This is not to say that smaller-sized, consumable pike are unattractive to recreational fishers. In fact, there are many consumptive pike fishers too (Pierce et al. 1995). However, both the theoretical finding by Johnston et al. (2013) and the empirical work by Beardmore et al. (2015) on angler satisfaction reinforce the point that it is the trophy (size) and challenge qualities that seem to attract many anglers to fish for pike.

The presence of different angler types with varying preferences (and behaviours) within an angler population leads to divergent views about how to best manage pike stocks (Arlinghaus 2005). For example, it is not possible to maximize the biomass yield of pike for consumptive anglers while at the same time maintaining large fish in the stock for trophy pike anglers to enjoy (Johnston et al. 2010, 2013, 2015, Gwinn et al. 2015). Hence, harvest and trophy experiences have to be differentially managed if optimal outcomes are desired for a single fishery. For example, reserving certain water bodies for consumptive fishers and managing these with very liberal harvest regulations could be coupled with the management of trophy lakes for trophy fishers where the harvesting opportunities are constrained. By developing fisheries to meet the expectations of specific angler types, these anglers types could best be served because it is usually impossible to maximize the quality of the fishing experience for all angler types jointly in a single fishery. When compromise solutions are sought for a given fishery, certain harvest regulations, such as harvest slots, that conserve both young and large, old fishes, may offer better outcomes than traditional minimum-size limits (Gwinn et al. 2015), but such tools can never be optimal for all angler types (Johnston et al. 2010, 2013, 2015).

12.3 THE SCIENCE OF PIKE ANGLING

In the last couple of years, the gear and lure market for pike fishing has exploded. One can now find a large variety of wobblers, crankbaits, spoons, spinners, spinnerbaits and soft plastic lures, alongside a rich array of tackle, rods, reel and equipment particularly tailored to pike fishing. In the social media, it is common to observe wild speculation among anglers about the importance of lure types, colours, shapes, etc. for affecting pike catch rates. Fortunately, there is an emerging science, summarized below, to support (and sometimes refute) many popular claims about what affects pike catch rates.

12.3.1 Lure Type

Although not significant on statistical grounds, Arlinghaus et al. (2008b) reported that the average catch rates offered by artificial lures were twice as high as the ones realized by natural bait in small, natural lakes in Germany, suggesting that bait type can impact catches by anglers. However, pike can learn to avoid artificial lures when regularly exposed to them, while no such learning effect has been found in relation to natural bait (Beukema 1970). It can thus be speculated that more natural lures may be more effective in the long term compared to less naturally appearing alternatives. Arlinghaus et al. (2017) recently conducted an angling experiment in a German lake where sites and gear as well as lure types were fully controlled. They found, in line with expectations, that the vulnerability of pike to soft plastic shads was greater than the vulnerability in relation to metal-based spoons. Soft plastics also tend to catch larger fish than other lures types of a similar size (Stålhammar et al. 2014). This work underscores recent developments in pike-angler communities that often strongly favour soft plastic lures given their more natural shape and action patterns. It is unknown whether and to what degree lure colours play a role in driving pike catchability, but recent research in largemouth bass (*Micropterus salmoides* Lacépède, 1802) – which also is a visual predator – has called the importance of lure colours into question (Moraga et al. 2015). Some fishes can sense UV light (Losey et al. 1999), thus, there is still room for further research to test for systematic effects of lure colours in relation to catchability in pike.

Different lures and hook types were found to be ingested differently by pike, with ingestion depth being a function of the size, shape, and texture of the bait, as well as the method of retrieval (active versus passive) (Dubois et al. 1994, Arlinghaus et al. 2008b, Stålhammar et al. 2014). One can reduce the incidence of catching small, undersized pike by using lures that are larger than 15 cm (Arlinghaus et al. 2008b). In addition, "softer" baits tend to be ingested more deeply (e.g. spin fly, Stålhammar et al. 2014, or small soft plastic shads, Arlinghaus et al. 2008b) than stiffer alternatives. However, Stålhammar et al. (2014) found that if soft plastic lures are large enough, deep hooking can be largely avoided. Larger fish with larger gapes were not found to ingest lures more deeply, but they bled more often, suggesting a more vigorous fight elevating hook wounds (Stålhammar et al. 2014). A particularly high incidence of deep hooking and subsequent hooking mortality was reported in a study in the USA where so called Swedish hooks were used with baitfish (Dubois et al. 1994). Work by Pullen et al. (2017) has revealed that pike can quickly get rid of lures after "break-offs" and that pike show little physiological, metabolic or behavioural reaction to embedded artificial lures in simulated break-offs.

12.3.2 Habitat

Obviously, one can only capture fish that encounter the lure. Pike have a high affinity for structured habitat (e.g. Kobler et al. 2008a, b), and can thus be expected to be closely associated with vegetated habitats of the littoral zone, which are used for shelter or as a hunting refuge (chapters 3, 5). This particularly applies for smaller sizes classes below 55 cm total length (Grimm and Klinge 1996). However, pike have difficulties hunting in structurally dense vegetation (Grimm 1994), preferring the intermediate zones among vegetated and the sublittoral for hunting or the interface of structure and open water (Chapman and Mackay 1984, Eklöv 1997). In line with these biological realities, Arlinghaus et al. (2017) found that pike catch rates were significantly elevated in areas offering medium levels of structural complexity in littoral zones, while catches were lower in the pelagic and in densely vegetated habitats. However, it has also been reported that pike form groups of so-called behavioural types that differ strongly in behaviour while being similarly sized (Kobler et al. 2009). Some behavioural types of pike readily search for food in the pelagic too (Kobler et al. 2009), and hence one can of course catch fish in areas with little structure as well, particularly when high turbidity offers protection and facilitates free roaming through the pelagic (Anderson et al. 2008). It is particularly the fish larger than 50 cm that can be expected to be regularly captured outside refuges or even in the pelagic, because larger pike increasingly lose their attachment to vegetation (Chapman and Mackay 1984, Vøllestad et al. 1986, Grimm and Klinge 1996, chapter 5). Indeed, Crane et al. (2015) recently reported that the maximum body size of fish captured by angling was substantially larger than those captured by electrofishing. Electrofishing captures fishes that are bound to structure in the littoral, suggesting that the fish that are angled have faster growth rates and achieve a larger terminal size. It has been reported that the more active behavioural types of pike also tend to grow more (Kobler et al. 2009), suggesting that one largely catches particular behavioural types in particular habitats (Pieterek 2014, Pieterek et al. 2016).

12.3.3 Abiotic Factors

Mogensen et al. (2014) reported that the catchability of pike was substantially increased at northern latitudes, most likely due to the more compact growing season elevating hunger and aggression. Pike are generally faster growing in Europe compared to the USA due to a combination of climate, latitude and the prey resource base (Rypel 2012). In a large among-lake comparative study in Wisconsin, angler catches of large pike were found to be greater in large, deep and hence cooler waters (Margenau et al. 2003). There was also a strong seasonal trend in pike catch rates by anglers, with catch rates being lowest during summer and highest in May and June (Margenau et al. 2003). Likewise, Jansen et al. (2013) found higher pike catch rates in May compared to July in a large lake in Denmark, but experimental work by Arlinghaus et al. (2017) found pike catch rates to be greater in autumn compared to May in a small German research lake. The caveat remains that Arlinghaus et al. (2017) only studied two one week-long fishing sessions. In the USA, open-water pike fishing generated higher catch rates than ice fishing (Margenau et al. 2003), but lakes offering high catch rates usually also produced, on average, smaller fishes in the catch (Margenau et al. 2003). Kuparinen et al. (2010) studied a full season of experimental pike angling data from a small, natural lake in Germany, reporting that the catch rates of pike increased at twilight, at low water temperatures, at times with greater wind speeds, and during full and new moon phases (Fig. 12.5). The authors did not detect any influence of barometric pressure or rainfall on pike catch rates, possibly due to correlations with the mentioned abiotic factors. The factor having the greatest relative effect on pike catch rates by anglers, however, was the level of fishing pressure during the days prior to a test angling event, suggesting that human-induced influences on pike catchability might override abiotic ones. Overall, 20% of the variation in catch rate was explained by abiotic and fishing-related factors (Kuparinen et al. 2010), suggesting a substantial amount of unexplained variance. Nevertheless, the importance of temperature for driving pike catch vulnerability is undisputed. For example, Casselman (1978) found that mesothermal pike are not overly active and hence reactive to gear at water temperatures beyond 21°C. The best pike fishing can indeed be expected in cooler periods, e.g. spring or autumn (Margenau et al. 2003). Water temperature also affects the depth of hooking, with deeper hooking happening at low water temperatures, suggesting intensive foraging in cold water (Stålhammar et al. 2014). At the same time, pike are particularly resilient to catch-related stressors at low water temperatures and show little to no mortality when angled and released under ice (Louison et al. 2017).

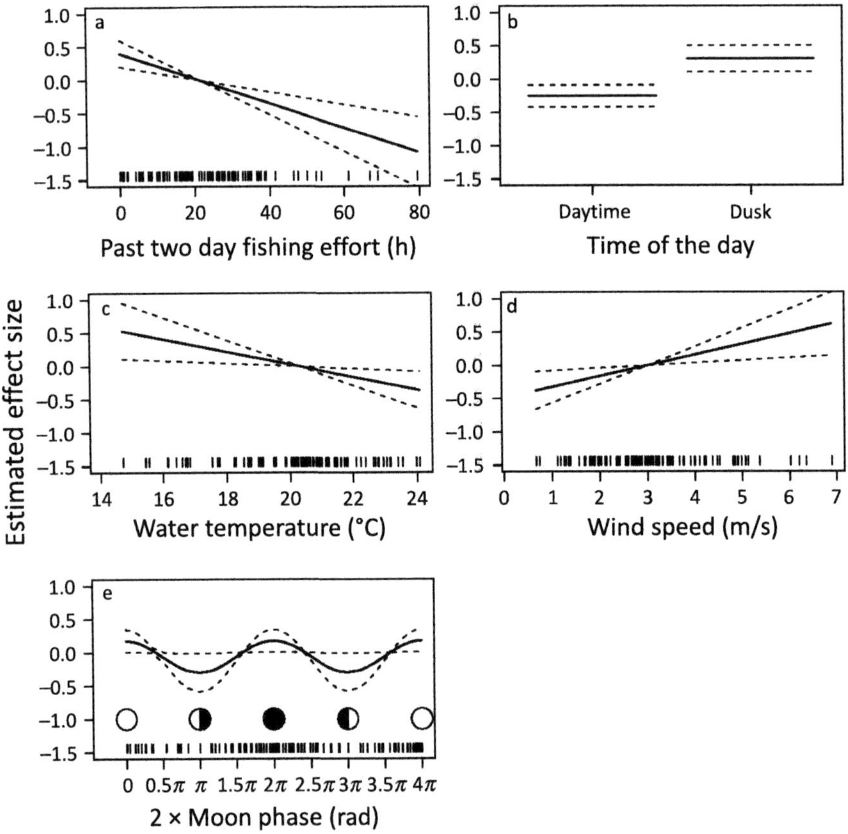

FIGURE 12.5 Effect size of various determinants of pike catch rates in a small lake in Germany (reprinted with permission by Elsevier from Kuparinen et al. 2010).

12.3.4 Individual Traits

In addition to gear, habitat and timing of fishing, a range of individual traits affect a pike's likelihood of capture. First and foremost, pike angling is positively size-selective, and larger fishes, which are often females (Casselman 1975), have a higher probability of capture compared to smaller ones (Pierce et al. 1995, Arlinghaus et al. 2009a, Pieterek 2014, Pieterek et al. 2016, Tianinen et al. 2017). Larger fish have higher absolute metabolic demands, are more active and possibly more aggressive, are less gape limitated, and may be more bold regularly roaming outside refuges, all of which might contribute to the positive size-selectivity observed in pike angling. Relatedly, it was found that faster growing individuals are more likely to be captured compared to slower growing conspecifics (Pieterek 2014, Crane et al. 2015). Pieterek (2014) reported that fish that grew fast when young were more readily captured independent of size at capture, suggesting that juvenile growth rate is a surrogate for a vulnerable behavioural phenotype. Lastly, using high resolution tracking data, Pieterek (2014) showed that more active pike (Fig. 12.6) and those with larger home ranges (Fig. 12.7) were more likely captured by angling than more sedentary individuals (see also Pieterek et al. 2016).

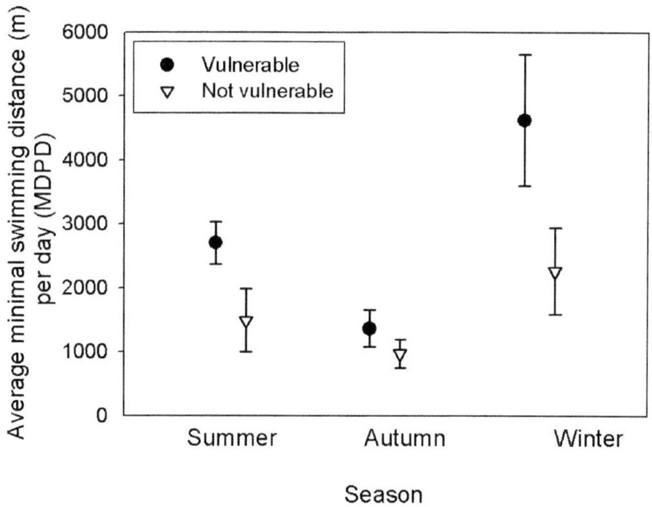

FIGURE 12.6 Minimum swimming distance of pike vulnerable and invulnerable to angling in Kleiner Döllnsee, Germany, in three seasons (modified from Pieterek et al. 2016). The pike were assigned as vulnerable or not based on the historic catch record whether they were ever captured by rod and reel or not (alternatively captured by electrofishing only).

The findings just described align with telemetry work by Kobler et al. (2009) who reported the presence of up to three distinct pike behavioural types in the same study lake that Pieterek (2014) examined. One of these types – the so called habitat opportunist – regularly visited the pelagic areas and was overall more active. The very same behavioural type also showed a tendency for faster growth, which could only have been achieved by higher prey consumption rates compared to less active conspecifics (Kobler et al. 2009). A higher prey consumption rate increases the likelihood of being captured. We can conclude that anglers seem to catch particular behavioural types (personality, Mittelbach et al. 2014) within pike populations, namely the active, exploratory, bold and possibly aggressive individuals. Conversely, it has become abundantly clear that some pike are highly invulnerable to angling, in particular sedentary behavioural types that seek

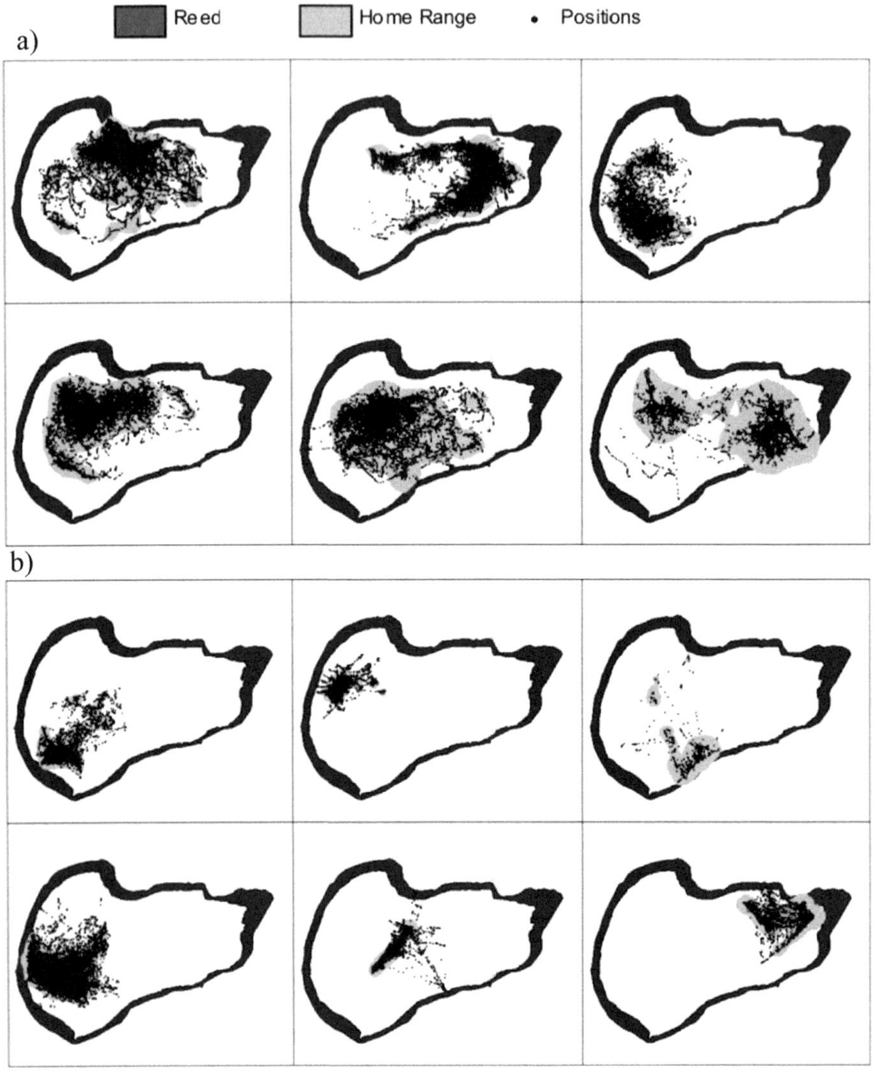

FIGURE 12.7 Individual-specific home ranges (95% kernel distribution) of a) vulnerable and b) invulnerable pike in Kleiner Döllnsee, Germany, during summer (July, 15 to August, 11, 2010) assessed using high resolution acoustic telemetry (modified from Pieterek et al. 2016).

dense refuge (e.g. the so-called reed selectors, Kobler et al. 2009). Thus, these less vulnerable individuals might survive even highly intensive angling pulses, in turn preserving the population against overexploitation through the process of fisheries-induced evolutionary change towards increased timidity (Arlinghaus et al. in press). However, at the same time angling is bound to induce evolutionary (i.e., genetic) changes in vulnerability-related behaviours (i.e., elevated timidity), which will cause ecological consequences for social groups, population dynamics and predator-prey interactions of insofar unknown relevance (Arlinghaus et al. 2016a, 2017).

12.3.5 Learning

As already mentioned, pike learn to avoid artificial lures (Kuparinen et al. 2010, Arlinghaus et al. 2017), but not natural bait (Beukema 1970). In addition, the behavioural selection just described might induce evolution of low vulnerability phenotypes over time (i.e., adaptive genetic change, Alós et al. 2012, Arlinghaus et al. 2016a, 2017). Overall, however, pike are poor learners compared to other fish species (Coble et al. 1985), which is largely why time series of heavily exploited pike stocks show a lot of stochasticity (see Fig. 1 in Kuparinen et al. 2010). By contrast, other species exploited by anglers show a rapid and consistent drop in catch rates after the onset of fishing (van Poorten and Post 2005, Klefoth et al. 2013). Nevertheless, field studies in pike fished with artificial lures have shown that some degree of hook avoidance due to learning happens also in this species, which reduces catch rates over period of days (Kuparinen et al. 2010, Arlinghaus et al. 2017), after which pike seem to recover their original vulnerability.

12.4 BIOLOGICAL IMPACTS OF RECREATIONAL FISHING ON PIKE

Pike have been found to be highly vulnerable to overexploitation (Weithman and Anderson 1976, Mosindy et al. 1987, Pierce et al. 1995, Post et al. 2002). For example, even very low angling effort (1.24 h/ha) removed about 50% of the total pike production in a small, low-productivity boreal lake in Ontario, Canada (Mosindy et al. 1987). Although this is an extreme example, the message is that pike angling can remove a large component of the vulnerable stock within a short period of time. Several factors play a role in explaining the high vulnerability of pike to angling. First, as a top predator natural selection has favoured high levels of aggression and low abilities for learning (Coble et al. 1985). This renders pike especially vulnerable to artificial lures because these lures often tap into the aggressive behaviour of predators. Moreover, the species readily associates with vegetation and other structure, which can be easily identified by anglers, in turn fostering repeated exposure to fishing gear. Finally, pike angling techniques are quite straightforward so that few anglers are constrained from effective pike angling once a basic knowledge of angling is attained.

One can broadly differentiate three forms of overfishing in (recreational) fisheries: growth overfishing, recruitment overfishing and of particular relevance for some angler types – size overfishing (Fig. 12.8). We will investigate whether pike angling can achieve these overfishing states and under which conditions. In addition, we will explore the potential for pike angling to induce evolutionary changes and examine how pike react behaviourally to exploitation. We start with growth overfishing, which is of key importance for fisheries management in general as it is related to the popular concept of maximum sustainable yield (MSY).

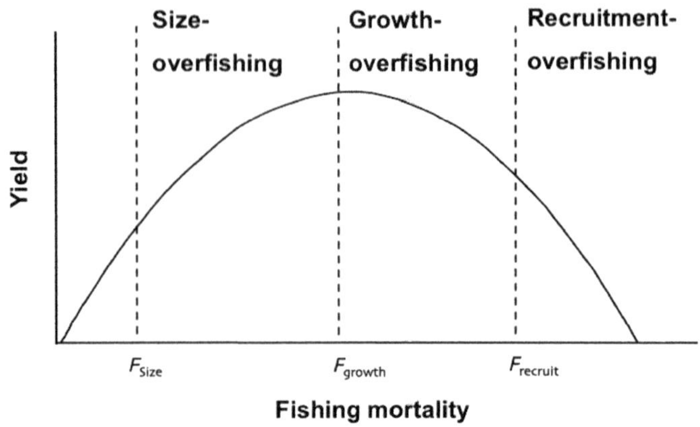

FIGURE 12.8. Conceptual sketch of three overfishing states in recreational fisheries (modified from Radomski et al. 2001). F = fishing mortality rate.

12.4.1 Growth Overfishing

Overfishing is commonly defined based on the concept of maximum sustainable (biomass) yield (MSY). Accordingly, overfishing is achieved when the production of new biomass of the exploited stock is no longer maximal and yield starts to decline as fishing effort, and hence fishing mortality, increases (Fig. 12.8). Individual growth, recruitment, and survival rates all influence the rate of biomass renewal (Ricker 1975), and are thus involved in determining the "turning point" used to judge growth overfishing, i.e., MSY (Fig. 12.8). Assuming that the exploited spawning stock is large enough that recruitment is still unconstrained (which is the defining marker of recruitment overfishing, see below), growth overfishing happens when the fish are taken too young and have not reached their maximum growth potential.

Growth overfishing can easily happen in pike fisheries. To illustrate this, we reanalysed published biomass, exploitation and release rate data from Pierce et al. (1995) (Table 12.1) to determine whether reported or theoretically possible removal rates of pike exceed MSY. Several rules of thumb have been developed to diagnose growth overfishing in relation to MSY. Applying these demands some basic knowledge of the instantaneous natural mortality rate, M, in the unexploited state [note and to avoid confusion: one can easily transform annual mortality rates into instantaneous ones using a simple rearrangement, Allen and Hightower 2010]. Although the unexploited natural morality rate M can only be measured with certainty in unexploited stocks, easily estimable parameters from the standard von Bertanlanffy growth equation can be used to approximate M, e.g. using the life-history invariant $M/k = 1.5$ or the temperature-dependent regression model by Pauly (1980). Alternatively, one can measure total mortality, Z, and the fishing mortality rate, e.g. using mark-recapture data or catch curves (Allen and Hightower 2010) and infer M using an additive model of mortality, which is reasonable for adult pike (Allen et al. 1998). Mark-recapture data can also be used to measure F and M jointly (Haugen et al. 2007). Assuming that the instantaneous fishing

TABLE 12.1 Summary information of data presented in Pierce et al. (1995) and associated re-calculations.

Lake	*Density* (#/ha)	*μ (annual exploitation rate in %* per yr)	*Harvest* (#/ha)	*Theoretical exploitation w/o catch-and-release* (#/ha)	*F* (per yr)	*Z* (per yr)	*M* (per yr)	*MSY (Garcia et al. 1989)* (#/ha)	*Ratio Harvest/MSY*	*Ratio Harvest/MSY assuming no catch-and-release*	*Effort* (h/ha)	*% pike anglers*	*Targeted pike effort* (h/ha)	*Catch rate* (#/h)	*Release rate* (%)
Julia	11.0	22	3.1	5.3	0.25	0.45	0.20	3.78	0.82	1.41	40.6	0.26	10.56	0.33	42
Medicine	25.4	16	8.8	28.4	0.17	0.69	0.52	9.89	0.89	2.87	79.6	0.17	13.53	0.91	69
North Twin	29.8	4	2.2	16.9	0.04	0.46	0.42	6.88	0.32	2.46	53.6	0.12	6.43	0.82	87
French	50.0	7	3.5	10.9	0.07	0.92	0.84	22.01	0.16	0.50	80.0	0.18	14.40	0.32	68
Sissabagmah	55.1	9	4.2	16.8	0.09	1.05	0.96	27.42	0.15	0.61	88.4	0.17	15.03	0.32	75
Wilkins	21.2	5	1.8	5.6	0.05	0.92	0.86	9.64	0.19	0.58	57.3	0.30	17.19	0.53	68
Coon	18.2	10	2.4	15	0.11	0.48	0.37	4.12	0.58	3.64	93.7	0.23	21.55	0.23	84

mortality that produces MSY, F_{MSY}, is equal to or smaller than M (Walters and Martell 2004, Zhou et al. 2012, Lester et al. 2014), Garcia et al. (1989) proposed the following simple equation to estimate MSY using a standard biomass-based Schaefer surplus production model:

$$MSY = (M \times B_Q)^2/(2 \times M \times B_Q - Y_Q) \quad (12.1)$$

where M is the instantaneous natural mortality rate of recruits (defined as fishes entering the fishery, which are usually adults in fishery managed by minimum-size limits), B_Q is the average fish standing stock (kg/ha) over the course of a year, and Y_Q is the actual fish yield (kg/ha). The model assumes that the fishery is not already overexploited (and hence the exploited biomass is reasonably high). An alternative rule of thumb approach to approximate MSY is based on the Gulland (1970) equation:

$$MSY = c \times M \times B_0 \quad (12.2)$$

where B_0 is the unexploited biomass per hectar and c is a parameter ranging between 0.1 and 0.5 depending on fish species. The expected unexploited pike biomass can be inferred from published regression models that relate the littoral area below a certain depth (Pierce and Tomcko 2005) and other habitat features (e.g. amount of area vegetated, Grimm 1989) to the expected pike biomass (Fig. 12.9). Applying these regressions to estimate pike biomass in the unexploited state is confined to similar water bodies from which the regression model was estimated.

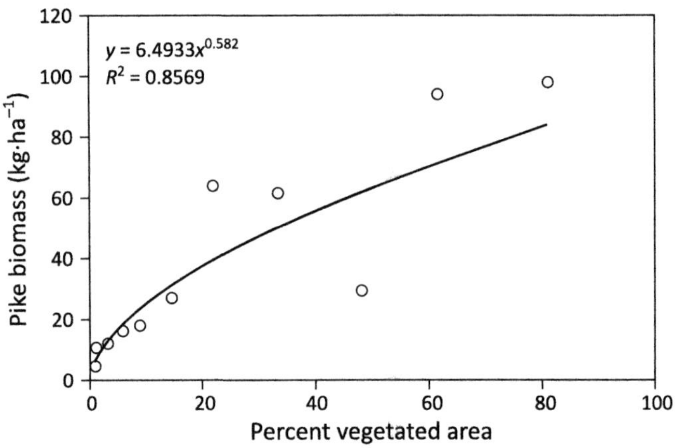

FIGURE 12.9 Regression of the percent vegetated area and pike biomass in small (< 40 ha) and shallow (mean depth < 2 m) standing waters (data taken from Grimm 1989, one outlier removed, which is no longer presented in the figure). Alternative models for deeper lakes also exist (see regression models of pike biomass as a function of littoral area reported in Pierce and Tomcko 2005). Application of these models to local conditions should always be constrained to similar water bodies from which the regression models were built.

Both models (equations 12.1 and 12.2) imply that fish stocks that have a high natural mortality rate are more productive in terms of production of pike biomass (and generation of yield) and thus they can support greater biomass extraction due to the higher biomass renewal rate. Pike is a species characterized by a reasonably high M (Raat 1988) and thus in principle is able to compensate well for fishing mortality without strongly reducing biomass (Tiainen et al. 2014). The coefficient c in equation (12.2) ranges from 0.1 to 0.5 but varies from stock to stock depending on natural productivity. However, it should rarely exceed 0.2 – 0.3 (Beddington and Cooke 1983). Therefore, in the absence of additional information, which is typical for data-poor situations in pike fisheries (Post et al. 2002), the MSY of pike can be approximated as $0.3 \times M \times B_0$. For example, assume the unexploited harvestable stock size is 50 kg/ha, and M of adults is 0.3, then the sustainable take should not exceed 4.5 kg/ha. Using the more sophisticated dynamic age-structured population model of Johnston et al. (2013) and simulating pike stocks from low to very highly productive (by varying parameters of the stock-recruitment relationship, Fig. 12.10a, which is conceptually similar to having different density-dependent natural mortality rates), we can see that the sustainable yield of pike ranges from 0.5 to 5 kg/ha (Fig. 12.10c), which is equivalent to a numerical yield of between 0.5 and 9 pike/ha (Fig. 12.10b) that can be removed annually without damaging the stock. Obviously, more productive stocks than those modelled in Figure 12.10 may exist and thus the yield can be higher than shown in Fig. 12.10 in reality (see Fig. 12.13).

The question now is: do pike-angling fisheries reach unsustainable levels of exploitation? Published annual exploitation rates, μ, in North America range from as low as 3% to a high of 50% (Snow 1978, Goedde and Coble 1981, Mosindy et al. 1987, Pierce et al. 1995). However, many pike in North America are captured and released and thus are not quantified in the harvest rates just mentioned (Pierce et al. 1995). When we use published data from Pierce et al. (1995) to estimate MSY (equation (12.2); Table 12.1) and relate the numerical removal to MSY we can see that the actual take in seven small Minnesota lakes is indeed sustainable and no growth overfishing occurred (Fig. 12.11). However, we can also see that the overfishing state was almost reached at annual exploitation rates, μ, above 16%. We also note that the pike fishery's sustainability was positively related to increasing catch-and-release rates (Fig. 12.12), which agrees with Jansen et al. (2013) and Johnston et al. (2015). Now assume that all fish captured in the lakes reported by Pierce et al. (1995) would have been removed (fourth data column in Table 12.1). Then four of the lakes would experience substantial growth overfishing (Fig. 12.13) at annual harvest rates between 5 and 28 pike/ha. Notably, the angling effort necessary to lead to growth overfishing varied widely in relation to the underlying productivity of the stocks and likely the angler types that are present (Johnston et al. 2013, 2015), ranging from 40 to slightly more than 90 annual angling h/ha (Fig. 12.13). Note also that the targeted pike angling effort was much lower than the absolute angling effort just mentioned (Table 12.1). This means that pike angling effort *per se* is a poor indicator of overfishing potential. Angling efforts reported for pike fisheries range from only a couple of h/ha (Mosindy et al. 1987) to over 300 h/ha (Kempinger and Carliner 1978). The message is: rather than absolute angling effort, what matters for sustainability is the mortality in relation to the underlying resiliency of the stock (represented by variation in M or variation in the slope of the stock recruitment relationship, Fig. 12.10) as well as which angler type is fishing, because anglers vary in their skill level and also their propensity to take fish home (Johnston et al. 2010, 2013, 2015).

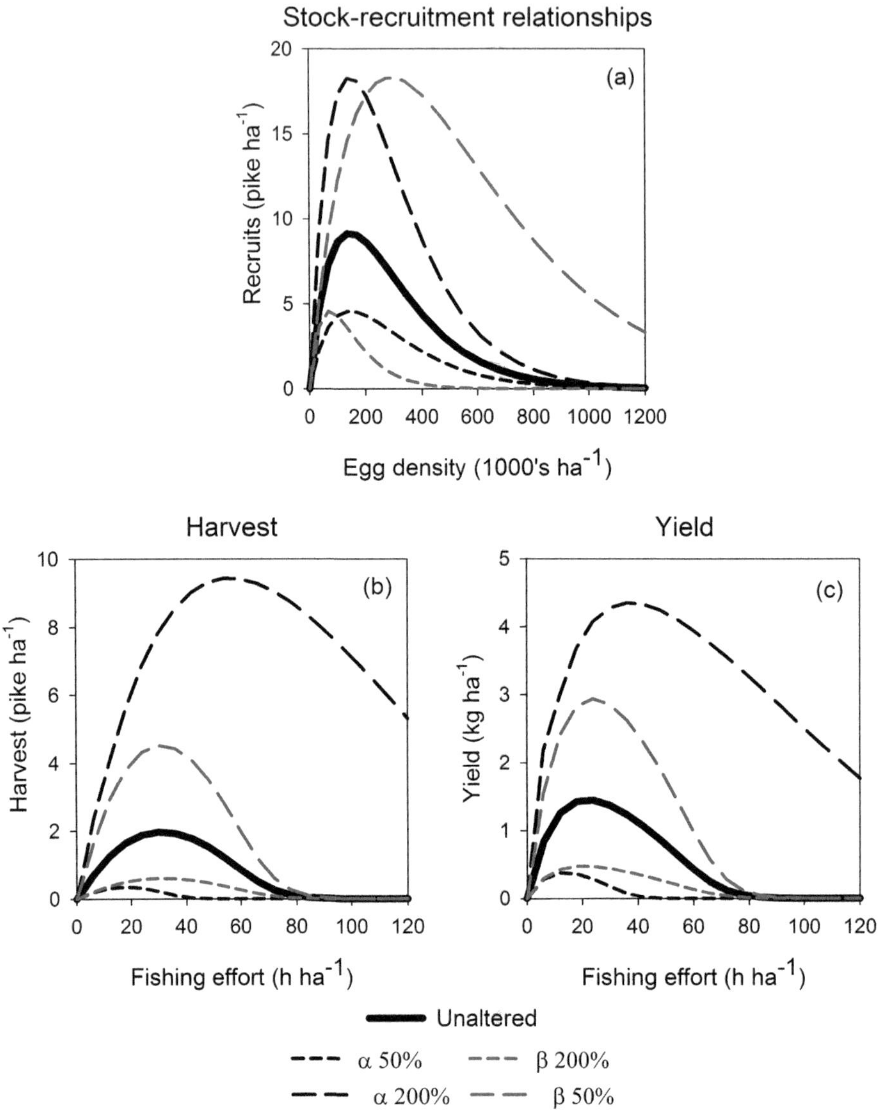

FIGURE 12.10 Yield curves (panel c) and numerical harvest (panel b) of pike as a function of different productivities (stock-recruitment, panel a) predicted by the pike model of Johnston et al. (2013) at the fishing effort displayed. The stock-recruitment curves follow a standard Ricker stock-recruitment curve $R = \alpha S e^{\beta S}$ (empirically supported in pike, Edeline et al. 2008), where α defines the maximum survival rate from spawning to recruitment R (i.e., age-1) when spawner density S is very low, and where β is the inverse of the spawner density that maximizes recruitment, and describes the strength of density-dependent interactions influencing the cohort's survival. Alternatively termed, β is the rate of decrease of recruits-per-spawner as spawner density increases, which is typical for cannibalistic species.

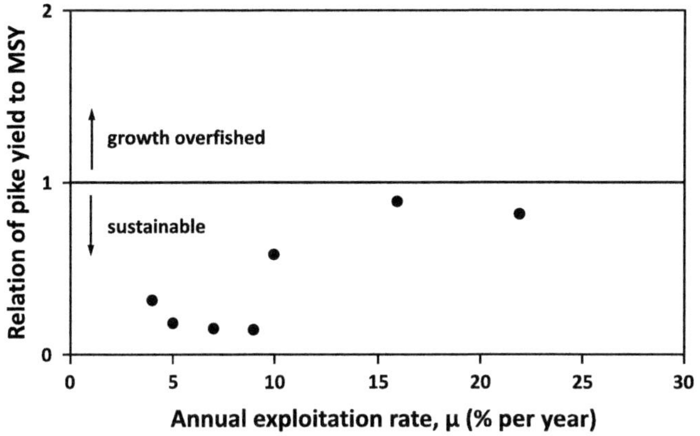

FIGURE 12.11 Sustainability assessment of pike fishing in Minnesota, U.S.A. (second data column in Table 12.1).

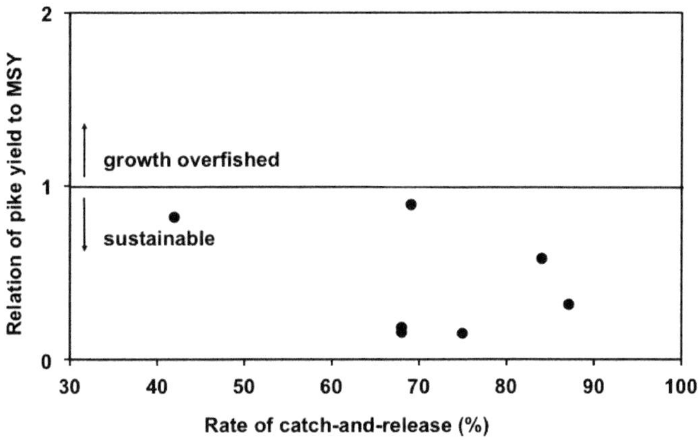

FIGURE 12.12 Sustainability assessment of pike fishing in Minnesota (USA) in relation to catch-and-release rates (raw data in Table 12.1).

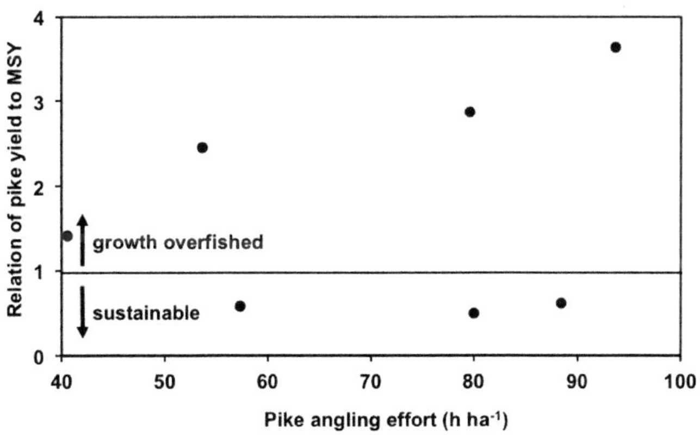

FIGURE 12.13 Sustainability assessment of pike fishing in Minnesota, USA, in relation to untargeted lake angling effort (exploitation data is theoretical exploitation of all captured fishes without catch-and-release, fourth data column in Table 12.1).

An alternative approach to judge growth overfishing is not to look at whether current take is larger than the long-term MSY, but rather to directly estimate current fishing mortality rates and compare them with a reference point of sustainable fishing mortality (Walters and Martell 2004). Lester et al. (2014) recently developed a theory for data-poor situations that accounts for survival and growth compensation and allows a quick appraisal of sustainable fishing rates that are "safe" (F_{safe}) against growth overfishing while promising to approach MSY. Assuming that fish are exploited at sizes larger than size at maturation (which is commonly the case in pike fisheries managed by minimum-length limits) and conservatively assuming no recruitment compensation (which in reality can stem from increased juvenile survival in the exploited state), F_{safe} can be estimated as (Lester et al. 2014):

$$F_{safe}/M = 0.75\ ((h_F/h_M) - 1) \tag{12.3}$$

where h_F is the juvenile growth rate in the exploited condition, h_M is the juvenile growth rate in the unexploited condition, and M is natural mortality of adults in an unexploited state. Note again that this equation is highly conservative because it assumes no recruitment compensation due to increased juvenile survival after biomass reduction. The literature on recruitment compensation in pike is controversial with some papers reporting substantial increase in juvenile density and hence recruitment after removal of adults (Sharma and Borgstrøm 2008, Edeline et al. 2008), while others report monotonic decreases in recruitment with reductions in the spawning stock (Langangen et al. 2011). Interestingly, Edeline et al. (2008) and Langangen et al. (2011) are studies from the same lake (Windermere in the UK). Without additional information, it is therefore prudent to assume no recruitment compensation, bearing in mind that F_{safe} then represents a particularly conservative estimate of sustainable pike fishing mortality and in reality higher fishing mortality rates may well be sustainable.

We directly assessed fishing mortality rates in a range of small lakes managed and exploited by anglers in north-western Germany and compared estimates of current fishing mortality with F_{safe} to diagnose the potential for growth overfishing. To that end, we tagged as many pike as possible with external tags that offered different reward to reporting anglers (to estimate tag reporting rates), and subsequently monitored recaptures using angler diaries over two years in 18 shallow and small (< 12 ha) lakes in north-western Germany. We estimated tag loss in ponds (Hühn et al. 2014) and the tag reporting rate using a reward level of 25 € assuming full reporting at this reward level. Methods outlined in Allen and Hightower (2010) were used to estimate annual fishing mortality rates (μ and correspondingly F, Table 12.2). To estimate F_{safe} we used survival information derived from telemetry data to estimate the M of adult pike larger than 50 cm from an unexploited lake in the same latitude, but situated north-western Germany ($M = 0.08$, 95% confidence interval 0 – 0.18, Pagel et al. unpublished data). The growth compensation of juveniles was approximated from a multi-lake meta-analysis of the Brodie coefficient k of the von Bertalanffy growth equation, which revealed a three-fold growth compensation potential in pike as a function of density (i.e., pike at low density have a three-fold higher k than the same genotypes of pike at high density, Pagel et al. unpublished data). Using the upper confidence interval for the telemetry-based M ($\approx$ 0.2) and assuming a three-fold growth compensation equates to an estimate of F_{safe} = 0.3 per year using equation (12.3), which can serve as a first reference point for sustainable pike fishing. We successfully estimated μ and hence F (= –ln(1 – μ)) in nine of 18 lakes we surveyed (Table 12.2). Some of the estimates were quite uncertain due to low sample size (Fig. 12.14). F varied widely among lakes and between years within lakes from as low as 0 to a high of 0.93 per year. Current F exceeded F_{safe} in Darnsee in both 2011 and 2012 and in Vörum 2 in 2011. Moreover, the fishing mortality exerted by anglers in two further lakes was close to the threshold value (0.27 Vockfeyer See in 2011 and 0.28 in Handorf in 2012). Put differently, of the 18 estimates of F in pike angling, 16% (N = 3) were considered unsustainable from a growth overfishing perspective, but the vast majority of stocks were sustainably exploited.

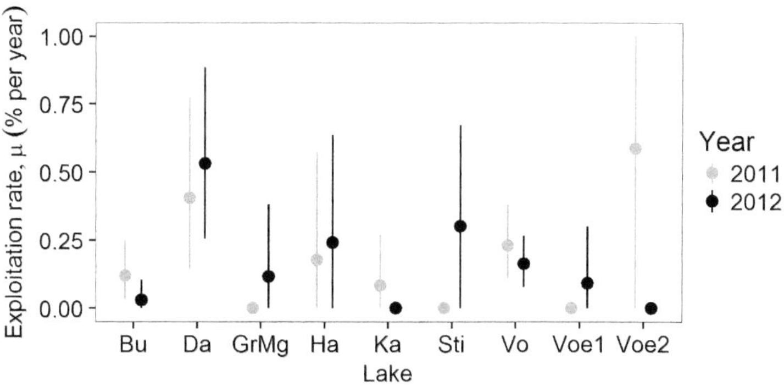

FIGURE 12.14 Estimated annual exploitation rate, μ, and corresponding 95% confidence intervals for pike stocks in nine German lakes in two years (see Table 12.2 for details). Abbreviations relate to lake names in Table 12.2.

TABLE 12.2 Annual exploitation rate, μ, (mean and SD of 1,000 bootstrapped iterations) and the corresponding instantaneous fishing mortality rate, F, of pike in 9 German lakes exploited by anglers in 2011 and 2012.

Gravel pit lake name	**$\mu \pm SD$ (%, 2011)**	**$\mu \pm SD$ (%, 2012)**	**F (yr^{-1}, 2011)**	**F (yr^{-1}, 2012)**
Buschmühlenteich	0.12 ± 0.06	0.03 ± 0.03	0.13	0.03
Darnsee	0.41 ± 0.16	0.53 ± 0.16	0.53	0.76
Große Mergelgrube	0.00 ± 0.00	0.12 ± 0.13	0.00	0.13
Handorf	0.19 ± 0.18	0.24 ± 0.17	0.21	0.28
Karpfenteich	0.09 ± 0.09	0.00 ± 0.00	0.09	0.00
Stiegerteich	0.00 ± 0.00	0.29 ± 0.16	0.00	0.34
Vockfeyer See	0.24 ± 0.07	0.17 ± 0.05	0.27	0.18
Vörum 1	0.00 ± 0.00	0.09 ± 0.09	0.00	0.09
Vörum 2	0.61 ± 0.28	0.00 ± 0.00	0.93	0.00

Fishing mortality rates were estimated taking tag loss, tagging mortality and reporting rate into account by following an R code provided by Daniel Gwinn (2015) (https://dgwinn.wordpress.com). The confidence intervals are visualized in Figure 12.14.

To conclude, there is abundant evidence that fishing mortality rates on pike vary widely and that many stocks are sustainably exploited. However, our analyses also show that growth overfishing is possible and is likely taking place in several fisheries (Post et al. 2002). This does not mean that growth overfished stocks are necessarily collapsed. In fact, quite the opposite can be expected because pike stocks react with a reduction in juvenile mortality and an increase in recruitment to intensive exploitation (Sharma and Borgstrøm 2008, Tiainen et al. 2014, 2017). This can stabilize biomass, at the cost of a strongly altered size and age distribution (Sharma and Borgstrøm 2008, Tiainen et al. 2014, 2017).

12.4.2 Recruitment Overfishing

If fishing mortality continues beyond reaching the growth overfishing threshold, recruitment overfishing can occur (Fig. 12.8). Models have suggested that recruitment overfishing is indeed possible in pike (Fig. 12.15; see Johnston et al. 2010, 2013, 2015 for details), in particular when harvest regulations are liberal or absent and effort is uncontrolled ("open access"). By contrast, when minimum-length limits are high enough and larger than size at maturation in females (e.g. > 50 cm, Fig. 12.15), recruitment overfishing can be effectively avoided in pike. The potential for recruitment overfishing is not confined to just harvest-oriented fishing. It can also occur when stocks are exploited by trophy anglers who occasionally harvest fish and in addition induce hooking mortality (Fig. 12.15). Pike are less resilient to recruitment overfishing than other fish species, such as brown trout or perch, but more resilient compared to slow growing, late maturing salmonids, i.e., bull trout (*Salvelinus confluentus* Suckley, 1859) (Fig. 12.15).

Recruitment overfishing is conventionally defined as when the spawning stock has been reduced to a point where recruitment is impaired. A typical reference point is a spawning potential ratio of 0.35 (Mace 1994). The stock-recruitment relationship described for pike has been shown to be of Ricker type in two studies (Minns et al. 1996, Edeline et al. 2008, but see Langangen et al. 2011 for an alternative view). A Ricker-type stock recruitment means that within certain limits, due to relaxed cannibalism, recruitment increases as the spawning stock declines due to fishing, while under low spawning stock values recruitment declines due to impaired total egg numbers released by the shrinking spawning stock (Fig. 12.10). Moreover, fishing mortality is additive to natural mortality in adult pike, while it is compensatory for juveniles (Allen et al. 1998). This means that fishing mortality acting on adults comes on top of a base natural morality level, which reduces the compensatory potential of the stock against exploitation. The threat of recruitment overfishing might be obscured to anglers and managers because the Ricker stock recruitment function means that initial removal of adult pike elevates recruitment. Indeed, several exploitation studies have found that hitting a spawning stock hard releases juveniles from cannibalism, fostering recruitment and increasing the abundance of small individuals (Sharma and Borstrøm 2008, Jolley et al. 2008). If fishing pressures continues, however, pike stocks will initially destabilize due to size truncation effects (van Kooten et al. 2007), and eventually collapse as shown

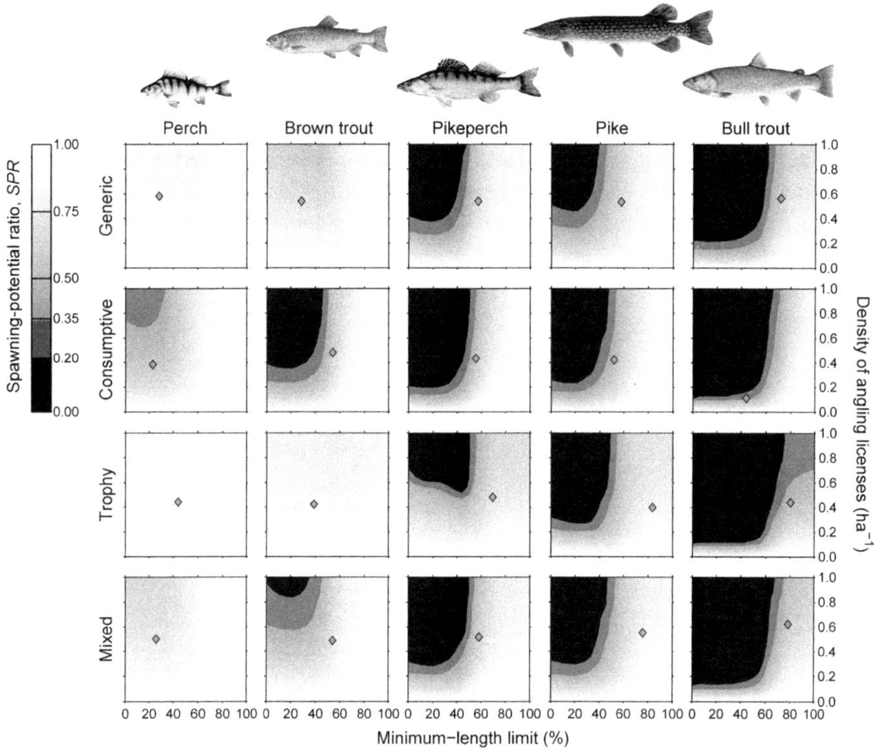

FIGURE 12.15 Spawning potential ratio (SPR) in relation to various minimum-length limits (for comparability across species expressed in % of the maximum theoretical size assumed in the model for each of the five species) and number of anglers (angler licenses) across five popular freshwater fish species. Recruitment overfishing (SPR < 0.35) is seen in the black colour. The diamonds show the optimal regulation that maximizes angler utility (modified from Johnston et al. 2013). Generic, consumptive and trophy in the panels of the first three rows represent three different angler types varying by level of specialization. In the first three rows it is assumed that the angler type represents the entire angler population that is fishing, while in the "mixed" scenario a mixture of the three angler types is assumed to represent the angler population. See Johnston et al. (2013) for further details.

in several areas in Canada (Post et al. 2002). Moreover, Langangen et al. (2011) presented data for Windermere pike that called into question the existence of a Ricker stock-recruitment relationship presented by Edeline et al. (2008), instead suggesting that recruitment monotonically increases (yet at diminishing rates) with an increase in the spawning stock. This would mean that any declines in the spawning stock would automatically reduce pike recruitment. Independent of whether the stock-recruitment relationship in pike is over Ricker or Beverton-Holt-type, the most effective way of avoiding problematic scenarios where recruitment is impaired due to fishing is to keep fishing mortality within sustainable bounds by using a combination of input (i.e., effort) and output controls (e.g. size-limits) (Fig. 12.15). Johnston et al. (2015) showed in a modelling study incorporating a mechanistic model of angler behaviour (of German anglers) that optimally controlling effort and minimum-length limits can effectively avoid recruitment overfishing in pike (Fig. 12.16). However, for this to happen compliance with regulations is of paramount importance, particularly when consumptive ("committed" in the jargon of Johnston et al. 2015) anglers are fishing and hooking mortality of protected fish sizes is high (Fig. 12.16). Note that with increasing hooking mortality the likelihood of recruitment overfishing increases, particularly when managers are not able to optimally control effort and set minimum-length limits (Fig. 12.16). By contrast, if managers can optimally set harvest controls, pike stocks are immune to recruitment overfishing even in the presence of non-compliance or substantial hooking mortality.

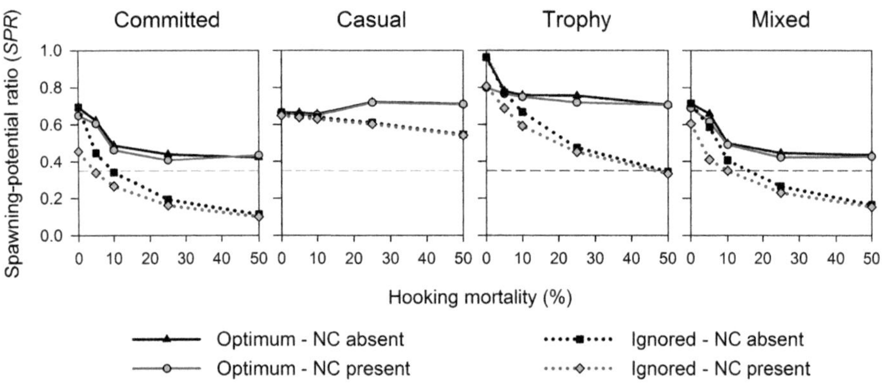

FIGURE 12.16 Impact of hooking mortality and the presence or absence of non-compliance (NC) on recruitment overfishing in pike (recruitment overfishing is indicated by values below the horizontal broken line) for situations where managers can control both effort and set minimum-length limits optimally (to maximize angler welfare) and were such optimal setting of regulations is not possible (ignored). For details on the angler types and the mixed angler population, see the figure legend of Figure 12.15 (modified from Johnston et al. 2015).

12.4.3 Size Overfishing

All pike anglers value the catch of reasonably large fishes (Arlinghaus et al. 2014, Beardmore et al. 2015; Fig. 12.3), but these fishes are quickly lost from exploited stocks (Olson and Cunningham 1989, Pierce 2010, Arlinghaus et al. 2010), leading to demographic truncation and something we here call "size overfishing" (Fig. 12.8). In fact, the average pike size in the exploited stock is a reasonable indicator of overfishing status, similar to other species (Goodyear 2015). Size overfishing can be felt well before MSY or even recruitment overfishing happens and starts at reasonably low fishing mortalities (Fig. 12.8). Radomski et al. (2001) call this reference point left of MSY (Fig. 12.8) "quality overfishing", which is conceptually similar to the idea of "utility overfishing" proposed by Johnston et al. (2010). The basic idea is that once size-truncation through the loss of fish over a minimum desired length has happened, the quality of the fishery for anglers will decline, i.e., overfishing as perceived by anglers has happened through the loss of large fish although the stock per se might not be at risk. Therefore, from both an ecosystem perspective as well as an angler well-being perspective, sustainable fishing mortality rates for pike are usually smaller than F_{MSY}, while F_{MSY} is a reasonable target reference point for commercial fisheries interested in maintaining high biomass yield. The need to compromise among numerical harvest to be distributed among as many consumptive anglers as possible while safeguarding the presence and capture probability of rare trophy pike is one important reason why harvest slots often seem to be superior harvest regulation for managing pike in recreational fisheries compared to minimum-length limits (Arlinghaus et al. 2010, Tiainen et al. 2014, 2017, Gwinn et al. 2015; see sections below for details).

12.4.4 Evolutionary Overfishing

All of the above considerations did not consider evolutionary effects of fishing. Ecological factors (e.g. density dependence) change rates of relevance for producing surplus that can be taken as yield (e.g. growth rates), and these changes are in principle reversible when fishing relaxes. By contrast, evolutionary facts change the heritable features of exploited stocks (e.g. the genetic capacity to growth) and may thus lead to a long-term, irreversible change in productivity-related traits. Intensive and selective exploitation can indeed foster evolutionary adaptation in life-history characteristics such as growth (Edeline et al. 2007, 2009), reproductive investment (Arlinghaus et al. 2009a) and size and age at maturation (Diana 1983, Matsumura et al. 2011) in pike because it usually pays off to develop a fast life history in the face of elevated adult mortality. Interestingly, however, the common intuition that due to the positive size-selectivity of angling, selection pressures should favour small growth, seems unsupported (Fig. 12.17). The actual selection responses depend on a complex interplay of the selectivity patterns on a range of adaptive traits, angling effort and natural selection pressures and can lead both to evolution of smaller adult body sizes when large minimum-length limits are in place or larger adult sizes when harvest slots are in place (Fig. 12.17; Matsumura et al. 2011). Moreover, natural selection operating in opposite directions to fisheries selection might weaken or even reverse fisheries-induced selection on growth rate (Carlson et al. 2007, Edeline et al. 2007). Based on the current understanding, however, it is reasonable to assume that under most situations fisheries selection should elevate reproductive investment and lower size at maturation, which will lower post maturation growth. Moreover, Pieterek's (2014) and Pieterek et al.'s (2016) work shows that angling removes the fish with the fastest juvenile growth rate, which could further reduce post maturation growth and have substantial effects on population dynamics of pike (Vindenes and Langangen 2015, Tibblin et al. 2016). However, if adult mortality is exceptionally large and future reproduction by investment into growth does not pay off, one can also accept the hypothesis that juvenile growth will be positively selected so fish can become as large as possible on the first opportunity to reproduce (Fig. 12.17, e.g. in the absence of regulations; Matsumura et al. 2011).

In addition to life history, pike angling may also lead to selection of inactive, shy behavioural phenotypes, with insofar unknown consequences for catchability and yield (Philipp et al. 2009, Arlinghaus et al. 2016a). A key conclusion is that

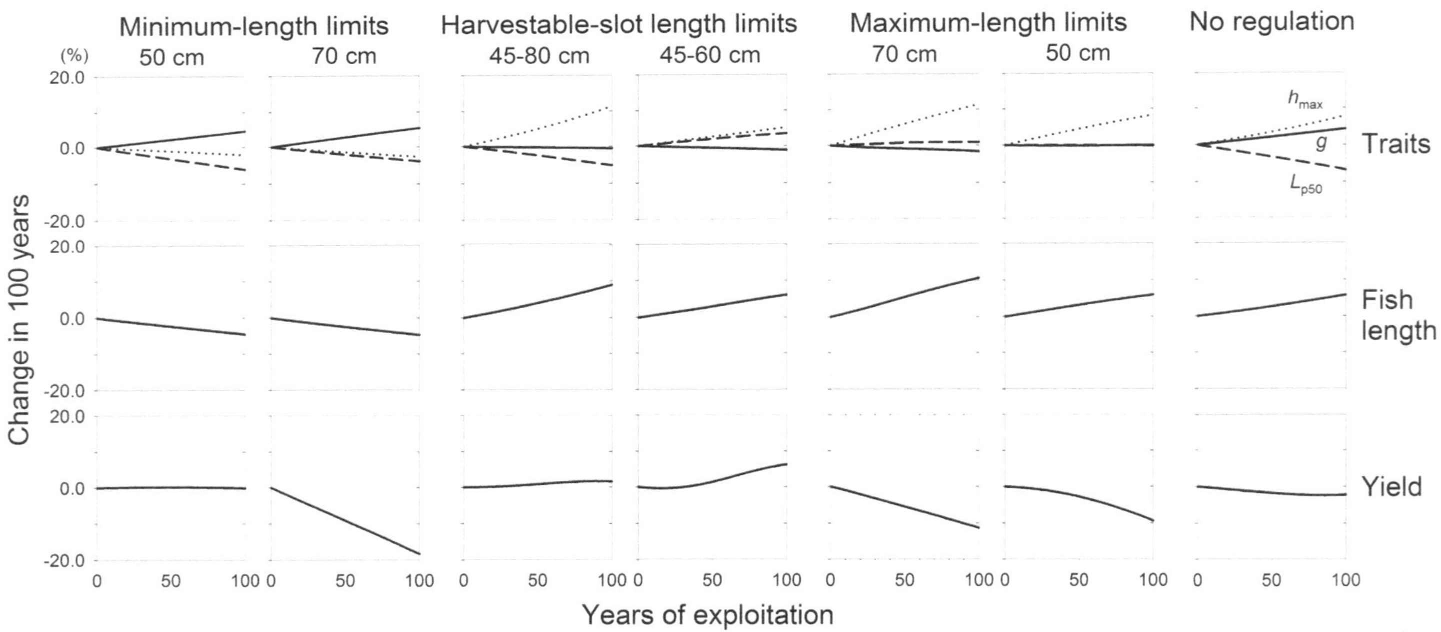

FIGURE 12.17 Predictions of evolutionary changes (in % over 100 years) caused by size-selective pike angling at 100 h/ha for different fishing regulations (data from Matsumura et al. 2011). h_{max} is the maximum juvenile growth rate, g is the reproductive investment and L_{p50} is the 50% probability to mature at a given size. The fish length is shown for an average age-3 pike. Yield is in biomass (kg/ha).

fisheries-induced evolution is probably inevitable (Matsumura et al. 2011), but wise choice of management actions will influence the evolutionary trajectory. When the goal is to avoid selection of lower growth rate, the suggestion is to implement dome-shaped selectivity patterns, which foster growth and yield (Matsumura et al. 2011) as well as lead to disruptive selection fostering trait variation (Edeline et al. 2009), which could be crucial to maintaining the ability of pike populations to adapt, e.g. to ongoing climate change (Vindenes et al. 2014). Evolutionary changes in life history and behaviour could lead to pike populations that do better in the face of exploitation, but are maladapted if fishing pressures cease hampering recovery (Uusi-Heikkilä et al. 2015). Moreover, as a top predator, changes in pike sizes are bound to have food-web effects and can lead to destabilization of pike population dynamics and trophic interactions (Kuparinen et al. 2016, chapter 8). In addition to managing size selectivity as expounded above, keeping fishing mortality within ecological bounds can be recommended as a clear-cut solution bearing in mind that some level of evolutionary adaptation caused by fishing seems inevitable (Matsumura et al. 2011). It is also likely that fisheries-induced evolution reduces catchability of pike to anglers (Philipp et al. 2009), which may be unavoidable and is a challenge to anglers.

12.4.5 Behavioural Overfishing and the Timidity Syndrome

Exploited pike also adapt physiologically (Arlinghaus et al. 2009b) and behaviourally (Klefoth et al. 2008, 2011, Baktoft et al. 2013) to fishing within the realm of phenotypic plasticity. Klefoth et al. (2011) found that targeting pike by boat angling and regular catch-and-release led the fish to reduce activity and increase their use of safe refuges. However, such effects were mild and quickly reversible within hours or days, similar to the findings of Baktoft et al. (2013) on the short-term effects of handling on behaviour in a small lake. Physiologically, pike have been found to be very resilient to catch, fight, air exposure and release, and to modify behaviour within only a few hours (Arlinghaus et al. 2009b). Even harsh treatments, e.g. the attachment of lures to the mouth of pike to simulate "break-offs", did not induce strong behavioural responses (Arlinghaus et al. 2008c). In fact, pike manage to get rid of crankbaits and spoons attached to the jaws within a few hours to a couple of days, with no long-term impacts reported (Pullen 2013, Pullen et al. 2017). The only exception is catch-and-release under high density conditions, where Klefoth et al. (2011) reported a growth depression, most likely resulting from lost foraging opportunity and the stress of capture. Therefore, although hooking mortality is usually far below 5% in esocids (Hühn and Arlinghaus 2011), catch-and-release is physiologically stressful and may affect growth (Klefoth et al. 2011). Catch-and-release, whether partial or total, is thus beneficial because fishing mortality is reduced (Jansen et al. 2013, Johnston et al. 2015), but it is not entirely harmless and must be conducted with care to avoid sublethal impacts and mortality (Arlinghaus et al. 2007, FAO 2012).

Although the short-term behavioural changes caused by pike angling seem minor, the potential for pike fisheries to selectively remove certain highly active and fast growing individuals continues to be prevalent. Behavioural selection might cause patterns that Arlinghaus et al. (2016a, 2017) recently coined a "timidity syndrome". The syndrome implies that catchability of pike might decline over time due to fisheries-induced evolution of behaviour (Philipp et al. 2009) and, additionally, learning to avoid capture with artificial lures (Beukema 1970, Kuparinen et al. 2010, Arlinghaus et al. 2017). Reductions in catchability would improve the population's resiliency to fishing, but might strongly affect the well-being of anglers who depend on catches for achieving satisfaction (Arlinghaus et al. 2014, Beardmore et al. 2015).

12.5 HARVEST REGULATIONS FOR MANAGING PIKE

As was elaborated before, pike are vulnerable to excessive fishing mortality, and quickly respond to exploitation through size and age truncation and increases in the abundance of small, usually anthropocentrically undesired individuals. To manage unsustainable fishing mortality, the manager can turn to a range of regulations in response. Regulations come in two broad variants: input controls manage the effort devoted to a given locality, while output controls are designed to manage and direct the removal rate of pike of different sizes. Typical input controls are direct effort limitations and indirectly protected seasons and areas. Output controls include bag limits of various sorts (daily, weekly, annual), removal tags, and variants of size-based harvest limits, which are commonly applied in pike management (Paukert et al. 2001, Diana and Smith 2008, Arlinghaus et al. 2010).

12.5.1 Input Controls

Controls on fishing effort are usually the least desired management tools by anglers because they directly constrain personal freedom of choice. Integrated models developed by Johnston et al. (2010, 2013, 2015) however clearly showed that to make anglers happy and achieve optimal social yield, controls on inputs are essential to avoid size, growth or recruitment overfishing. In reality, however, many managers are still reluctant to implement effort controls (Cox and Walters 2002), and instead turn to size-based harvest limits and other tools to less directly manage fishing mortality. Alternative input tools that constrain spatial or temporal effort include the implementation of protected seasons (usually during spawning periods) and protected areas. Although strongly resonating with common sense, there is no published study with appropriate controls that has investigated whether protected seasons or sites have their intended effect of maintaining or increasing recruitment, abundance and catch rates of pike. Because pike usually develop confined home ranges (Kobler et al. 2008a, b, but see chapter 5 for exceptions), appropriately spaced protected zones may well have intended effects by saving at least some fish from harvest by anglers, but it is also known that pike engage in large scale habitat change when the fitness landscape changes (Haugen et al. 2006). A life-history model published by Minns et al. (1996) also suggests that protected zones should target juvenile refuge habitat because recruitment is usually determined in this life stage (and not in the egg or the adult life stages).

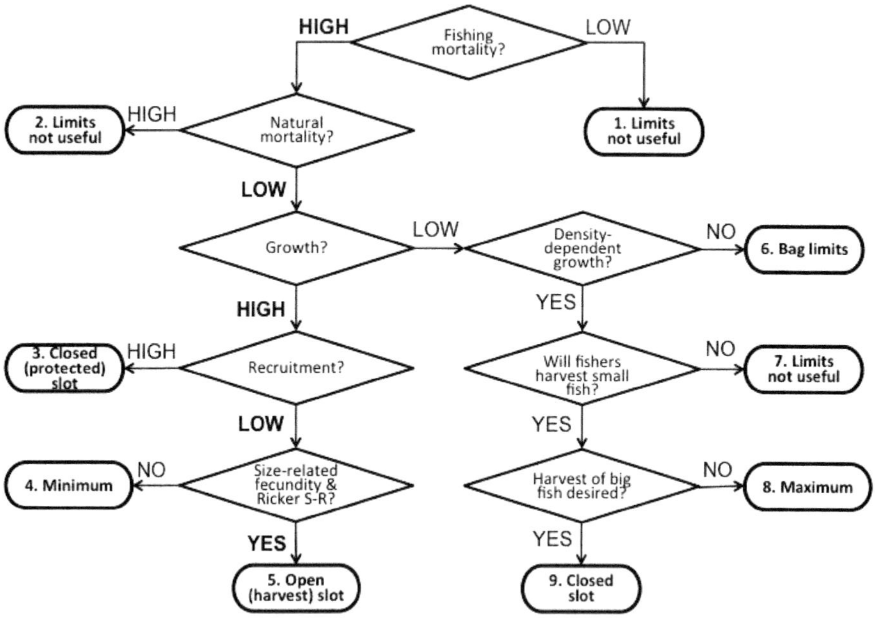

FIGURE 12.18 Decision tree to identify suitable harvest regulations (modified from FAO 2012 and Arlinghaus et al. 2016b). The most likely scenario for most pike fisheries is indicated in bold-face letters.

12.5.2 Output Controls

In central Europe almost all fisheries have some output controls in place to manage pike. The situation is more complex in the USA, where many fisheries lack any form of pike regulation and where a greater variety of harvest regulations are present (Paukert et al. 2001, Diana and Smith 2008). Arlinghaus et al. (2016b) proposed a decision-tree to help managers identify suitable harvest regulations for recreational fisheries in general. Application of this tree to the life-history of pike (Fig. 12.18) suggests that harvest regulations are only an option if (*i*) fishing mortality is high and (*ii*) natural mortality is moderate or low, because otherwise few fish would reach recruitment age where they enter the catch despite the presence of harvest regulations.

When the decision is made that harvest regulations are a possible way forward, a manager can choose among several variants. Size-based harvest limits commonly applied to pike management include minimum-length limits (Min-LL) (Paukert et al. 2001), maximum-length limits (Max-LL, Pierce 2010), harvest slots (HS, also known as kitchen window, open slot or inverse slot limits, Arlinghaus et al. 2010) and the reverse – the protected slot-length limit (Pierce 2012). Protected slots are for example used to cull high density, slow growing, small-bodied pike populations experiencing density-dependent growth (Fig. 12.18, chapter 6). However, few anglers are incentivized to harvest enough small pike (Pierce and

Tomcko 1998), which is why many protected slot length limit function as large minimum-length limits. For reasons of space and practicality, we will thus confine the discussion about the advantages and disadvantages of size-based harvest limits to Min-LL, Max-LL and HS.

12.5.2.1 Min-LL

The rationale for Min-LL is to maintain sufficient recruitment by allowing each spawner to spawn at least once prior to harvest and thereby maintain recruitment and avoid recruitment overfishing and collapse (Fig. 12.18). Indeed, Min-LL larger than size at maturation of females (e.g. 45 cm or higher) are usually effective at preventing recruitment overfishing (Johnston et al. 2013, 2015). However, low Min-LL limits come at the significant costs of leading to a pervasive truncation of the age and size structure (juvenescence effect) (e.g. Arlinghaus et al. 2010, Tiainen et al. 2014, 2017). Pierce (2010) conducted a comprehensive regulation assessment in Minnesota and found that Min-LL were indeed ineffective at maintaining large pike over 76 cm in the stock. Modelling results (Arlinghaus et al. 2010) and whole-lake experiments (Tiainen et al. 2014, 2017) support this important meta-analytical result. Thus, while Min-LL can in fact avoid recruitment overfishing, this comes at the cost of producing a highly unnatural size distribution with many small and young fishes and few large specimens remaining in the stock.

If a manager is interested in not only maintaining recruitment, but also wants to maximize biomass yield, classical age-structured yield/recruit models suggest that Min-LL must be set rather high at about 2/3 of maximum theoretical length to leave the fish enough time to accumulate biomass prior to harvest (Froese 2004). Such high Min-LL however induce very strong selectivity and in evolutionary models have been found to exert particularly strong selective pressures favouring increased reproductive investment and smaller size at maturation, leading to the downsizing of adults and to reductions in long-term yield (Matsumura et al. 2011). Therefore, from an evolutionary perspective, large Min-LLs seem unwise.

12.5.2.2 Max-LL

If a manager wants to maintain large fishes in the stock to benefit reproduction and angler well-being, implementing Max-LLs are an option and have indeed been found to improve the size structure of pike stocks (Pierce 2010). Notably, maintaining large fishes in lakes did not affect the relative abundance of smaller-sized fishes, suggesting predation pressure by the few trophy specimens on smaller size classes is less than previously believed (Pierce 2010). However, Max-LL are only a suitable option if anglers have no interest in keeping large fish and if pike stocks suffer from intensive competition for food in the younger age classes, justifying thinning out (Fig. 12.18). Again, as anglers might not be inclined to harvest very small pike, a Max-LL might ultimately work similar to a harvest slot (Arlinghaus et al. 2010, Gwinn et al. 2015), where both small immature and very large fishes are protected from harvest, the former due to voluntary choice by anglers, the latter due to mandatory release.

12.5.2.3 HS

Harvest slots (HS) have been proposed as a compromise between consumptive and trophy objectives (Gwinn et al. 2015), with the interesting side effect of maintaining a more natural age structure and large fishes in stock (Figs. 12.20 and 12.21; Arlinghaus et al. 2010), thereby fostering an ecosystem approach to fisheries management (Francis et al. 2007). Harvest slots are recommended when growth rates are sufficiently high, recruitment is impaired or is to be safeguarded, and for species, such as pike, that often (yet not necessarily, Langangen et al. 2011) follow a stock-recruitment pattern that is Ricker-like (Edeline et al. 2008; Figs. 12.10 and 12.18). The idea is that large, highly fecund pike serve as a fecundity reserve and contribute to a broader age class distribution in the spawning stock, which is known to stabilize recruitment and dampen population fluctuations (e.g. Hsieh et al. 2006, van Kooten et al. 2007, Anderson et al. 2008, Ohlberger et al. 2014). Moreover, since more of the fecundity of the entire stock is confined to fewer, large, fecund fishes, recruitment is boosted due to reduced degree of cannibalism. Note that the

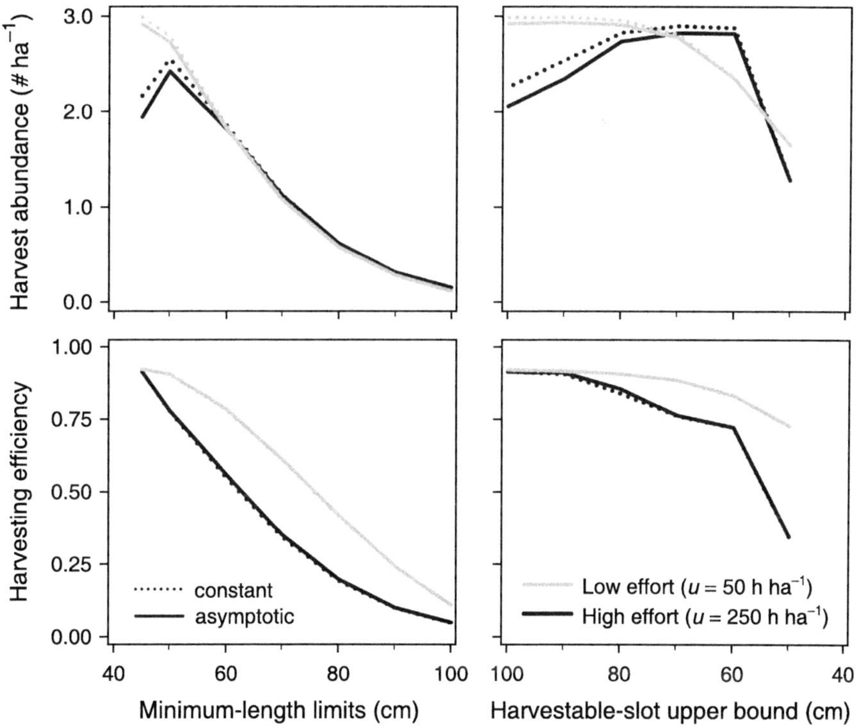

FIGURE 12.19 Impact of minimum-length limits and harvest slots on numerical pike yield and the efficiency of harvesting (at 1 all dead fish are taken home rather than dying from catch-and-release) at two pike angling intensities, u, "low" and "high" (modified from Arlinghaus et al. 2010). The lines indicate simulations with and without size-dependent maternal effects on offspring survival (none = constant, asymptotic = asymptotic increase of offspring fitness with size as documented by pond experiments, Arlinghaus et al. 2010).

mere presence of highly fecund large fishes has been found to be sufficient to justify the superiority of HS over Min-LL to manage pike (Arlinghaus et al. 2010). If, in addition, size-dependent maternal effects on offspring quality and survival are present (as documented in several studies, Arlinghaus et al. 2010, Kotakorpi et al. 2013), the positive effects of saving large fish from harvest would be amplified (Fig. 12.19; Arlinghaus et al. 2010). However, recent research has revealed that maternal effects are of limited importance for natural pike population dynamics when compared with other ecological factors, particularly temperature (Pagel et al. 2015, Vindenes et al. 2016). This finding does not reduce the importance of saving large pike through HS for reason of the fecundity reserve. In fact, simulation models by Gwinn et al. (2015) and Arlinghaus et al. (2010) that omitted size-dependent maternal effects have clearly shown that HS still produces better compromises by generating a maximized numerical yield of medium-sized fishes while maintaining the presence and catch probability of large fishes (Figs 12.19, 12.20 and 12.21) – effects that were never achieved by Min-LLs. Although the model by Gwinn et al. (2015) suggests that maximized biomass yield should be expected with a Min-LL regulation rather than HS, further work has now shown that this prediction only holds for Beverton-Holt-like stock recruitment (Arlinghaus et al. unpublished data). However, when fishes recruit following a Ricker stock-recruitment model, like pike, the biomass yield is also predicted to be maximized using HS as opposed to Min-LL (Arlinghaus et al. unpublished data). Further work has also revealed that HS increase the resilience of exploited stocks by maintaining a buffering capacity, while Min-LL foster recovery after severe overexploitation because the stock is made up primarily of fast-growing, small-sized fishes (Le Bris et al. 2015).

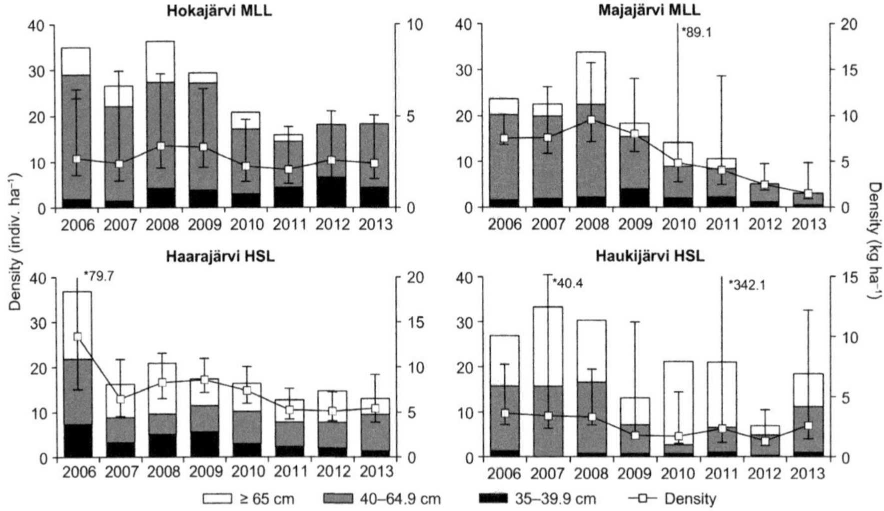

FIGURE 12.20 Pike population number estimates (solid line), total biomasses (columns) as well as biomasses of three sizes classes in Finnish research lakes in years 2006-2013 (from Tiainen et al. 2017, reprinted with permission by Boreal Environment Research Publishing Board). HSL = harvest slot lakes, MLL = lakes exploited by minimum-length limits. Each year 50% of the vulnerable biomass was removed.

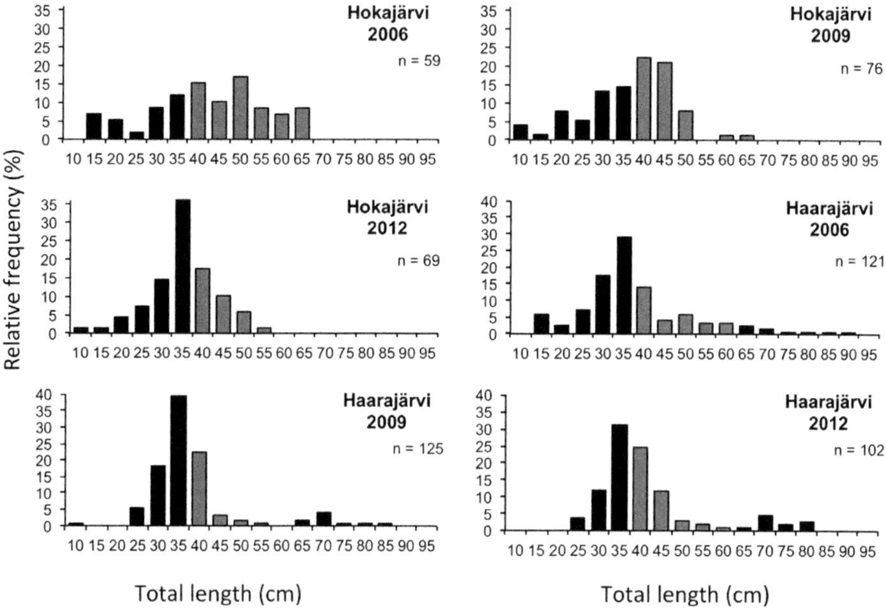

FIGURE 12.21 Length structure over time in one lake experimentally fished with a minimum-length limit (left panels) and one experimentally fished with a harvest slot (right panels). Grey bars indicate the vulnerable size classes that were fished with an annual biomass removal of 50% of the vulnerable biomass (modified from Tiainen et al. 2014, reprinted with permission).

An ongoing experiment at whole-lake level by Tiainen et al. (2014, 2017) has recently confirmed the theoretical model predictions about the superiority of HS over Min-LL for managing pike (Figs 12.20 and 12.21). The authors exposed two lakes to HS exploitation and another two lakes to more standard Min-LL exploitation. In agreement with the theoretical model of Arlinghaus et al. (2010) for pike, the population biomass declined in one of the MinLL-managed lakes, while the biomass and abundance density was stable in both HS-exploited lakes, even at annual removal rates of 50% of the vulnerable stock (Fig. 12.20). Moreover, large fish were entirely lost from the pool of fishes in the Min-LL lakes, while they were maintained at near-natural relative abundances in the HS-managed lakes despite the exploitation (Figs 12.20 and 12.21). HS were also predicted to foster increases in pike length and yield from an evolutionary perspective, while Min-LL were predicted to lead to smaller adult pike sizes (Matsumura et al. 2011). It is also noteworthy that Pierce (2010) did not detect any correlation of the abundance of large fishes and the catch per unit effort of smaller conspecifics, thereby providing evidence that predation by larger fish that are increased under Max-LL and HS regulations must not necessarily constrain the abundance of smaller pike, possibly because small and very large pike use different habitats (chapter 5). The situation might be different in very small lakes that lack refuges.

12.5.2.4 Comparing HS with Min-LL

When evaluating all the evidence (summarized in Table 12.3), from a purely ecological perspective HS certainly promises to provide a more sustainable means of management for pike stocks because HS regulations maintain a more natural stock that also carries large fishes (Francis et al. 2007). In addition, as shown by Arlinghaus et al. (2014) many anglers enjoy occasionally catching large fish, which is only possible in stocks managed either with HS (Fig. 12.21) or with Max-LL (Pierce 2010). The only caveat is that under such regulations the harvest of trophy fish would no longer be possible, which can cause conflicts with anglers who desire to put trophies on their walls or to remove large fishes for consumption. In addition, in some countries where voluntary catch-and-release of trophies is resented on moral grounds, e.g. in Germany, it is our experience that managers are reluctant to implement HS because this might be perceived as fostering apparently illegal catch-and-release of trophies (Arlinghaus 2007). Recent surveys, however, show that the German population morally accepts the release of large fishes as long as this has an underlying ecological motivation (Riepe and Arlinghaus 2014). Ironically, from an ethical perspective HS are superior regulations to large Min-LL because fewer fish are "wasted" to unwanted hooking mortality (Fig. 12.19). Table 12.3 summarizes the benefits and disadvantages of Min-LL and HS and concludes that in most situations HS seem to be the more suitable tool when the stock needs protection from too high fishing mortality. As a general rule the lower bound of a HS should be set above the length at maturation and the upper bound at 2/3 of maximum theoretical length of females to reach optimal compromises (Gwinn et al. 2015). At extreme fishing mortality levels, the upper bound might be moved downwards to about 0.5 of L_{inf} (Gwinn et al. 2015).

12.5.2.5 Other Output Controls

Other commonly applied tools to manage pike stocks include daily bag limits. There is no rigorous evaluation available on the usefulness of bag limits in pike fisheries. Research from other species might provide some interesting insights. Although daily bag limits rarely control the total fishing mortality, they can affect individual's daily take. However, because few anglers reach their limits, bag limits may not lead to more equitable harvest distribution among anglers (Seekell 2011, Seekell et al. 2011). However, this depends on how the bag limits are set in relation to typical daily catch rates. Bag limits of pike in central Europe often involve two or three fish per day (while daily bag limits are often absent or higher in the US, Paukert et al. 2001). Typical catch rates in pike recreational angling in Europe are often well below 0.5 pike per hour (Arlinghaus et al. 2008b, Beardmore et al. 2011, Arlinghaus et al. 2017, Table 12.1). Average catch rates of trophy anglers are closer to 0.5 fish per hour while the less specialized anglers are characterized by lower catch rates (Beardmore et al. 2011). Assuming an average fishing duration of five pike angling-hours per day (Beardmore et al. 2011), average daily captures of a maximum of three pike per day is the norm in many situations. Hence, typical central European daily bag limits of two or three fishes will not substantially affect the harvest by individual anglers, and might not constrain total take. The situation

TABLE 12.3 Overview about the benefits and disadvantages of minimum-length limits (Min-LL) and harvest slots (HS) for manging pike.

Issue	*Min-LL*	*Harvest slots*	*Key reference(s)*
Avoiding recruitment overfishing	Suitable	Suitable	Arlinghaus et al. (2010), Tiainen et al. (2014, 2017)
Avoiding growth overfishing	Suitable if *F* is not too high and limit properly set	Suitable if *F* is not too high and limit properly set	Arlinghaus et al. (2010)
Avoiding size overfishing	Unsuitable	Suitable	Arlinghaus et al. (2010), Pierce (2010) (for maximum size limit), Tiainen et al. (2014, 2017)
Fisheries-induced evolution	Sharp directional selection, selects for small-growing fish and depresses yield	Disruptive selection, selects for fast growing fish, maintains trait variance and increases yield	Edeline et al. (2009), Matsumura et al. (2011)
Behavioural change	Selection for shy and low active and low aggressive fishes	Possibly selection for fast growing, aggressive and active fish or at least reduced selection strength due to disruptive selection	Hypothesis stage, but see Matsumura et al. (2011), Pieterek (2014) for partial support
Maintenance of near natural age structure	Not possible, sharp juvenescence, artificial age and size structure far away from natural status	Possible, old fish maintained if set properly, albeit at smaller abundance, more in line with an ecosystem approach to fisheries management	Francis et al. (2007), Arlinghaus et al. (2010), Tiainen et al. (2014, 2017)
Recovery	Faster because stock consists of mainly young, fast growing fish	Slower because stock is more balanced and less composed of fast growing young fishes	Le Bris et al. (2015)
Resilience	Low, increased stock variability	High due to high buffering effect (age class diversity portfolio effect)	Le Bris et al. (2015)
Numerical harvest	Medium	High	Arlinghaus et al. (2010), Gwinn et al. (2015)
Biomass yield	High (Beverton-Holt stock-recruitment) if set properly	High (Ricker stock-recruitment)	Arlinghaus et al. (unpublished data), Gwinn et al. (2015)
Catch of trophies	Low	High	Arlinghaus et al. (2010), Tiainen et al. (2014, 2017)
Harvesting efficiency	Low when minimum-size limits are high	High as the abundant intermediate size classes are heavily cropped	Arlinghaus et al. (2010)
Suggested procedure	Set > $L_{maturation}$ to avoid recruitment overfishing Set at 2/3 L_{inf} for maximized biomass yield	Lower bound > $L_{maturation,}$ upper bound at about 2/3 L_{inf}	Gwinn et al. (2015)

will be different in high catch rate pike fisheries (cf Rypel 2015). Generally, daily bag limits seem to exert their greatest effect socially by reminding people that fish stocks are limited (Radomski et al. 2001), and hence overly high bag limits are often disliked by anglers (e.g. Dorow et al. 2010) because they are perceived to foster overfishing or the excessive take of fishes by just a few anglers.

Daily bag limits are often combined with Min-LL in the management of pike. There are very few thorough assessments that have examined impacts of combined regulations. Recently, Oele et al. (2016) showed that increased Min-LL coupled with low bag limits are indeed effective at improving the size structure of pike. However, for such effects to happen, Min-LL must be very high (> 80 cm) and daily bag limits very low (one pike per day), i.e., the combined regulations must be very conservative.

12.5.2.6 Compliance and Best Practise in Catch-and-Release

What unites all harvest regulations is that they can only be effective if anglers comply with the regulations (Fig. 12.16; Johnston et al. 2015). To foster compliance with rules, it is recommended to engage in some base level of enforcement. A study in Alberta, Canada, showed that as little as a 3% enforcement rate might be sufficient to achieve a sufficient level of rule compliance in pike fishing (Walker et al. 2007). Of similar importance is to work on awareness of regulations and to carefully communicate the objectives and intentions of any regulatory change to avoid conflict and achieve a smooth implementation.

Similar to compliance, it is important that pike survive the catch-and-release event and that sublethal impacts on behaviour, growth and reproduction are minimized (Arlinghaus et al. 2007). Catch-and-release is an inevitable by-product of any regulation for protected sizes and thus always occurs when harvest regulations are in place. Fortunately, maybe, pike are very resilient to catch-and-release related stressors in both warm (Arlinghaus et al. 2009b) and cold water (Louison et al. 2017) and often show very low hooking mortality levels < 5% (Hühn and Arlinghaus 2011). Johnston et al. (2015) showed that such low levels of hooking mortality do not significantly affect the conservation value of output regulations. Nevertheless, it is always advisable to engage in best practices to further minimize injury and impacts on the fish that is to be released (EIFAC 2008, FAO 2012, Brownscombe et al. 2017). Figure 12.22 provides a general roadmap to best practice in catch-and-release. To identify proper behaviours for the specific target species it useful to further break apart the catch-and-release event into the following six steps: 1) Hooking, 2) Retrieval, 3) Landing, 4) Unhooking, 5) Documentation, 6) Assessment, Recovery and Release.

In relation to the first *hooking* component, it is of importance to minimize the hooking of non-targeted organisms that have to be released, e.g. undersized fishes, and to change fishing sites if many undersized fishes occur in the catch. Arlinghaus et al. (2017) found that larger-sized pike were predominantly found in littoral or pelagic sites, while the smaller conspecifics were predominantly captured in sparsely vegetated, suboptimal habitats, hence choice of habitat can minimize the catch of small, undersized fishes that have to be released. Anglers can also influence the size of fish captured by the choice of lure types or baits, particularly by using large sized baits (Arlinghaus et al. 2008b). The use of artificial baits limits deep hooking and bleeding, which is more pronounced when natural baits are used in pike angling (Arlinghaus et al. 2008b). To further reduce injury, the choice of proper lure types (e.g. avoiding using spin flys who hook more deeply, Stålhammar et al. 2014) and attention to the size of hooks is of relevance, with smaller hooks producing smaller wounds (e.g. Rapp et al. 2008). Recent tackle innovations in predator fishing, such as the "release rig" where the hook is no longer directly attached to the lure, not only increase the landings rates but due to the reliance on smaller treble hooks on artificial lures also minimize injury during the hooking process (Bursell and Arlinghaus, unpublished data).

The *retrieval* process can be optimized by minimizing fight times through matching the line strength and the choice of rod and reel in relation to the fish size that is to be expected. Also, the use of steel leaders is essential to avoid "break-offs" during the retrieval process (Pullen et al. 2017).

The *landing* can be conducted in the most fish-friendly manner by either using wet hands or by (large) knotless rubber nets that strongly reduce mucus abrasion (Colotelo and Cooke 2011). The use of lip gripping devices does not damage the skin (Colotelo and Cooke 2011) but could produce injury to the mouth region and the jaws (Gould and Grace 2009), although these effects may be milder than in other, less "toothy" fish species (Danylchuk et al. 2008). Specialized pike anglers increasingly land pike by sliding one hand along the operculum and lifting the pike out of the water without the use of any technical assistance tool. While this technique avoids mucus abrasion (Colotelo and Cooke 2011) that is almost inevitable when landings nets are used, it is important to support the body of the fish to avoid misaligning vertebra as shown in barrumundi (*Lates calcarifer* Bloch, 1790) held vertically on lip gripping devices (Gould and Grace 2009). It is also important to avoid that the pike "rolls" over the line during the landing process to minimize epithelian damage (Colotelo and Cooke 2011).

The *unhooking* process should ideally be done underwater and as quickly as possible to avoid air exposure. All efforts should be done to avoid the pike getting contract with carpeted boats or stony bottom substrate because this damages the skin (Colotelo and Cooke 2011). Although pike are rather resilient to air exposure (Arlinghaus et al. 2009b, Louison et al. 2017), it is still advisable to keep air exposure as brief as possible if the intention is to make a quick photograph. The "release-rig" recently developed in Denmark for artificial lures allows unhooking underwater even for deeply hooked pike (Bursell and Arlinghaus, unpublished data), but if the fish has to be air exposed it is useful to use special unhooking mats common in carp angling (*Cyprinus carpio* Linnaeus, 1758) (Arlinghaus 2007) to avoid the fish flopping on stones or on the floor of the fishing boat (Colotelo and Cooke 2011). Pike anglers should also always carry pliers and side cutters to facilitate a rapid hook removal, e.g. by carefully cutting hook shanks via the operculum for hooks embedded in the gills carefully (Arlinghaus et al. 2008b). For fishes that are deeply hooked in the stomach, it is advisable to cut the leader rather than to try to operate a treble hook out of the stomach (e.g. Fobert et al. 2009), although delayed mortality is then very likely especially when natural

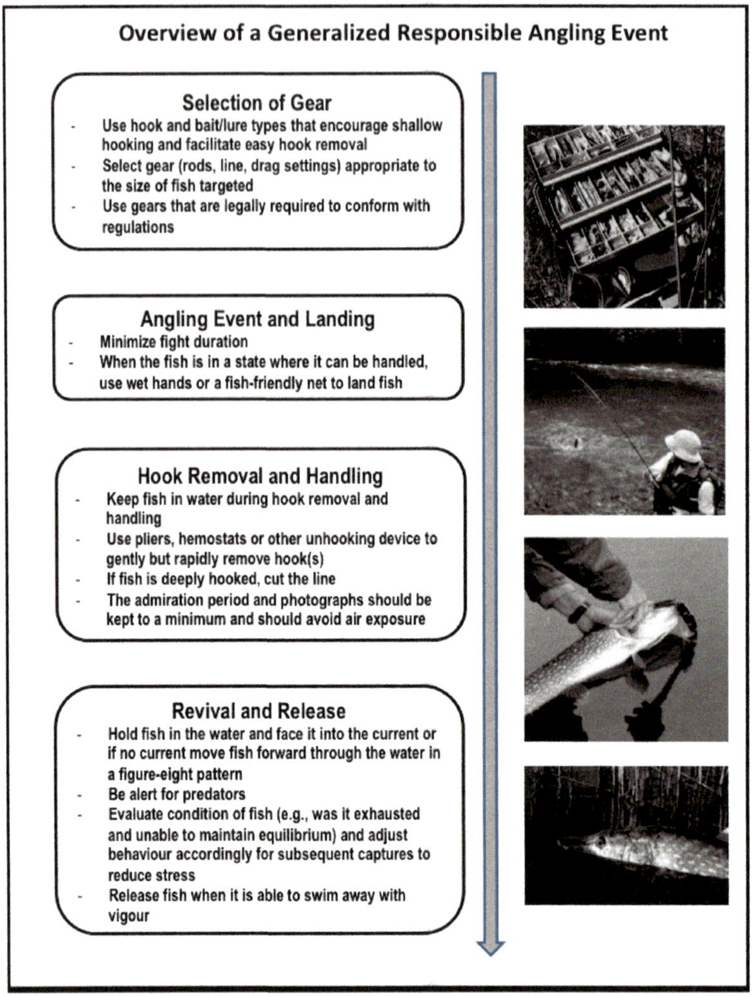

FIGURE 12.22 Best practice recommendations for minimizing impacts on fishes during catch-and-release (modified from FAO 2012 with permission to use the right bottom picture from Philipp Freudenberg).

bait is used (Margenau 2007). Specialized anglers use special handling through one operculum, which facilitates opening the mouth of the toothy predator and allows rapid dehooking without mucus abrasion through flopping on some surface. There is no research available that has studied whether the handling technique is harmful to pike or not, but it is probably safe to recommend one should always support the body and not hold the pike vertically.

The *documentation* phase is meant to check whether the fish is legal to keep, and anglers may take a quick picture, e.g. with trophy fish, as a memory. Ideally, the pike are measured as quickly as possible for length and the photo taken in a few seconds to minimize air exposure and facilitate rapid release. Obviously, the fish is best off if not exposed to air at all.

In the final *assessment* phase that may lead to a release event, the chances that the fish will survive the catch-and-release event is assessed. To that end, a range of reflex impairment measures can be used to assess the viability of the target organism (Brownscombe et al. 2017). Strongly bleeding pike have a reduced likelihood of survival (Arlinghaus et al. 2008b), and as a general rule warmer water is more stressful and problematic for the fishes that are to be released (Arlinghaus et al. 2007). Barotrauma is possibly an issue in pike captured from great depth, although no research is available on this topic. If the fish is to be revived, it should be moved in a "figure-eight" mode through well oxygenated water (Fig. 12.22) and quickly released when the fish is able to swim away on its own.

12.6 CONCLUSION AND RESEARCH NEEDS

Pike are relevant fisheries resources for recreational fisheries in the northern hemisphere. They are vulnerable to overexploitation and even mild fishing pressure results in demographic size truncation. Pike have a high natural mortality and thus have the ability to withstand substantial fishing pressure before collapsing. Pike can be managed sustainably using a combination of input and output controls, supplemented by gear management, while management based on pike stocking seems to be unproductive in many situations (see chapter 9). Harvest slots seem to be particularly suited to the management of pike fisheries, because they create win-win situations for angler well-being and the conservation of near-natural age structures at high standing biomasses. A largely unresolved issue remains with respect to the potential for intensive pike fishing to alter the behavioural composition of exploited stocks in terms of favouring low activity, slow growing and possibly shy phenotypes. Whether behavioural selection fosters "timidity" syndromes (Arlinghaus et al. 2016a, 2017) and affects the ecology of exploited lakes constitutes an exciting research area for the future. Moreover, there is also a need to properly evaluate harvest regulations and to study how anglers respond to changes in the regulatory and ecological environment.

Acknowledgements

Own work presented in this chapter has been funded by the Gottfried-Wilhelm-Leibniz-Community through the grant Adaptfish to RA, the Federal German Ministry of Education and Research (BMBF) through the grant Besatzfisch to RA (grant no. 01UU0907, www.besatz-fisch.de) and the Academy of Finland through a grant to AK. The finalization of this manuscript was made possible with funding within the Baggersee project funded by the BMBF and the German Federal Ministry for the Environment, Nature Conservation, Building and Nuclear Safety (BMUB). The BMBF funded Baggersee as Research for Sustainable Development (FONA); www.fona.de (grant number 01LC1320A to RA, www.baggersee-forschung.de). All involved assistants, students and national and international co-workers are cordially thanked for their input over the years, particularly Ulf Dieckmann, Mike Allen, Jürgen Meyerhoff and Daniel Gwinn. Mikko Olin gave valuable advice on some portion of this manuscript and fostered the possibility to use Figures 12.20 and 12.21.

REFERENCES CITED

Allen, M.S., L.E. Miranda and R.E. Brock. 1998. Implications of compensatory and additive mortality to the management of selected sportfish populations. *Lakes Reserv. Res. Manage.* 3:67-79.

Allen, M.S. and J.E. Hightower. 2010. Fish population dynamics: mortality, growth, and recruitment. pp 43-79. *In*: W.A. Hubert and M.C. Quist (eds.). *Inland fisheries management in North America*. Bethesda, Maryland: American Fisheries Society.

Alós, J., M. Palmer and R. Arlinghaus. 2012. Consistent selection towards low activity phenotypes when catchability depends on encounters among human predators and fish. *Plos One* 7:1-9.

Anderson, C.N.K., C.H. Hsieh, S.A. Sandin, R. Hewitt, A. Hollowed, J. Beddington, R.M. May and G. Sugihara. 2008. Why fishing magnifies fluctuations in fish. *Nature* 452:835-839.

Arlinghaus, R. 2005. A conceptual framework to identify and understand conflicts in recreational fisheries systems, with implications for sustainable management. *Aquat. Res., Culture and Development* 1:145-174.

Arlinghaus, R. 2007. Voluntary catch-and-release can generate conflict within the recreational angling community: a qualitative case study of specialised carp, *Cyprinus carpio*, angling in Germany. *Fish. Manag. Ecol.* 14:161-171.

Arlinghaus, R., S.J. Cooke, J. Lyman, D. Policansky, A. Schwab, C. Suski, S.G. Sutton and E.B. Thorstad. 2007. Understanding the complexity of catch-and-release in recreational fishing: an integrative synthesis of global knowledge from historical, ethical, social, and biological perspectives. *Rev. Fish. Sci.* 15:75-167.

Arlinghaus, R., M. Bork and E. Fladung. 2008a. Understanding the heterogeneity of recreational anglers across an urban-rural gradient in a metropolitan area (Berlin, Germany), with implications for fisheries management. *Fish. Res.* 92:53-62.

Arlinghaus, R., T. Klefoth, A. Kobler and S.J. Cooke. 2008b. Size-selectivity, capture efficiency, injury, handling time and determinants of initial hooking mortality of angled northern pike (*Esox lucius* L.): the influence of bait type and size. *N. Am. J. Fish. Manage.* 28:123-134.

Arlinghaus, R., T. Klefoth, A.J. Gingerich, M.R. Donaldson, K.C. Hanson and S.J. Cooke. 2008c. Behaviour and survival of pike, *Esox lucius*, with a retained lure in the lower jaw. *Fish. Manag. Ecol.* 15:459-466.

Arlinghaus, R., S. Matsumura and U. Dieckmann. 2009a. Quantifying selection differentials caused by recreational fishing: development of modeling framework and application to reproductive investment in pike (*Esox lucius*). *Evol. Appl.* 2:335-355.

Arlinghaus, R., T. Klefoth, S.J. Cooke, A. Gingerich and C. Suski. 2009b. Physiological and behavioural consequences of catch-and-release angling on northern pike (*Esox lucius*). *Fish. Res.* 97:223-233.

Arlinghaus, R., S. Matsumura and U. Dieckmann. 2010. The conservation and fishery benefits of protecting large pike (*Esox lucius* L.) by harvest regulations in recreational fishing. *Biol. Conserv.* 143:1444-1459.

Arlinghaus, R., B. Beardmore, C. Riepe, J. Meyerhoff and T. Pagel. 2014. Species-specific preferences of German recreational anglers for freshwater fishing experiences, with emphasis on the intrinsic utilities of fish stocking and wild fishes. *J. Fish Biol.* 85:1843-1867.

Arlinghaus, R., K. Lorenzen, B.M. Johnson, S.J. Cooke and I.G. Cowx. 2016. Management of freshwater fisheries. pp 557-579. *In*: J.F. Craig (ed.). *Freshwater fisheries ecology*. Chichester, U.K.: John Wiley & Sons, Ltd.

Arlinghaus, R., J. Alós, T. Klefoth, K. Laskowski, C.T. Monk, S. Nakayama and A. Schröder. 2016a. Consumptive tourism causes timidity, rather than boldness, syndromes: A response to Geffroy et al. *Trends Ecol. Evol.* 31:92-94.

Arlinghaus, R., J. Alós, T. Pieterek and T. Klefoth. 2017. Determinants of angling catch of northern pike (*Esox lucius*) as revealed by a controlled whole-lake catch-and-release angling experiment—the role of abiotic and biotic factors, spatial encounters and lure type. *Fish. Res.* 186:648-657.

Arlinghaus, R., K.L. Laskowski, J., Alós, T. Klefoth, C.T. Monk, S. Nakayama, A. Schröder. 2017. Passive gear-induced timidity syndrome in wild fish populations and its potential ecological and managerial implications. *Fish Fish.* 18:360–373.

Baktoft, H., K. Aarestrup, S. Berg, M. Boel, L. Jacobsen, A. Koed, M.W. Pedersen, J.C. Svendsen and C. Skov. 2013. Effects of angling and manual handling on pike behaviour investigated by high-resolution positional telemetry. *Fish. Manag. Ecol.* 20:518-525.

Beddington J.R. and J.G. Cooke. 1983. *The potential yield of fish stocks.* FAO Fisheries Technical Paper, 242.

Beardmore, B., W. Haider, L.M. Hunt and R. Arlinghaus. 2011. The importance of trip context for determining primary angler motivations: Are more specialized anglers more catch-oriented than previously believed? *N. Am. J. Fish. Manage.* 31:861-879.

Beardmore, B., L.M. Hunt, W. Haider, M. Dorow and R. Arlinghaus. 2015. Effectively managing angler satisfaction in recreational fisheries requires understanding the fish species and the anglers. *Can. J. Fish. Aquat. Sci.* 72:500-513.

Beukema, J.J. 1970. Acquired hook-avoidance in the pike *Esox lucius* L. fished with artificial and natural baits. *J. Fish Biol.* 2:155-160.

Brownscombe, J.W, A.J. Danylchuk, J.M. Chapman, L.F.G. Gutowsky, and S.J. Cooke. 2017. Best practices for catch-and-release recreational fisheries – angling tools and tactics. *Fish. Res.* 186:693-705.

Bryan, H. 1997. Leisure value systems and recreational specialization: the case of trout fishermen. *J. Leis. Res.* 9:174-187.

Carlson, S. M., E. Edeline, L.A. Vøllestad, T.O. Haugen, I.J. Winfield, J.M. Fletcher, J.B. James and N.C. Stenseth. 2007. Four decades of opposing natural and human-induced artificial selection acting on Windermere pike (*Esox lucius*). *Ecol. Lett.* 10:512-521.

Casselman, J.M. 1975. Sex ratios of Northern Pike, *Esox lucius* Linnaeus. *Trans. Am. Fish. Soc.* 104:60-63.

Casselman, J.M. 1978. Effects of environmental factors on growth, survival, activity and exploitation of Northern Pike. pp 114-128. *In*: R.L. Kendall (ed.). *Selected coolwater fishes of north america.* Bethesda, Maryland: American Fisheries Society, Special Publication 11.

Chapman, C.A. and W.C. Mackay. 1984. Versatility in habitat use by a top aquatic predator, *Esox Lucius* L.. *J. Fish Biol.* 25:109-115.

Coble, D.W., G.B. Farabee and R.O. Anderson. 1985. Comparative learning ability of selected fishes. *Can. J. Fish. Aquat. Sci.* 42:791-796.

Colotelo, A.C., and S.J. Cooke. 2011. Evaluation of common angling-induced sources of epithelial for popular freshwater sport fish using fluorescein. *Fish. Res.* 109:217-224

Cox, S. and C. Walters. 2002. Maintaining quality in recreational fisheries: how success breeds failure in management of open-access sport fisheries. pp 107-119. *In*: T.J. Pitcher and C.E. Hollingworth (eds.). *Recreational fisheries: Ecological, economic and social evaluation.* Oxford: Blackwell Scientific Publications.

Crane, D.P., L.M. Miller, J.S. Diana, J.M. Casselman, J.M. Farrell, K.L. Kapuscinski and J.K. Nohner. 2015. Muskellunge and northern pike ecology and management: important issues and research needs. *Fisheries* 40:258-267.

Danylchuk, A.J., A. Adams, S.J. Cooke, and C.D. Suski. 2008. An evaluation of the injury and short-term survival of bonefish (*Albula* spp.) as influenced by a mechanical lip-gripping device used by recreational anglers. *Fish. Res.* 93:248-252.
Diana, J.S. 1983. Growth, maturation, and production of northern pike in three Michigan lakes. *Trans. Am. Fish. Soc.* 112:38-46.
Diana, J.S. and K. Smith. 2008. Combining ecology, human demands, and philosophy into the management of northern pike in Michigan. *Hydrobiologia* 601:125-135.
Donaldson, M.R., C.M. O'Connor, L.A. Thompson, A.J. Gingerich, S.E. Danylchuk, R.R. Duplain and S.J. Cooke. 2011. Contrasting global game fish and non-game fish species. *Fisheries* 36:385-397.
Dorow, M., B. Beardmore, W. Haider and R. Arlinghaus. 2010. Winners and losers of conservation policies for European eel, *Anguilla anguilla*: an economic welfare analysis for differently specialised eel anglers. *Fish. Manag. Ecol.* 17:106-125
Dubois, R.B., T.L. Margenau, R.S. Stewart, P.K. Cunningham and P.W. Rasmussen. 1994. Hooking mortality of northern pike angled through ice. *N. Am. J. Fish. Manage.* 14:769-775.
Edeline, E., S.M. Carlson, L.C. Stige, I.J. Winfield, J.M. Fletcher, J.B. James, T.O. Haugen, L.A. Vøllestad and N.C. Stenseth. 2007. Trait changes in a harvested population are driven by a dynamic tug-of-war between natural and harvest selection. *Proc. Nat. Acad. Sci. USA* 104:15799-15804.
Edeline, E., T.B. Ari, L.A. Vøllestad, I.J. Winfield, J.M. Fletcher, J.B. James, and N.C. Stenseth. 2008. Antagonist selection from predators and pathogens alters food-web structure. *Proc. Nat. Acad. Sci. USA* 105:19792-19796.
Edeline, E., A. Le Rouzic, I.J. Winfield, J.M. Fletcher, J.B. James, N.C. Stenseth and L.A. Vøllestad. 2009. Harvest-induced disruptive selection increases variance in fitness-related traits. *Proc. R. Soc. B* 276:4163-4171.
EIFAC. 2008. *Code of practice for recreational fisheries*. EIFAC Occasional Paper 42, FAO, Rome, Italy.
Eklöv, P. 1997. Effects of habitat complexity and prey abundance on the spatial and temporal distributions of perch (*Perca fluviatilis*) and pike (*Esox lucius*). *Can. J. Fish. Aquat. Sci.* 54:1520-1531.
Ensinger, J., U. Brämick, E. Fladung, M. Dorow and R. Arlinghaus 2016. *Charakterisierung und Perspektiven der Angelfischerei in Nordostdeutschland.* Schriftenreihe des Instituts für Binnenfischerei e.V., Band 44, Potsdam-Sacrow, Germany (available at www.ifishman.de).
FAO, 2012. *Recreational Fisheries.* FAO Technical Guidelines for Responsible Fisheries 13, FAO, Rome, Italy.
Fobert, E., P. Meining, A. Colotelo, C. O'Connor and S.J. Cooke. 2009. Cut the line or remove the hook? An evaluation of sublethal and lethal endpoints for deeply hooked freshwater recreational fish. *Fish. Res.* 99:38-46. .
Francis, R.C., M.A. Hixon, M.E. Clarke, S.A Murawski and S. Ralston. 2007. Ten commandments for ecosystem-based fisheries scientists. *Fisheries* 32:217-233.
Froese, R. 2004. Keep it simple: three indicators to deal with overfishing. *Fish Fish.* 5:86-91.
Garcia, S., P. Sparre and J. Csirke. 1989. Estimating surplus production and maximum sustainable yield from biomass data when catch and effort time series are not available. *Fish. Res.* 8:13-23.
Goedde, L.E. and D.W. Coble. 1981. Effects of angling on a previously fished and an unfished warmwater fish community in two Wisconsin lakes. *Trans. Am. Fish. Soc.* 110:594-603.
Goodyear, C.P. 2015. Understanding maximum size in the catch: Atlantic blue marlin as an example. *Trans. Am. Fish. Soc.* 144:274-282.

Gould, A. and B.S. Grace. 2009. Injuries to barramundi *Lates calcarifer* resulting from lip-gripping devices in the laboratory. *N. Am. J. Fish. Manage.* 29:1418-1424.

Grimm, M.P. 1989. Northern pike (*Esox lucius*) and aquatic vegetation, tools in the management of fisheries and water quality in shallow waters. *Hydrobiol. Bull.* 23:59-65.

Grimm, M.P. 1994. The influence of aquatic vegetation and population biomass on recruitment of 0+ and 1+ northern pike (*Esox lucius* L.). pp 226-234. *In*: I.G. Cowx (ed.). *Rehabilitation of freshwater fisheries*. Oxford: Fishing News Books, Blackwell Science.

Grimm, M.P. and M. Klinge. 1996. Pike and some aspects of its dependence on vegetation. pp 125-156. *In*: J.F. Craig (ed.). *Pike: Biology and exploitation*. London: Chapman & Hall.

Gulland, J.A. 1970. *The fish resources of the ocean*. Fisheries Technical Paper 97. FAO, Rome, Italy.

Gwinn, D.C., M.S. Allen, F.D. Johnston, P. Brown, C.R. Todd and R. Arlinghaus. 2015. Rethinking length-based fisheries regulations: the value of protecting old and large fish with harvest slot. *Fish Fish.* 16:259-281.

Haugen, T.O., I.J. Winfield, L.A. Vøllestad, J.M. Fletcher, J.B. James and N.C. Stenseth. 2006. The ideal free pike: 50 years of fitness-maximizing dispersal in Windermere. *Proc. R. Soc. B.* 273:2917-2924.

Haugen, T.O., I.J. Winfield, L.A. Vøllestad, J.M. Fletcher, J.B. James and N.C. Stenseth. 2007. Density dependence and density independence in the demography and dispersal of pike over four decades. *Ecol. Monogr.* 77:483-502

Hsieh C.H., C.S. Reiss, J.R. Hunter, J.R. Beddington, R.M. May and G. Sugihara. 2006. Fishing elevates variability in the abundance of exploited species. *Nature* 443:859-862.

Hühn, D. and R. Arlinghaus. 2011. Determinants of hooking mortality in freshwater recreational fisheries: a quantitative meta-analysis. *Amer. Fish. Soc. Symp.* 75:141-170.

Hühn, D., T. Klefoth, T. Pagel, P. Zajicek and R. Arlinghaus. 2014. Impacts of external and surgery-based tagging techniques on Small Northern Pike under field conditions. *Amer. J. Fish. Manage.* 34:322-334.

Hunt, L., W. Haider and K. Armstrong. 2002. Understanding the fish harvesting decisions by anglers. *Hum. Dimens. Wildl.* 7:75-89.

Jansen, T., R. Arlinghaus, T.D. Als and C. Skov. 2013. Voluntary angler logbooks reveal long-term changes in a lentic pike, *Esox lucius*, population. *Fish. Manag. Ecol.* 20:125-136.

Johnston, F.D., R. Arlinghaus and U. Dieckmann. 2010. Diversity and complexity of angler behaviour drive socially optimal input and output regulations in a bioeconomic recreational-fisheries model. *Can. J. Fish. Aquat. Sci.* 67:1507-1531.

Johnston, F.D., R. Arlinghaus and U. Dieckmann. 2013. Fish life history, angler behaviour and optimal management of recreational fisheries. *Fish Fish.* 14:554-579.

Johnston, F.D., B. Beardmore and R. Arlinghaus. 2015. Optimal management of recreational fisheries in the presence of hooking mortality and noncompliance-predictions from a bioeconomic model incorporating a mechanistic model of angler behavior. *Can. J. Fish. Aquat. Sci.* 72:37-53.

Jolley, J.C., D.W. Willis, T.J. DeBates and D.D. Graham. 2008. The effects of mechanically reducing northern pike density on the sport fish community of West Long Lake, Nebraska, USA. *Fish. Manag. Ecol.* 15:251-258.

Kempinger, J.J. and R.F. Carline. 1978. Dynamics of the northern pike population and changes that occurred with a minimum size limit in Escanaba Lake, Wisconsin. *Am. Fish. Soc. Spec. Publ.* 11:382-389.

Klefoth, T., A. Kobler and R. Arlinghaus. 2008. The impact of catch-and-release on short term behaviour and habitat choice of northern pike (*Esox lucius* L.). *Hydrobiologia* 601:99-110.

Klefoth, T., A. Kobler and R. Arlinghaus. 2011. Behavioural and fitness consequences of direct and indirect non-lethal disturbances in a catch-and-release northern pike (*Esox lucius*) fishery. *Knowl. Manag. Aquat. Ecosyst.* 403:11.

Klefoth, T., T. Pieterek and R. Arlinghaus. 2013. Impacts of domestication on angling vulnerability of common carp, *Cyprinus carpio*: the role of learning, foraging behaviour and food preferences. *Fish. Manag. Ecol.* 20:174-186.

Kobler, A., T. Klefoth, C. Wolter, F. Fredrich and R. Arlinghaus. 2008a. Contrasting pike (*Esox lucius* L.) movement and habitat choice between summer and winter in a small lake. *Hydrobiologia* 601:17-27.

Kobler, A., T. Klefoth and R. Arlinghaus. 2008b. Site fidelity and seasonal changes in activity centre size of female pike (*Esox lucius*) in a small lake. *J. Fish Biol.* 73:584-596.

Kobler, A., T. Klefoth, T. Mehner and R. Arlinghaus. 2009. Co-existence of behavioural types in an aquatic top predator: a response to resource limitation? *Oecologia* 161:837-847.

Kotakorpi, M., J. Tiainen, M. Olin, H. Lehtonen, K. Nyberg, J. Ruuhijärvi and A. Kuparinen. 2013. Intensive fishing can mediate stronger size-dependent maternal effect in pike (*Esox lucius*). *Hydrobiologia* 718:109-118.

Kuparinen, A., T. Klefoth and R. Arlinghaus. 2010. Abiotic and fishing-related correlates of angling catch rates in pike (*Esox lucius*). *Fish. Res.* 105:111-117.

Kuparinen, A., A. Boit, F.S. Valdovinos, H. Lassaux and N.D. Martinez 2016. Fishing-induced life-history changes degrade and destabilize harvested ecosystems. *Scient. Rep.* 6:22245.

Langangen, Ø., E. Edeline, J. Ohlberger, I.J. Winfield, J.M. Fletcher, J.B. James, N.C. Stenseth and L.A. Vøllestad. 2011. Six decades of pike and perch population dynamics in Windermere. *Fish. Res.* 109:131-139.

Le Bris, A., A.J. Pershing, C.M. Hernandez, K.E. Mills and G.D. Sherwood. 2015. Modelling the effects of variation in reproductive traits on fish population resilience. *ICES J. of Mar. Sci.* 72:2590-2599.

Lester, N.P., B.J. Shuter, P. Venturelli and D. Nadeau. 2014. Life-history plasticity and sustainable exploitation: a theory of growth compensation applied to walleye management. *Ecol. Appl.* 24:38-54.

Louison, M. J., C.T. Hasler, M.M. Fenske, C.D. Suski and J. A. Stein. 2017. Physiological effects of ice-angling capture and handling on northern pike, *Esox lucius*. *Fish. Manag. Ecol.* 24:10-18.

Losey, G.S., T.W. Cronin, T.H. Goldsmith, D. Hyde, N.J. Marshall and W.N. McFarland. 1999. The UV visual world of fishes: a review. *J. Fish Biol.* 54:921-943.

Mace, P.M. 1994. Relationships between common biological reference points used as thresholds and targets of fisheries management strategies. *Can. J. Fish. Aquat. Sci.* 51: 110-122.

Margenau, T.L., S.J. Gilbert and G.R. Hatzenbeler. 2003. Angler catch and harvest of northern pike in northern Wisconsin lakes. *N. Am. J. Fish. Manage.* 23:307-312.

Margenau, T.L. 2007. Effects of angling with a single-hook and live bait on muskellunge survival. *Env. Biol. Fish.* 79:155-162.

Matsumura, S., R. Arlinghaus and U. Dieckmann. 2011. Assessing evolutionary consequences of size-selective recreational fishing on multiple life-history traits, with an application to northern pike (*Esox lucius*). *Evol. Ecol.* 25:711-735.

Mickiewicz, M. and A. Wołos. 2012. Economic ranking of the importance of fish species to lake fisheries stocking management in Poland. *Arch. Pol. Fish.* 20:11-18.

Minns, C.K., R.G. Randall, J.E. Moore and V.W. Cairns. 1996. A model simulating the impact of habitat supply limits on northern pike, *Esox lucius*, in Hamilton Harbour, Lake Ontario. *Can. J. Fish. Aquat. Sci.* 53:20-34.

Mittelbach, G.G., N.G. Ballew and M.K. Kjelvik. 2014. Fish behavioral types and their ecological consequences. *Can. J. Fish. Aquat. Sci.* 71:927-944.

Mogensen, S., J.R. Post and M.G. Sullivan. 2014. Vulnerability to harvest by anglers differs across climate, productivity, and diversity clines. *Can. J. Fish. Aquat. Sci.* 71:416-426.

Moraga, A.D., A.D.M. Wilson and S.J. Cooke. 2015. Does lure colour influence catch per unit effort, fish capture size and hooking injury in angled largemouth bass? *Fish. Res.* 172:1-6.

Mosindy, T.E., W.T. Momot and P.J. Colby. 1987. Impact of angling on the production and yield of mature walleyes and northern pike in a small boreal lake in Ontario. *N. Am. J. Fish. Manage.* 7:493-501.

Oele, D.L., A.L. Rypel, J. Lyons, P. Cunningham and T. Simonson. 2016. Do higher size and reduced bag limits improve northern pike size structure in Wisconsin Lakes? *N. Am. J. Fish. Manage.* 36:982-994.

Ohlberger, J., S.J. Thackeray, I.J. Winfield, S.C. Maberly and L.A. Vøllestad. 2014. When phenology matters: age-size truncation alters population response to trophic mismatch. *Proc. R. Soc. B.* 281:20140938.

Olson, D.E. and P.K. Cunningham. 1989. Sport-fisheries trends shown by an annual Minnesota fishing contest over a 58-year period. *N. Am. J. Fish. Manage.* 9:287-297.

Pagel, T., D. Bekkevold, S. Pohlmeier, C. Wolter and R. Arlinghaus. 2015. Thermal and maternal environments shape the value of early hatching in a natural population of a strongly cannibalistic freshwater fish. *Oecologia* 178:951-965.

Paukert, C.P., J.A. Klammer, R.B. Pierce and T.D. Simonson. 2001. An overview of northern pike regulations in North America. *Fisheries* 26:6-13.

Pauly, D. 1980. On the interrelationships between natural mortality, growth parameters, and mean environmental temperature in 175 fish stocks. *J. Cons. Int. Explor. Mer.* 39:175-192.

Philipp D.P., S.J. Cooke, J.E. Claussen, J.B. Koppelman, C.D. Suski and D.P. Burkett. 2009. Selection for vulnerability of angling in largemouth bass. *Trans. Am. Fish. Soc.* 138:189-199.

Pierce, R.B., C.M. Tomcko and D.H. Schupp. 1995. Exploitation of northern pike in seven small north-central Minnesota lakes. *N. Am. J. Fish. Manage.* 15:601-609.

Pierce, R.B. and C.M. Tomcko. 1998. Angler noncompliance with slot length limits for northern pike in five small Minnesota lakes. *N. Am. J. Fish. Manage.* 18:720-724.

Pierce, R.B. and M.F. Cook. 2000. Recreational darkhouse spearing for northern pike in Minnesota: historical changes in effort and harvest and comparisons with angling. *N. Am. J. Fish. Manage.* 20:239-244.

Pierce, R.B. and C.M. Tomcko. 2005. Density and biomass of native northern pike populations in relation to basin-scale characteristics of north-central Minnesota lakes. *Trans. Am. Fish. Soc.* 134:231-241.

Pierce, R.B. 2010. Long-term evaluations of length limit regulations for northern pike in Minnesota. *N. Am. J. Fish. Manage.* 30:412-432.

Pierce, R.B. 2012. Northern Pike: Ecology, Conservation, and Management History. Minneapolis, Minn.: University of Minnesota Press.

Pieterek, T. 2014. *Determinanten der anglerischen Fangbarkeit von Hechten* (*Esox lucius*). M.Sc. Thesis, Humboldt-Universität zu Berlin, Germany. http://besatz-fisch.de/images/stories/Papers/thesis_msc_pieterek.pdf.

Pieterek, T., T. Klefoth, J. Alós and R. Arlinghaus. 2016. *Einfluss verschiedener Faktoren auf die anglerische Fangbarkeit von Hechten* (*Esox lucius*). Schriftenreihe des Deutschen Angelfischereiverbands, Gewässer- und Naturschutzseminar 2015 1:17-26.

Post, J.R., M. Sullivan, S. Cox, N.P. Lester, C.J. Walters, E.A. Parkinson, A.J. Paul, J. Jackson and B.J. Shuter. 2002. Canada's recreational fisheries: the invisible collapse? *Fisheries* 27:6-17.

Pullen, C.E. 2013. *The consequences of retained lures on free swimming fish: physiological, behavioural and fitness perspectives*. M.Sc Thesis, Carleton University, Ontario, Canada.
Pullen, C.E., K. Hyes, C.M. O'Connor, R. Arlinghaus, C.D. Suski, J.D. Midwood and S.J. Cooke. 2017. Consequences of oral lure retention on the physiology and behaviour of adult northern pike (*Esox lucius* L.). *Fish. Res.* 186:601-611.
Raat, A.J.P. 1988. Synopsis of biological data on the northern pike (*Esox lucius* Linnaeus, 1758). *FAO Fisheries Synopsis 30, Review 2*. FAO, Rome, Italy.
Radomski, P.J., G.C. Grant, P.C. Jacobson and M.F. Cook. 2001. Visions for recreational fishing regulations. *Fisheries* 26:7-18.
Rapp, T., S.J. Cooke and R. Arlinghaus. 2008. Exploitation of specialised fisheries resources: The importance of hook size in recreational angling for large common carp *Cyprinus carpio*. *Fish. Res.* 94:79-83.
Ricker, W E. 1975. Computation and interpretation of biological statistics of fish populations. *B. Fish. Res. Bd. Can.* 191:1-382.
Riepe, C. and R. Arlinghaus, R. 2014. *Einstellungen der Bevölkerung in Deutschland zum Tierschutz in der Angelfischerei*. Ber. IGB 27:1-198.
Rypel, A.L. 2012. Meta-analysis of growth rates for a circumpolar fish, the northern pike (*Esox lucius*), with emphasis on effects of continent, climate and latitude. *Ecol. Freshw. Fish* 21:521-532.
Rypel, A.L. 2015. Effects of a reduced daily bag limit on bluegill size structure in Wisconsin lakes. *N. Am. J. Fish. Manage.* 35:388-397.
Schroeder, S.A. and D.C. Fulton. 2013. Comparing catch orientation among Minnesota walleye, northern pike, and bass anglers. *Hum. Dimens. Wildl.* 18:355-372.
Schroeder, S.A. and D.C. Fulton. 2014. Fishing for northern pike in Minnesota: comparing anglers and dark house spearers. *N. Am. J. Fish. Manage.* 34:678-691.
Seekell, D.A. 2011. Recreational freshwater angler success is not significantly different from a random catch model. *N. Am. J. Fish. Manage.* 31:203-208.
Seekell, D.A., C.J. Brosseau, T.J. Cline, R.J. Winchombe and L.J. Zinn. 2011. Long-term changes in recreational catch inequality in a trout stream. *N. Am. J. Fish. Manage.* 31:1110-1115.
Sharma, C.M. and R. Borgstrøm. 2008. Increased population density of pike *Esox Lucius* – a result of selective harvest of large individuals. *Ecol. Freshw Fish* 17:590-596.
Snow, H.E. 1978. Response of northern pike to exploitation in Murphy Flowage, Wisconsin. *Am. Fish. Soc. Spec. Publ.* 11:320-327.
Stålhammar, M., T. Fränstam, J. Lindström, J. Höjesjö, R. Arlinghaus and P.A. Nilsson. 2014. Effects of lure type, fish size and water temperature on hooking location and bleeding in northern pike (*Esox lucius*) angled in the Baltic Sea. *Fish. Res.* 157:165-169.
Tiainen, J., M. Olin and H. Lehtonen. 2014. The effects of size-selective fishing on pike populations. pp 3-6. *In*: N. Valkonen Joensuu (ed.). *Perspectives on sustainable fisheries management – case examples from Sweden and Finland*. Finland: Future Missions Oy.
Tiainen, T., M. Olin, Lehtonen, H., Nyberg, K. and Ruuhijärvi, J. 2017. The capability of harvestable slot-length limit regulation in conserving large and old northern pike (*Esox lucius*). *Bor. Env. Res.* 22 (Special Issue 1):169-186.
Tibblin, P., A. Forsman, T. Borger and P. Larsson. 2016. Causes and consequences of repeatability, flexibility and individual fine-tuning of migratory timing in pike. *J. Anim. Ecol.* 85:136-145.
Uusi-Heikkilä, S., A.R. Whiteley, A. Kuparinen, S. Matsumura, P.A. Venturelli, C. Wolter, J. Slate, C.R. Primmer, T. Meinelt, S.S. Killen, D. Bierbach, G. Polverino, A. Ludwig and R. Arlinghaus. 2015. The evolutionary legacy of size-selective harvesting extends from genes to populations. *Evol. Appl.* 8:597-620.

van Kooten, T., L. Persson and A.M. de Roos. 2007. Size-dependent mortality induces life-history changes mediated through population dynamical feedbacks. *Am. Nat.* 170:258-270.

van Poorten, B.T. and J.R. Post. 2005. Seasonal fishery dynamics of a previously unexploited rainbow trout population with contrasts to established fisheries. *N. Am. J. Fish. Manage.* 25:329-345.

Vindenes, Y., E. Edeline, J. Ohlberger, Ø. Langangen, I.J. Winfield, N.C. Stenseth and L.A. Vøllestad. 2014. Effects of climate change on trait-based dynamics of a top predator in freshwater ecosystems. *Am. Nat.* 183:243-256.

Vindenes, Y. and Ø. Langangen. 2015. Individual heterogeneity in life histories and eco-evolutionary dynamics. *Ecol. Let.* 18:417-432.

Vindenes, Y., Ø. Langangen, I.J. Winfield and L.A. Vøllestad. 2016. Fitness consequences of early life conditions and maternal size effects in a freshwater top predator. *J. Anim. Ecol.* 85(3):692-704.

Vøllestad, L.A., J. Skurdal and T. Qvenild. 1986. Habitat use, growth, and feeding of pike (*Esox lucius* L.) in SE Norway. *Arch. Hydrobiol.* 108:107-117.

Walker, J.R., L. Foote and M.G. Sullivan. 2007. Effectiveness of enforcement to deter illegal angling harvest of northern pike in Alberta. *N. Am. J. Fish. Manage.* 27:1369-1377.

Walters, C.J. and S.J.D. Martell. 2004. *Fisheries Ecology and Management.* Princeton, USA: Princeton University Press,

Weithman, A.S. and R.O. Anderson. 1976. *Angling vulnerability of esocidae.* Proceedings of the Annual Conference of the South-Eastern Association of Fisheries and Wildlife Agencies 30:99-102.

Wolter, C., R. Arlinghaus, U.A. Grosch and A. Vilcinskas. 2003. *Fische und Fischerei in Berlin.* VNW Solingen, Germany: Verlag Natur & Wissenschaft,.

Zhou, S., S. Yin, J.T. Thorson, A.D.M. Smith and M. Fuller. 2012. Linking fishing mortality reference points to life history traits: an empirical study. *Can. J. Fish. Aquat. Sci.* 69:1292-1301.

Author Address

**Corresponding author*: [1]Department of Biology and Ecology of Fishes, Leibniz-Institute of Freshwater Ecology and Inland Fisheries, Müggelseedamm 310, 12587 Berlin, Germany;
Email: arlinghaus@igb-berlin.de
and
Division of Integrative Fisheries Management, Faculty of Life Sciences & Integrative Research Institute on Transformations of Human-Environment, Systems (IRI THESys), Humboldt-Universität zu Berlin, Invalidenstrasse 42, 10115 Berlin, Germany

[2]Department of Biology and Ecology of Fishes, Leibniz-Institute of Freshwater Ecology and Inland Fisheries, Müggelseedamm 310, 12587 Berlin, Germany; Email: alos@igb-berlin.de

[3]Wisconsin Department of Natural Resources, 101 S. Webster St. Madison, Wisconsin 53707, USA; Email: Alan.Beardmore@wisconsin.gov

[4]Division of Integrative Fisheries Management, Faculty of Life Sciences & Integrative Research Institute on Transformations of Human-Environment, Systems (IRI THESys), Humboldt-Universität zu Berlin, Invalidenstrasse 42, 10115 Berlin, Germany; Email: angela.martindiaz@mail.uca.es

[5]Institute of Inland Fisheries in Potsdam-Sacrow, Im Königswald 2, 14469 Potsdam, Germany; Tel.: +49-(0)33201 406 33, Email: daniel.huehn@ifb-potsdam.de
and
Department of Biology and Ecology of Fishes, Leibniz-Institute of Freshwater Ecology and Inland Fisheries, Müggelseedamm 310, 12587 Berlin, Germany; Email: huehn@igb-berlin.de

(Contd.) . . .

Author Address (*Contd.*)

[6]Department of Biology and Ecology of Fishes, Leibniz-Institute of Freshwater Ecology and Inland Fisheries, Müggelseedamm 310, 12587 Berlin, Germany; Email: fdjohnst@googlemail.com

[7]Angling Association of Lower Saxony (Anglerverband Niedersachsen e.V.), Brüsseler Strasse 4, 30539 Hannover, Germany; Email: t.klefoth@av-nds.de

[8]The Department of Biological and Environmental Science, P.O. Box 35 (Survontie 9 C), 40014 University of Jyväskylä, Finland; Email: anna.k.kuparinen@jyu.fi

[9]Faculty of Applied Biological Sciences, Gifu University, Yanagido 1-1, Gifu, 501-1193, Japan; Email: matsumur@gifu-u.ac.jp

[10]Department of Biology and Ecology of Fishes, Leibniz-Institute of Freshwater Ecology and Inland Fisheries, Müggelseedamm 310, 12587 Berlin, Germany; Email: pagel@igb-berlin.de

[11]Division of Integrative Fisheries Management, Faculty of Life Sciences & Integrative Research Institute on Transformations of Human-Environment, Systems (IRI THESys), Humboldt-Universität zu Berlin, Invalidenstrasse 42, 10115 Berlin, Germany; Email: pippek@msn.com

[12]Department of Biology and Ecology of Fishes, Leibniz-Institute of Freshwater Ecology and Inland Fisheries, Müggelseedamm 310, 12587 Berlin, Germany; Email: riepe@igb-berlin.de

Chapter 13

Northern Pike Commercial Fisheries, Stock Assessment and Aquaculture

Anna Kuparinen[*1] *and Hannu Lehtonen*[2]

13.1 INTRODUCTION

Few fish are as significant to mankind as the northern pike (*Esox lucius* Linnaeus, 1758, hereafter referred to as pike). In most areas of the species distribution range, it is targeted by both commercial and sport fishing, and it is also protected by closed seasons and minimum size restrictions as well as regulations to avoid bycatch. In some situations and places, pike may be regarded as a nuisance because of their excessive predation on other species of fish or other fauna of commercial or conservation value (chapter 14). However, elsewhere, pike have an important economic value, attracting large numbers of recreational anglers and thereby supporting local communities (chapter 12).

The pike is a cannibalistic top predator that can be found across the northern hemisphere in freshwater and brackish water ecosystems (Raat 1988). It is a commonly targeted species, both in recreational, subsistence, and commercial fisheries throughout its range (Paukert et al. 2001, Arlinghaus & Mehner 2004, chapter 12). Due to its large potential size, wide distribution, locally high abundance and economic value, the pike has been of importance in various cultures. This cultural significance is exemplified by the common occurrence of pike bones recovered in archeological excavations around the world. The older remains date back between 8000 years before present (Mathiassen 1943) and

**Corresponding author*: [1]The Department of Biological and Environmental Science, P.O. Box 35 (Survontie 9 C), 40014 University of Jyväskylä, Finland; Email: anna.k.kuparinen@jyu.fi

[2]Department of Environmental Sciences, P.O. Box 65 (Viikinkaari 1), 00014 University of Helsinki, Finland; Email: hannu.lehtonen@helsinki.fi

to Roman times (Jones 1988). The pike is also a central part of some national epics. For example, in Finnish folklore and the national epic Kalevala, the main hero known as Väinämöinen made a sweet-sounding zither (a traditional Finnish musical instrument) from a giant pike's jawbone. Of further significance, pike were historically dried together with furs, and these products were used as currency for paying Swedish taxes (Hansen 2009). The current significance of pike is quite remarkable in many parts of the world. For instance, commercial pike gillnet fisheries in Waterhen Lake in Manitoba, Canada, became the first freshwater pike fishery in the Western Hemisphere to attain certification as a sustainable, well-managed fishery according to the global Marine Stewardship Council (MSC) standard. Annual catches here have averaged 26,740 kg over recent years (Marine Stewardship Council 2014).

This chapter considers three important aspects of human interactions with pike. First, a range of information is presented on the extent of pike fisheries. Second, methods for assessing stock densities of pike are summarized and contrasted, and, finally, an overview is presented on the potential problems relating to pike aquaculture.

13.2 COMMERCIAL FISHERIES

According to Mann (1996), commercial pike fisheries include those where pike is the prime target species and those where it is only a small contribution to the total fish catch. The European Inland Fisheries Advisory Commission (EIFAC 2007) defines 'commercial fisheries' as fisheries where a stock is exploited for commercial gain either through the provision of food fish, fish for stock enhancement, or seed or brood stock for aquaculture production purposes. In this chapter, commercial pike fishing is further defined as including all fishing where the catch is sold.

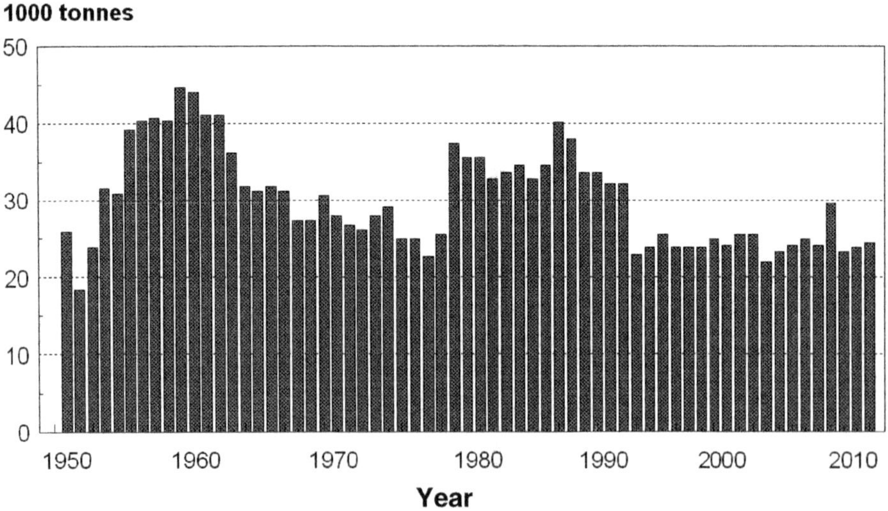

FIGURE 13.1 Global commercial catch of pike (in 1000 tonnes), redrawn from FAO (2015).

Commercial and recreational fishers may compete for the same resource. In particular, harvesting of pike by commercial fisheries may reduce the value for recreational fisheries, due to reductions in pike abundance in general and in the abundance of large trophy pikes in particular. As a consequence, the objectives for managing fisheries have largely changed over the past decades, as recreational fisheries have proliferated. While pike fisheries were historically managed primarily as commercial fisheries, management for recreational fisheries is increasingly considered (chapter 12). As a result, regulatory changes such as restricting the use of commercial gear is now common in many waters (e.g. Colby 2012).

TABLE 13.1 Commercial pike catch in 2013 by country in tonnes (FAO 2014).

Tonnes	*Country*
11991	Russia
1886	Canada
638	Kazakhstan
364	Finland
347	Germany
317	Poland
294	Estonia
223	Serbia
213	Turkey
145	Sweden
138	Hungary
129	Czech Republik
114	Switzerland
90	Romania
53	Slovakia
49	Croatia
39	Latvia
38	Belarus
17	France
15	Lithuania
9	Denmark
8	Greece
7	Belgium
7	Slovenia
6	Ukraine
6	Uzbekistan
3	United Kingdom
2	Bulgaria
2	Macedonia
1	Moldova

The global annual commercial capture rate of pike has, in recent decades, ranged between approximately 20,000-40,000 tonnes (FAO 2015, Fig. 13.1). In the 1980s, total commercial pike harvest in Europe was almost 20,000 tonnes, although in North America it was only 3,200 tonnes (Marine Stewardship Council 2014). In North America, commercial fisheries are mainly restricted to the northern parts of the distribution range, whereas in their southern range pike are primarily managed for recreational fisheries (Fsheries and Oceans Canada 2012). European commercial catch data by country from 2013 indicated that landings were the greatest in Russia, Kazakshtan, Finland, Germany and Poland (Table 13.1). Long-term catch data is also available from Lake Superior (USA), where the highest landings were in the early 1900's (Fig. 13.2). In Polish Lake Mamry, commercial pike catches declined rapidly during the 1970's, and catch levels have since been about one third of the catch rates from the 1960's (Fig. 13.3).

Fishing methods include gillnets, fyke nets, trap nets, traps, long lines, hooks, and also active methods such as trawling, seining, angling and spearing (Raat 1988). Spatial and temporal differences in the use of fishing gear are evident. Gillnets and hooks are used almost everywhere in all seasons, fyke net, trap net and trap fisheries are practiced primarily during the spawning period when pike gather in shallow waters. During winter, hook fishing is a traditional commercial pike fishing method in northern areas. Hooks and long lines are also used during the open water season. Fishing practices and gear choice are commonly local decisions and fishing methods have sometimes not changed for decades (or even centuries). The raw materials used for gear manufacturing have, however, changed, and natural fibers have now been replaced by synthetic nylon fibers. As pike live mainly in shallow near-shore areas, small rowboats or outboard motorboats are commonly used to access fishing areas (Raat 1988, Mann 1966, Ernst and Young 2011).

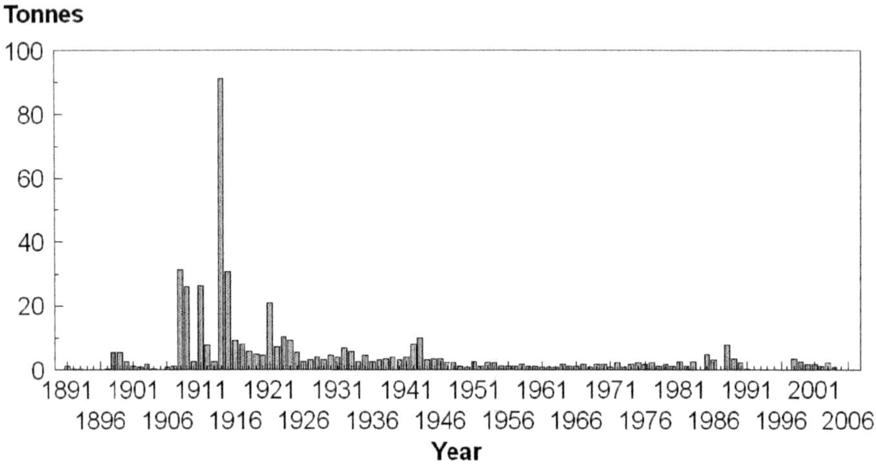

FIGURE 13.2 Commercial pike catch from the Canadian part of Lake Superior in 1891-2004 (redrawn from Baldwin et al. 2009).

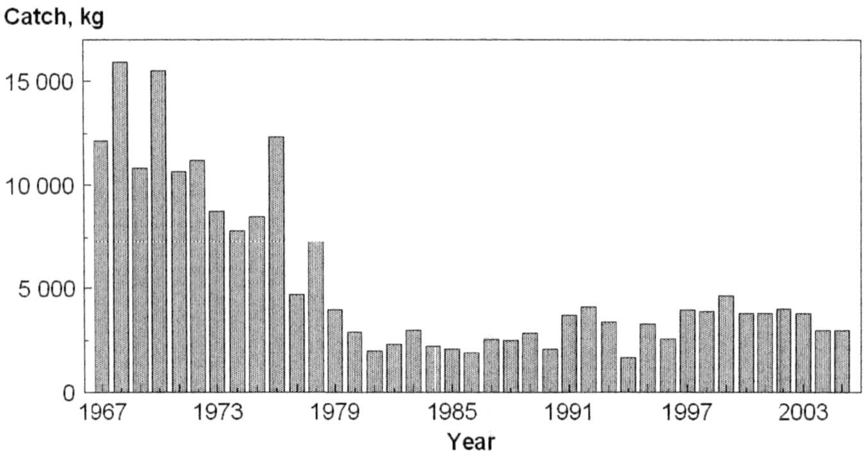

FIGURE 13.3 Commercial pike catch in Lake Mamry in Poland (redrawn from Wolos et al. 2009).

The European commercial harvest of pike remains important in many countries (Ernst and Young 2011). However, the existing data demonstrate considerable differences in the intensity and catch levels of pike fishing. This is possibly due to difference in catch statistics reported by different countries, e.g. reported statistics include recreational catches or not (Ernst and Young 2011, FAO 2014). Similarly, the total commercial pike catch in Canada in 2013 (1,886 tonnes, valued at 1,335,000 Canadian dollars) differed between provinces. For example, in Manitoba, the annual catch was 1,533 tonnes, but in Saskatchewan it was 302 tonnes, while in other provinces the catch volume was considerably smaller still (Table 13.2). In some places, considerable changes in landings have occurred. For example, the Great Lakes pike fishery declined from 1,600 tonnes year^{-1} in the early 1900s to less than 50 tonnes year^{-1} by the late 1960s (Baldwin et al. 2009). This drastic decline was attributed to overfishing, and the response was to close the fishery during the 1970s (Baldwin et al. 1979).

TABLE 13.2 Commercial and recreational pike catch in kg in some countries and provinces.

Country	*Year*	*Commercial*	*Recreational*	*References*
Sweden	2013	43,757	3,134,000	https://havbi.havochvatten.se/analytics/saw.dll?PortalPages https://www.havochvatten.se/download/18.203ea9d8149410b71c2c7c54/1416390851137/officiell-statistik-JO57SM1401.pdf http://www.scb.se/Statistik/JO/JO1104/2013A01/JO1104_2013A01_SM_JO57SM1401.pdf
Canada	2013	1,886,000	20,499,919 (number caught)	http://www.dfo-mpo.gc.ca/stats/commercial/land-debarq/freshwater-eaudouce/2013-eng.htm http://www.dfo-mpo.gc.ca/stats/rec/can/2010/RECFISH2010_ENG.pdf
Finland	2013	364,000	5,742,000	http://www.stat.luke.fi
Denmark	1973	19,624	13,453	http://www.fao.org/docrep/009/t0377e/t0377e08.htm
Estonia	2011	32,070	-	http://issuu.com/eurofish/docs/estonian_fishery_2011_web?e=1376257/3387243#search
Manitoba	2013	1,533,000	-	https://www.gov.mb.ca/waterstewardship/fisheries/commercial/history.pdf
Quebec	2013	4,000	-	http://www.dfo-mpo.gc.ca/stats/commercial/land-debarq/freshwater-eaudouce/2013-eng.htm
Ontario	2013	26,000	-	http://www.dfo-mpo.gc.ca/stats/commercial/land-debarq/freshwater-eaudouce/2013-eng.htm
Saskatchewan	2013	302,000	-	http://www.dfo-mpo.gc.ca/stats/commercial/land-debarq/freshwater-eaudouce/2013-eng.htm
Alberta	2013	8,000	-	http://www.dfo-mpo.gc.ca/stats/commercial/land-debarq/freshwater-eaudouce/2013-eng.htm
NW-territories	2013	12,000	-	http://www.dfo-mpo.gc.ca/stats/commercial/land-debarq/freshwater-eaudouce/2013-eng.htm

The most extensive historical record of pike fisheries is likely from Finland, where most commercial pike fishing uses gill nets and trap nets (Table 13.3). In Finland pike are also common by-catch in trawl and seine fishing for vendace (*Coregonus albula* Linnaeus, 1758). In 2012, the total Finnish pike catch was 6,081 tonnes, of which the majority stems from recreational fisheries. The commercial harvest was only 339 tonnes (5.6%). Altogether, 65% of the commercial catch, but only 18% of recreational catch was taken from the Baltic Sea. The total commercial pike catch in Finland has decreased since the early 1980s (Fig. 13.4). The reason is probably the decline in appreciation of pike as a food fish. This becomes evident also from the consumer price index of pike from 1980, which shows a decline of almost 50% adjusted to 2013 prices (Finnish Game and Fisheries Research Institute 2013).

TABLE 13.3 Principal methods used for the capture of pike in Finnish commercial fisheries in the Baltic Sea area and lakes and rivers in 2013 (the Baltic Sea) and in 2012 (freshwater) in per cent of harvest (Finnish Game and Fisheries Research Institute 2013, 2014).

Fishing gear	*Baltic Sea*	*Freshwaters*
Trap net	26	31
Gill net	67	51
Hook	7	3
Trawl	0	5
Winter seine	0	6
Summer seine	0	4

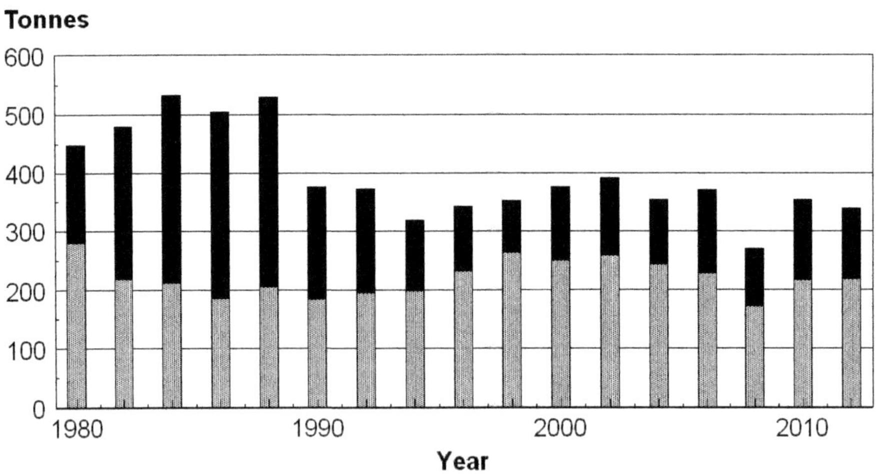

FIGURE 13.4 Finnish commercial pike catch in 1980-2013 from the Baltic Sea (light) and inland waters (black) (redrawn from Finnish Game and Fisheries Research Institute 2014).

There is a clear seasonality in commercial pike catches. On the Finnish Baltic Sea coast, the highest catches in 2013 were taken during the spawning period in May and during the ice period in winter when fishing takes place mainly by hook and line. During the summer months the pike catch is smaller (Fig. 13.5). In Finnish commercial inland fisheries the pike catches during the open water period in 2012 consisted of 68% of the annual catch, while fishing from ice comprised 32% (Finnish Game and Fisheries Research Institute 2013).

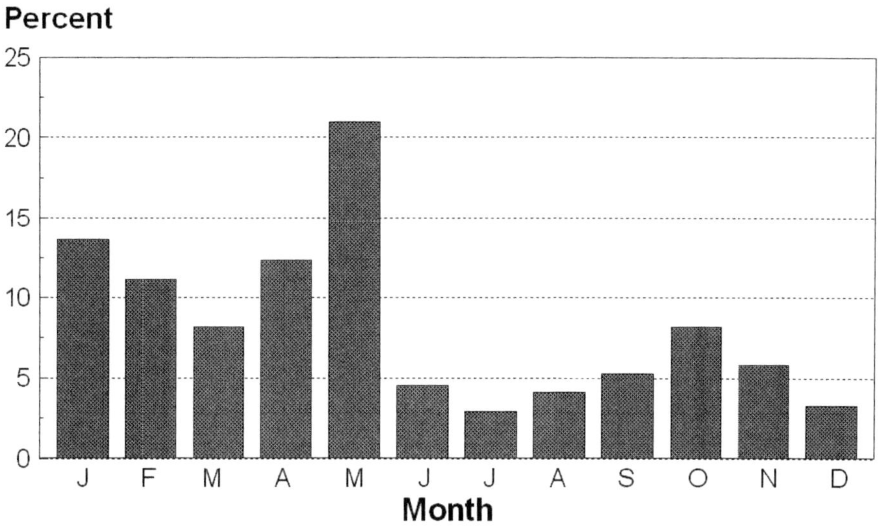

FIGURE 13.5 Annual distribution of pike catches in commercial fishery on the Baltic coast of Finland in 2013 (redrawn from Finnish Game and Fisheries Research Institute 2014).

The commercial market for pike in North America is primarily as fresh and/or frozen fish products with minimal value-added processing (i.e., fillets, minced product). Pike is predominantly sold as frozen fillets in the European Union, primarily in France (Marine Stewardship Council 2014). In other parts of Europe, such as in Russia, the catch is primarily sold fresh.

13.3 STOCK ASSESSMENT

Stock assessment usually involves quantifying the numbers (absolute or relative) and size and age distributions of pike within a defined area, waterbody or population. Population assessments are often used by managers to balance the interests of recreational and commercial fisheries, to set fisheries regulations (i.e., gear restrictions, catch quotas, closed seasons and more), and to evaluate the conservation status of the populations. However, systematic pike stock assessments are rare.

Published pike population abundance estimates are most often used for management of recreational fisheries and evaluation of stocking programmes (Arlinghaus et al. 2014). Ecological studies, such as food web manipulation experiments and habitat monitoring have also explored trends in pike population abundances (e.g. Prejs et al. 1997, Brosse et al. 2001) but without direct links to fisheries management. Additionally, scientific fishing programs, such as in Lake

Windermere in United Kingdom, provide insights into the long term development of pike population sizes as well as factors affecting pike population dynamics (Haugen et al. 2007). In the following discussion, data and methods applied to evaluate pike population abundances are categorized into two main groups: (1) metrics considered as indirect indices of population abundance, and (2) attempts to estimate population size directly.

13.3.1 Abundance Indicators

Direct catch-per-unit-effort (CPUE) observations are generally considered a well-established yet sometimes problematic indicator of fish density (Maunder et al. 2006). Depending on species mobility, a complete picture might require systematic sampling of all suitable habitats, and selectivity of sampling gears must be carefully considered. For pike, Craig (1996) suggested that seine and trawl catches are generally more reliable than gillnet catches in CPUE analyses because they catch large pike more efficiently. Regardless, gill net and angling are currently the most common pike fishing methods and, despite possible limitations, catches by these methods have provided the basis for stock assessments (reviewed in Craig 1996). A recent experimental manipulation of pike density in a lake in Minnesota (USA) revealed that gillnet catches tracked changes in pike density well, whereas no association was found between density and angling catch rates (Pierce and Tomcko 2003).

Differences between angling and gill netting as CPUE indicators probably reflect inherently different processes. Gill net catches reflect fish movement activity (e.g. Biro and Post 2008) and angling catches reflect fish attack rate at lure or bait which may depend on feeding motivation (Kuparinen et al. 2009). Furthermore, catchability of pike by angling is likely to be influenced by availability of other prey fish, weather conditions, seasonality (Kuparinen et al. 2010), and angler skills (Pierce and Tomcko 2003), which all can mask variations in CPUE arising from population abundance. Therefore, at least finer scale variations in abundance over small temporal scales may not be detectable based on angling records alone. It should be noted, however, that the power of gillnetting-based CPUE to reflect abundance variations can also depend on the quality of the sampling. In an experiment conducted by Pierce and Tomcko (2003), fishing pressure was standardized across pike age classes and sizes by applying multiple mesh sizes. Less standardized gillnet catch records might only track the component of the population selected by the applied gear. Moreover, factors that affect pike movement rates, such as general food availability and visibility of nets may confound the relationship between CPUE from gill net catch and pike density.

Some of the most extensive time-series data are from log book records of pike caught by commercial and recreational angling (Lehtonen et al. 2009, Jansen et al. 2013). Hence, although angling-based data may have variation around any trend between CPUE and fish density, this may be countered by the length of the data set, assuming that there is no systematic bias over time. As an example, log book information from a regulated and monitored fishing area in the Western Gulf of Finland yielded a CPUE time series spanning 68 years from 1939 to 2007 with CPUE expressed as the number of pike caught by angling or line per fishing day

(Lehtonen et al. 2009). Since the 1980's, the CPUEs were 3-4% of the highest records which likely reflects a strong decline in the abundance of the local pike population. Variations in CPUE were found to be correlated with an increase in nutrients in the Baltic Sea, but no association was found between CPUE and salinity or temperature (Lehtonen et al. 2009). Interestingly, however, recorded pike weights were negatively correlated with CPUE. A similar log book-based long time series of pike CPUE was recorded in Lake Esrom that is among the largest lakes in Denmark. These data revealed that pike CPUE was positively related to catch-and-release practices of recreational anglers but negatively to the occurrence of commercial fishing (Jansen et al. 2013). The authors emphasize the utility of log book based CPUE records as the best available proxy of population abundance in the absence of expensive fishing surveys.

Given that pike growth can be density dependent (chapter 6) growth-history based indices also have potential to reflect underlying variations in population abundance. A comparison of pike growth patterns, relative size-class densities, and population size estimates among 12 lakes in Minnesota and 17 lakes in Wisconsin revealed a consistent inverse relationship between body size and population abundance (Pierce et al. 2003). This finding suggests that pike growth could provide insights into population abundance, yet such patterns are likely to be very location specific as ecosystem productivity, prey availability, temperature, and other local environmental variables can also substantially modify pike growth (chapters 2, 3, 4, 6, 10). Therefore, changes in the local environment can also mask density-dependent variations in growth.

An interrelated size-based approach was suggested by Jansen et al (2013), who used log book data to explore temporal variations in the sizes of the annual record pike caught by anglers, i.e., the annual variation in trophy pike. Analyses of log book data conducted by Jansen et al. (2013) suggested that the abundance of trophy pike was related to occurrence of commercial fishing (fyke net and seining) and increased when commercial fishing at the lake stopped. This illustrates that a very minimalistic approach in terms of data requirement can provide useful information about population structure, e.g. records of trophy fish can be available through angling clubs and fishing competitions (see also chapter 12).

13.3.2 Abundance Estimates

While well-chosen abundance indices can provide insights into underlying population densities and size distributions, the core objective of fisheries stock assessment is often to provide estimates of the population size itself, ideally with limits of certainty. In rare special cases pike populations can be observed visually, given that the abundance of structured habitats (e.g. submerged vegetation) is fairly limited and water sufficiently clear (Turner and Mackay 1985, Brosse et al. 2001). However, such approaches are very location specific and require intensive fisheries-independent sampling. More typically, the assessments are conducted indirectly, by coupling catch and potentially also survey data with a method that infers the population size based on certain underlying assumptions about the population dynamics.

In terms of sampling requirements, the best methods to estimate population abundance are typically age- or length-based cohort analyses, but these require data for a large proportion of the entire annual catch, and assume that age and length distributions of the catch correspond to that of a cohort (Sparre and Venema 1998). Therefore, the population size must remain fairly stable. For pike, there exist attempts to assess pike stocks size using Jones' length-based cohort analysis (Cubuk et al. 2005). This method back-calculates ages of pike based on the population's von Bertalanffy growth curve. Unfortunately, the study does not provide any discussion about the suitability of the method for pike, and further research on this would be beneficial.

Within the context of wildlife conservation and management, mark-recapture studies have long been used to monitor and census the size of the study population (Lebreton et al. 1992). For pike, mark-recapture studies in the literature are mostly from scientific monitoring experiments, although rare mark-recapture monitoring specific to recreational interests do exist (Dye et al. 2002). A long-term mark-recapture pike study has been ongoing for six decades in Lake Windermere (Haugen et al. 2007), and shorter-term studies have also been conducted in, for example, several lakes in UK (Hawkins and Armstrong 2008), Finland (Kuparinen et al. 2012), and in the USA (Pierce and Tomcko 2003). It should be noted, however, that mark-recapture based stock assessments could also be conducted in collaboration with recreational catch-and-release fisheries as ecological awareness of anglers is increasing, and anglers could be recruited to report recaptures (chapter 12).

The most commonly applied method to estimate the population size from mark-recapture data is to utilize the Chapman modification of Petersen's population size estimate (Ricker 1975). This method, however, requires temporally very close captures by the same capture methods. In other words, the population should stay stable during the recaptures (no recruitment, mortality, or immigration/emigration). The size of the catchable population (N) is estimated by

$$N = ((T + 1)(n + 1))/(m + 1) \tag{13.1}$$

where T is the total number of marked individuals in the population, n is the size of the recapture sample, and m is the number of marked individuals in the recapture sample. Confidence intervals for N can be produced by Poisson approximation (Ricker 1975).

Petersen's population size estimates were utilized, for example, by Pierce and Tomcko (2003), though they further accounted for handling mortality arising from captures. Petersen's method assumes that the probability of being captured does not depend on the previous capture history or whether an individual carries a mark or not. This is probably a reasonable working assumption because behavioural and stress consequences of catch-and-release of pike have been found to be short term, often lasting no more than a couple of days (Klefoth et al. 2008, Arlinghaus et al. 2009, chapter 12).

Given the limitations associated with Petersen's method, it is not best suited for mark-recapture monitoring spanning across several years. While in theory Petersen's estimates could be corrected for recruitment and natural mortality, these rates are typically unknown and, therefore, population size estimates would remain highly sensitive to the uncertainty associated with these values. One way to address this challenge is to analyse mark-recapture data within the hierarchical Bayesian framework. Such a modelling approach has been applied to pike mark-recapture observations collected over four years in four lakes within the same geographic region in Southern Finland (Kuparinen et al. 2012). The advantage of this approach is that it effectively utilizes information from the several populations (at different sampling times) simultaneously and can account for differing capture methods as well as other potentially relevant covariates. The approach estimates the rates of natural mortality (M) in addition to the census population sizes. Through data augmentation, the method also accounts for the uncatchable component of the population (i.e., individuals in the population that were never captured).

Derived in this manner, Bayesian population size estimates were typically lower than Petersen's estimates and uncertainty about them was much lower than that associated with Petersen's estimates (Kuparinen et al. 2012, Fig. 13.6). Given that

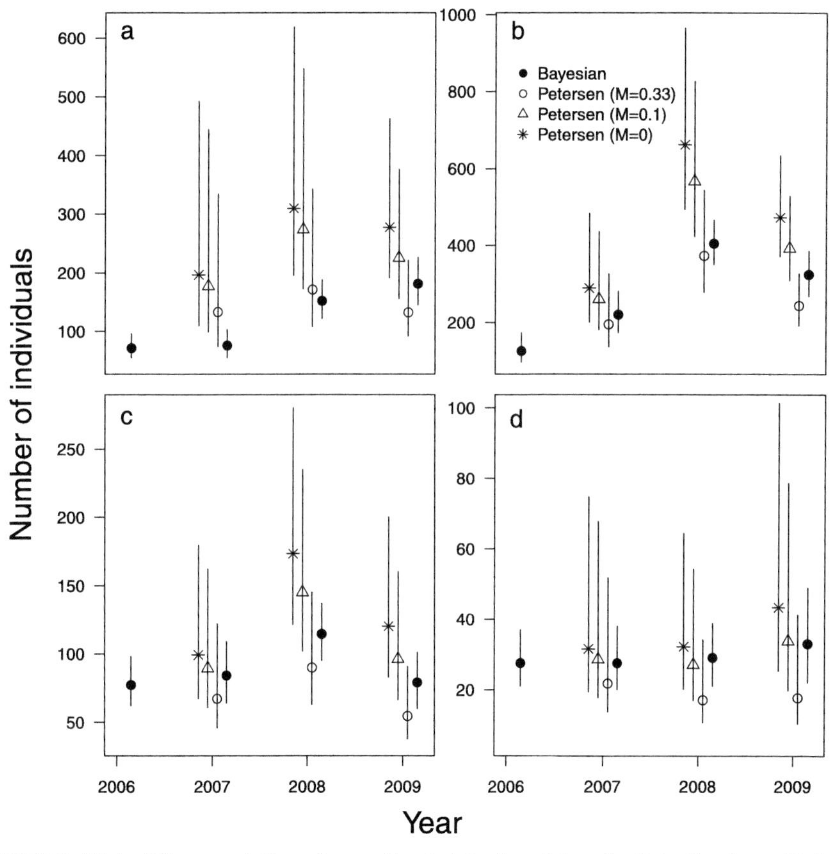

FIGURE 13.6 Pike population sizes estimated in four lakes (a-d) in Southern Finland based on a mark-recapture experiment conducted in years 2006-2009. Population sizes are estimated using Bayesian hierarchical method and Chapman's modification of Petersen's method with three alternative natural mortality (M) scenarios. (Reproduced from Fig. 2 in Kuparinen et al. 2012).

sampling was conducted in four consecutive years, Petersen's assumption about stable population size was not met. To correct for this factor, baseline population size estimates assuming $M = 0$ were contrasted to $M = 0.1$ as estimated by catch-curve method (Robson and Chapman 1961), and $M = 0.33$, which was the Bayesian estimate of the average natural mortality across study sites and years. The closest match with the Bayesian estimates was often (but not always) provided by Petersen's estimates corrected by the Bayesian natural mortality estimate. This study emphasizes the importance of estimating M and illustrates that long-term mark-recapture monitoring of pike requires analysis tools that go beyond Petersen's estimates.

13.3.3 Lake Windermere Monitoring Program

Apart from the extensive mark-recapture program, the pike of Lake Windermere have been monitored through annual experimental gillnet surveys mimicking commercial fishing, as well as through reported fisheries catches and efforts. Together, these data provide a basis for a pike stock assessment comparable to those conducted in marine commercial fisheries. Langangen et al. (2011) utilized an age-structured modelling approach typical of commercial fisheries stock assessments (Megrey 1989) to describe pike population dynamics and to estimate key demographic parameters through numeric optimization. The modelling approach further incorporated environmental covariates and information about the abundance of European perch (*Perca fluviatilis* Linnaeus, 1758), a species that constitutes both key predator and prey of pike in Lake Windermere. With these ingredients, the authors managed to construct time series of pike abundance together with its 95% confidence intervals (Fig. 13.7). The natural mortality estimates provided by the model matched closely to those estimated independently, based on mark-recapture

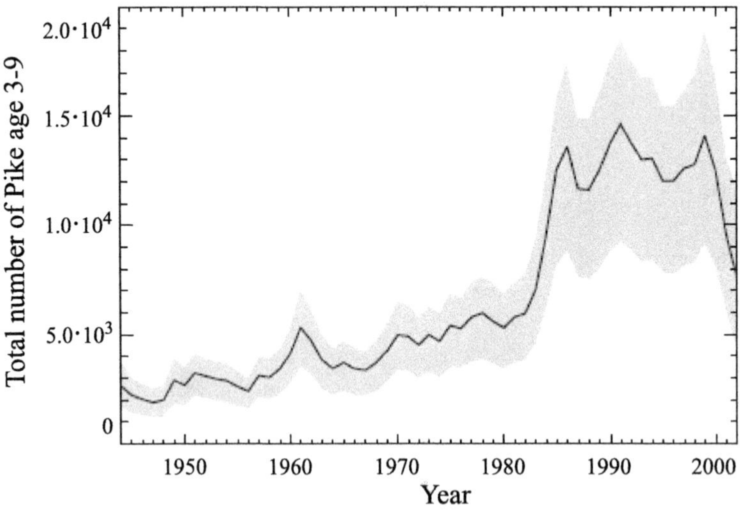

FIGURE 13.7 Temporal development of pike population abundance in Lake Windermere (UK). Population sizes comprise 3-9 year old pike; grey area encompasses the 95% confidence interval about the mean (solid line). (Reproduced from Fig 3. in Langangen et al. 2011).

data (Haugen et al. 2007). The model was further validated by comparing modelled and reported pike catches. The estimated pike stock-recruitment relationship was fairly linear, with weak compensation (i.e., negative density-dependence at low abundances) and positive temperature effect.

The assessment of Lake Windermere population provides the most complete insights into pike population dynamics and the long-term development of population abundance. Furthermore, it sets guidelines for sampling and conditions required to conduct pike stock assessment in a manner comparable to commercial fishery assessments. Interestingly, the study also suggests that pike population dynamics cannot be evaluated in isolation from its key competitors, such as the perch in the case of Lake Windermere (Langangen et al. 2011). This conclusion is also supported by the analyses of pike CPUE in Lake Windermere, suggesting that prey availability rather than direct environmental impacts regulate pike population growth (Winfield 2008).

13.4 AQUACULTURE

Pike is a relatively difficult fish species from an aquacultivator's point of view (Huet 1975). Problems arise because pike are voracious carnivores, and cannibalism is the major limiting factor in intensive farming (Grimm 1981, Bry et al. 1995). Low survival is common even under good conditions with plenty of food. Pike cannibalism has been reported in fish as young as four to five weeks old (Fago 1977, Giles et al. 1986).

Evidence regarding historic propagation and rearing of pike for direct human consumption is scant, and nowadays pike are primarily produced for stocking of natural waters. In nearly all cases, eggs and milt are collected from wild fish. These are readily caught on their spawning grounds with either trap nets or pound nets. Sometimes electrofishing or gillnetting is also used. Natural spawning of pike occurs in spring, normally at temperatures from 4 to 11°C (Scott and Crossman 1973). Males arrive on the spawning grounds before females (chapter 5, Fig. 5.2); usually at the time of ice-out. Eggs hatch about 120 degree days (accumulated product of time and temperature) after fertilization. After hatching, the fry have a hanging or attached stage (Leslie and Gorrie 1985) (see chapters 3 and 9 for definitions of early life stages, i.e., fry, fingerlings and juveniles). In aquaculture, it is often useful to improve the substrate by placing aquatic plants or conifer branches into the ponds. In small rearing tanks the newly hatched larvae may also attach to the walls of the tanks. The optimal water temperature is from 7 to 12°C, but a range of 5 to 21°C can produce viable fry (Hokanson et al. 1973, Stickney 1986). Lethal temperatures for pike embryos and yolk sac fry have been reported outside the 5-21°C range at 3 and 24°C (Hokansson et al. 1973). After hatching, another 160 to 180 degree days are required for yolk-sac absorption. The time from initiation of incubation to feeding is from 300 to 360 degree days. After that period the fry swim to the surface in order to fill their gas bladders and adopt a horizontal swimming position (Raat 1988). At this stage the yolk resorption is almost complete, and the fry are stocked either in natural water or in culture ponds to grow them to fingerlings.

Due to the high degree of cannibalism, pike for human consumption are not commonly cultivated in ponds in monoculture (Helfrich et al. 2009). Other fish species are usually more economical to farm if the goal is to produce biomass. This is why the principal goal of pike culture is to produce fry and fingerlings for stocking purposes (chapters 9, 11). Pike have been propagated and reared only to a minor extent for human consumption in some areas of Europe (FAO 2015). The global annual aquaculture production of pike was highest in the 1990s, and, after a decrease, again increased during the 2000s (Fig. 13.8).

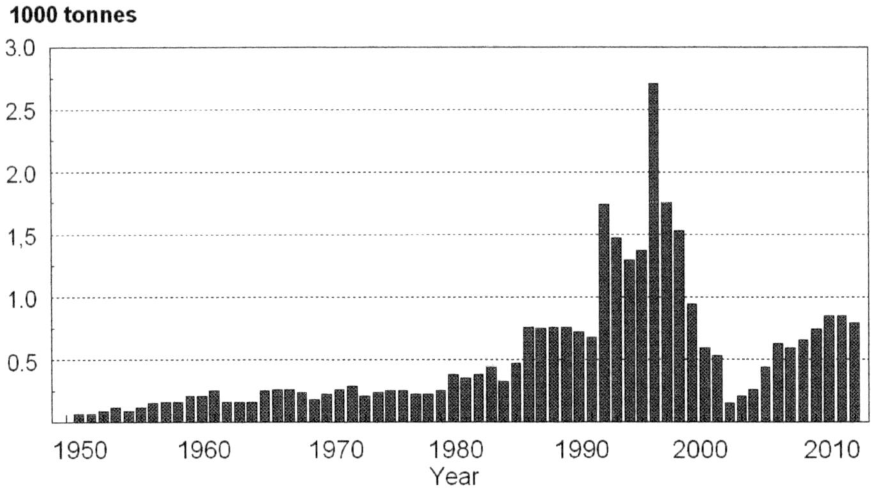

FIGURE 13.8 Global aquaculture production of pike in tonnes (redrawn from FAO 2015).

The rearing of pike in ponds has more often been successful in polyculture where pike are fed with live fish as prey (Bry et al. 1995). For example, in France, roach (*Rutilus rutilus* Linnaeus, 1758) and tench (*Tinca tinca* Linnaeus, 1758), and in North America, bluegills (*Lepomis macrochirus* Rafinesque, 1819) are used as prey (Bardach et al. 1972, Scott and Crossman 1973). In these systems, pike seldom exceed 10% of the total fish biomass. Growth of pike in such conditions is generally good and their weight may increase from 100 g to 800 g in one summer (Bardach et al. 1972). This is an expensive and very laborious cultivation method. More recently, artificial pellets have been successfully used in hatchery systems (Westers 1986, Westers and Stickney 1993), and may boost interest in pike culture for human consumption.

It was formerly thought that supplementation with forage fish was necessary beyond the pike length of 60 mm (Hiner 1961). However, the prolonged holding of young pike in ponds may be possible in the presence of large-sized invertebrate prey, such as macrocrustacea (Bry et al. 1995). Pike have been reared to sizes of up to 100 mm on pellets (Westers 1986). According to Bry et al. (1995), size at harvest may vary from 40 mm (the production of early juveniles) to 200 mm total length, depending on rearing practice. They suggest that an optimal harvest of pike is between 70 and 100 mm (weight 2-5 g).

In most cases, the goal is to produce only 4-6 cm long fingerlings, as cannibalism increases significantly at larger sizes. The duration of culture varies widely but is usually 3-5 weeks. The time depends primarily on the productivity of the pond or the amount of food given and the desired length of the fingerlings at harvest. Pike are usually reared to grow in natural ponds but sometimes also in temporarily flooded shallow vegetated areas. Ponds are always shallow and their area is normally not more than 1-3 ha. The presence of aquatic vegetation is a prerequisite for high pike production (chapters 3, 10).

In intensive culture, the size of tanks is usually only a few square meters, and the water depth is 40-50 cm. Both round and rectangular tanks are in use. The tanks are usually stocked with 2000-4000 newly hatched fry/m^2. The duration of the production period depends on the stocking density and feeding rate. In the beginning, the larvae are usually fed once per day with live zooplankton. Feeding rate is later increased to 2 or 3 times per day (Westers and Stickney 1993). Collecting the prey animals is a cumbersome task. In some cases, the ponds are fertilized to increase the food production.

Commercial trout fodder has proved suitable for pike juveniles (Hevroy et al. 2004). In spite of this, there is a need to develop a special pike fodder. Kucska et al. (2005) observed in their experiment that pike fingerlings fed with artificial pellets had nearly 100% survival and all mortalities occurred during the first week of the experiment only. They believe that intensive culture by pellet feeding can be a profitable method for the production of large pike fingerlings in the future.

Pike grow much more rapidly than salmonids. Therefore, it is necessary to adjust feeding levels more frequently than is common for salmonid culture. Preferred feeding temperatures of pike range from 7°C to 15°C (Bry et al. 1991). Fish hatcheries are often equipped with water temperature control devices and it is best to work with a relatively constant temperature. Although temperatures higher than recommended may produce somewhat more rapid rates of growth, it is advisable to operate below optimum growth temperatures (23 to 24°C) as a means of avoiding problems with pathogens which have comparable optimal temperature ranges (Stickney 1992).

Often, the aim of pike culture is to obtain larger juveniles (10 cm or more). The major problem with this is, again, cannibalism which is difficult to avoid even when an abundant quantity of forage fish is present. According to Stickney (1986) the predator-prey ratio of 1 : 10 is enough to prevent cannibalism. Forage fish must be continually added in high numbers. When the ratio drops below the critical level, considerable losses due to cannibalism will begin to occur. Because of the need for large numbers of forage fish of right sizes, the production costs of large pike fingerlings may be very high.

13.5 CONCLUDING REMARKS

In spite of the fact that pike today are mainly caught by recreational fishers, commercial pike fisheries can still be economically important. This has caused competition between different fisher groups. Pike fisheries were historically managed primarily as commercial fisheries. However, management for recreational fisheries is today increasingly considered. As a result, regulatory changes such as restricting the use of commercial gears is now common in many water areas.

Stock assessment of pike usually involves quantifying the numbers (absolute or relative) and size and age distributions of pike within a defined area, waterbody or population. Population assessments are often used by managers to balance the interests of recreational and commercial fisheries, to set fisheries regulations and to evaluate the conservation status of populations. However, systematic pike stock assessments are rare due to difficulties to use traditional methods for a sedentary species. Ecological studies, such as food web manipulation experiments and habitat monitoring have also explored trends in pike population abundances, but without direct links to fisheries management.

As a voracious carnivore, pike is a relatively difficult fish species from an aquacultivator's point of view. Also, cannibalism is the major limiting factor in intensive farming. Low survival is common even under good conditions with plenty of food. Rearing of pike in ponds has more often been successful in polyculture, where pike are fed with live fish as prey. Despite these difficulties, pike are locally, and rarely, reared for direct human consumption, while the principal goal of current pike culture is to produce fry and fingerlings for stocking purposes.

Acknowledgements

Writing of this chapter was facilitated though Academy Fellowship funding and the Natural Sciences and Engineering Research Council of Canada Discovery Grant to Anna Kuparinen and inspired by KESKALA research program funded by Bergsrådet Bror Serlachius Stiftelse and led by Hannu Lehtonen. We thank the Great Lakes Fishery Commission for providing us with pike catch records.

REFERENCES CITED

Arlinghaus, R. and T. Mehner. 2004. Testing the reliability and construct validity of a simple and inexpensive procedure to measure the use value of recreational fishing. *Fish. Man. Ecol.* 11:61-64.

Arlinghaus, R., T. Klefoth, S.J. Cooke, A. Gingerich and C. Suski. 2009. Physiological and behavioural consequences of catch-and-release angling of northern pike (*Esox lucius* L.). *Fish. Res.* 97:223-233.

Arlinghaus, R. B. Beardmore, C. Riepe, J. Meyerhoff and T. Pagel. 2014. Species-specific preferences of German recreational anglers for freshwater fishing experiences, with emphasis on the intrinsic utilities of fish stocking and wild fishes. *J. Fish Biol.* 85: 1843-1867.

Baldwin, N.S., R.W. Saalfeld, M.A. Ross and H.J. Buettner. 1979. *Commercial fish production in the Great Lakes 1867-1977.* Technical Report No. 3, Great Lakes Fishery Commission, Ann Arbor, MI.

Baldwin, N.A., R.W. Saalfeld, M.R. Dochoda, H.J. Buettner and R.L. Eshenroder. 2009. *Commercial Fish Production in the Great Lakes 1867-2006.* Available from http://www.glfc.org/databases/commercial/commerc.php.

Bardach, J.E., J.H. Ryther and W.O. McLarney. 1972. *Aquaculture, the Farming and Husbandry of Freshwater and Marine Organisms.* John Wiley & Sons, New York.

Biro, P.A. and J.R. Post. 2008. Rapid depletion of genotypes with fast growth and bold personality traits from harvested fish populations. *Proc. Nat. Acad. Sci. USA* 105:2919-2922.

Brosse, S., P. Laffaille, S. Gabas and S. Lek. 2001. Is scuba sampling a relevant method to study fish microhabitat in lakes? Examples and comparisons for three European species. *Ecol. Freshw. Fish.* 10:138-146.

Bry, C., M.G. Hollebecqa, V. Ginotb, G. Israela and J. Manelphea. 1991. Growth patterns of pike (*Esox lucius* L.) larvae and juveniles in small ponds under various natural temperature regimes. *Aquaculture* 97:155-168.

Bry, C., F. Bonamyj, J. Manelphe and B. Duranthon. 1995. Early life characteristics of pike, Esox lucius, in rearing ponds: temporal survival pattern and ontogenetic diet shifts. *J. Fish. Biol.* 46:99-113.

Craig, J. (ed.). 1996. *Pike: Biology and exploitation.* Chapman & Hall, Cambridge.

Colby, P.J. 2012. *Sustainability of Commercial Fisheries at Selected Lakes in Alberta's Commercial Fishery Zone E: Final Assessment.* Available from http://esrd.alberta.ca/fish-wildlife/fisheries-management/documents/SustainabliityOfCommericalFisheriesZoneE-FinalAssesment-Oct-2012.pdf

Cubuk, H., Ü Balik, R Uysal and R. Özkök. 2005. Some Biological Characteristics and the Stock Size of the Pike (*Esox lucius* L., 1758) Population in Lake KaramÝk (Afyon, Turkey). *Turk. J Vet. Anim. Sci.* 29:1025-1031.

Dye, J., M. Wallendorf, G.P. Naughton and A.D. Gryska. 2002. *Stock assessment of Northern pike in Lake Aleknagik, 1998-1999.* Fishery Data Series No. 02-14, Alaska Department of Fish and Game.

Ernst and Young, 2011. *EU intervention in inland fisheries. Framework contract N°FISH/2006/09(LotN°3).* Available from http://ec.europa.eu/fisheries/documentation/studies/inland_fisheries_en.pdf

Fago, D. M. 1977. Northern pike production in managed spawning and rearing marshes. Wisconsin. *Dept. Natural Resources Tech. Bull.* 96.

FAO, 2014. *Global capture production (FishStat).* Available from http://data.fao.org/dataset-data-filter?entryId=af556541-1c8e-4e98-8510-1b2cafba5935&tab=data&type=Dimensionmember&uuidResource=7b7377c8-d56f-4661-b74a-b6a1115f42f0).

FAO, 2015. *Species fact sheets, Esox lucius L. 1758.* Available from http://www.fao.org/fishery/species/2942/en.

Finnish Game and Fisheries Research Institute. 2013. Commercial Inland Fishery 2012. *Riista- ja kalatalous – Tilastoja 6/2013. Official Statistics of Finland – Agriculture, Forestry and Fishery.* Finnish Game and Fisheries Research Institute.

Finnish Game and Fisheries Research Institute. 2014. Commercial Marine Fishery 2013. *Riista- ja kalatalous - Tilastoja 3/2014. Official Statistics of Finland – Agriculture, Forestry and Fishery.* Finnish Game and Fisheries Research Institute.

Fisheries and Oceans Canada. 2012. Survey of Recreational Fishing in Canada 2010. Available from http://www.dfo-mpo.gc.ca/stats/rec/can/2010/RECFISH2010_ENG.pdf.

Giles, N., R.M. Wright and M.E. Nord. 1986. Cannibalism in pike fry, *Esox lucius* L.: some experiments with fry densities. *J. Fish Biol.* 29:107-113.

Grimm, M.P. 1981. Intraspecific Predation as a principal factor controlling the biomass of northern pike (*Esox lucius* L.). *Fish. Manage.* 12:77-79.

Hansen, L.I. 2009. Varanger-siidaen – et handelssentrum. pp 349-367. *In*: E. Niemi and C. Smith-Simonsen. (eds.). *Det Hjemlige og det Globale. Festskrift til Randi Rønning Balsvik.* Oslo: Akademisk Publisering.

Haugen, T.O., J.I. Winfield, L.A. Vollestad, J.M. Fletcher, B. James and N.C. Stenseth. 2007. Density dependence and density independence in the demography and dispersal of pike over four decades. *Ecol. Monogr.* 77:483-502.

Hawkins, L. and J. Armstrong 2008. *Population size estimates of pike, Esox Lucius: a brief review*. Fisheries Research Services Internal Report No 15/08.

Helfrich, L.A., D.J. Orth and R.J. Neves. 2009. *Freshwater Fish Farming in Virginia: Selecting the Right Fish to Raise*. Virginia Cooperative Extension, Publ. 420-010. Available from https://pubs.ext.vt.edu/420/420-010/420-010.html.

Hevroy, E.M., K. Sandnes and G.-I.Hemre 2004. Growth, feed utilisation, appetite and health in Atlantic salmon (*Salmo salar* L.) fed new type of high lipid fish meal, Sea GrainR, processed from various pelagic marine fish species. *Aquaculture* 235:371-392.

Hiner, L.E. 1961. Propagation of northern pike. *Trans. Am. Fish. Soc.* 90:298-302.

Hokanson, K.E.F., J.H. McCormick and B.R. Jones 1973. Temperature requirements for embryos and larvae of the northern pike, *Esox lucius* (Linnaeus). *Trans. Amer. Fish. Soc.* 102:89-100.

Huet, M. 1975. *Textbook of fish culture. Breeding and cultivation of fish.* Fishing News Books Ltd. Page Bros: Norwich.

Jansen T., R. Arlinghaus, T.D. Als and C. Skov. 2013. Voluntary angler logbooks reveal long-term changes in a lentic pike, *Esox lucius*, population. *Fish. Manag. Ecol.* 20:125-136.

Jones, A.K.G. 1988. Provisional remarks on fish remains from archeological deposits in York. In: *The exploitation of wetlands,* P. Murphy and C. French, C. Br. Archaeol (eds.). Rep. British Ser. 186:113-127.

Leslie, J.K. and J.F. Gorrie 1985. Distinguishing features for separating protolarvae of three species of esocids. pp 1-20. *In*: A.W. Kendall Jr. and J.B. Marliave (eds.). *Descriptions of early life history stages of selected fishes: Third international symposium on the early life history of fishes*. Canadian Technical Report of Fisheries and Aquatic Sciences, 1359.

Klefoth, T., A. Kobler and R. Arlinghaus. 2008. The impact of catch-and-release angling on short-term behaviour and habitat choice of northern pike (*Esox lucius* L.). *Hydrobiologia* 601:99-110.

Kucska, B., T. Müller, J. Sári, M. Bódis and M. Bercsényi. 2005. Successful growth of pike fingerlings (*Esox lucius* L.) on pellet at artificial condition. *Aquaculture* 246:227-230.

Kuparinen, A., S. Kuikka and J. Merilä 2009. Estimating fisheries-induced selection: traditional gear selectivity research meets fisheries-induced evolution. *Evol. Appl.* 2:234-243.

Kuparinen, A., J. Alho, M. Olin and H. Lehtonen. 2012. Estimation of northern pike population sizes via mark-recapture monitoring. *Fish. Manag. Ecol.* 19:323-332.

Langangen, Ø, E. Edeline, J. Ohlberger, I.J. Winfieldc, J.M. Fletcherc, J.B. James, N.C. Stenseth and L.A. Vøllestad. 2011. Six decades of pike and perch population dynamics in Windermere. *Fish. Res.* 109:131-139.

Lebreton, J.D., K.P. Burnham, J. Clobert and D.R. Anderson. 1992. Modelling survival and testing biological hypotheses using marked animals: a unified approach with case studies. *Ecol. Mon.* 62:67-118.

Lehtonen, H., E. Leskinen, R. Selen and M. Reinikainen. 2009. Potential reasons for the changes in the abundance of pike, *Esox lucius*, in the western Gulf of Finland, 1939-2007. *Fish. Manag. Ecol.* 16:484-491.

Mann, R.H.K. 1996. Fisheries and economics. pp 219-241. *In*: J.F. Craig (ed.). *Pike: Biology and exploitation*. London: Chapman & Hall.

Marine Stewardship Council. 2014. *Waterhen Lake walleye and Northern pike Commercial Gillnet. MSC status*. Available from https://www.msc.org/newsroom/news/manitoba-freshwater-fishery-enters-msc-assessment-waterhen-lake-walleye-and-northern-pike-commercial-gillnet-fishery

Mathiassen, T. 1943. *Stenalderbopladser i Aamosen.* Kommission hos Gyldendalske boghandel: Nordisk forlag.

Maunder, M.N., J.M. Sibert, A. Fonteneau, J. Hampton, P. Kleiber and S.J. Harley. 2006. Interpreting catch per unit effort data to assess the status of individual stocks and communities. *ICES J. Mar. Sci.* 63:1373-1385.

Megrey, B.A. 1989. Review and comparison of age-structured stock assessment models from theoretical and applied points of view. *In*: E. Edwards and B. Megrey (eds.). Mathematical analysis of fish stock dynamics: Reviews and current applications. *Amer. Fish. Soc. Symp.* 6:8-48.

Paukert C.P., J.A. Klammer, R.B. Pierce and T.D. Simonson. 2001. An overview of northern pike regulations in North America. *Fisheries* 26:6-13.

Pierce, R.B., C.M. Tomcko and T.L. Margenau. 2003. Density Dependence in Growth and Size Structure of Northern Pike Populations. *N. Am. J. Fish. Man.* 23:331-339.

Pierce, R.B. and C.M. Tomcko. 2003. Variation in gillnet and angling catchability in with changing density in Northern pike in a small Minnesota lake. *Trans. Am. Fish. Soc.* 132:771-779.

Prejs, A., J. Pijanowska, P. Koperski, A. Martyniak, S. Boron and P. Hliwa. 1997. Food-web manipulation in a small, eutrophic LakeWirbel, Poland: long-term changes in fish biomass and basic measures of water quality. A case study. *Hydrobiologia* 342/343:383-386.

Raat, A.J.P. 1988. Synopsis of biological data on the northern pike *Esox lucius* Linnaeus, 1758. FAO Rome, *FAO Fish. Synop. No 30.*

Ricker, W.E. 1975. Computation and interpretation of biological statistics of fish populations. *Bull. Fish. Res. Bd. Can.* 191:1-382.

Robson, D.S. and D.G. Chapman. 1961. Catch curves and mortality rates. *Trans. Am. Fish. Soc.* 90:181-189.

Scott, W.B. and E.J. Crossman. 1973. Freshwater fishes of Canada. *Fish. Res. Bd. Can. Bull.* 184.

Sparre, P. and S.C. Venema. 1998. *Introduction to Tropical Fish Stock Assessment - Part 1: Manual.* . FAO Fisheries Technical Paper 306/1: Rome.

Stickney, R.R. 1986. Culture of Non-Salmonid Freshwater Fishes. CRC Press, Boca Raton. FL, 201 pp

Stickney, R.R. 1992. *Culture of Nonsalmonid Freshwater Fishes.* CRC Press: USA.

Turner, L.J. and W. Mackay. 1985. Use of Visual Census for Estimating Population Size in Northern Pike. (*Esox lucius*). *Can. J Fish. Aquat. Sci.* 42:1835-1840.

Winfield, I.J., J.B. James and J.M. Fletcher. 2008. Northern pike (*Esox lucius*) in a warming lake: changes in population size and individual condition in relation to prey abundance. *Hydrobiologia* 601:29-40.

Westers, H. 1986. Northern pike and muskellunge. pp 91-101. *In*: R.R. Stickney (ed.). *Culture of nonsalmonid freshwater fishes*. Boca Baton, USA: CRC Press.

Westers, H. and R.R. Stickney. 1993. Northern pike and muskellunge. pp 199-214. *In*: R.R. Stickney (ed.). *Culture of nonsalmonid freshwater fishes*. Boca Baton, USA: CRC Press.

Wolos, A., B. Zdanowski and M. Wierzchowska. 2009. Long-term changes in commercial fish catches in Lake Mamry Pólnocne (northeastern Poland) on the background of physical, chemical, and biological data. *Arch. Pol. Fish.* 17:195-210.

Chapter 14

The Northern Pike, A Prized Native but Disastrous Invasive

Kristine Dunker[*1], Adam Sepulveda[2], Robert Massengill[3] and David Rutz[4]*

14.1 NORTHERN PIKE AS AN INVASIVE SPECIES

As the chapters in this book describe, the northern pike *Esox lucius* Linneaus, 1758 is a fascinating fish that plays an important ecological role in structuring aquatic communities (chapter 8), has the capacity to aid lake restoration efforts (chapter 11), and contributes substantially to local economies, both as a highly-sought after sport fish (chapter 12) and as a commercial fishing resource (chapter 13). However, despite the magnificent attributes of this fish, there is another side to its story. Specifically, what happens when northern pike, a highly efficient predator, becomes established outside its natural range? To explore this question, this chapter will investigate observed consequences from many locations where northern pike (hereafter referred to as "pike") have been introduced and discuss potential reasons why pike, under the right circumstances, can be considered an invasive species.

The concept of 'invasive species' is becoming increasingly familiar among scientists, policy makers and the general public. Personal experiences with invasive species, from pulling garden weeds to more extreme circumstances involving property damage or lost livelihoods, have led to this broad familiarity. Most people

Corresponding author: [1]Alaska Department of Fish and Game, Division of Sport Fish, 333 Raspberry Road, Anchorage, AK 99518, Email: kristine.dunker@alaska.gov

[2]United States Geological Survey, Biological Resources Division, Northern Rocky Mountain Science Center, 2327 University Way, Suite 2, Bozeman, MT 59715, Email: asepulveda@usgs.gov

[3]Alaska Department of Fish and Game, Division of Sport Fish, 43961 Kalifornsky Beach Road, Suite B, Soldotna, AK 99669, Email: robert.massengill@alaska.gov

[4]Alaska Department of Fish and Game, Division of Sport Fish *(Retired)*

with basic knowledge about invasive species would agree they can be harmful. However, not as many might recognize that seemingly beneficial species might also have the capacity to become invasive. This can be the case with the pike which is typically prized by anglers for its aggressiveness and excellent food quality. To understand how such a revered species can fall into the invasive category, it is important to have a fundamental understanding of what an invasive species is.

By definition, an invasive species is a species that has been introduced to an environment where it is non-native, or alien, and whose introduction causes environmental or economic damage or harm to human health (IUCN 2015). Not all non-native organisms become invasive, and the degrees of impact from invading organisms on native biota are difficult to predict and dependent on multiple factors (Havel et al. 2015, Ricciardi and Cohen 2007). To become invasive, a species must first successfully colonize their new environment, and successful invasive colonizers tend to share certain biological attributes (Ricciardi et al. 2013). For example, among vertebrates, they tend to have large native ranges, high dispersal capabilities, broad diets and environmental tolerances, short generation times, exhibit distinctive functional traits from the native community, are abundant, have a history of successful colonizations, and are associated with humans (Havel et al. 2015, Ehrlich 1989).

Pike have many of these traits. They have one of the largest geographic distributions of all freshwater fish species and can move significant distances to spawn or overwinter (Tyus and Beard 1990; chapter 5). Once transplanted to a new region, they can colonize connected waters where they have the capacity to alter the trophic structure of their new environments. As apex ambush predators with high trophic adaptability, pike are capable of shifting their piscivorous diets to include amphibians, small mammals, waterfowl, invertebrates, and conspecifics when preferred prey resources are scarce (Solman 1945, Sepulveda et al. 2013; chapter 2). Pike also have broad physico-chemical tolerances and can survive hypoxic, saline, and other variable environmental conditions (Raat 1988, Craig 1996; chapter 3). They have high reproductive potential, can hybridize with other escocids (Raat 1988) and can achieve high population densities (Mann 1980; chapter 6). Pike are utilized by humans for sport (chapter 12), commercial harvest (chapter 13) and subsistence and have been widely introduced outside of their native range where they have established populations and are now considered an invasive species in many of these locations (Zelasko et al. 2016, Bradford et al. 2009, ADFG 2007, Marchetti et al. 2004).

The ecological and economic ramifications stemming from non-native pike establishment can be severe. Their high predation rates can reduce and often extirpate populations of native fish leading to declines in aquatic biodiversity, diversions of energy from native predators, and, in the case of salmon predation, lost nutrient inputs into riparian environments. Pike populations are also highly susceptible to stunting when optimal prey resources are depleted (Ventruelli and Tonn 2006), and these smaller fish are not popular among anglers (Pierce et al. 1995, Paukert et al. 2001). This can lead to potential economic consequences because industries that are built around healthy fisheries can become threatened and occasionally collapse altogether following the establishment of an introduced pike population. Alarmingly, many of these consequences occur across relatively short temporal scales and require substantial resources by local governments to remediate. Throughout this chapter, the circumstances in which pike are an invasive species will be explored by highlighting their patterns of colonization, discussing the effects of non-native pike establishment on salmonids and other fishes, addressing the anthropomorphic consequences of these effects, identifying management strategies, and suggesting future needs for research to better inform the management of this species when it becomes invasive.

14.2 NON-NATIVE PIKE INTRODUCTIONS

Pike have a circumpolar holarctic distribution including most of the north temperate zone above 40 degrees latitude, but they have been widely introduced outside of their native range in North America, Europe, and Africa. Introductions have included legitimate stockings by government entities to establish sport fisheries and illegal stockings by private citizens (Welcomme 1988, Lever 1996, Brautigam and Lucas 2008; Fig. 14.1).

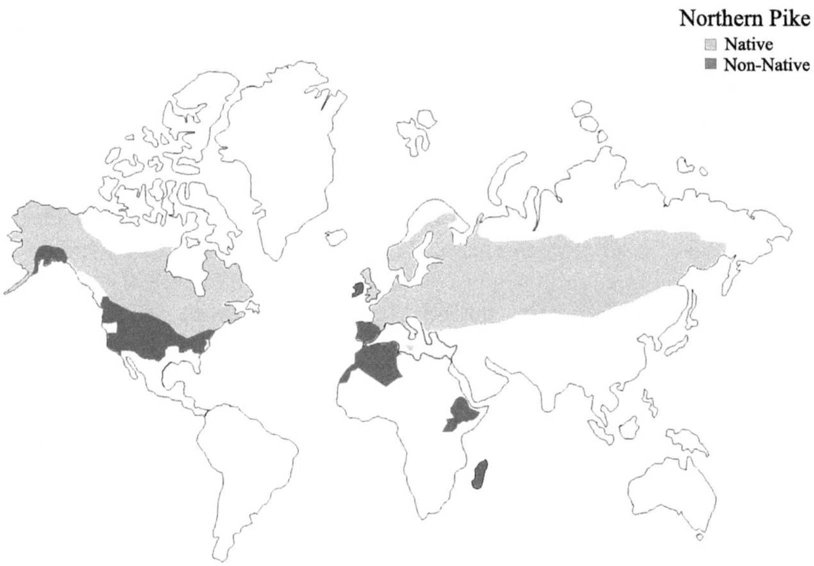

FIGURE 14.1 Global native and non-native pike distribution.

While the degree of scientific investigation, planning, and oversight in authorized stockings of pike has varied greatly with respect to when and where the stockings occurred, there is no oversight at all when pike are stocked illegally. Consequently, the implications for native aquatic communities can be more pronounced in these cases (Johnson et al. 2009, Hickley and Chare 2004, McMahon and Bennet1996).

In Eurasia, pike are native throughout central Asia and Siberia eastward to the Anadyr drainage in the Bering Sea basin. Their historic distribution also includes drainage basins of the Aral, Caspian, Black, White, Barents, Baltic, and North seas south to the Adour drainage (Kottelat and Freyhof 2007). The natural range of pike excludes Greenland, Iceland, western Norway and Northern Scotland (Fig. 14.1). Their historical distribution in Ireland is a topic of debate (Ensing 2014, Pedreschi et al. 2013, 2015; chapter 7), but some think that Ireland was the site of the first anthropomorphic introduction of this species, occurring sometime between the 12th and 16th centuries. Within the Mediterranean Basin, pike are native to the Rhone River drainage and portions of northern Italy, but they were historically absent from central Italy, the eastern Adriatic basin, southern and western Greece, and the Iberian Peninsula. Beginning in the late 1940's pike were introduced to the Iberian Peninsula to provide sport fishing opportunities (Leunda 2010) and today are widespread throughout the major drainages of Spain and Portugal (Elvira 1995a). In Africa, pike have been introduced to Morocco, Algeria, Tunisia, Ethiopia, Uganda, and Madagascar for the purposes of sport fishing or aquaculture (Welcomme 1988, Lever 1996), but their present status in these countries is largely unknown (Lever 1996).

The historical North American range for pike included northern and western Alaska, Canada south of the Arctic Circle, the lower Missouri River drainage to the upstream confluence of the Mississippi River, the Ohio River drainage, the Great Lakes drainage and the Saskatchewan River drainage (Scott and Crossman 1973). However, non-native pike populations outside of these drainages have been documented in 38 of the 50 United States (Fuller 2016; Fig. 14.2) and in southeast British Columbia (Heise et al. 2016), and many present-day concerns with pike functioning as an invasive species are centered in western North America.

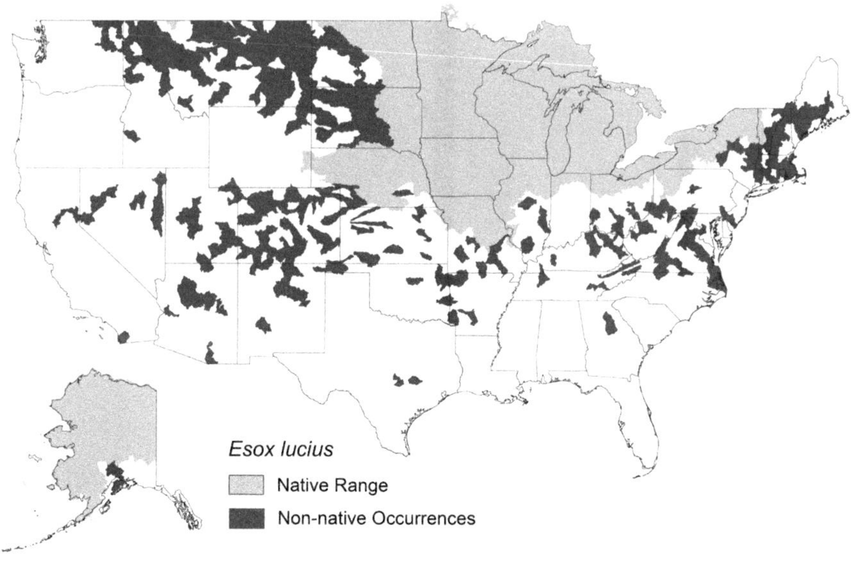

FIGURE 14.2 Pike distribution in the United States (Data provided by the US Geological Survey Nonindigenous Aquatic Species Database (https://nas.er.usgs.gov/).

In many states in the western US, pike were intentionally stocked by government agencies to provide recreational fishing opportunities (Fuller 2003) or for biocontrol of invasive common carp *Cyrinus carpio* Linnaeus, 1758

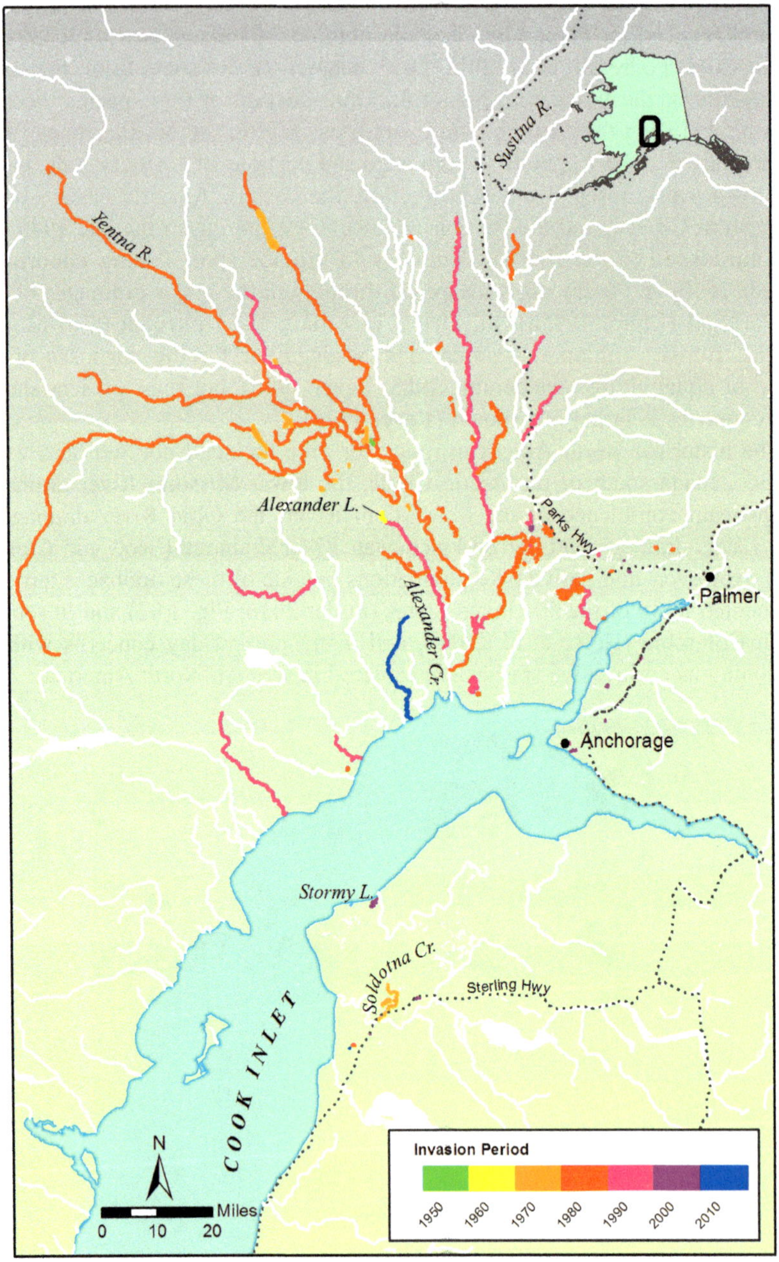

FIGURE 14.3 Expansion of pike throughout Southcentral Alaska since the late 1950's from illegal introductions and subsequent dispersal.

(Cook and Bergersen 1988). This has been especially prevalent in modified river systems containing reservoirs where, consequently, pike and other introduced fishes tend to thrive (Wydoski and Bennett 1981, Olden et al. 2006, Scarnecchia et al. 2014). In some cases, such as in the Upper Colorado River drainage in Colorado or the Upper Missouri River drainage in Montana, pike migrated from their original legal stocking locations to colonize unintended waters. Further, some legally stocked pike populations provided sources for misguided anglers to illegally transport them elsewhere. Unauthorized, illegal pike introductions have been an extensive problem in parts of the western United States. In Alaska, in particular, the situation is complex because pike are a native species in drainages north and west of the Alaska Mountain Range which has served as a natural barrier for resident freshwater fishes since the late Wisconsinan Glaciation (Oswood et al. 2000, Seeb et al. 1987). Beginning in the 1950s, however, pike in Alaska were illegally transported south from the Minto Flats in the Yukon River drainage, where they are native, to a lake in the Susitina River drainage, where they are not (ADFG, Unpublished data). Over the course of five decades, pike dispersal and continued illegal stockings resulted in their establishment in over 120 lakes and rivers in the northern Cook Inlet and Kenai Peninsula regions of the state (Fig. 14.3). In nearly all cases in western North America where pike have established following illegal introductions or escaped from authorized stockings, significant conservation concerns have resulted and have spurred efforts by local governments to curtail the damages (McMahon and Bennett 1996).

14.3 ECOLOGICAL EFFECTS OF INVASIVE PIKE INTRODUCTIONS

Globally, the introduction and proliferation of invasive species is recognized as one of the greatest threats to biodiversity in terrestrial and aquatic communities (Mooney et al. 2005). Effects on the native biota can be direct or indirect. Typically, disease transmission, hybridization, competition, prey substitution, and predation on native species are the primary drivers that lead to alterations of the invaded communities (Mooney et al. 2005). The introduction of an apex predator to a novel environment can have dramatic effects on native communities because of the top-down effects the predator has on the abundance of lower-trophic levels (Byström et al. 2007, chapter 8). Ecological disruptions can be amplified when the invasive predator is also a generalist or opportunistic feeder because alternative prey species can support the new predator after preferred prey species have declined, thus allowing the predator to increase in abundance and distribution while reducing or extirpating native species (Albins and Hixon 2008, Ogutu-Ohwayo 1990). It is in this capacity that introduced pike, an opportunistic apex predator, has had its greatest impact on native fish communities.

The risks to native fishes tend to be greatest when there is a high degree of habitat overlap with pike. As discussed in chapter 3, optimal pike habitat includes shallow, slow moving waters with abundant aquatic vegetation (Cook and Bergersen 1988; Fig. 14.4).

FIGURE 14.4 Alexander Lake in Southcentral Alaska. This lake provides ideal habitat conditions for pike and has an abundant invasive pike population.

Pike are not well adapted for strong currents (Inskip 1982), and high velocity waters and steep channel slopes can impede their movement (Spens et al. 2007, Inskip 1982). Adequate abundance of aquatic macrophytes is a critical habitat component for all life stages of pike (Inskip 1982). Shallow vegetated marshes, flooded terrestrial vegetation, weedy bays of lakes, and backwater sloughs of rivers are prime habitats for pike spawning and rearing. The limiting factor in these areas is the degree of water-level fluctuation, as this can affect nearshore vegetation and potentially desiccate eggs or larval fish (Inskip 1982). Young pike eventually disperse from spawning and rearing areas, but pike of all ages continue to frequent vegetated shallows (Chapman and Mackay 1984). For salmonids, where pike have been introduced, this is problematic because some species of juvenile salmonids rear in the same shallow, vegetated, calm waters in which pike thrive (Sepulveda et al. 2013).

14.3.1 Native Fish Impacts

14.3.1.1 Pacific Salmon

Pacific salmon are keystone species that play vital roles in the functioning of riparian and freshwater coastal ecosystems in the north Pacific because they fuel aquatic food webs, are prey for multiple terrestrial species, and supply marine-derived nutrients to these environments (Cederholm et al. 1999, Naiman et al. 2002). A myriad of environmental and anthropomorphic factors affect Pacific salmon population dynamics, but the introduction of an invasive predator can adversely complicate these factors and lead directly to declines (Sepulveda et al. 2014). Though pike are opportunistic predators, they predominately forage on soft-rayed fusiform fishes when they are available and exhibit strong preferences for juvenile salmonids (Sepulveda et al. 2013, Rutz 1999, Eklöv and Hamrin 1989). Compensatory effects in ocean survival can be a significant driver of salmon abundance patterns (Zabel et al. 2006, Greene and Beechie 2004), but reduced survival of juvenile salmon from pike predation has the potential to result in fewer adults returning to their natal streams to spawn. Marine-derived nutrients supplied by spawning salmon are critical components of optimal ecosystem functioning in coastal freshwaters. For example, marine-derived nitrogen provided through salmon carcasses contributes greater than 20% of the nitrogen taken up by trees in coastal riparian forests near salmon streams (Helfield and Naiman 2001). This can serve as a positive feedback mechanism by which nutrients deposited from salmon support ecological processes that, in turn, improve spawning and rearing conditions for future generations of salmon (Helfield and Naiman 2001). In this regard, reduced salmon abundance from pike predation can have ecological consequences, extending beyond food web disruptions, and contributing to declining habitat conditions which can accelerate losses in salmon production. Confounding ecological effects as illustrated in this example demonstrate the complexity of impacts when a non-native predator, such as pike, becomes established. In general, impacts resulting from non-native predators can be difficult to predict and depend on multiple factors such as environmental variability and the degree of invasion resistance from the incipient biota (Ricciardi and Cohen 2007).

For the majority of the following discussion, the focus will be on the direct top-down effect of reduced salmon abundance from pike predation where salmon populations are most vulnerable. Reductions in Pacific salmon, specifically Chinook salmon *Oncorhynchus tshawytscha* Walbaum, 1792, coho salmon *O. kisutch* Walbaum, 1792, and sockeye salmon *O. nerka* Walbaum, 1792 have, in some southcentral Alaskan waters, provided striking examples of the negative consequences of non-native pike introductions. There is, however, a great deal of variability in how different drainages have responded to the establishment of this species, and the degree of impact to salmon appears to be largely habitat-dependent (Sepulveda et al. 2013, 2014).

Pike occupy the same predator niche both within and outside of their native ranges. In interior and western Alaska, where pike are native, it is unknown how fish communities might differ in their absence, but pike naturally occur in the same drainages, albeit typically in different habitats, with salmon and resident species like arctic grayling *Thymallus arcticus* Pallas, 1776, Arctic char *Salvelinus alpinus* Linnaeus, 1758, whitefish *Coregonus* spp. Linnaeus, 1758, burbot *Lota lota* Linnaeus, 1758 and Alaska blackfish *Dallia pectoralis* Bean, 1880. In the Bristol Bay region of Alaska, the world's largest sockeye salmon runs occur alongside native pike populations. In contrast, in the southcentral part of Alaska, where pike have been part of the icthyofauna for less than 60 years, there are multiple cases where formerly robust salmon populations have dwindled in response to pike establishment (Smukall 2015, Sepulveda et al. 2014, Patankar 2006) and more cases where salmon populations are threatened by their potential spread (Massengill 2017b, ADFG 2007). Further research aimed at exploring this distinction is needed, but the prevailing hypotheses are that, in a broad sense, the degree of predation risk is mitigated by life histories of salmon with brief periods of freshwater juvenile residence (i.e., chum salmon *O. keta* Walbaum, 1792 and pink salmon *O. gorbuscha* Walbaum, 1792) and habitat heterogeneity within waters that allow juvenile salmon to rear in habitats separate from pike. In southcentral Alaska, anadromous salmon species that are most impacted spend between one and three years rearing in freshwaters before smolting and are available as pike prey for longer periods. In addition to this temporal availability, these species often share complete spatial habitat overlap with invasive pike populations (Sepulveda et al. 2013).

A mechanism that is hypothesized to facilitate coexistence between pike and salmon is the availability of spatial refugia (i.e., areas of suboptimal habitat conditions for pike within a lake or river) that results in habitat segregation between pike and salmon (Sepulveda et al. 2013, Hein et al. 2014). In invaded waters that are dominated by ideal pike spawning and rearing conditions, impacts from invasive pike on native fishes are predicted to be larger than in invaded waters that provide refuge for native fish via high velocity waters, channel slopes, or greater depths (Sepulveda et al. 2014, Hein et al. 2014). In boreal lakes in Sweden, it has been reported that coexistence between salmonids and pike is rarely possible (Spens and Ball 2008). In Alaska, however, the coexistence of pike and salmon is observed in waters with sufficient habitat complexity to allow their segregation (Sepulveda et al. 2013). Where pike are native and salmon populations are robust, such as in Southwestern Alaska, the landscape includes large, complex, deep water bodies such as the Wood River and Illiamna Lake systems that drain into Bristol Bay. These drainages contain a mixture of habitats that support both pike and other species (Figs. 14.5 and 14.6).

In the slow backwaters of these systems, pike are the dominant species, but as the habitat shifts toward deeper and more pelagic conditions or higher velocity, salmon and resident species become the dominant taxa (J. Dye, Personal communication, Hauser 2011). In particular, with the exception of shallow lake populations, juvenile sockeye salmon are largely pelagic foragers and spend little time near the vegetated banks were pike occur (Chihuly 1976). When they do interact with habitats that support pike, predation is a natural mechanism of their mortality. However, predation primarily occurs during the smolting period and has less of a regulating effect on their populations than in homogenous habitats were pike and juvenile sockeye share the same habitat for the entire period between emergence and smolting (Glick and Willette 2016).

As discussed in chapters 6-8 of this book, pike play an important role in structuring fish communities within their native range, particularly in homogenous habitats that are shallow, low gradient, and have high connectivity (Spens and Ball 2008). Comprehensive fish surveys of tributaries in Alaska's Lower Yukon River indicate that pike are primarily found in low relief areas within medium to large drainages that are subject to warmer temperatures and lower velocities. In contrast,

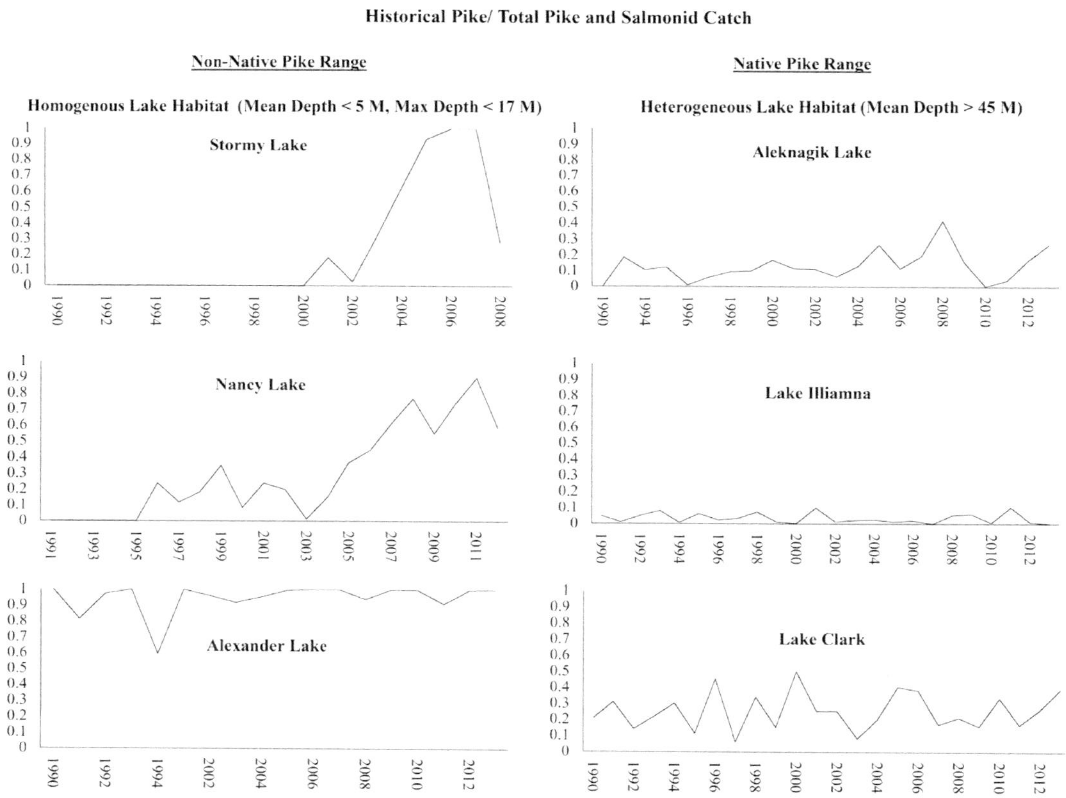

FIGURE 14.5 Angler catch patterns in Alaska in select waters where pike were introduced to shallow, more homogenous habitats vs. where pike are native and occupy deeper, more variable habitats. In the former, pike replace salmonids in proportional catch (Y axis = proportion of pike catch to the total catch of pike and salmonids), whereas in the latter, proportional catch rates between pike and salmonids do not differ.

Chinook and coho salmon were most often detected in medium to small streams within high relief areas rarely inhabited by pike (Buckwalter et al. 2010). Fish surveys conducted in a mixture of open and closed low-elevation lakes across six

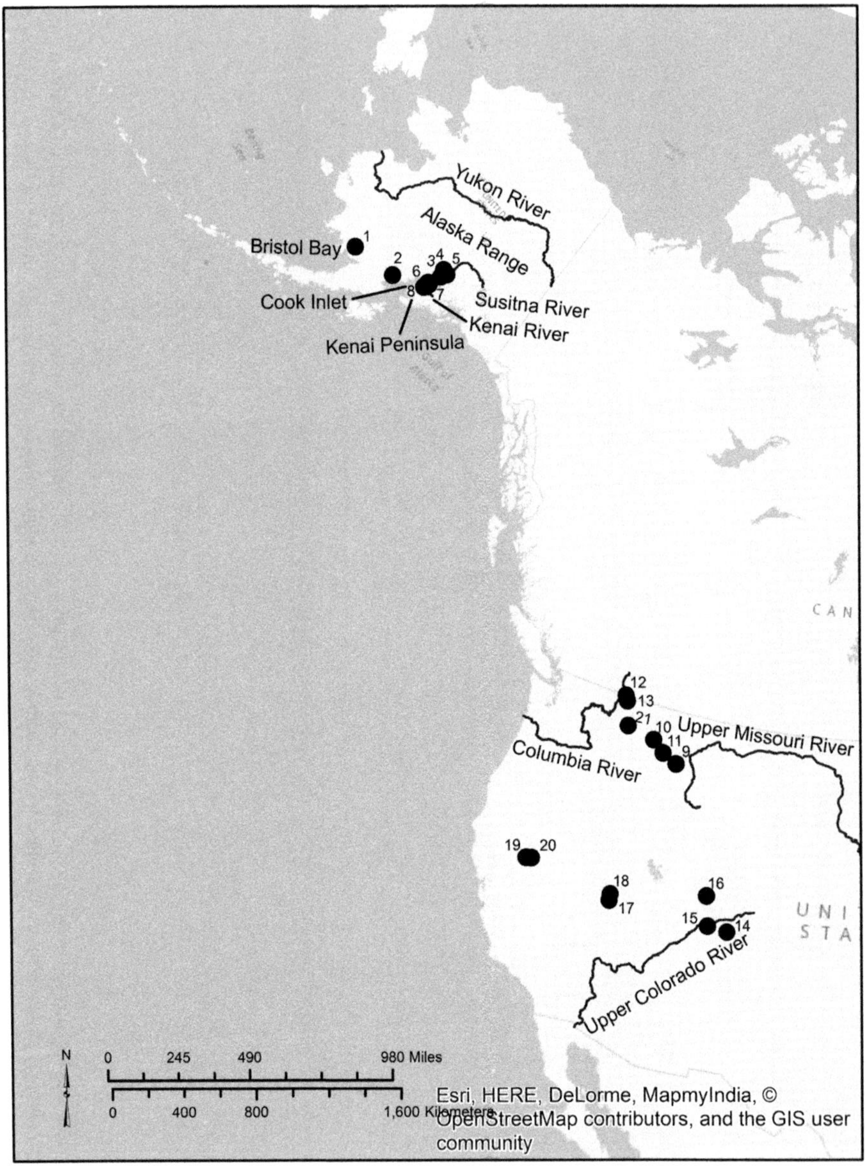

FIGURE 14.6 North American drainages and northern pike populations discussed in this chapter (1. Wood River, 2. Illiamna Lake, 3. Alexander Creek, 4. Alexander Lake, 5. Deshka River, 6. Swanson River, 7. Stormy Lake, 8. Soldotna Creek, 9. Clark Fork River, 10. Flathead River, 11. Milltown Reservoir, 12. Pend Oreille River, 13. Box Canyon Reservoir, 14. Paonia Reservoir, 15. Gunnison River, 16. Yampa River, 17. Comins Lake, 18. Basset Lake, 19. Lake Davis, 20. Frenchman Reservoir, 21. Coeur D'Alene Lake).

national wildlife refuges within the Yukon River drainage found that pike were the most frequently collected species, and no salmon were detected in any lake despite some having linkages to anadromous tributaries. Species such as grayling, lake trout *Salvelinus namaycush* Walbaum, 1792, slimy sculpin *Cottus cognatus* Richardson, 1836 and round whitefish *Prosopium cylindraceum* Pennant, 1784 were only found in higher elevation foothill lakes (Glense 1986). This pattern is corroborated by the work of Spens and Ball (2008), who found that if pike were absent from low elevation, interconnected lakes in the northern boreal region of Sweden, fish communities is this landscape would likely be dominated by salmonids. This aligns with observations of natural fish communities in boreal lakes in southcentral Alaska. This region has similar low-relief interconnected waters but most of these areas, unless pike have invaded, are occupied by juvenile salmonids (Johnson and Coleman 2014).

The greatest losses of salmon that are clearly attributed to invasive pike establishment in southcentral Alaska have occurred in waters with a high proportion of shallow, vegetated homogenous habitats that lacked refugia for other species. A particularly interesting case study, and perhaps the greatest example of this, is the loss of Chinook salmon in a Susitna River tributary known as Alexander Creek. Alexander Creek flows 64 km from Alexander Lake to the confluence with the Susitna River. Alexander Lake, at the headwaters, is shallow with a maximum depth of less than 3 m. The lake is approximately 356 hectares and is heavily vegetated throughout (Fig. 14.4). The main stem of Alexander Creek is mostly low gradient, slow velocity with instream vegetative mats, and surrounded by numerous vegetated side-channel sloughs that are less than 1.5 m deep. The majority of Alexander Creek flows through expansive interconnecting wetland areas that remain flooded throughout the spring, coinciding with the pike spawning period. The entire Alexander system is prime pike habitat for all life stages of the species.

Alexander Lake represents one of the first known introductions of pike into Southcentral Alaska (Fig. 14.3). Pike are believed to have been introduced to the lake sometime in late 1960s. All five Pacific salmon species found in Alaska historically spawned either in the lake or tributaries flowing directly into Alexander Lake which provided nursery habitat for salmon as well as other native resident fishes. Today, Alexander Lake is an abundant monoculture of pike, and other formally abundant fish species are now rare within its boundaries. Pike dispersal out of the lake and into all reaches of Alexander Creek occurred over the following three decades, and pike are now widespread throughout the system today. Historically, Alexander Creek supported one of the most productive Chinook salmon runs in the Susitna River basin with average escapements (the number of salmon needed to escape fisheries and spawn to sustain their population) of 3,500 (based on aerial index surveys) between 1979 and 1999. By the late 1990s, pike were well established throughout the entirety of Alexander Creek, and Chinook salmon escapements declined. Between 2000 and 2008, escapements fell to average indices of only 1,600 fish, prompting the closure of Chinook salmon fisheries on the river. This closure did not help. By 2010, Chinook salmon escapement in Alexander Creek had declined to an estimated 177 fish (Fig. 14.7), and other salmonid populations similarly crashed.

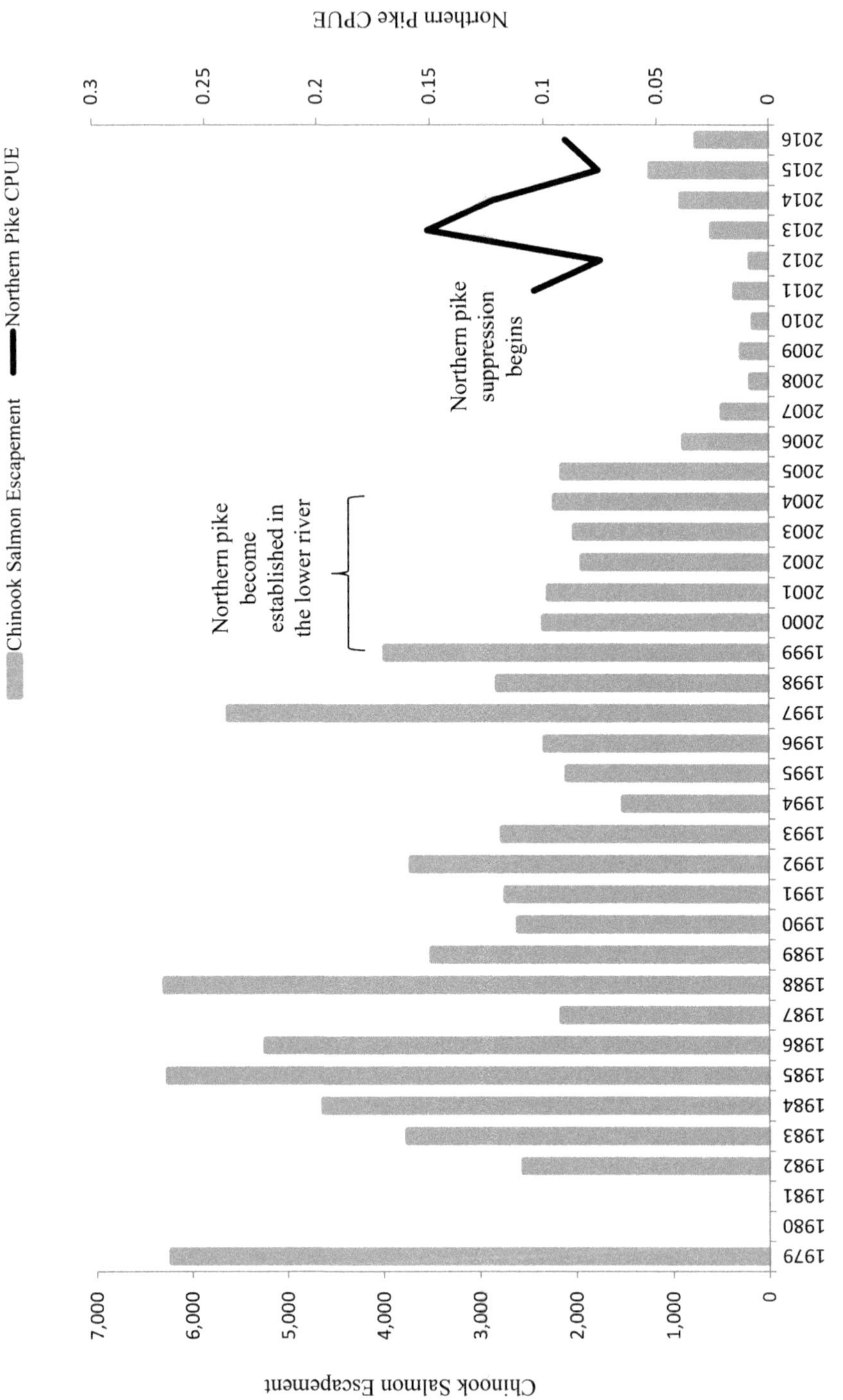

FIGURE 14.7 Aerial Escapements of Chinook salmon in Alexander Creek in southcentral Alaska.

In neighboring systems that either lacked invasive pike or had minimal pike habitat, Chinook salmon escapements were either stable or increasing during this same period (Oslund and Ivey 2010), suggesting that in the homogenous Alexander system, the thriving population of predatory pike coupled with the lack of refugia for salmonids in the system were the driving mechanisms behind the lost salmon production.

There have been several investigations of pike diets in southcentral Alaska (Glick and Willette 2016, Sepulveda et al. 2013, Haught and von Hippel 2011). Sepulveda et al. (2013) were among the first to quantify these patterns and found that juvenile salmon are the dominant prey for invasive pike, both where salmon are abundant and where they are rare (Fig. 14.8).

Their study compared pike diets in Alexander Creek and in the nearby Deshka River. The Deshka River contains pike, but has fewer slow-moving sloughs and backwaters, most of which are in the lower river, and these are far less plentiful than in Alexander Creek. The mainstem of the Deshka River is deeper, higher velocity, and is dominated by mid-channel gravel bars and riffles. Pike are thought to have been present in the Deshka River since approximately 1970, but they were not prevalent until the early 1990s (Whitmore and Sweet 1998). Despite the persistence of pike in the Deshka River, Chinook salmon abundance remains stable, presumably because of the river's habitat heterogeneity that provides salmonids with refuge from pike predation.

FIGURE 14.8 Approximately 100 juvenile Chinook and chum salmon in the stomach contents of a pike from Alexander Creek in Southcentral Alaska.

The Deshka River was selected for Sepulveda et al. (2013)'s study because it offered an opportunity to study pike diets where salmonids are abundant within the non-native range of pike. Conversely, Alexander Creek offered an opportunity to assess pike diets where salmonid abundance is critically low. They found that in the Deshka River, pike diets were dominated by Chinook and coho salmon < 100 mm in length, and that smaller pike (< 400 mm) consumed more of these fish than larger pike. In Alexander Creek, pike diets reflected the distribution of spawning salmonids, which decreased with distance upstream. In the lowest study reach of Alexander Creek, pike diets were dominated by salmonids despite their rarity. Pike diets in the middle and upper reaches of Alexander Creek were dominated by Arctic Lamprey *Lampetra camtschatica* Tilesius, 1811 and slimy sculpin, respectively, as salmonids became increasingly unavailable as prey. These data demonstrate the trophic adaptability of pike to shift their diets toward more available prey when preferred resources are scarce. The largest implication of this trophic adaptability is that populations of multiple species can decline unless management actions are undertaken to reverse the trend (Sepulveda et al. 2013). Even when management actions are implemented, bioenergetics modeling suggests that in homogenous systems of optimal pike habitat like Alexander Creek, nearly all pike need to be removed to restore salmon stocks to sustainable levels (Sepulveda et al. 2014).

14.3.1.2 Trout and Char

Pacific salmon species are not the only salmonids affected by invasive pike. Trout populations are also highly vulnerable, and significant impacts to native trout and char populations have been observed. One of the greatest conservation challenges resulting from invasive pike establishment in the western United States is their effect on native bull trout *Salvelinus confluentus* Suckley, 1859 and westslope cutthroat trout *Oncorhynchus clarki lewisi* Suckley, 1856 populations in Montana, Idaho, and Washington.

Westslope cutthroat trout and bull trout have both declined throughout their historic range. Habitat fragmentation and interactions with other invasive species contribute to population declines of these species, but pike predation is also a significant factor (Muhlfeld et al. 2008). Pike were illegally introduced in the upper Flathead River in Montana, where they have incidentally become a popular sport fish, but also contribute greatly to declines in westslope cutthroat trout and bull trout populations in this river and in downstream waters in Idaho and Washington. As with Pacific salmon in Alaska, spatial and temporal habitat overlap between juveniles of these species and invasive pike in the Flathead River system increases their predation risk. This dynamic is corroborated by bioenergetics modeling of pike diets in the Upper Flathead River. Bioenergetics models indicate that a population of 1,200-1,300 pike in this system has the potential to annually consume 800 kg (3,500 individuals) of bull trout and 700 kg (13,000 individuals) of westslope cutthroat trout. These quantities translate numerically to 5% of the overall prey base consumed by pike, even though these two species numerically comprise only 1.7% of the prey fishes in the available community (Muhlfeld et al. 2008). This again demonstrates how adept pike are at seeking out preferred salmonid prey, even when salmonids are in low abundance (Sepulveda et al. 2013).

In the Coeur d'Alene Lake system in northern Idaho, a similar pattern exists where invasive pike predation is thought to be a causative mechanism behind declining westslope cutthroat trout populations (Walrath et al. 2015). In this system, shallow, vegetated backwaters are common where tributaries enter Coeur d'Alene Lake. Juvenile westslope cutthroat trout migrate to Coeur d'Alene Lake in the spring. During this migration they must pass through the vegetated tributaries where pike are concentrated, making them highly vulnerable to predation during this period (Walrath et al. 2015). Predation risk in this scenario is well documented. Where pike are native in Europe, several studies have shown pike predation to be a primary cause of mortality for migrating salmonids (Kekäläinen et al. 2008, Jepsen et al. 1998, Larsson 1985). A bioenergetics model of the dietary patterns of pike in the Coeur d' Alene Lake system demonstrated that pike primarily feed on non-native kokanee *O. nerka* Walbaum, 1792, westslope cutthroat trout, and yellow perch *Perca flavescens* Mitchill, 1814 and that westslope cutthroat trout can comprise up to 30% of the biomass consumed by small pike (Walrath et al. 2015). Not surprisingly, the highest occurrence of pike predation on westslope cutthroat trout occurs during the spring when juveniles are migrating through the tributaries (Walrath et al. 2015).

To the west in northeastern Washington, invasive pike threaten to undermine current and future recovery efforts for bull trout, westslope cutthroat trout, other native fishes in the Pend Oreille River watershed (Bean 2014). Pike in the Pend Oreille River originated from illegal introductions in the Clark Fork River drainage, which includes the Flathead River in Montana, where they spread downstream through Idaho and into Washington. Pike are now firmly established in the Box Canyon Reservoir and its tributaries and are causing declines in native trout and introduced game species (J. Connor, Personal communication). Fish surveys reveal that the pike population in the Box Canyon reservoir grew exponentially from less than 400 individuals in 2006 to an estimated 10,000 individuals in 2014 (Bean 2014; J. Connor, Personal communication). One potential explanation for the growth in pike abundance was the creation of more favorable overwintering habitat by reductions in winter water releases from the reservoir (Olson and Connor 2009). Native trout survival is now reduced throughout the system due to the high degree of spatial habitat overlap with pike. If left unchecked, pike in Box Canyon Reservoir and the Pend Oreille River pose significant risks to the anadromous salmon and steelhead *O. mykiss* Walbaum 1792 populations of the Columbia River.

As with bull trout and westslope cutthroat trout, rainbow trout populations are also highly susceptible to reductions from invasive pike predation (Flinders and Bonar 2008). Similar to juvenile coho salmon, rainbow trout fry prefer slower velocity water (Raleigh et al. 1984), and in lacustrine areas, they inhabit lake margins (Ford et al. 1995) where they have increased exposure risk to pike predation. Stormy Lake in the Swanson River drainage on Alaska's northwestern Kenai Peninsula provides a notable example of the deleterious impacts illegally-introduced pike have on native rainbow trout and Arctic char (Massengill 2017a, Fig. 14.5). Prior to the illegal pike introduction, Stormy Lake was a consistent producer of large rainbow trout and Arctic char and resultantly supported popular fisheries for these species. A decade after the 2001 discovery of pike into this lake, few native fishes remained, and no catches of either species were reported in angler surveys. Further, over 2,000 hours of gillnetting effort conducted during fish surveys in 2009 and 2010 only captured three rainbow trout and two Arctic char in the lake while 188 pike were captured (Massengill 2017a).

14.3.1.3 Extirpation of Native Fishes

In extreme cases following introduced pike establishment, populations of native fish have not only been reduced, but have been extirpated. On the Iberian Peninsula in Europe, piscivorous fishes were absent from freshwater communities until recreational species like pike were intentionally introduced and subsequently spread (Elvira 1995a). It is postulated that the lack of native predatory fish leaves native species more prone to predation and replacement by non-native species because they lack anti-predator behavioral responses (Nicholson et al. 2015, Leunda 2010), while non-native fishes that have established in the region may have greater predator avoidance adaptations (Jacobsen and Perrow 1998, Eklov and Harmin 1989, Wahl and Stein 1988). Studies of pike diets in Spain describe their opportunistic predatory nature, but these studies found a preference for soft-rayed species like native cyprinids (Dominguez and Pena 2000 and Rincon et al. 1990). In the Ruidera Lakes region in central Spain where pike were stocked for recreation in the 1950s, many native fish species have disappeared except for areas of the drainage where pike are absent (Almodovar and Elvira 1994; Fig. 14.9).

Pike are believed to have extirpated the native fish populations there, consequently resulting in a shift of pike diets to non-native red swamp crayfish *Procambarus clarkii* Girard, 1852 and other non-native fishes in the lakes (Elvira et al. 1996). This is, perhaps, now a benefit of the pike introduction but does not negate their original effect of eliminating native fishes from the region, and thus creating the pathway for non-native fishes to replace the native species. A similar negative pattern of succession following introduced pike establishment was reported in wetlands within the Daimiel National Park. In this region, the pike population eliminated the native fish community, and then, despite the presence of red swamp crayfish as available prey, died out in the mid-1980s leaving only two non-native fish species, the common carp and eastern mosquitofish *Gambusia holbrooki* Girard, 1859, to comprise the fish community (Elvira and Barrachina 1996, Elvira 1995b). For reasons like these, pike are considered a significant threat to the fragile endemic ichtyofauna of the eastern Mediterranean basin, which is already highly vulnerable from other anthropomorphic stressors (i.e., other non-native fish introductions, pollution, agriculture, and reservoir construction; Ribeiro and Leunda 2012).

In the western United States, non-native pike have been implicated in the extirpation of populations of rare prairie stream fishes such as the pearl dace *Margariscus margarita* Cope, 1867, and there are great concerns for fishes listed as endangered under the US Endangered Species Act (Johnson et al. 2008, Hawkins et al. 2005, Mueller 2005). Extirpation of native fishes has been observed in numerous lakes throughout southcentral Alaska where robust populations of salmonids and threespine sticklebacks *Gasterosteous aculeatus* Linnaeus, 1758 have been completely eliminated (Patankar et al. 2006, Haught and von Hippel 2011, McKinley 2013). It is hypothesized that in these locations, salmonid prey is depleted first, which shifts predation pressure onto spiny-rayed sticklebacks until their populations eventually disappear (Haught and von Hippel 2011). In one lake in the Susitna River drainage, introduced pike caused the local extinction

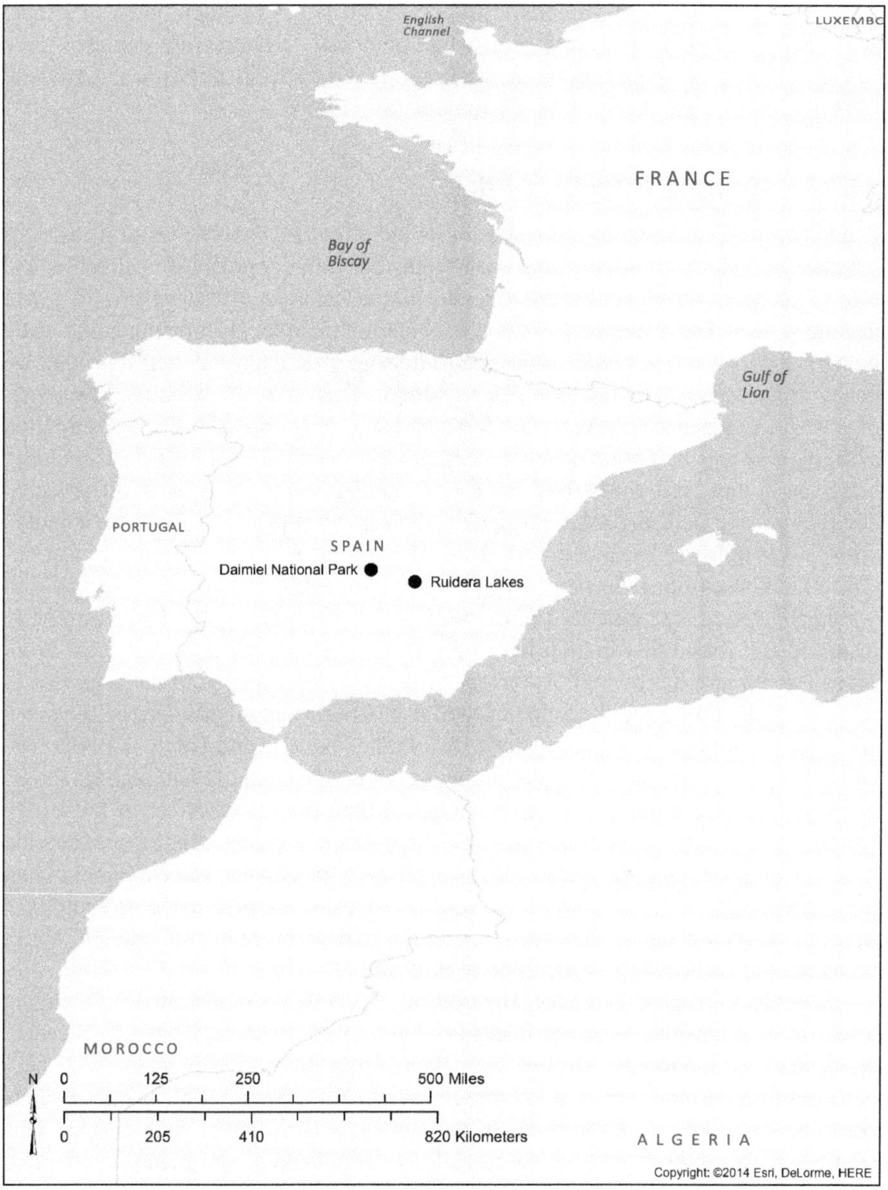

FIGURE 14.9 Iberian Peninsula locations discussed in this section.

of a genetically-unique form of threespine stickleback in a lake that formerly lacked piscivores. The lack of piscivores in the lake allowed for an adaptation for weaker armoring within the stickleback population. This rendered the stickleback population highly vulnerable to predation, and the population was extinct within six years of pike establishment (Patankar et al. 2006). Pike are opportunistic predators so amphibians and other aquatic organisms (i.e., invertebrates, waterfowl, and small mammals) are also vulnerable to population losses (Haught and von Hippel 2011, Solman 1945), and eventually, when other prey resource are depleted, pike can shift their diets to subsist on invertebrates and conspecifics (Glick and Willette 2016, Venturelli and Tonn 2006, Chapman et al. 1989).

14.3.2 Indirect Effects of Pike

In addition to pike predatory impacts, non-lethal effects, such as behavioral cascades (chapter 8), can occur where pike are introduced. Indirect or non-lethal ecological effects of pike introductions often relate to the behavior of organisms at lower trophic levels. For instance, a study in Alberta, Canada, documented that male flathead minnows *Pimephales promelas* Rafinesque, 1820 altered their reproductive behavior in response to predation risk from pike. This study demonstrated that costs of predator avoidance behaviors can include lost mating opportunities or usurpation of nests and egg predation when predation risk from pike inhibits nest guarding behavior (Jones and Paszkowsi 1997). A recent study in Alaska documented declines in age and size of reproducing threespine sticklebacks and their clutch sizes in response to a pike introduction, illustrating non-consumptive predation-risk effects resembling nutrient deprivation (Heins et al. 2016). An experiment in Wisconsin introduced pike to a piscivore-free lake to measure the fish community response and found that in addition to direct declines in prey fish abundance from predation, emigration of prey fishes out of the lake, in response to the presence of pike, accounted for a rapid decline of total fish density and greatly altered the mean sizes and community composition of fishes in the lake (He and Kitchell 1990). The authors of this study state that the indirect effects due to the behavioral responses of prey fish were rapid and more pronounced then the direct effect of predation. Others have shown that species can suffer high fitness costs associated with sharing habitat with pike. This was demonstrated by a study negatively associating duckling production with pike presence in northern Sweden, where pike acted as ecological traps for breeding ducks that used vegetation and prey abundance cues to select breeding and rearing habitat (Dessborn et al. 2011).

Specific to trophic cascades, Byström et al. (2007) demonstrated a cascading effect from a presumed natural invasion of pike into a subarctic lake in Sweden. Here, the fish community shifted from Arctic char as the top predator with high densities of ninespine stickleback *Pungitus pungitus* Linnaeus, 1758 as intermediate consumers to pike as the top predator and very low densities of sticklebacks. This change in the top predator species cascaded to zooplankton and macroinvertebrate primary consumers, which increased in abundance. In this scenario, Byström et al. (2007) suggest that intraguild predation was the causal mechanism behind the replacement of char in the lake and resulting food web alteration (see also chapter 8). Other studies have corroborated these findings and also demonstrate strong effects of pike on prey fish communities with cascading effects down to both primary consumers and producers (i.e., Persson et al. 1996, Findlay et al. 2005). Byström et al. (2007) postulate that future invasions of pike into cold alpine lakes may be facilitated by climate change and could have great potential to completely alter the trophic structures of their invaded communities.

14.3.3 Effects of Climate Change on Native Species Interactions with Pike

Changing climate patterns have great potential to exacerbate ecological impacts of invasive and non-native species (Havel et al. 2015), including pike. A study by Öhlund et al. (2015) investigated the role of temperature on the attack rates of pike on brown trout (*Salmo trutta* Linnaeus, 1758). In Sweden, the two species can co-exist in large cold lakes that provide pelagic refugia for brown trout, but not in small warm lakes where habitat overlap with pike leads to their extinction (Hein et al. 2014, Spens and Ball 2008). In their experiments, Öhlund et al. (2015) found that the average attack swimming speed of pike was an order of magnitude lower than the escape swimming speed of brown trout at 5°C but the two swimming speeds were approximately equal at temperatures above 11°C, suggesting that as temperatures warm, pike will become more efficient predators on brown trout. This study demonstrates that thresholds in temperature dependence on ecological rates can create tipping points in the responses of ecosystems to increasing temperatures and have broad implications for the effects of warming on predator-prey dynamics and ecosystem stability.

In another study, Hein et al. (2014) modeled brown trout and pike coexistence patterns across lakes in Sweden using lakes where the two species coexist and others where anthropomorphic introductions of pike have led to extirpation of brown trout. The authors found that patterns of coexistence were best described by air temperature and lake area. All lakes where pike were introduced and caused subsequent trout extirpations were warm and small. Their model suggested that with predicted rates of warming, many small, cold lakes that currently contain both species will not facilitate coexistence in the future. To predict how expected climate patterns influence this, Hein et al. (2014) applied recognized climate scenarios to the coexistence interactions. Factoring in warming air temperatures and dispersal limitations (i.e., barriers, high velocities and gradients), pike are expected to expand their range in Sweden and invade 9,100 more lakes by the end of the century (Hein et al. 2011). Model results from Hein et al. (2014) suggest that by 2100, only 15 of the 9,100 lakes that pike are expected to invade might allow for the continued existence of brown trout. When considered across the broader geographic range for which pike are not indigenous, the implications of these predictions are daunting and highlight the importance of minimizing human-mediated introductions to avoid further ecological consequences, especially in the face of warming climate trends that could make current negative interactions exponentially worse.

14.4 SOCIOECONOMIC IMPACTS FROM INVASIVE PIKE INTRODUCTIONS

In addition to the ecological ramifications of invasive pike introductions, the economic and recreational impacts can be substantial. One of the trickiest dynamics for fisheries managers is the balance between providing fishing opportunities for pike while at the same time culling their populations to protect other fisheries.

Indeed, it is the desire for quality pike fisheries that motivated most stockings of the species, either legally or illegally, in the first place. Over the years, many government agencies that conduct authorized fish stockings have become more restricted in their practices and thorough permitting and review processes are increasingly required prior to these activities. However, the popularity of non-native species like pike and walleye *Stizostedion vitreum* Mitcill, 1818 as local sport fish has greatly increased in the last two decades in the western United States, and websites and media that tout the opportunities to catch trophy-size fish have led to an increase in the incidences of illegal stockings by anglers for these species (McMahon and Bennet 1996). This has created a difficult and sometimes political dynamic where fishery managers must weigh the potential recreational and economic benefits derived from pike introductions against potential long-term ecosystem effects (Johnson et al. 2009). This is fraught with complex biological and social considerations and often leads to tension between conservationists, fishery managers and resource users (McMahon and Bennet 1996). This problem is not unique to any one geographic area. Illegal introductions of sport fish are a growing problem worldwide (Hickley and Chare 2004), and government agencies everywhere are increasingly engaged in public education campaigns to raise awareness of the consequences of illegal stocking practices and develop stronger laws aimed at discouraging such activity (Johnson et al. 2009, Hickley and Chare 2004, McMahon and Bennet 1996).

In Alaska, all non-native pike populations originated from illegal introductions, and there is an obvious trend of pike waters being road or float-plane accessible (ADFG 2007). Unfortunately, where salmonid populations resultantly decline following these introductions, so do the fisheries that depend upon them. In the United States, anglers annually spend approximately \$ 50 billion on expenditures related to fishing trips and equipment, and the US recreational fishing industry supports over 326,000 jobs (National Marine Fisheries Service 2011). An economic analysis in 2007 found that in Alaska, alone, recreational fisheries contributed approximately \$ 1.4 billion to the state's economy with 73% of those expenditures (\$ 989 million dollars) occurring within the range where pike are invasive (Southwick Associates Inc. et al. 2008). In addition, over 11,500 jobs in Southcentral Alaska are supported through the recreational fishing industry which primarily targets the state's internationally-renown salmonid fisheries. However, harvest trends in Alaska typically show that after pike become established, salmonid harvests precipitously decline (Fig. 14.5), rendering profound economic effects. In theory, these economic losses can be mitigated to some degree by the economic boost that pike fisheries can provide. For example, a recent economic study estimated that pike angling in Ireland annually contributes \$ 102 million to the country's economy (National Strategy for Angling Development 2015). Within the native range for pike in Alaska, an economic analysis in 2001 listed ~ \$ 3.7 million as the estimated total annual net economic value of sport fishing for pike, burbot, and lake trout combined (Duffield et al. 2001). In Southcentral Alaska, where pike are not native, the estimated annual value of pike, burbot, and lake trout fisheries was estimated at ~ \$ 588 thousand (Duffield et al. 2001). In this region of Alaska, pike populations easily stunt when prey sources are depleted, and this may limit some of the angling interest for all but the most enthusiastic pike anglers. Overall in southcentral Alaska, angler harvest surveys tend to show declines in recreational fishing effort in waters following pike establishment, especially when salmonids are no longer present (Jennings et al. 2015).

Salmonid fisheries in Alaska are managed for harvest of surpluses generated by natural production and hatchery releases. Reductions in surpluses due to pike predation can diminish the number of fish available for commercial, recreational, and subsistence fisheries. Many lodges and guiding services that accommodate anglers participating in salmonid fisheries operate in areas where pike have been introduced. In addition, multimillion-dollar commercial fishing operations rely on harvesting salmon that now, in some cases, originate in waters with invasive pike populations (Glick and Willette 2016). Local businesses also depend on the sport and commercial aspects of salmon resources. Though not easily measurable, the economic losses caused by pike predation can be substantial.

In Alaska, the best example of economic losses directly attributable to pike is, again, within Alexander Creek. Prior to the early 2000s, before pike were well-established throughout the entire river, Alexander Creek supported a vibrant Chinook salmon fishery. At its peak, 13 fishing lodges were in operation. Nine of these operated throughout the ice-free months, and four operated only during the Chinook season. Another six local businesses operated charter fishing trips to the river. In addition, multiple air charters and boat rental facilities operated in the area to service Chinook anglers. This all amounted to a thriving multi-million dollar industry. By 2008, when the rapidly declining Chinook fishery was closed to harvest, this industry collapsed and today no longer exists.

In addition to lost fishing industries like the Alexander Creek example illustrates, costs for pike control can be excessive. In Alaska, pike control and eradication efforts have totaled over $ 3.7 million within the period between 2011 and 2016, alone. The largest pike eradication effort to date, which took place in Lake Davis, California, cost $ 33 million to implement (K. Thomas, Personal communication). All governments involved in pike management incur significant costs, and the financial burden is often perpetuated by the lengthy timeframes necessary to achieve project goals as well as the ever-present risk that management efforts could be foiled by future invasions.

14.5 MANAGEMENT STRATEGIES AND STRATEGY SUCCESS

Management of introduced pike populations is complicated, and there have been many different approaches to the problem. With unauthorized fish stockings in general, there are four recognized response options: educate the public on the negative consequences of transferring fish to new water bodies with the hope of deterring future introductions, pass legislation to make unauthorized stockings illegal and impose fines and penalties that serve as deterrents, institute control or eradication programs when feasible to mitigate losses from introductions, and if all else fails, accept the introductions and associated consequences as part of the biota (Rahel 2004). With introduced populations of pike, all of these management approaches have been applied to some degree. Increasing public outreach and legally prohibiting unauthorized stockings is, today, standard practice among most fisheries management authorities. Decisions to engage in remediation activities, however, are typically less standard, and the chosen responses vary considerably with the perceived threat from the introduced pike population, the feasibility of implementing control or eradication strategies, the anticipated benefits to native species, the availability of financial and personnel resources, and the degree of public or political support. The following discussion will describe some of the approaches that have been applied throughout western North America to reduce or eradicate populations of invasive pike (Table 14.1).

14.5.1 Prevention and Early Detection

With the financial burdens and uncertainty of success, attempts to remove invasive pike populations can be daunting and complex. Overall, the most successful and cost-efficient management strategy is to prevent pike introductions in the first place (Park 2004). Where prevention fails, early detection can yield a greater likelihood of future eradication success. An overarching principle in invasion biology is that a lag time exists between an invader's initial colonization and its exponential increase in abundance, at which point it becomes cost-prohibitive to control (Mehta et al. 2007). If an invasive species is detected early during this lag phase, eradication is typically more feasible.

For early detections of pike introductions, environmental DNA (eDNA) is one relatively new tool that is now available. The basic concept behind eDNA testing is that organisms shed cells, gametes, scales, excrement, etc. into the water column, and DNA from these sources collected in water samples can be amplified in the lab to test for the presence of a target species (Darling and Mahon 2011, Minamoto et al. 2012). eDNA can be a more sensitive tool than traditional fisheries techniques for detecting taxa in low abundance (Ficetola et al. 2008), making it a potentially useful tool for detecting new invasions of pike or evaluating the success of pike eradication efforts. Genetic markers designed for pike eDNA detection were developed in 2014 (Olson et al. 2015), and pike surveys utilizing this technology are now ongoing (Dunker et al. 2016).

14.5.2 Eradication

Complete eradication of established invasive pike populations may have the greatest benefit to native fish populations in the long-term because this eliminates predation pressure and removes source populations from which pike can spread. Drawbacks to eradication projects are that they are often cost-prohibitive, publically contentious, not always feasible pending the scale of infestation, and potentially prohibited by local environmental regulations (Lee 2001). Few tools are available to fisheries managers to completely eradicate unwanted fish, and nothing currently exists that is species-specific to pike. The only two proven methods for invasive fish eradication include using piscicides or dewatering (Finlayson et al. 2010). There is, however, a growing field in the development of genetics techniques such as triploidy

Table 14.1 Invasive pike management areas in western North America.

Location	*Drainage*	*Water*	*Introduction*	*~Date*	*Species Affected*[1]	*Management Actions*
New Mexico	Six Mile Creek	Eagle Nest Lake	Illegal	2011	Rainbow Trout (S) Kokanee (S)	Gillnets No Harvest Limits
Arizona	Walnut Creek	Rainbow Lake	Illegal	1990s	Rainbow Trout (S)	Gillnets, Stock Large Trout, Introduce Carp
Arizona	Walnut Creek	Lake O'Woods	Illegal	1990s	Rainbow Trout (S)	Gillnets
Arizona	Show Low Creek	Fool's Hollow Lake	Illegal	1990s	Rainbow Trout (S)	Stock Large Trout
Arizona	San Cruz River	Parker Canyon Lake	Illegal	1990s	Rainbow Trout (S)	Surveys
California	Sacramento-San Joaquin	Frenchman Reservoir	Illegal	1989	Rainbow Trout (S)	Rotenone
California	Sacramento-San Joaquin	Feather River	Illegal	1980s	Rainbow Trout (S)	Rotenone
California	Sacramento-San Joaquin	Lake Davis	Illegal	2007	Rainbow Trout (S)	Rotenone
Nevada	Tailings Creek	Basset Lake	Legal	1990's	Recreational Species[2] (S)	No Harvest Limits, Rotenone
Nevada	Steptoe Creek	Comins Lake	Legal/ Illegal	1970's/ 1990's	Rainbow Trout (S)	Rotenone No Harvest Limits, Discontinue Trout Stockings
Utah	Provo River	Utah Lake	Illegal	2010s	Recreational Species[2] (S), June Sucker (E, EN)	Surveys, No Harvest Limit, Mandatory Kill Regulation
Colorado	Gunnison River	Paonia Reservoir	Legal	1960s	Colorado pikeminnow (E), Humback chub (E), Razorback sucker (E), Bonytail (E)	Rotenone, No Harvest Limits
Colorado	Yampa River	Elkhead Reservoir	Legal	1977	Colorado pikeminnow (E), Humback chub (E), Razorback sucker (E), Bonytail (E)	No Harvest Limits, Barrier

Table 14.1 (*Contd...*)

Table 14.1 Invasive pike management areas in western North America. (*Contd...*)

Location	***Drainage***	***Water***	***Introduction***	***~ Date***	***Species Affected***[1]	***Management Actions***
Colorado	Yampa River	Yampa River	Spread	1979	Colorado pikeminnow (E), Humback chub (E), Razorback sucker (E), Bonytail (E)	Gillnets, Fyke nets, Electrofishing, Trammel nets, Seines, Surveys, No Harvest Limits
Colorado	Yampa River	Green River	Spread	1990s	Colorado pikeminnow (E), Humback chub (E), Razorback sucker (E), Bonytail (E)	Gillnets, Fyke nets, Electrofishing, Trammel nets, Seines, Surveys, No Harvest Limits
Wyoming	Yampa River	Little Snake River	Spread	2012	Colorado pikeminnow (E), Humback chub (E), Razorback sucker (E), Bonytail (E)	Surveys
Montana	Clark Fork River	Milltown Reservoir	Spread	1990s	Bull trout (SC) Westslope Cutthroat Trout (SC)	Dam Removal
Montana	Upper Missouri River	Big Muddy River	Legal	~1990s	Pearl Dace	Surveys
Montana	Upper Missouri River	Toston Reservoir	Illegal	2000s	Bull trout (SC) Westslope Cutthroat Trout (SC)	Habitat Alteration from Natural Flood Event
Idaho	Spokane River	Coer d'Alene Lakes	Illegal	1970s	Bull trout (SC) Westslope Cutthroat Trout (SC)	Surveys, No Harvest Limits
Idaho	Pend Oreille River	Lake Pend Oreille	Spread	2000s	Bull trout (SC) Westslope Cutthroat Trout (SC)	Surveys, No Harvest Limits
Washington	Pend Oreille River	Box Canyon Res-ervoir	Spread	~2004	Bull trout (SC) Westslope Cutthroat Trout (SC)	Gillnets, No Harvest Limits
British Columbia	Columbia River	Pend Oreille River	Spread	2009	Pacific Salmonids (N)[3] White Sturgeon (SC)	Gillnets, No Harvest Limits

Table 14.1 (*Contd...*)

Table 14.1 Invasive pike management areas in western North America. (*Contd...*)

Location	*Drainage*	*Water*	*Introduction*	*~ Date*	*Species Affected*[1]	*Management Actions*
Alaska	Alexander Creek	Alexander Lake	Illegal	1960s	Pacific Salmonids (N)[3]	Slot Limit Evaluation
Alaska	Alexander Creek	Alexander Creek	Illegal	1990s	Pacific Salmonids (N)[3]	Gillnets, No Harvest Limits
Alaska	Cottonwood Creek	Anderson Lake	Illegal	1980s	Pacific Salmonids (N)[3]	Fyke net, No Harvest Limits
Alaska	Susitna River	Deshka River	Illegal	1990s	Pacific Salmonids(N)[3]	Gillnets, No Harvest Limits
Alaska	Yentna River	Moose Creek	Illegal	1980s	Pacific Salmonids (N)[3]	Gillnets, No Harvest Limits
Alaska	Yentna River	Indian Creek	Illegal	1980s	Pacific Salmonids (N)[3]	Gillnets, No Harvest Limits
Alaska	Kenai River Drainage	Soldotna Creek	Spread	1970s	Pacific Salmonids N)[3]	Rotenone
Alaska	Kenai River	6 Soldotna Creek Lakes	Illegal/ Spread	1970s	Pacific Salmonids (N)[3]	Rotenone
Alaska	Swanson River	Stormy Lake	Illegal	2001	Coho Salmon (N) Rainbow trout (N) Arctic Char (N) Longnose Suckers (N)	Rotenone
Alaska	Kenai Peninsula	Arc Lake	Illegal	2000	Coho Salmon (S)	Rotenone
Alaska	Kenai Penin-sula	Scout Lake	Illegal	2002	Coho Salmon (S)Rainbow Trout (S)	Rotenone
Alaska	Kenai Peninsula	6 Tote Road Lakes	Illegal	1980s	Sticklebacks	Rotenone
Alaska	Anchorage	Cheney Lake	Illegal	2000	Rainbow Trout (S)	Rotenone/ Gillnets
Alaska	Anchorage	Sand Lake	Illegal	1994	Rainbow Trout (S)	Rotenone
Alaska	Anchorage	Campbell Lake	Illegal	2000	Pacific Salmonids (N)[3]	Gillnets, Winterkill
Alaska	Anchorage	Otter Lake	Illegal	2000	Rainbow Trout (S) Coho Salmon (N)	Rotenone

[1](S) – Stocked for recreational purposes, (N) - Native, (EN) - Endemic, (E) - Endangered Species, (SC) - Species of Special Concern
[2]Largemouth bass, Yellow perch, Channel catfish, Walleye, Bluegill, Crappie, Bullhead, White bass
[3]Pacific salmonids include: Chinook salmon, Coho salmon, Sockeye salmon, Pink salmon, Chum salmon, Rainbow trout, Arctic grayling, Dolly varden.

(releasing triploid males into the population to slow reproductive potential), Trojan Y chromosome (using female fish with two Y chromosomes to shift the sex-ratio of the population), recombinant approaches such as sterile-feral technologies (creation of an inheritable gene that renders both males and females sterile) and autocidal technologies (species-specific diseases, viruses, and "self-killing" genetic manipulations) to slowly cause the extirpation of invasive fishes (Schill et. al. 2016, Thresher et al. 2014). Much of this work is still in development, and it is unknown how publically acceptable implementation of some of these techniques might be, but this body of work has shown potential promise in adding to the tools available for eradication of invasive fishes (Schill et. al. 2016, Thresher et al. 2014). As yet, genetics techniques have not been applied for pike eradications, and the remainder of this discussion will focus on what has been implemented.

Chemical treatments using piscicides, particularly liquid and powdered formulations of rotenone, have been used for invasive pike eradication most commonly in California, Nevada and Alaska. Rotenone has also been used to manage pike fisheries in reservoirs in other states and to remove pike for fisheries management purposes in Scandinavia. Rotenone is a highly regulated, restricted-use pesticide for fisheries management. It is botanical in origin and is lethal to fish because it is rapidly absorbed into the blood stream of gilled organisms where it inhibits oxygen uptake during cellular respiration. In California, rotenone treatments in Frenchman Reservoir, Sierra Valley tributary streams, and Lake Davis ultimately resulted in the complete eradication of this species from the state. These efforts, however, were not without complications. Lake Davis is the drinking water source for the town of Portola, and the town's economy is closely tied to fisheries in the lake. There was tremendous public contention and litigation over the state's decision to use rotenone there. Despite the controversy, rotenone was applied to Lake Davis in 1994. Either through a failure of this treatment or a subsequent illegal introduction, pike were rediscovered in Lake Davis in 1997. The state of California initiated a second attempt, this time extensively involving the public in project planning, and Lake Davis was re-treated in 2007. This second attempt was successful, and California currently remains pike free.

Rotenone is routinely used to remove unwanted fishes from reservoirs (Table 14.1). For example, Paonia Reservoir was treated in Colorado specifically to remove a source population of pike to the Gunnison River. In Nevada, rotenone has been used to remove pike from Basset and Comins Lakes, which are both manmade reservoirs. Pike had been stocked into Comins and nearby Basset Lakes in the late 1960's for biocontrol of carp and Utah chub *Gila altaria* Girard, 1856 and to create a sport fishery. Pike were effective in their role as biocontrol agents, and the pike populations resultantly crashed in the 1980s when the prey base became insufficient to support them (Korell 2015). In 1989, following the collapse of the pike fishery, Comins lake was treated with rotenone. The following year, the reservoir was drained for irrigation purposes and no fish were discovered, confirming that the treatment was successful. In 1996, the reservoir was stocked with trout to reestablish a sport fishery. Unfortunately, northern pike were illegally reintroduced to the reservoir in the late 1990s and confirmed to be spawning by 2001. Nearby Basset Lake was suspected as the source population for this illegal introduction

(Korell 2015). In 2015, both reservoirs were treated with rotenone to remove their pike populations and reduce the risk of reintroduction. Extensive post-treatment surveys involving both gillnetting and electrofishing confirmed the absence of live fish in the reservoirs, and both treatments were considered successful (Korell 2015). However, the history of reintroductions in both the examples from California and Nevada underscore the precarious nature or non-native fish eradication programs, where continued illegal introductions and/ or unanticipated survival of the target species can undermine the management objectives and considerable resources these projects entail.

In Alaska, rotenone use has been employed for invasive pike eradication since 2008 (Table 14.1). Initial treatments were conducted in small closed urban lakes that had formerly been stocked with rainbow trout, coho salmon or both. In these treatments, pike were the only harvestable species remaining in the lakes. In addition to redeveloping popular urban fisheries, these initial treatments served as public education opportunities to demonstrate, on a small scale, a potential strategy for tackling Alaska's mounting issues with invasive pike. In 2012, the complexity of rotenone treatments increased substantially with the rotenone treatment of Stormy Lake on the Kenai Peninsula to prevent pike from spreading to the nearby Swanson River (Massengill 2017b). This was the first rotenone treatment in a pike lake where native fish were still present. A large effort was undertaken to rescue native fish from Stormy Lake prior to the treatment to preserve their genetics. These fish were held offsite in net pens until the rotenone in Stormy Lake degraded. Additionally, broodstock were collected from the few Arctic char survivors in the lake, and juveniles were raised in a hatchery to ensure enough would be available for reintroduction. The Stormy Lake rotenone treatment was successful (Massengill 2017b). Pike have not been rediscovered in the lake, and native fishes in the lake are quickly rebounding. Catch per Unit Effort during a post-treatment survey in 2014 was 1.9 salmonids per net hour, and monitoring in the fall of 2016 indicated an increase in salmonid CPUE to 5.7 fish per net hour (ADFG, Unpublished data). This is in stark contrast to the previously-mentioned total salmonid catch of five (CPUE = 0.0025) during the 2010-2011 pre-treatment surveys (Massengill 2017a).

Between 2014 and 2017, Alaska conducted its largest invasive pike eradication effort, as of the writing of this chapter, in the Soldotna Creek tributary of the Kenai River. This multi-year effort was intended to prevent pike from spreading to vulnerable Kenai River tributaries. This project included a significant native fish restoration component and required rotenone treatments of six lakes, 35 km of stream, and 58 hectares of connected wetlands. Rotenone treatments completed between 2014 and 2016 were evaluated with gillnet and eDNA surveys, and no pike were detected using either method (Dunker et al 2016). Completion of this project, followed by rotenone treatments of six small landlocked lakes outside the Kenai River drainage, should result in the complete eradication of pike from Alaska's Kenai Peninsula.

Throughout the American West, dewatering can be an option for removing populations of unwanted fish in reservoirs where water levels can be manipulated through dam releases. Depending on the surrounding bathymetry, water-level fluctuations through dam releases can either create or decrease optimal habitat conditions for pike. The complete dewatering and removal of a dam at Milltown Reservoir in Montana, however, is one notable example of a pike population being virtually eliminated through a dewatering effort. Milltown reservoir was routinely drawn down in an effort to strand young of the year pike so they could more easily be trapped (D. Schmetterling, Personal communication). The decision to remove the dam and reservoir altogether was greatly influenced by the threat pike imposed to native trout and the desire to substantially reduce pike density in the Middle Clark Fork River. In 2008 the dam was removed. With the resulting lack of optimal habitat, pike are now rarely reported from the vicinity whereas prior to the dam removal, pike densities were greater than 70 fish/km (D. Schmetterling, Personal communication).

14.5.3 Suppression

Where pike eradication is not possible due to habitat complexity, drainage size, cost, or other factors, population suppression can be the next-best option. Pike suppression can utilize many different strategies with mechanical removal being the most widely-used approach in North America. Pike suppression projects are currently employed in Alaska, Washington, British Columbia, Idaho, Montana, Colorado, New Mexico, and Arizona (Table 14.1).

In Alaska, the largest invasive pike suppression effort takes place in Alexander Creek, where side-channel sloughs are annually gillnetted throughout the pike spawning period. Each year, approximately 60 sloughs connected to the mainstem of the river are netted. Within each slough, nets are fished until their catch rates decrease to less than 85% of the highest CPUE of the season (Rutz and Dunker 2017). These annual efforts are attempting to reduce predation pressure on juvenile Chinook salmon and, thus, increase their survival probability. Since this project's inception in 2011, over 17,300 pike have been removed, but it remains to be seen if this strategy will prove successful in the long-run. Gillnets are not as effective in selecting small young-of-the-year pike (Pierce and Tomcko 2003), and the removal of the largest individuals has the potential to omit self-regulating, density-dependent mechanisms in pike populations like cannibalism (Haugen et al. 2007). Other potential drawbacks are that size selectivity from the gillnets and exploitation could increase the rate of stunting in the pike population, causing earlier maturation and increasing the density of smaller age classes of pike (Ylikarjula 1999, Goedde and Coble 1981, Kipling and Frost 1970) that most often target juvenile salmonids (Glick and Willette 2016, Sepulveda 2014). Further, it is unclear if reducing the abundance of pike will equate to greater salmonid survival. Decreasing the amount of intraspecific competition among pike for prey could have the potential to render the surviving pike more efficient predators (Eklöv 1992, Nilsson and Brönmark 1999). However, despite these potential and recognized limitations, adaptive management is warranted to try and lessen the impact to Alexander Creek salmonids. Though the long-term success of the project is unknown, favorable patterns are emerging. Aerial indices of Chinook salmon are increasing during a period where salmonid productivity overall has been in a state of decline (ADFG Chinook Salmon Research Team 2013; Fig. 14.7). Also noteworthy is that both juvenile salmonids and adult spawners have reestablished throughout Alexander Creek, whereas juvenile and spawning salmonids were only observed in the lower river prior to suppression activities (Rutz and Dunker 2017). Finally, radio telemetry studies in the drainage demonstrate that few radio-tagged adult pike > 440 mm leave Alexander Lake and those that do are killed downstream in gillnets. Taken together, these are positive signs that pike suppression activities are beneficial to salmonid recovery in the system. However, fishery managers remain cautiously optimistic. Bioenergetics modeling predicts that pike must be virtually eradicated for salmonids to rebound (Sepulveda et al. 2014). In Alexander Creek, as with most suppression projects for pike, it is important for management to remain adaptive.

Other large-scale pike suppression initiatives are currently underway in the Upper Columbia River drainage in Washington and British Columbia and the Yampa River drainage in Colorado, the latter associated with endangered species recovery efforts. In the Box Canyon Reservoir of the Pend Oreille River in eastern Washington, over 16,200 pike were removed with gillnets between 2012 and 2014 (Bean 2014). As with Alexander Creek, Box Canyon Reservoir pike suppression primarily takes place during the pike spawning period. Gillnet mesh size specifically targets pike greater than 35 cm to reduce bycatch of smaller bodied fishes and other aquatic organisms. Based on pike abundance levels observed in 2010 and 2011, a target reduction of 87% was identified as the necessary level to crash the pike population in the reservoir (Bean 2014). Achievement of this target reduction is evaluated annually by conducting surveys to measure inter-annual reductions in the pike population. In 2014, a pike suppression project began in the Pend Oreille and Columbia Rivers in British Columbia to join the effort of trying to protect native fish in the region and prevent pike from spreading further downstream in the Columbia River (Heise et al. 2016). Pike suppression in British Columbia employs both targeted spring gillnetting and angler incentives to harvest pike. As with Alexander Creek in Alaska, fisheries managers in this region also recognize the potential for compensatory responses from surviving pike to inhibit the effectiveness of removal efforts and are taking an adaptive approach with future implementation of pike removal efforts.

As part of the Upper Colorado River Endangered Fish Recovery Program, mechanical suppression of pike, smallmouth bass *Micropterus dolomieu* Lacépède, 1802, and channel catfish *Ictalurus punctatus* Rafinesque, 1818 populations is conducted annually in the Yampa River, with the objective of reducing predation risk for endangered fishes (Hawkins et al. 2005). Boat-mounted electrofishing gear, fyke nets, and gill nets are used, and removal efforts are conducted in coordination with mark-recapture studies (Zelasko et al. 2016). Among the many challenges with predatory fish reductions in this region is the strong opposition by local anglers who advocate for continued sport fisheries for the targeted species. In an attempt to meet these social needs and reduce predation risk, non-native predatory fishes were translocated from endangered species critical habitat to ponds where they could be harvested by anglers (Webber 2009), but this practice of translocating pike was officially ended in 2014 (Battige 2014). While the effort to provide harvest opportunities offered a unique compromise between fisheries managers and anglers, it ultimately had too many risks. There was frequent recapture evidence that translocated pike were escaping back into the river from these ponds and that these ponds were serving as source populations for anglers to illegally translocate pike elsewhere (Webber 2009). The decision was eventually made to euthanize all pike removed from the Yampa River during control efforts.

One of the critiques of mechanical suppression programs for predatory fishes in the Colorado River basin is that efforts are based on available resources as opposed to specific removal levels (Mueller 2005). In Yellowstone Lake, the US National Park Service used gillnets to remove nearly 450,000 invasive lake trout from 1995 through 2009. A review of this program in 2010 found no evidence of overharvest despite more than a decade of fish removal at an annual cost of $ 2 million (Syslo et al. 2011). The difficulties of achieving overharvest and the high cost of suppression work indicate that monitoring and evaluation are critical for cost-effective suppression (Zelasko et al. 2016). However, following the review of the Yellowstone Lake Trout Removal program and implementation of recommended targets, the program has achieved greater success (Koel et al. 2015). As with the former pike suppression projects discussed, the Colorado River Endangered Fish Recovery Program is adaptive and must remain so to address the multiple causative mechanisms involved in native fish declines in the drainage (i.e. dynamic hydrography, multiple non-native species, etc.). Overall, individual suppression programs for invasive predators can have varying levels of success, but if carefully executed and well evaluated, they have great potential for benefiting native species (Zelasko et al. 2016).

14.5.4 Fishing Regulations and Harvest Strategies

There are substantial differences in invasive pike management approaches, ranging from managing recreational fisheries for them to spending millions on their eradication. One commonality, however, is that it is illegal to stock pike into waters without permits, and legal ramifications can be severe including potential misdemeanor convictions, large fines and potential liability for remediation. Harvest regulations are less consistent. Many places still have size or possession limits for non-native pike, whereas harvest limits are unrestricted in others. In some southcentral Alaskan waters with salmon conservation concerns, it is illegal to return pike to waters alive, and methods of pike harvest are greatly liberalized compared with legal methods for other species. Where pike fishing regulations have been liberalized, the general theory is that angler harvest will aid in reducing abundance. To that end, fishing organizations or municipalities have held fishing derbies for pike. For example, the Kalispel Native American tribe hosts a large annual derby as a component of the overall pike suppression effort in Box Canyon Reservoir. Pike fishing was formerly illegal in the Columbia River in British Columbia, but with the implementation of pike suppression programs there, pike harvest is now legal with no catch limits. Rewards are even offered for anglers who turn in pike carcasses with tags from mark-recapture studies (Heise et al. 2016). However, the effectiveness of using angler harvest to reduce the overall population size of pike is often limited because the removal of the largest pike can lead to stunting of the population, the smaller fish contribute greater predation pressure on juvenile salmonids (Sepulveda et al. 2014), and anglers tend to lose interest when large fish are not available (Pierce et al. 1995).

In Arizona, fishery managers have tried reducing pike impacts to stocked rainbow trout fisheries by stocking rainbows in excess of 30 cm. Producing these fish is more expensive and there are fewer of them to stock, so possession limits for larger rainbow trout are often restricted. Further, Arizona is experimenting with grass carp *Ctenopharyngodon idella* Valenciennes, 1844 introductions to reduce macrophyte cover for pike through herbivory (A. Lopez, Personal communication). Currently there are no limits for grass carp in Arizona, but this could change if this practice becomes more widely employed. Overall, legal considerations with introduced pike populations are variable and reflect the complex nature of managing a species that is both highly popular with anglers and an invasive species capable of significant ecological and economic damage.

14.6 RESEARCH RECOMMENDATIONS

The complexity of the issues associated with invasive pike certainly warrant further investigation, and there are many topics for which future research could be helpful. Where pike are considered invasive in North America, many of the investigations are based on correlative fisheries data. While there is nothing wrong with this *per se*, continued focus on peer-reviewed research into the cause and effect relationships of pike introductions are important for clearly understanding the implications. Bioenergetics modeling, in particular, is extremely useful in predicting impacts and prioritizing management responses (Johnson et al. 2008). First and foremost, however, there is a need to discover novel approaches for eradication and suppression of invasive fish (Britton et al. 2011). As demonstrated earlier, the use of piscicides can be an effective means of eradicating invasive pike populations. However, aside from draining a waterbody, it is currently the only means of eradicating them. Though effective, it is costly and, at times, extremely contentious (Lee 2001). In the United States, piscicide use requires extensive permitting that often takes years to acquire, thus limiting the ability for a rapid response to a new invasion. As discussed earlier, mechanical removal using gillnets and traps is widely used, yet the effectiveness is often limited by the resources that management agencies can allocate to their use (Mueller 2005, Syslo et al. 2011). Identifying new and efficient approaches to pike population suppression is an area of research that would be immensely useful to current management of invasive populations. Another area of research that is needed is the continued investigation of effective techniques to prevent pike from spreading. For example, the use of vertical drop barriers, electric barriers, and the like are intriguing, but there are limited data to support or refute their effectives (Layhee et al. 2016). In locations where eradication is not possible, preventing pike from spreading in conjunction with suppressing their populations may be the best management approach, so identifying the most effective means to prevent their spread in varying habitat conditions is needed.

Another area of research that would benefit ongoing invasive pike management includes developing accurate assessments of existing efforts and identifying biologically meaningful targets (Syslo et al. 2011, Bean 2014). For example, with suppression projects, assessments that identify the spatial extent of suppression needed to achieve recovery of native fishes are vital to project planning and effectiveness. Also, continued research into the habitat characteristics associated with the coexistence of introduced pike and native fishes are beneficial, particularly for setting management targets on suppression activities or predicting the impacts that development or habitat restoration projects might have on fish communities affected by pike. Conducting surveys of native fishes in waters proximate to invasive populations would also be useful to quantify losses in production of native species should pike establishment occur. These survey data would also facilitate the development of biologically meaningful targets for recovery if pike suppression efforts would be necessary. Resource agencies typically lack funding to conduct surveys of native biota without a justified reason for doing so. Unfortunately, this lack of baseline data makes it very difficult to set native species recovery targets if future invasive pike management is needed. Proactively quantifying native fish assemblages prior to invasion ultimately improves the success of management activities down the line.

Finally, fishery managers could be aided by the development of predictive models that consider broader geographical ranges of pike expansions to assess potential interactions between pike and other invasive species and how changes in climate might affect pike predation risk (Byström et al. 2007, Hein et al. 2014, Rahel et al. 2008). If fishery managers could better predict which areas are most vulnerable to invasion and from where, proactive efforts to prevent pike from spreading could be more efficiently implemented, ultimately resulting in substantial cost savings to resource agencies in avoiding the hefty costs of suppression or eradication. Also, the potential for interactions between pike and other invasive species is a daunting scenario. For example, pike are only one of several aquatic invasive species that threaten salmonid populations. Other invasive species such as lake trout, smallmouth bass and whirling disease parasites *Myxobolus cerebralis* Hofer, 1903 in the American West are impacting the quality of salmonid habitats. Invasive aquatic plant species such as waterweed *Elodea canadensis* Planchon, 1848 and Eurasian water milfoil *Myriophyllum spicatum* Linnaeus, 1753 have dense growth patterns that can degrade salmon spawning and rearing habitat conditions. Plants like these also have the potential to facilitate pike establishment by providing additional rearing and spawning habitat, thereby increasing pike consumption of salmonids. This potentially leads to a process termed 'invasional meltdown' (Simberloff and Von Holle 1999, Ricciardi 2001). As habitat conditions change, whether through new invaders to the landscape, changing climate patterns, development, or combinations of these, interactions between invasive pike and native fishes are likely to change as well. Predicting how species associations might differ under different climate scenarios will help fishery managers prioritize management activities and better prepare for the future when management of invasive pike might only become more complicated.

14.7 PIKE MANAGEMENT CONSIDERATIONS

The discussions in this chapter have focused on invasive pike predation as a primary causal mechanism behind native fish declines in areas where spatial habitat overlap is prevalent and have addressed some of the management efforts implemented to reverse these effects. The issues surrounding invasive pike management, however, are broad, and the story is often far more complex than discussed here. Environmental drivers such as habitat degradation (i.e., Collares-Pereira et al. 2000), interactions with other invasive species (Johnson et al. 2008), changing climate patterns (Havel 2015, Pauchard et al. 2016), and countless others all play a role. Habitat alterations that are associated with the development of reservoirs around the globe are particularly susceptible to invasive species introductions (Havel 2015), and many of the problems associated with pike populations described in this chapter occur in these modified habitats. Political and economic factors such as the quantity of resources made available for management or restoration activities, or changes in public policy that either shift priorities toward or away from progressive invasive species management, further complicate the issues.

Fish communities across the globe have been homogenized (Britton et al. 2011). For centuries fish have been transplanted from all over to provide recreation, support commercial enterprises, or control other invasive species. Unintentional introductions have also been prevalent, from accidental hitch hikes through bait buckets and ballast water to the well-meaning child setting his gold fish free. No matter the cause, the end result is that there are few places on Earth where current freshwater fish communities entirely mirror those provided by nature. This sets up a management framework where prioritizing the needs of one species over another is commonplace and complicates fisheries management strategies, especially when predatory and potentially invasive species like pike are involved.

Knowledge and hindsight exists today that perhaps did not sixty years ago. Introducing predators, like pike, has consequences, and today those consequences are better recognized and understood when deciding the courses of management actions. Most governing bodies no longer intentionally introduce predatory fishes without stringent controls (i.e., sterilization), but the effects of past introductions persist. Moving forward, the focus must now be on curbing illegal introductions of pike or any other species by well-intentioned but misinformed anglers or other members of the public who wish to try their hands at fish stocking. This requires consistent engagement with the public through directed outreach and the development of key messages regarding laws and regulations for pike introductions and the basic biological reasons behind these laws. It also remains helpful to continue public outreach on the more general principles of invasive species and the practices people can employ to avoid being vectors. Where pike are not native, public awareness of the consequences of these fish in their backyards, in the context of the broader invasive species concept, will hopefully lessen the incidences of uninformed bucket biology and more readily engage the public in management decisions where public support is crucial to the success of management actions.

To that end, management approaches for invasive pike across jurisdictional boundaries are highly variable, and a higher degree of interjurisdictional coordination is warranted. In the western United States, there are vast differences in how states, or even regions within states that have non-native pike populations, manage the species. Where pike are not native, there are also differences in how they are categorized. In the western United States and southeastern British Columbia, non-native pike are frequently considered "invasive", yet in other countries (i.e. Spain) that have documented impacts from non-native populations, pike tend to be referred to only as "introduced". One approach that could potentially help foster greater consistency could be the development of an inter-agency task force on invasive pike, much like the task force that currently exists for invasive Asian Carp Management in the United States (http://www.asiancarp.org/). Greater communication, networking, and interaction among fisheries professionals tasked with managing this species would be of tremendous value in identifying efficiencies with management strategies that have been found effective in one place and could potentially benefit another. Such coordination could also be of great value in areas where pike are problematic in drainages that cross jurisdictional boundaries. To a large extent coordination in those scenarios already occurs, but in cases where pike threaten to spread, proactive coordination before this happens can only aid prevention success down the line.

Preventing further spread of invasive pike populations is of the highest importance. This, above all else, is likely to yield the greatest benefit. The costs of invasive pike management can be staggering. Despite the difficulties of putting a price on something that doesn't exist, the cost-savings associated with not having to implement pike control or eradication activities are potentially immense. However, in areas currently grappling with invasive pike to which prevention is no longer a feasible option beyond containment, there is a need to identify clear priorities. Financial resources for invasive species management can often be scarce or require proposed projects to compete for funding. Setting clear priorities at the local and interjurisdictional levels based on criteria that ensure the best possible outcomes for preferred fish populations and the people who depend upon them ultimately increases the likelihood that prioritized management efforts are successful (Pauchard et al. 2016). Implementing strategies that have the greatest likelihood of success in the locations with the most potential benefit is key, and careful prioritization of efforts will help ensure this. There is also a need to continue learning which strategies are the most effective. Continued research is needed on novel approaches for pike suppression, eradication, and detection, especially approaches that can reduce the collateral damage to co-occurring species, thus making them more ecologically sound and more easily accepted by the public (Britton et al. 2011). Further study of the dynamics between invasive pike and co-occurring species at the local level is continually needed to inform better management and the prioritization of these efforts. Also, anticipating future scenarios that may come about through climate change is prudent for identifying long-term management strategies with this species. Finally, well-planned and prioritized management projects need to be implemented to protect the species that are at risk. Taken together these principles of *Prevention*, *Prioritization*, and *Protection* offer the greatest opportunity to remediate the damages from pike where they have become an invasive species.

Acknowledgments

The authors would like to recognize and sincerely thank the following individuals for their contributions to the discussion is this chapter: Kathrin Sundet, Gayle Neufeld, Jason Graham, Ryan Ragan, Jason Dye, and Tammy Davis (Alaska Dept. Fish and Game), Peter Westley (University of Alaska Fairbanks), Kevin Thomas

(California Dept. of Fish and Game), Anna Senecal and Bobby Compton (Wyoming Game and Fish Dept.), Koreen Zelasko (Colorado State University), Jason Connor (Kalispel Tribe), Bruce Bolding (Washington Dept. of Fish and Wildlife), Rick Boatner (Oregon Dept. of Fish and Wildlife), David Schmetterling (Montana Fish, Wildlife and Parks), Antonio Lopez (Arizona Game and Fish Dept.), Michael Slater (Utah Div. Wildlife Resources), Heath Korell (Nevada Dept. of Wildlife) Eric Frey (New Mexico Dept. Game and Fish), and Pam Fuller and Amy Benson (U.S. Geological Survey Nonindigenous Aquatic Species Program). The authors also wish to thank James Hasbrouck with the Alaska Dept. of Fish and Game, Michael Carey with the United States Geological Survey, and two anonymous reviewers for their improvements to this chapter. *Any use of trade, product, or firm names is for descriptive purposes only and does not imply endorsement by the US Government.*

REFERENCES CITED

ADFG. 2007. Management Plan for Invasive Northern Pike in Alaska. Div. of Sport Fish. http://www.adfg.alaska.gov/static/species/nonnative/invasive/pike/pdfs/invasive_pike_management_plan.pdf.

ADFG Chinook Salmon Research Team. 2013. Chinook salmon stock assessment and research plan, 2013. AK Dept. of Fish and Game SP No. 13-01, Anchorage, USA.

Albins, M.A. and M.A. Hixon. 2008. Invasive Indo-Pacific lionfish *Pterois volitans* reduce recruitment of Atlantic coral-reef fishes. *Mar. Ecol. Prog. Ser.* 367:233-238.

Almodovar, A. and B. Elvira. 1994. Further data on the fish faunan catalogue of the Natural Park of Ruidera Lakes (Guadiana River basin, central Spain). *Verh. Int. Ver. Limnol.* 25:2173-2177.

Battige, K. 2014. Middle Yampa River northern pike removal and evaluation: smallmouth bass removal and evaluation. 2014 APR to the Bureau of Reclamation. Project No. 13AP40029-98a.

Bean, N. 2014. Box Canyon Reservoir Northern Pike Suppression Project. APR. Kalispel Tribe. Prepared for Avista Corporation. Project # R-39234.

Bradford, M.J., P.C. Tovey , L.M. and Herborg. 2009. Biological risk assessment for northern pike (*Esox lucius*), pumpkinseed (*Lepomis gibbosus*), and walleye (*Sander vitreus*) in British Columbia. Dept. of Fisheries and Oceans, Ottawa, ON (Canada), Canadian Science Advisory Secretariat, Ottawa, ON (Canada. DFO, Ottawa, ON (Canada), 2009.

Brautigam, F. and J. Lucas. 2008. Northern Pike Assessment. Maine Dept. of Inland Fisheries and Wildlife, Div. of Fisheries and Hatcheries, Region A. http://www.maine.gov/ifw/fishing/species/management_plans/northernpike.pdf

Britton, J.R., R.E. Gozlan and G.H. Copp. 2011. Managing non-native fish in the environment. *Fish Fish.* 12:256-274.

Buckwalter, J.D., J.M. Kirsch and D.J. Reed. 2010. Fish inventory and anadromous cataloging in the lower Yukon River drainage, 2008. AK Dept. of Fish and Game, FDS No. 10-76, Anchorage, USA.

Bystrom, P., J. Karlsson, P. Nilsson, T. Van Kooten, J. Ask. and F. Olofsson. 2007. Substitution of top predators: effects of pike invasion in a subarctic lake. *Freshw. Biol.* 52:1271-1280.

Cederholm, C.J., M.D. Kunze, T. Murota and A. Sibatani. 1999. Pacific salmon carcasses: essential contributions of nutrients and energy for aquatic and terrestrial ecosystems. *Fisheries* 24:6-15.

Chapman, C.A. and W.C. Mackay. 1984. Direct observation of habitat utilization by northern pike. *Copeia*. 1984:255-258.

Chapman, L.J., W.C. Mackay and C.W. Wilkinson. 1989. Feeding flexibility in northern pike (*Esox lucius*): fish versus invertebrate prey. *Can. J. Fish. Aquat. Sci.* 46:666-669.

Chihuly, M. 1976. Biology of northern pike (*Esox lucius*) in the Wood River Lake system Bristol Bay, Alaska. MS. Fairbanks, AK: University of Alaska.

Collares-Pereira, M.J., I.G. Cowx, F. Riberiro, J.A. Rodrigues and L. Rogardo. 2000. Threats imposed by water resource development schemes on the conservation of endangered fish species in the Guardiana River Basin in Portugal. *Fisheries Manag. Ecol.* 7:167-178.

Cook, M.F. and E.P. Bergersen. 1988. Movements, habitat selection, and activity periods of northern pike in Eleven Mile Reservoir, Colorado. *Trans. Am. Fish. Soc.* 117:495-502.

Craig, J.F. (ed.). 1996. *Pike: Biology and Exploitation*. London: Chapman & Hall.

Darling, J.A. and A.R. Mahon. 2011. From molecules to management: adopting DNA-based methods for monitoring biological invasions in aquatic environments. *Environ. Res.* 111:978-988.

Dessborn, L., J. Elmberg and G. Englund. 2011. Pike predation affects breeding success and habitat selection of ducks. *Freshwater Biol.* 56:579-589.

Dominguez, J. and J.C. Pena. 2000. Spatio-temporal variation in the diet of the northern pike (*Esox lucius*) in a colonized area (Esla Basin, NW Spain). *Limnetica* 19:1-20.

Duffield, J.W., C.J. Neher and M.F. Merritt. 2001. Alaska angler survey: Use and valuation estimates for 1998 with a focus on burbot, pike, and lake trout fisheries in Region III. AK Dept. of Fish and Game, SP No. 01-3, Anchorage.

Dunker, K.J, A.J. Sepulveda, R.L. Massengill, J.B.Olsen, O.L. Russ, J.K. Wenburg and A. Antonivich. 2016. Potential of environmental DNA to evaluate northern pike (*Esox lucius*) eradication efforts: an experimental test and case study. *PLoS ONE* 11(9):e0162277. doi:10.1371/journal.pone.0162277.

Ehrlich, P.R. 1989. Attributes of invaders and the invading processes: vertebrates. pp 315-328. *In*: J.A. Drake, H.A. Mooney, F. di Castri, R.H. Groves, F.J. Kruger, M. Rejmánek and M. Williamson (eds.). *Biological invasions: A global perspective.* New York: John Wiley & Sons.

Eklöv, P. and S.F. Hamrin. 1989. Predatory efficiency and prey selection: interactions between pike *Esox lucius*, perch *Perca fluviatilis* and rudd *Scardinus erythrophthalmus*. *Oikos* 56:149-156.

Eklöv, P. 1992. Group foraging versus solitary foraging efficiency in piscivorous predators: the perch, Perca fluviatilis, and pike, *Esox lucius*, patterns. *Anim. Behav.* 44:313-326.

Elvira, B. 1995a. Freshwater fishes introduced in Spain and relationships with autochthonous species. pp 262-265. *In*: D.P. Philipp, J.M. Epifania, J.E. Marsden and J.E. Claussen (eds.). *Protection of aquatic biodiversity, Proc. of the world fisheries congress, Theme 3.* New Delhi: Oxford & IBH Publishing.

Elvira, B. 1995b. Native and exotic freshwater fishes in Spanish river basins. *Freshwater Biol.* 33:103-108.

Elvira, B. and P. Barrachina. 1996. Peces. *In*: Cirujano, S. (ed.). *Una Limnol. de las Tablas de Daimiel*. Madrid: ICONA, Coleccion Technica.

Elvira, B., G.G. Nicola and A. Almodovar. 1996. Pike and reds swamp crayfish, a new case on predator-prey relationship between aliens in central Spain. *J. Fish Biol.* 48:437-446.

Ensing, D. 2014. Pike (*Esox lucius*) could have been an exclusive human introduction to Ireland after all: a comment on Pedreschi et al. (2014). *J. Biogeogr.* 42:604-607.

Ficetola, G.F., C. Miaud, F. Pompanon and P. Tarberlet. 2008. Species detection using environmental DNA from water samples. *Biol. Lett.* 4:423-425.

Findlay, D.L., M.J. Vanni, M. Paterson, K.H. Mills, S.E.M. Kasian, W.J. Findlay and A.G. Salki. 2005. Dynamics of a Boreal Lake Ecosystem during a Long-Term Manipulation of Top Predators. *Ecosystems* 8:603-618.

Finlayson, B., R. Schnick, D. Skarr, J. Anderson, L. Demong, D. Duffield, W. Horton and J. Stinkjer. 2010. *Planning and Standard Operating Procedures for the Use of Rotenone in Fish Management.* Bethesda: American Fisheries Society.

Flinders, J.M. and S.A. Bonar. 2008. Growth, condition, diet, and consumption rates of northern pike in three Arizona reservoirs. *Lake and Res. Manage.* 24:99-111.

Ford, B.S., P.S. Higgins, A.F. Lewis, K.L. Cooper, T.A. Watson, C.M. Gee, G.L. Ennis and R.L. Sweeting. 1995. Literature reviews of the life history, habitat requirements and mitigation/compensation strategies for thirteen sport fish species in the Peace, Liard and Columbia river drainages of British Columbia. *Can. Man. Rep. Fish. Aquat. Sci.* 2321.

Fuller, P.L. 2003. Freshwater aquatic vertebrate introductions in the United States: patterns and pathways. pp 123-151. *In*: G.M. Ruiz and J.T. Carlton (eds.). *Invasive species vectors and management strategies*. Washington D.C.: Island Press.

Fuller, P. and M. Neilson. 2016. Esox lucius. USGS Nonindigenous Aquatic Species Database, Gainesville, FL. http://nas.er.usgs.gov/queries/FactSheet.aspx?SpeciesID=676 Revision Date: 12/16/16.

Glense, R.S. 1986. Lake fishery habitat survey and classification on interior Alaska National Wildlife Refuges, 1984 and 1985. PR. U.S. Fish and Wildlife Service, Fisheries Resources, Fairbanks, AK. http://www.fws.gov/alaska/fisheries/fish/Progress_Reports/p_1986_07A.pdf

Glick, W.J. and T.M. Willette. 2016. Relative abundance, food habits, age, and growth of northern pike in 5 Susitna River drainage lakes, 2009-2012. AK Dept. of Fish and Game, FDS No. 16-34, Anchorage.

Goedde, L.E. and D.W. Coble. 1981. Effects of angling on a previously fished and unfished warmwater fish community in two Wisconsin lakes. *Trans. Am. Fish. Soc.* 108:14-25.

Greene, C.M. and T.J. Beechie. 2004. Consequences of potential density-dependent mechanisms on recovery of ocean-type Chinook salmon (*Oncorhynchus tshawytscha*). *Can. J. Fish. Aquat. Sci.* 61:590-602.

Haugen, T.O., I.J. Winfield, L.A. Vøllestad, J.M. Fletcher, J.B. James and N.C. Stenseth. 2007. Density dependence and density independence in the demography and dispersal of pike over four decades. *Ecol. Monogr.* 77:483-502.

Haught, S. and F.A. von Hippel. 2011. Invasive pike establishment in Cook Inlet Basin lakes, Alaska: diet, native fish abundance and lake environment. *Biol. Invasions* 13:2103-2114.

Hauser, W.J. 2011. *Fishes of the Last Frontier: Life Histories, Biology, Ecology and Management of Alaska Fishes*. Anchorage: Publication Consultants.

Havel, J.E., K.E. Kovalenko, S.M. Thomaz, S. Amalfitano and L.B. Kats. 2015. Aquatic invasive species: challenges for the future. *Hydrobiologia* 750:147-170.

Hawkins, J., C. Walford and T. Sorensen. 2005. Northern pike management studies in the Yampa River, 1999-2002. FR of Larval Fish Laboratory, Colorado State University, Fort Collins to Upper Colorado River Endangered Fish Recovery Program, US Fish and Wildlife Service, Denver, Colorado. http://www.coloradoriverrecovery.org/documents-publications/technical-reports/nna/YampaPike1999-2002.pdf.

He, X. and J.F. Kitchell. 1990. Direct and indirect effects of predation on a fish community: a whole-lake experiment. *Trans. Am. Fish. Soc.* 119:825-835.

Hein, C.L., G. Öhlund and G. Englund. 2011. Dispersal through stream networks: modeling climate-driven range expansions of fishes. *Divers. Distrib.* 17:641-651.

Hein, C.L., G. Öhlund and G. Englund. 2014. Fish introductions reveal the temperature dependence of species interactions. Proc. of the Royal Society B: *Biol. Sci.* 281:20132641.

Heins, D.C., H. Knoper and J.A. Baker. 2016. Consumptive and non-consumptive effects of predation by introduced northern pike on life-history traits in threespine stickleback. *Evol. Ecol. Res.* 17:355-372.

Heise, B., D. Doutaz, M. Herborg, M. Neufeld, D. Derosa and J. Baxter. 2016. Ecology and control of invasive northern pike in the Columbia River, Canada. PPT presented at the 19th International conference on Aquatic Invasive Species, Winnipeg, April 10-14, 2016. http://www.icais.org/pdf/2016abstracts/ICAIS%20Wednesday%20PM%20Session%20B/400_Heise.pdf.

Helfield, J.M. and R.J. Naiman. 2001. Effects of salmon-derived nitrogen on riparian forest growth and implications for stream productivity. *Ecology* 82:2403-2409.

Hickley, P. and S. Chare. 2004. Fisheries for non-native species in England and Wales: angling or the environment. *Fish. Manag. Ecol.* 11:203-212.

Inskip, P.D. 1982. Habitat suitability index models: northern pike. U.S. Dept. of Interior, USFWS. FWS/OBS-82/10.17.

International Union for Conservation of Nature (IUCN). 2015 Invasive Species Definition. https://www.iucn.org/theme/species/our-work/invasive-species.

Jacobsen, L. and M.R. Perrow. 1998. Predation risk from piscivorous fish influencing the diel use of macrophytes by planktivorous fish in experimental ponds. *Ecol. Fresh. Fish* 7:78-86.

Jennings, G.B., K. Sundet and A.E. Bingham. 2015. Estimates of participation, catch, and harvest in Alaska sport fisheries during 2011. AK Dept. of Fish and Game, FDS No. 15-04, Anchorage.

Jepsen, N., K. Aaestrup, F. Økland and G. Rasmussen. 1998. Survival of radio-tagged Atlantic salmon (*Salmo salar* L.) and trout (*Salmo trutta* L.) smolts passing a reservoir during seaward migration. *Hydrobiologia* 371.372:347-353.

Johnson, B.M., P.J. Martinez, J.A. Hawkins and K.R. Bestgen. 2008. Ranking predatory threats by nonnative fishes in the Yampa River, Colorado, via bioenergetics modeling. *North Am. J. Fish. Manage.* 28:1941-1953.

Johnson, B.M., R. Arlinghaus and P.J. Matrinez. 2009. Are we doing all we can to stem the tide of illegal fish stocking? *Fisheries* 34:389-393.

Johnson, J. and J. Coleman. 2014. Catalog of waters important for spawning, rearing, or migration of anadromous fishes – Southcentral Region, Effective June 1, 2014. AK Dept. of Fish and Game, SP No. 14-03, Anchorage, USA.

Jones, H.M. and C.A. Paszowski. 1997. Effects of northern pike on patterns of next use and reproductive behavior of male fathead minnows in a boreal lake. *Behav. Ecol.* 8:655-662.

Kekäläinen, J., T. Niva and H. Huuskonen. 2008. Pike predation on hatchery-reared Atlantic salmon smolts in a northern Baltic river. *Ecol. Fresh. Fish* 17:100-109.

Kipling, C. and W.E. Frost. 1970. A study of the mortality, population numbers, year class strengths, production and food consumption of pike, *Esox lucius*, in Windermere from 1944 to 1962. *J. Anim. Ecol.* 39:115-157.

Koel, T.M., J.L. Arnold, P.E. Bigelow, C.R. Detjens, P.D. Doepke, B.D. Ertel and M.E. Ruhl. 2015. Native Fish Conservation Program, Fisheries & Aquatic Sciences, Yellowstone Nat. Park: AR, 2012-2014. Nat. Park Service, Yellowstone Center for Resources, Yellowstone Nat. Park, Wyoming, YCR-2015-01.

Korell, H.J. 2015. Comins and Bassett Lakes Restoration Project Field Trip Report. Nevada Dept. of Wildlife. *Unpublished.*

Kottelat, M. and J. Freyhof. 2007. *Handbook of European freshwater fishes*. Berlin: Kottelat, Cornol, Switzerland and Freyhof.

Larsson, P.O. 1985. Predation on migrating smolt as a regulating factor in Baltic salmon, *Salmo salar* L., populations. *J. Fish Biol.* 26:391-397.

Layhee, M., A.J. Sepulveda, A. Shaw, M. Smukall, K. Kappenman and A. Reyes. 2016. Effects of electric barrier on passage and physical condition of juvenile and adult rainbow trout. *J. Fish Wild. Manage.* 7:28-35.

Lee, D.P. 2001. Northern pike control at Lake Davis, California. pp 55-61. *In* R.L. Caileteux, L. Demong, B.J. Finlayson, W. Horton, W. McClay, R.A. Schnick and C. Thompson (eds.). *Rotenone in fisheries: Are the rewards worth the risks?* Bethesda, Maryland: American Fisheries Society.

Leunda, P.M. 2010. Impacts of non-native fishes on Iberian freshwater ichthofauna: current knowledge and gaps. *Aquat. Invasions* 5:239-262.

Lever, C. 1996. *Naturalized Fishes of the World.* San Diego: Academic Press.

Mann, R.H.K. 1980. The numbers and production of pike (*Esox lucius*) in two Dorset Rivers. *J. Anim. Ecol.* 49:899-915.

Marchetti, M.P., P.B. Moyle and R. Levine. 2004. Invasive species profiling? Exploring the characteristics of non-native fishes across invasion stages in California. *Freshwater Biol.* 49:646-661.

Massengill, R.M. 2017a. Stormy Lake northern pike movement study. AK Dept. of Fish and Game, SP, Anchorage, USA.

Massengill, R.M. 2017b. Stormy Lake Restoration: Northern Pike Eradication, 2012. AK Dept. of Fish and Game, SP Anchorage, USA.

McKinley, T.R. 2013. Survey of northern pike in lakes of Soldotna Creek drainage, 2002. AK Dept. of Fish and Game, SP No. 13-02, Anchorage, USA.

McMahon, T.E. and D.H. Bennett. 1996. Walleye and northern pike: boost or bane to northwest fisheries? *Fisheries* 21:6-13.

Mehta, S.V., R.G. Agiht, F.R. Homans, S. Polasky and R.C. Venette. 2007. Optimal detections and control strategies for invasive species management. *Ecol. Econ.* 61:237-245.

Minamoto, T., H. Yamanaka, T. Takahara, M.N. Honjo and Z.I. Kawabata. 2012. Surveillance of fish species composition using environmental DNA. *Limnology* 13:193-197.

Mooney, H.A., R. Mack, J.A. McNeely, L.E. Neville, P.J. Schei and J.K. Waage (eds.). 2005. *Invasive Alien Species: A New Synthesis.* Washington D.C.: Island Press.

Mueller, G.A. 2005. Predatory fish removal and native fish recovery in the Colorado River Mainstem: what have we learned? *Fisheries* 30:10-19.

Muhlfeld, C.C., D.H. Bennett, R.K. Steinforst, B. Marotz and M. Boyer. 2008. Using bio-energetics modeling to estimate consumption of native juvenile salmonids by nonnative northern pike in the Upper Flathead River System, Montana. *North Am. J. Fish. Manage.* 28:636-648.

Naiman, R.J., R.E. Bilby, D.E. Schindler and J.M. Helfield. 2002. Pacific salmon, nutrients, and the dynamics of freshwater and riparian ecosystems. *Ecosystems* 5:399-417.

National Marine Fisheries Service. 2011. Fisheries economics of the United States, 2010. U.S. Dept. of Commerce, NOAA TM NMFS-F/SPO-120. https://www.st.nmfs.noaa.gov/Assets/economics/documents/feus/2010/FEUS_2010-FINAL.pdf

National Strategy for Angling Development. 2015. The economic contribution of pike angling in Ireland.

Nicholson, M.E., M.D. Rennie and K.H. Mills. 2015. Apparent extirpation of prey fish communities following the introduction of Northern Pike (*Esox lucius*). *Can. Field-Natur.* 129:165-309.

Nilsson, P.A. and C. Brönmark. 1999. Foraging among cannibals and kleptoparasites: effects of prey size on pike behavior. *Behav. Ecol.* 10:557-566.

Ogutu-Ohwayo, R. 1990. The decline of the native fishes of the lakes Victoria and Kyoga (East Africa) and the impact of introduced species, especially the Nile perch, *Lates niloticus*, and the Nile tilapia, *Oreochromis niloticus*. *Environ. Biol. Fish.* 27:81-96.

Öhlund, G., P. Hedström, S. Norman, C.L. Hein and G. Englund. 2015. Temperature dependence of predation depends on the relative performance of predators and prey. *Proc. R. Soc. B.* 282:20142254.

Olden, J.D., N.L. Poff and K.R. Bestgen. 2006. Life-history strategies predict fish invasions and extirpations in the Colorado River Basin. *Ecol. Monogr.* 76:25-40.

Olson, J.A. and J.M. Connor. 2009. Resident fish stock status above Chief Joseph and Grand Coulee Dams. Box Canyon Reservoir Fisheries Survey. 2009 AN. Kalispel Tribe. Prepared for U.S. Dept. of Energy, Bonneville Power Administration, Div. of Fish and Wildlife, Project #97208-3621.

Olson, J.B., C.J. Lewis, R.L. Massengill, K.J. Dunker and J.K. Wenburg. 2015. An evaluation of target specificity and sensitivity of three qPCR asssays for detecting environmental DNA from Northern Pike (*Esox lucius*). *Conserv. Genet. Resour.*:1-3. DOI:10.1007/s12686-015-0459-x.

Oslund, S. and S. Ivey. 2010. Recreational fisheries of northern Cook Inlet, 2009-2010: report to the Alaska Board of Fisheries, February 2011. AK Dept. of Fish and Game, FMR No. 10-50, Anchorage, USA.

Oswood, M.W., J.B. Reynolds, J.G. Irons III and A.M. Milner. 2000. Distributions of freshwater fishes in ecoregions and hydroregions of Alaska. *J. North Amer. Benthol. Soc.* 19:405-418.

Park, K. 2004. Assessment and management of invasive alien predators. *Ecol. Society* 9: 12. http://www.ecologyandsociety.org/vol9/iss2/art12

Patankar, R., F. Von Hippel and M. Bell. 2006. Extinction of a weakly armored threespine stickleback (*Gasterosteus aculeatus*) population in Prator Lake, Alaska. *Ecol. Fresh. Fish* 15:482-487.

Paukert, C.P., J.A. Klammer, R.B. Pierce and T.D. Simonson. 2001. An overview of northern pike regulations in North America. *Fisheries* 26:6-13.

Pauchard, A., A. Milbau, A. Albihn, J. Alexander, T. Burgess, C. Daehler, G. Englund, F. Essl, B. Evengard, G.B. Greenwood, Sy, Haider, J. Lenoir, K. McDougall, E. Muths., M.A. Nunez, J. Olofsson, L. Pellissier, W. Rabitsch, L.J. Rew, M.Robertson, N.Sanders and C. Kueffer. 2016. Non-native and native organisms moving into high elevation and high latitude ecosystems in an era of climate change: new challenges for ecology and conservation. *Biol. Invasions* 18:345-353.

Pedreschi, D., M.K. Quinn, J. Caffrey, M.O'Grady and S. Mariani 2013. Genetic structure of pike (*Esox lucius*) reveals a comples and previously unrecognized colonization history of Ireland. *J. Biogeogr*. DOI: 10.1111/jbi.12220.

Pedreschi, D. and S. Mariani. 2015. Towards a balanced view of pike in Ireland: a reply to Ensing. *J. Biogeogr.* 42:607-609.

Persson, L., J. Anderson, E. Wahlström and P. Eklöv. 1996. Size-specific interactions in lake systems: predator gape limitation and prey growth rate and mortality. *Ecology* 77:900-911.

Pierce R.B. and C.M. Tomcko. 2003. Variation in Gill-Net and Angling Catchability with Changing Density of Northern Pike in a Small Minnesota Lake. *Trans. Am. Fish. Soc.* 132:771-779.

Pierce R.B., C.M. Tomcko and D.H. Schupp. 1995. Exploitation of northern pike in seven small North-Central Minnesota Lakes. *N. Am. J. Fish. Manage.* 15:601-609.

Raat, A.J.P. 1988. *Synopsis of biological data on the northern pike Esox lucius Linneaeus, 1758*. FAO Fisheries Synopsis No. 30. Rev. 2. Rome: Food and Agriculture Organization of the U.N.

Rahel, F.J. 2004. Unauthorized fish introductions: fisheries management of the people, for the people, or by the people? *Amer. Fish. Soc. Sympos*. 44:431-443.

Rahel, F.J., B. Bierwagen and Y. Taniguchi. 2008. Managing aquatic species of conservation concern in the face of climate change and invasive species. *Conserv. Biol.* 22:551-561.

Raleigh, R.F., T. Hickman, R.C. Solomon and P.C. Nelson. 1984. Habitat suitability information: rainbow trout. United States Fish and Wildlife Service. FWS/OBS-82/10.60.

Ribeiro, F. and P.M. Leunda. 2012. Non-native fish impacts on Mediterranean freshwater ecosystems: current knowledge and research needs. *Fish. Manag. Ecol.* 19:142-156.

Ricciardi, A. 2001. Facilitative interactions among aquatic invaders: is an "invasional meltdown" occurring in the Great Lakes?" *Can. J. Fish. Aquat. Sci.* 58:2513-2525.

Ricciardi, A. and J. Cohen. 2007. The invasiveness of an introduced species does not predict its impact. *Biol. Invasions* 9:309-315.

Ricciardi, A., M.F. Hoopes, M.P. Marchetti and J.L. Lockwood. 2013. Progress toward understanding the ecological impacts of nonnative species. *Ecol. Monogr.* 8:263-282.

Rincon, P.A., J.C. Velasco, N. Gonzalez-Sanchez and C. Pollo. 1990. Fish assemblages in small streams in western Spain: The influence of an introduced predator. *Archiv f. Hydrobiol.* 118:81-91.

Rutz, D.S. 1999. Movements, food, availability and stomach contents of northern pike in selected river drainages, 1996-1997. AK Dept. of Fish and Game, FDS Report No. 99-5. Anchorage, USA.

Rutz, D. and K.J. Dunker. 2017. Invasive northern pike suppression in Alexander Creek, 2011-2016. AK Dept. of Fish and Game, FDS Report, Anchorage, USA. *In Review.*

Scarnecchia, D.L., Y. Lim, S.P. Moran, T.D. Tholl, J.M. Dos Santos and K. Breidinger. 2014. Novel Fish Communities: native and non-native species trends in two run-of-the-river reservoirs, Clark Fork River, Montana. *Rev. Fish. Sci. Aquacult.* 22:97-111.

Schill, D.J., J.A, Heindel, M.R. Campbell, K.A. Meyer and E.R.J.M. Mamer. 2016. Production of a YY Male Brook Trout Broodstock for Potential Eradication of Undesired Brook Trout Populations. *N. Am. J. Aquacult.* 78:72-83.

Scott, W. B. and E. J. Crossman. 1973. Freshwater Fishes of Canada. Bulletin 184. Dept. of Fisheries and Oceans. Ottawa: Scientific Information and Publications Branch.

Seeb, J.E., L.W. Seeb, D.W. Oates and F.M. Utter. 1987. Genetic variation and postglacial dispersal of populations of northern pike (*Esox lucius*) in North America. *Can. J. Fish. Aquat. Sci.* 44:556-561.

Sepulveda, A., A. Ray, R. Al-Chokhachy, C. Muhlfeld, R. Gresswell, J. Gross and J. Kershner. 2012. Aquatic invasive species: lessons from cancer research. *Amer. Sci.* 100:234-242.

Sepulveda, A.J., D.S. Rutz, S.S. Ivey, K.J. Dunker and J.A. Gross. 2013. Introduced northern pike predation on salmonids in Southcentral Alaska. *Ecol. Freshwater Fish.* 22:268-279.

Sepulveda, A.J., D.S. Rutz, A.W. Dupuis, P.A. Shields and K.J. Dunker. 2014. Introduced northern pike consumption of salmonids in Southcentral Alaska. *Ecol. Freshwater Fish.* DOI: 10.1111/eff.12164.

Simberloff, D. and B. Von Holle. 1999. Positive interactions of nonindigenous species: invasional meltdown? *Biol. Invasions* 1:21-32.

Smukall, M.J. 2015. Northern pike investigations FR 2012-2014. Cook Inlet Aquaculture Association. Solman, V.E. 1945. The ecological relations of pike, *Esox Lucius*, L., and waterfowl. *Ecology*. 26:157-170.

Southwick Associates, Inc., W.J. Romberg, A.E. Bingham, G.B. Jennings and R.A. Clark. 2008. Economic impacts and contributions of sport fishing in Alaska, 2007. AK Dept. of Fish and Game, Professional Publication No. 08-01, Anchorage, USA.

Spens, J., G. Englund and H. Lundqvist. 2007. Network connectivity and dispersal barriers: using geographical information system (GIS) tools to predict landscape scale distribution of a key predator (*Esox lucius*) among lakes. *J. Appl. Ecol.* 44:1127-1137.

Spens, J. and J.P. Ball. 2008. Salmonid or nonsalmonid lakes: predicting the fate of northern boreal fish communities with hierarchical filters relating to a keystone piscivore. *Can. J. Fish. Aquat. Sci.* 65:1945-1955.

Syslo, J.M., C.S. Guy, P.E. Bigelow, P.D. Doepke, B.D. Ertel and T.M. Koel. 2011. Response of non-native lake trout (*Salvelinus namaycush*) to 15 years of harvest in Yellowstone Lake, Yellowstone NP. *Can. J. Fish. Aquat. Sci.* 68:2132-2145.

Thresher, R.E., K. Hayes, N.J. Bax, J. Teem, T.J. Benfey and F. Gould. 2014. Genetic control of invasive fish: technological options and its role in integrated pest management. *Biol. Invasions* 14:1201-1216.

Tyus, H.M. and J.M. Beard. 1990. *Esox lucius* (Esocidae) and *Stizostedion vitreum* (Percidae) in the Green River Basin, Colorado and Utah. *Great Basin Natur.* 50:33-39.

Venturelli, P.A. and W.M. Tonn. 2006. Diet and growth of northern pike in the absence of prey fishes: initial consequences for persisting in disturbance-prone lakes. *Trans. Am. Fish. Soc.* 135:1512-1522.

Wahl, D.H. and R.A. Stein. 1988. Selective predation by three esocids: the role of prey behavior and morphology. *Trans. Am. Fish. Soc.* 117:142-151.

Walrath, J.D., M.C. Quist and J.A. Firehammer. 2015. Trophic ecology of nonnative northern pike and their effect on conservation of native westslope cutthroat trout. *N. Am. J. Fish Manage.* 35:158-177.

Webber, A. 2009. Translocation of northern pike from the Yampa River upstream of Craig, Colorado. Colorado River Recovery Program FY2009 APR. Project No. 98b.

Welcomme, R.L. 1988. International introductions of inland aquatic species. FAO Fisheries TP No. 294. Food and Agriculture Organization of the United Nations, Rome, Italy.

Whitmore, C. and D. Sweet. 1998. Area management report for the recreational fisheries of Northern Cook Inlet, 1997. AK Dept. of Fish and Game, FMR No. 98-4, Anchorage, USA.

Wydoski, R.S. and D.H. Bennett. 1981. Forage species in lakes and reservoirs of the western United States. *Trans. Am. Fish. Soc.* 110:764-771.

Ylikarjula, J., M. Heino and U. Dieckmann. 1999.Ecology and adaptation of stunted growth in fish. *Evol. Ecol.* 13:433. doi:10.1023/A:1006755702230

Zabel, R.W., M.D. Scheurell, M.M. McClure and J.G. Williams. 2006. The interplay between climate variability and density dependence in the population viability of Chinook salmon. *Conserv. Biol.* 20:190-200.

Zelasko, K.A., K.R. Bestgen, J.A. Hawkins and G.C. White. 2016. Evaluation of a long-term predator removal program: abundance and population dynamics of invasive northern pike in the Yampa River, Colorado. *Trans. Am. Fish. Soc.*: 145:1153-1170.

Index